THEORETICAL MECHANICS

The M Co

ELEMENTS OF

THEORETICAL MECHANICS

BY

ALEXANDER ZIWET

JUNIOR PROFESSOR OF MATHEMATICS IN THE UNIVERSITY OF MICHIGAN

REVISED EDITION OF
"AN ELEMENTARY TREATISE ON THEORETICAL MECHANICS,"
ESPECIALLY DESIGNED FOR STUDENTS OF ENGINEERING

New York
THE MACMILLAN COMPANY
LONDON: MACMILLAN & CO., LTD.
1908

New edition, set up, electrotyped and printed January, 1904
Reprinted February, 1906; February, 1908.

Norwood Press:
Berwick & Smith, Norwood, Mass., U.S.A.

PREFACE.

THE present edition differs considerably from the original edition of 1893–94, especially in the third part. It represents essentially the required course in theoretical mechanics as given in the Engineering Department of the University of Michigan. In order to keep within the bounds of a three-hour course extending through a year and within the reach of the mathematical attainments of a second or third year's college student it seemed best to confine the treatment largely to problems in one and two dimensions (except in Statics). Thus the motion of a rigid body about a fixed point had to be omitted, in spite of its importance. But rectilinear motion and rotation about a fixed axis have received more ample treatment, and at least some illustrations of plane motion have been given. It is hardly necessary to say that the text has been carefully revised throughout, and that the exercises have in part been modified and increased in number.

For the sake of completeness, certain fundamental subjects, such as simple and compound harmonic motion, the determination of centroids, motion under central forces, the theory of moments of inertia and principal axes, have been retained in greater fullness than might be thought necessary in so elementary a work. Where a shorter course is required the teacher will find no difficulty in retrenching. Thus, Arts. 129–148, 155–156, 170–174, 176–178, 234–236, 239–240, 248–249, 312–313, 324, 334, 382–388, 392, 411–417, 479–495, 509–511, 556–571, 637–663, 702–704 may be omitted, as well as many of the numerous applications and the more difficult exercises.

The work is *not* a treatise on applied mechanics, the applications being merely used to illustrate the general principles and to give the student an idea of the uses to which mechanics can be put. It is intended to furnish a safe and sufficient basis, on the one hand for the more advanced study of the science, on the other for the study of its more simple applications. I wish in particular that it may serve to stimulate the study of theoretical mechanics in engineering schools. At a future time, I hope to embody in a more advanced treatise, together with other matter, those portions of the old edition which could not find a place in the present volume.

To Professor E. R. Hedrick, of the University of Missouri, who had the kindness of reading the manuscript and proofs of a large part of the work, I am greatly indebted for pertinent criticism and many valuable suggestions. My thanks are also due to Mr. Wm. Marshall, of the University of Michigan, who has ably assisted me in reading the proofs and eliminating errors.

ALEXANDER ZIWET.

UNIVERSITY OF MICHIGAN,
January, 1904.

CONTENTS.

PAGE

INTRODUCTION 1

PART I: GEOMETRY OF MOTION; KINEMATICS.

CHAPTER I.

GEOMETRY OF MOTION.

I. LINEAR MOTION; TRANSLATION AND ROTATION . . . 3

II. PLANE MOTION 7

III. SPHERICAL MOTION 16

IV. SCREW MOTION 18

V. COMPOSITION AND RESOLUTION OF TRANSLATIONS; VECTORS. 22

CHAPTER II.

KINEMATICS.

I. TIME 29

II. LINEAR KINEMATICS:
1. Uniform rectilinear motion; velocity 30
2. Variable rectilinear motion; acceleration 34
3. Applications 37
4. Rotation; angular velocity; angular acceleration . . 49

III. PLANE KINEMATICS:
1. Velocity; composition of velocities; relative velocity . . 52
2. Applications 58
3. Acceleration in curvilinear motion 62
4. Applications 68
5. Velocities in the rigid body 109
6. Applications 117

PART II: Introduction to Dynamics; Statics.

CHAPTER III.

Introduction to Dynamics.

I. Mass; Moments of Mass; Centroids:
1. Mass; density 129
2. Moments and centers of mass 133
3. Centroids of particles and lines 137
4. Centroids of areas 141
5. Centroids of volumes 153

II. Momentum; Force; Energy 159

CHAPTER IV.

Statics.

I. Forces Acting on the Same Particle 170

II. Concurrent Forces; Moments 175

III. Parallel Forces 183

IV. Theory of Couples 201

V. Plane Statics:
1. The conditions of equilibrium 208
2. Stability 217
3. Jointed frames 219
4. Graphical methods 224
5. Friction 230

VI. Solid Statics:
1. The conditions of equilibrium 243
2. Constraints 255

VII. The Principle of Virtual Work 258

PART III: Kinetics.

CHAPTER V.

Kinetics of the Particle.

I. Impulses; Impact of Homogeneous Spheres . . . 275

II. Rectilinear Motion of a Particle 293

III. Free Curvilinear Motion:
1. General principles 325
2. Central forces 337

IV. Constrained Motion:
1. Introduction 362
2. Motion on a fixed curve 365
3. Motion on a fixed surface 375

CHAPTER VI.

Kinetics of the Rigid Body.

I. General Principles 379

II. Moments of Inertia and Principal Axes:
1. Introduction 392
2. Ellipsoids of inertia 399
3. Distribution of principal axes in space 410

III. Rigid Body with a Fixed Axis 416

IV. Plane Motion 450

THEORETICAL MECHANICS.

INTRODUCTION.

The science of theoretical mechanics has for its object the mathematical study of motion.

The idea of motion is intimately related to the fundamental ideas of **space, time,** and **mass.** It will be convenient to introduce these consecutively. Thus we shall begin with a purely geometrical study of motion, without regard to the time consumed in the motion and to the mass of the thing moved, the moving object being considered as a mere geometrical configuration. This introductory branch of mechanics may be called the **Geometry of motion.**

The introduction of the idea of time will then lead us to study the velocity and acceleration of geometrical configurations. This constitutes the subject-matter of **Kinematics** proper. The term kinematics is often used in a less restricted sense, so as to include the geometry of motion.

Finally, endowing our geometrical points, lines, and other configurations with mass, we are led to the ideas of momentum, force, energy, etc. This part of our subject, the most comprehensive of all, has been called **Dynamics,** owing to the importance of the idea of force in its investigation. For the sake of convenience it is usually divided into two branches, **Statics** and **Kinetics.** In statics those cases are considered in which no change of motion is produced by the acting forces, or, as it is commonly expressed, in which the forces are in equilibrium. The investigations of statics are therefore independent of the element of time. Kinetics treats of motion in the most general way.

PART I:

GEOMETRY OF MOTION; KINEMATICS

CHAPTER I.

GEOMETRY OF MOTION.

I. *Linear Motion; Translation and Rotation.*

1. Motion consists in change of position.

We begin with the simple case of a point moving in a straight line. The position of a point P in a line is determined by its distance $OP = x$ from some fixed point or origin, O, assumed in the line, the length x being taken with the proper sign to express the *sense* (say forward or backward, to the right or to the left) in which it is to be measured on the line. This sense is also indicated by the order of the letters, so that $PO = -OP$, and $OP + PO = 0$.

The position of a point in a line is thus fully determined by a single algebraical quantity or co-ordinate; viz. by its abscissa $x = OP$.

2. Let the point P move in the line from any initial position P_0 (Fig. 1) to any other position P_1, and let $OP_0 = x_0$, $OP_1 = x_1$.

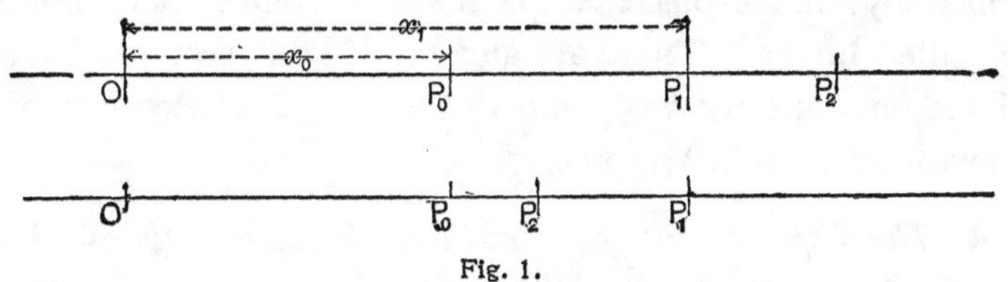

Fig. 1.

This change of position, or *displacement*, is fully determined by the distance $P_0P_1 = x_1 - x_0$ traversed by the point.

Now let this displacement P_0P_1 be followed by another displacement in the same line, from P_1 to P_2, in the same sense as the former, or in the opposite sense. In either case the total, or **resultant**, displacement P_0P_2 is the algebraic sum of the two displacements P_0P_1, P_1P_2, which are called its **components**; *i. e.*, we have $P_0P_2 = P_0P_1 + P_1P_2$, or $P_0P_1 + P_1P_2 + P_2P_0 = 0$, whatever may be the positions of the points P_0, P_1, P_2 in the line.

This reasoning is easily extended to any number of component displacements; that is, *the resultant of any number of consecutive displacements of a point in a line is a single displacement in the same line equal to the algebraic sum of the components.*

Similar considerations apply to the motion of a point in a curved line provided the displacements be always measured along the curve.

3. Let us next consider the motion of a rigid body. The term **rigid body**, or simply **body**, is used in kinematics to denote a figure of invariable size or shape, or an aggregate of points whose distances from each other remain unchanged. Examples are: a segment of a straight line, a triangle, a cube, an ellipsoid, etc.

Imagine such a body brought in any manner from some initial position into any other position. This change of position is called the *displacement* of the body. We shall see (Arts. 30–37) that, even in the most general case, the displacements of *three* points of the body determine those of all other points, and consequently the displacement of the whole body.

There are, however, two special cases of motion, *translation* and *rotation* in which the displacement of the body is fully determined by the displacement of a *single* point; such motions can be called **linear.** There are also motions determined by the displacements of only *two* points of the body; an example of this is **plane** motion (see Art. 11).

4. *The displacement of a rigid body is called a* **translation** *when the displacements of all of its points are parallel and equal.* It is evident that in this case the displacement of any one point of the body fully represents the displacement of the whole body. The translation of a rigid body from one position to another is therefore measured by the segment P_0P_1 of the straight line joining the initial and final positions of any point P of the body.

Two or more consecutive translations of a rigid body in the same direction produce a resultant translation in the same direction equal to the algebraic sum of the component translations.

5. When a rigid body has two of its points fixed, the only motion it can have is a rotation about the line joining the fixed points as axis. *In a motion of* **rotation** *all points of the body excepting those on the axis describe arcs of circles whose centers lie on the axis while the points on the axis are at rest.*

Thus any point P, not on the fixed axis, is carried from its initial position P_0 to its final position P_1 along a circular arc whose center C lies on the axis. The rotation is evidently measured by the angle P_0CP_1 subtended at C by this arc P_0P_1 (Fig. 2).

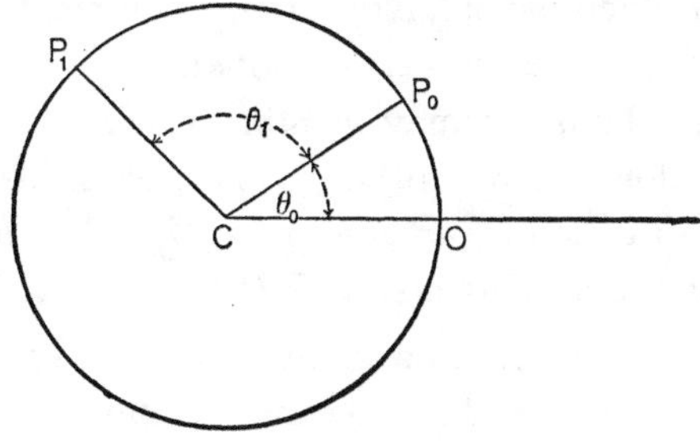

Fig. 2.

The position of any point P (not on the axis) fully determines the position of the whole body and is given by the angle θ made by CP with some initial line CO, passing through C at right angles to the axis. If $OCP_0 = \theta_0$, $OCP_1 = \theta_1$, the angle $P_0CP_1 = \theta_1 - \theta_0$ measures the rotation, just as (Art. 2) the distance $P_0P_1 = x_1 - x_0$ measures the displacement of a point, and hence (Art. 4) the translation of a rigid body.

Two or more consecutive rotations of a rigid body about the same axis give a resultant rotation about the same axis whose angle is the algebraic sum of the angles of the component rotations.

6. The particular case when the rigid body is a plane figure whose motion is confined to its plane deserves special mention. If one point, C, of such a figure be fixed, the figure can only have a motion of rotation about C, the *center of rotation*, every other point of the figure describing an arc of a circle whose center is C (Fig. 2).

7. We have seen that a translation as well as a rotation is measured by a single algebraical quantity, the translation by a distance, the rotation by an angle. This is the reason why such motions may be called *linear*. The two fundamental forms of motion, translation and rotation, are thus seen to correspond to the two fundamental magnitudes of metrical geometry, viz., distance and angle.

It is to be noticed that both for translations in the same direction and for rotations about the same axis the resultant displacement is found by algebraic addition of the components, not only when the components are *consecutive* motions, but even when they are *simultaneous*. Thus we may imagine a point P displaced by the amount P_1P_2 along a straight line, while this line itself is moved along in its own direction by an amount Q_1Q_2. The resultant displacement of P is the algebraic sum $P_1P_2 + Q_1Q_2$.

8. Translations being measured by distances or lengths and rotations by angles, we need in mechanics a unit of length and a unit of angle.

The two most important systems of measurement are the C. G. S. (*i. e.* centimeter-gram-second) system and the F. P. S. (*i. e.* foot-pound-second) system. The former is frequently called the scientific system; it is based on the international or metric system of weights and measures. The F. P. S. or British system is still used in England and the United States almost universally in engineering practice.*

* For fuller information on all questions relating to standards and units see J. D. Everett, *Illustrations of the C. G. S. system of units, with tables of physical constants*; London, Macmillan, 1902.

9. The unit of length in the C. G. S. system is the **centimeter** (cm.), *i. e.* $\frac{1}{100}$ of the meter. The original standard meter is a platinum bar preserved in the *Palais des Archives* in Paris; two carefully compared copies, known as *prototypes*, are kept at the National Bureau of Standards, in Washington, D. C. The meter can be defined as the distance between two marks on the standard meter when at a temperature of 0° C.

In the F. P. S. system, the unit of length is the **foot,** *i. e.* $\frac{1}{3}$ of the standard yard. The original British standard yard is a bronze bar preserved in London.

The relation between these two fundamental units of length is, according to the *United States Coast and Geodetic Survey Bulletin,* No. 9, 1889,

$$1 \text{ cm.} = 0.032\,808\,2 \text{ ft.}$$

For practical use we have the following approximate relations:

$$1 \text{ m.} = 3.2808 \text{ ft.}, \qquad 1 \text{ ft.} = 30.48 \text{ cm.}$$

$$1 \text{ cm.} = 0.3937 \text{ in.} \qquad 1 \text{ in.} = 2.54 \text{ cm.}$$

10. The unit of angle is either the **degree,** *i. e.* $\frac{1}{360}$ of one revolution, or the **radian,** *i. e.* the angle subtended by a circular arc equal in length to the radius.

If α be any angle expressed in radians, and α°, α', α'' the same angle expressed respectively in degrees, minutes, seconds, we have the relations

$$\alpha = \frac{\pi}{180}\cdot\alpha^\circ = \frac{\pi}{10800}\cdot\alpha' = \frac{\pi}{648000}\cdot\alpha''.$$

or $$\alpha = 0.017\,453\alpha^\circ = 0.000\,291\alpha' = 0.000\,004\,85\alpha''.$$

II. *Plane Motion.*

11. The position of a plane figure in its plane is fully determined by the positions of any two of its points since every other point of the figure forms with these two points an invariable triangle. But the position of the figure can of course be determined in other ways; for instance, by the position of one point and that of a line of the figure passing through the point; or by the position of two lines of the figure.

12. Let us now consider the motion of a plane figure in its plane from any initial position to any other position. This displacement can be brought about in various ways. Thus, it would be sufficient to bring any two points A, B of the figure from their initial positions A_0, B_0 (Fig. 3) to their final positions A_1, B_1. This can be done, for instance, by first giving the whole figure a translation

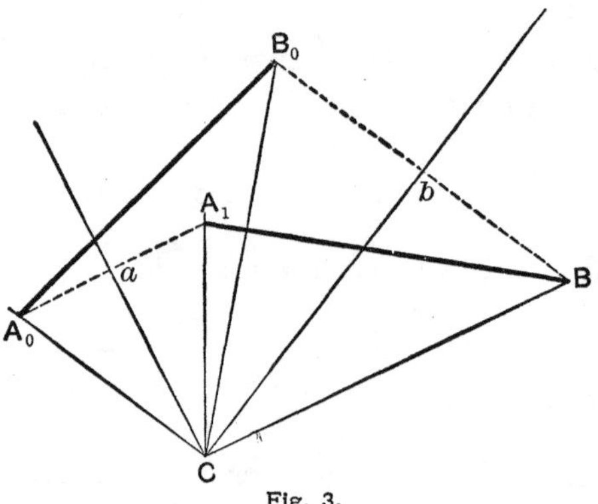

Fig. 3.

through a distance A_0A_1 and then a rotation through an angle equal to the angle between A_0B_0 and A_1B_1; or by such a rotation followed by the translation.

Instead of A we might have selected any other point of the figure. But it is important to notice that the angle of rotation required for a given displacement is always the same, while the translation will differ according to the point selected as center.

13. This leads us to inquire whether the center of rotation cannot be so selected as to reduce the translation to zero. Now any rotation that is to bring A from A_0 to A_1 must have its center on the perpendicular bisector of A_0A_1; similarly for B. Hence the intersection C of the perpendicular bisectors of A_0A_1 and B_0B_1 is the only point by rotation about which both A and B can be brought from their initial to their final positions. That they actually are so brought follows at once from the equality of the angles A_0CB_0 and A_1CB_1 (and hence of the angles A_0CA_1 and B_0CB_1) which are corresponding angles in the equal triangles A_0CB_0 and A_1CB_1.

We thus have the proposition: *Any displacement of an invariable plane figure in its plane can be brought about by a single rotation about a certain point* which we may call the *center* of the displacement.

14. The construction of the center C given in the preceding article becomes impossible when the bisectors coincide (Fig. 4) and when they are parallel (Fig. 5). In the former case, C is readily found as the intersection of A_0B_0 and A_1B_1. In the latter, *i. e.* whenever $A_0A_1 = B_0B_1$, the center lies at infinity, and the rotation becomes a translation.

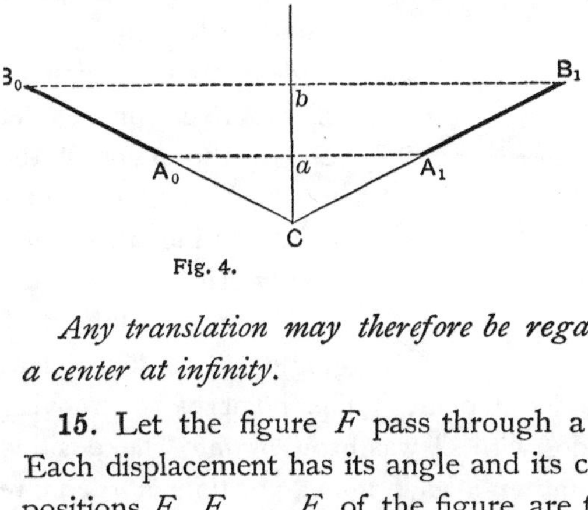

Fig. 4.

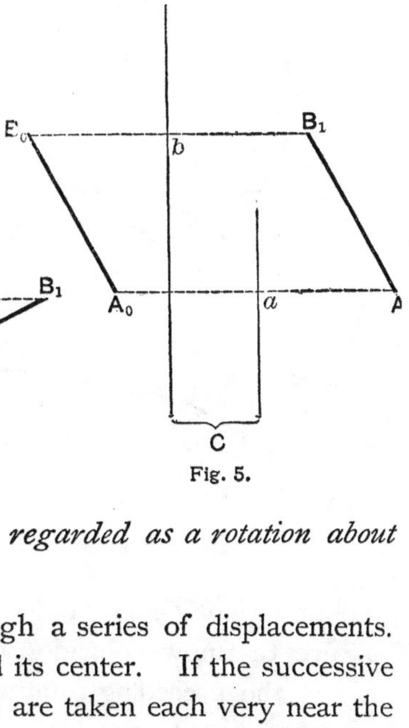

Fig. 5.

Any translation may therefore be regarded as a rotation about a center at infinity.

15. Let the figure F pass through a series of displacements. Each displacement has its angle and its center. If the successive positions $F_0, F_1, \ldots F_n$ of the figure are taken each very near the preceding one, the angles of rotation will be very small, and the successive centers $C_1, C_2, \ldots C_n$ will follow each other very closely. In the limit, *i. e.* when the series of finite displacements passes into a continuous motion of the figure, the centers C will form a continuous curve (c) and the successive angles of rotation approach zero while the radii of the successive arcs described by any point pass into the normals to the continuous path of that point. The point C about which the figure rotates in any one of its positions during the motion is now called the **instantaneous center**; the locus of the centers, that is the curve (c), is called the **fixed centrode,** or path of the center. It is apparent that in any position of the moving figure *the normals to the paths of all its points must pass through the instantaneous center*, and *the direction of motion of any such point is therefore at right angles to the line joining it to the center.*

16. The centers C are fixed points of the fixed plane in which the figure F moves. But in any position F_1 of this figure some point C'_1 of F will coincide with the point C_1 of the fixed plane. Thus, in the case of finite displacements (Fig. 6), let the figure F begin its motion with a rotation of angle θ_1 about a point C_1 of the fixed plane; let C'_1 be the point of the moving figure that coincides during this rotation with C_1.

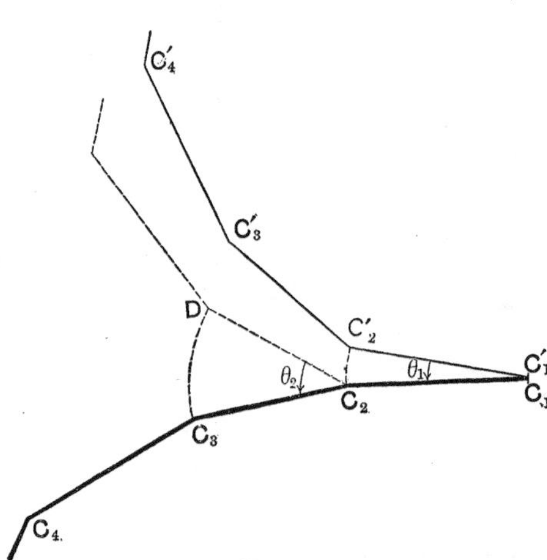

Fig. 6

The next rotation, of angle θ_2, takes place about a point C_2 of the fixed plane. The point of the moving figure that now coincides with C_2 was brought into the position C_2 by the preceding rotation. Its original position is therefore obtained by turning C_1C_2 back by an angle $-\theta_1$ into the position $C_1C'_2$. The rotation of angle θ_2 about C_2 brings a new point C'_3 of the moving figure to coincidence with the fixed center C_3; and the original position C'_3 of this point can be determined by first turning C_2C_3 back about C_2 by an angle $-\theta_2$ into the position C_2D, and then turning the broken line

C_1C_2D by a rotation of angle $-\theta_1$ about C_1 back into the position $C'_1C'_2C'_3$. Continuing this process we obtain, besides the broken line $C_1C_2C_3\ldots$ formed by joining the successive centers of rotation in the fixed plane, a broken line $C'_1C'_2C'_3\ldots$ in the moving figure formed by joining those points of this figure which in the course of the motion come to coincide with the fixed centers. The whole motion may be regarded as a kind of rolling of the broken line $C'_1C'_2C'_3\ldots$ over the broken line $C_1C_2C_3\ldots$

17. In the case of continuous motion each of the broken lines becomes a curve, and we have actual rolling of the curve (c'), or **body centrode,** over the curve (c), or **fixed centrode.** *The continuous motion of an invariable plane figure in its plane may therefore always be produced by the rolling* (without sliding) *of the body centrode over the fixed centrode.* The point of contact of the two curves is of course the instantaneous center.

18. It appears from the preceding articles that the continuous motion of a plane figure in its plane is fully determined if we know the center of rotation for every position of the figure. This center can be found as the intersection of the normals of the paths of any two points of the figure, so that the motion of the figure will be known if the paths of any two of its points are given. This, however, is only one out of many ways of determining plane motion by two conditions.

19. When a circle rolls (without slipping) over a straight line the path of any point on the circumference of the circle is called a *cycloid*, that of any other point rigidly connected with the rolling circle is called a *trochoid.* When a circle rolls over another circle three cases may be distinguished: (*a*) when each circle lies outside the other, the corresponding paths are called *epicycloid* and *epitrochoid;* (*b*) when a smaller circle rolls over the inside of a fixed larger circle, the paths are called *hypocycloid* and *hypotrochoid;* (*c*) when a larger circle rolls over an enclosed fixed smaller circle, the paths are called *pericycloid* and *peritrochoid.*

The following examples illustrate the method of finding the centrodes and the path of any point of the moving figure in the motion.

20. Elliptic motion: *Two points of a plane figure move along two fixed lines that are at right angles to each other.*

Let A, B (Fig. 7) be the points moving on the lines Ox, Oy; the perpendiculars to these lines erected at A and B intersect at the instantaneous center C. Denoting by $2a$ the invariable distance of A and B, we have $OC = AB = 2a$ for all positions of the moving figure. The fixed centrode (c) is therefore a circle of radius $2a$ described about the intersection O of the fixed lines.

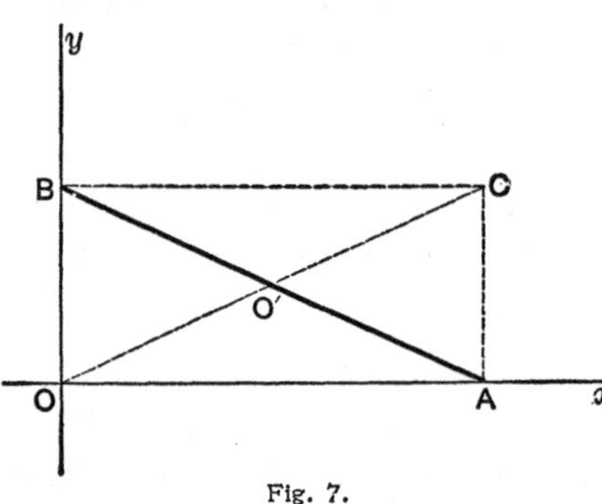

Fig. 7.

To find the body centrode (c') we must construct the triangle ABC for all possible positions of AB. As BCA is always a right angle, the body centrode will be a circle described on AB as diameter. Hence the whole motion can be produced by the rolling of a circle of radius a within a circle of radius $2a$.

The student is advised to carry out carefully the constructions indicated in this as well as the following problems. Thus, in the present case, draw the moving figure, *i. e.*, the segment AB, in a number of its successive positions in each of the four quadrants, and construct the instantaneous center C in every case. This gives a number of points of the fixed centrode. Then take any one position of AB and transfer to it as base

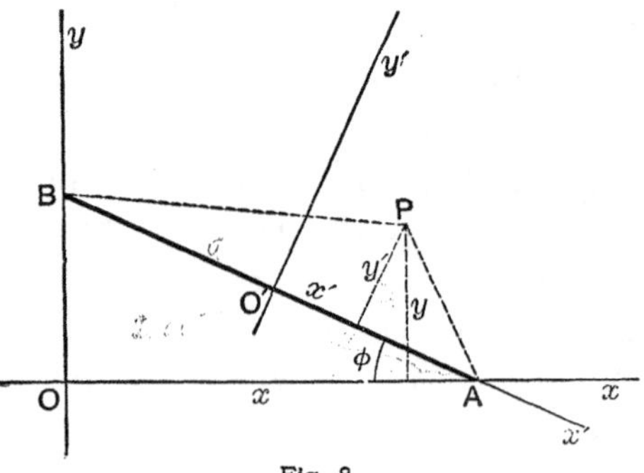

Fig. 8.

all the triangles ABC previously constructed. The vertices of these triangles all lie on the body centrode.

21. To find the equation of the path of any point P of the moving figure, let this point be referred to a co-ordinate system fixed in, and moving with, the figure (Fig. 8); let the middle point O' of AB be the origin, and $O'A$ the axis $O'x'$, of this system. Then the co-ordinates x', y' of P in this moving system are connected with its co-ordinates x, y in the fixed system Ox, Oy by the equations

$$x = (a + x')\cos\varphi + y'\sin\varphi,$$
$$y = (a - x')\sin\varphi + y'\cos\varphi,$$

where φ is the angle OAB that determines the instantaneous position of AB. Solving these equations for $\sin\varphi$ and $\cos\varphi$, squaring and adding, we find for the equation of the path of P

$$\left(\frac{y'x - (a + x')y}{x'^2 + y'^2 - a^2}\right)^2 + \left(\frac{y'y - (a - x')x}{x'^2 + y'^2 - a^2}\right)^2 = 1,$$

or $[(a-x')^2+y'^2]x^2 - 4ay'xy + [(a+x')^2+y'^2]y^2 = (x'^2+y'^2-a^2)^2$, which represents an ellipse, since

$$[(a - x')^2 + y'^2]\cdot[(a + x')^2 + y'^2] - 4a^2y'^2$$
$$= (x'^2 + y'^2 + a^2)^2 - 4a^2(x'^2 + y'^2) = (x'^2 + y'^2 - a^2)^2$$

is necessarily positive.

In general, therefore, the points of the figure describe ellipses; O' describes a circle; A and B describe straight lines passing through O, and so does every point on the circle of diameter AB. It is this fact that by rolling a circle within a circle of double diameter the points of the smaller circle are made to describe segments of straight lines, which makes this form of motion of practical importance: it may serve to transform circular into rectilinear motion.

22. Elliptic Motion (continued): *Two points A, B of a plane figure move along two fixed lines OA, OB, inclined to each other at an angle* ω (Fig. 9).

This case is readily reduced to the preceding one. For, let the circle through O, A, B intersect at B' the perpendicular erected at O to OA, and imagine AB rigidly connected to AB'; then the points A and B will move, by Art. 21, along OA and OB as desired, provided A and B' be made to move along the perpendicular lines OA and OB'.

The figure shows that, since $\measuredangle AB'B = \measuredangle AOB = \omega$, the diameter of the rolling circle is $OC = AB' = AB/\sin\omega$.

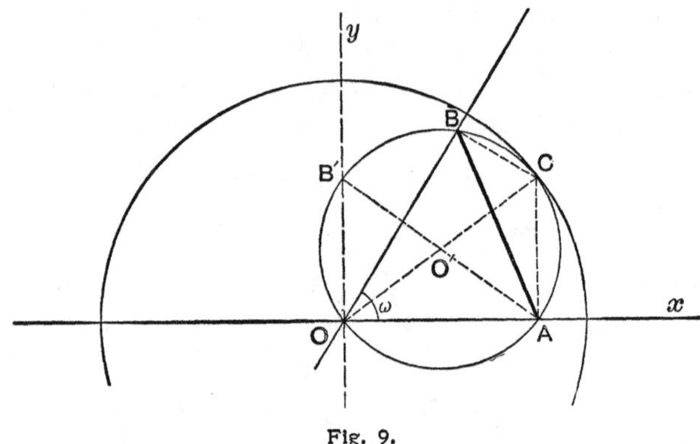

Fig. 9.

23. Connecting Rod Motion: *One point A of the figure describes a circle, while another point B moves on a straight line passing through the center O of the circle* (Fig. 10).

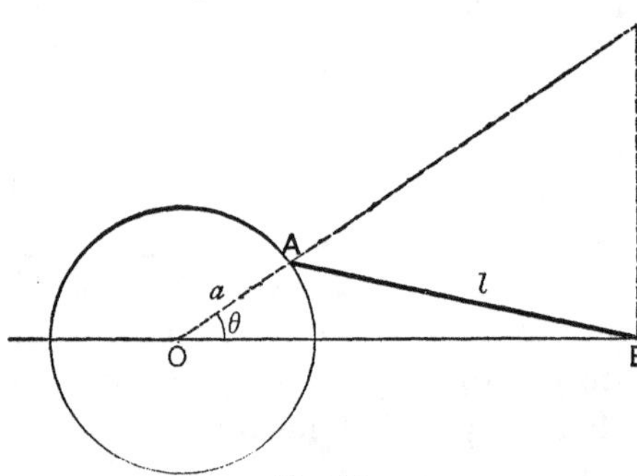

Fig. 10.

The two centrodes are readily constructed by points for a given ratio l/a, say 4, 3, or 2. If $l > a$ the fixed centrode consists of two branches having a common asymptote; the body centrode has two branches with a common tangent at A. For $l = a$ both centrodes become circles, one of radius $2a$ about O, the other of radius a about the point A. For $l < a$, the point B can describe only a portion of the crank circle, and the centrodes become closed curves.

24. Conchoidal Motion: *A point A of the figure moves along a fixed straight line l, while a line of the figure, l', containing the point A, always passes through a fixed point B* (Fig. 11).

The fixed point B may be regarded as a circle of infinitely small radius, which the line l' is to touch. The instantaneous center is therefore the intersection C of the perpendiculars erected at A on l and at B on l'.

The fixed centrode is a parabola whose vertex is B. To prove this we take the fixed line l as axis of y, the perpendicular OB to it drawn through the fixed point B as axis of x. Then, putting $\measuredangle OBA = \varphi$ and $OB = a$, we have for the co-ordinates of C

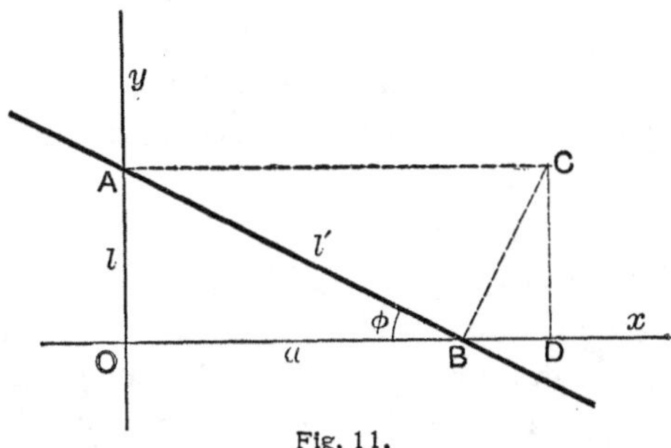

Fig. 11.

$$x = a + y \tan\varphi,$$

$$y = a \tan\varphi;$$

hence $x - a = y^2/a$, or, for B as origin and parallel axes, $y^2 = ax$. The proportion $y/x = a/y$ also follows directly from the similar triangles BDC and AOB.

The equation of the body centrode, for A as origin, AB as polar axis, is $r \cos^2\theta = a$, or in cartesian co-ordinates $a^2(x^2 + y^2) = x^4$.

The points of l' can easily be shown to describe conchoids, whence the name of this form of plane motion.

25. The results obtained in the preceding articles for the motion of a plane figure in its plane apply directly to the motion of a rigid body, if any one point of the body describes a plane curve while a line of the body remains parallel to itself. For in this case all points of the body move in parallel planes, and the motion in any one of these planes determines the motion of the whole figure.

The only modifications required would be that instead of an instantaneous center we should have an **instantaneous axis,** viz. the perpendicular to the plane of motion of any point through the center of motion of this point, and that the centrodes are now not curves, but cylindrical surfaces rolling one upon the other.

26. Exercises.

(1) Show how to find the direction of motion of any point P rigidly connected with the connecting rod of a steam engine.

(2) A wheel rolls on a straight track; find the direction of motion of any point on its rim. What are the centrodes in this case?

(3) Show how to construct the normal at any point of a conchoid.

(4) Find the equations of the centrodes when a line l' of a plane figure always touches a fixed circle O, while a point A of l' moves along a fixed line l.

(5) Show that, in (4), the centrodes are parabolas when the fixed circle touches the fixed line.

(6) Two straight lines l', l'' of a plane figure constantly pass each through a fixed point O', O''; investigate the motion.

(7) Four straight rods are jointed so as to form a plane quadrilateral $ABDE$ with invariable sides and variable angles. One side AB being fixed, investigate the motion of the opposite side; construct the centrodes graphically.

(8) A right angle moves so that one side always passes through a fixed point A, while a point B on the other side, at the distance a from the vertex, moves along a fixed line from which the fixed point A has the distance a; determine the centrodes.

(9) If the quadrilateral of Ex. (7) be a parallelogram, show that any point rigidly connected with the side opposite the fixed side describes a circle.

(10) In the problem of Art. 23 investigate the centrodes and the path of the middle point of AB analytically.

(11) Explain how elliptic motion (Art. 20) can be regarded as a limiting case of connecting rod motion (Art. 23).

(12) Explain how the paths described by various points of the rolling circle in elliptic motion (Art. 20) can be regarded as special hypocycloids and hypotrochoids (Art. 19).

III. *Spherical Motion.*

27. The motion of a spherical figure of invariable form on its sphere presents a close analogy to plane motion; in fact, plane motion is but a limiting case of spherical motion, since a plane may be regarded as a sphere of infinite radius.

By a generalization similar to that of Art. 25, the study of the motion of a spherical figure on its sphere leads directly to the laws of motion of a rigid body having one fixed point. For the motion of such a body is evidently determined by the spherical motion on any sphere described about the fixed point.

28. Let us consider any two positions F_0 and F_1 of a spherical figure F on its sphere, and let O be the center of the sphere. Just as in the case of plane motion (Art. 13) the displacement can always be brought about by a single rotation about a point C on the sphere, or what amounts to the same, by a single rotation about the *axis* OC. The proof is strictly analogous to that given in Art. 13. We first remark that the position of the figure on the sphere is fully determined by the position of two of its points (not on the same diameter), say A and B (Fig. 12), since any third point forms with these an invariable spherical triangle. Let A_0, B_0 be the positions of A, B in F_0; A_1, B_1 their positions in F_1; and draw the great circles A_0A_1 and B_0B_1. Their perpendicular bisectors intersect in two points C, D which are the ends of a diameter of the sphere. CD is the axis of the displacement and the angle A_0CA_1, or B_0CB_1, gives the angle of the displacement.

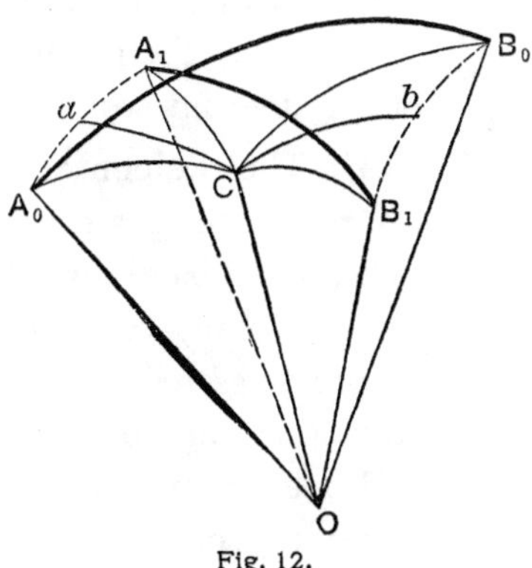

Fig. 12.

29. If we consider a series of positions of the moving figure, F_0, F_1, F_2, ..., we obtain a series of axes of rotation, say c_1, c_2, ...; and in the limit when the motion becomes continuous, the axes c_1, c_2, ... will form a cone fixed in space, with the vertex at the center O of the sphere. The points C_1, C_2, ... where these axes intersect the sphere form a curve (c) on the fixed sphere, while the points C'_1, C'_2, ... of the moving figure which come to coincide with these fixed points form a spherical curve (c') invariably connected with the moving figure. The whole motion may be produced by the rolling of the curve (c') over the curve (c), or also by the rolling of the corresponding cones one over the other. We have thus the proposition that *any continuous motion of a rigid body having a fixed point can be produced by the rolling of a cone fixed in the body over a fixed cone, the vertices of both cones being at the fixed point.*

IV. *Screw Motion.*

30. The position of a rigid body in space is fully determined by the position of any three of its points not situated in the same straight line. For any fourth point of the body will form an invariable tetrahedron with these three points. As two points determine a straight line, the position of a rigid body may also be given by the position of a point and line or by the positions of any two lines of the body.

31. The position of a point being determined by its three co-ordinates requires three conditions to be fixed. A point is therefore said to have three *degrees of freedom* when its position is not subject to any conditions. One conditional equation between its co-ordinates restricts the point to the surface represented by that equation; the point is then said to have two degrees of freedom and one *constraint.* Two conditions would restrict the point to a line, the curve of intersection of the two surfaces represented by the equations of condition; the point has then but one degree of freedom and two constraints.

A rigid body that is perfectly free to move has six degrees of freedom. For we have seen that its position is fully determined when three of its points not in the same line are fixed. The nine co-ordinates of these points are, however, not independent; they are connected by the three equations expressing that the three distances between the three points are invariable. Thus the number of independent conditions is $9 - 3 = 6$.

A rigid body with one fixed point has three degrees of freedom and therefore three constraints. For it takes two more points, *i.e.*

six co-ordinates, to fix the position of the body; and the distances of these two points from each other and from the fixed point being invariable, there are again three conditional equations to which the six co-ordinates are subject. The three co-ordinates of the fixed point may be regarded as the three constraints.

A rigid body with two fixed points, i. e., *with a fixed axis, has one degree of freedom, and five constraints.* Indeed, the six co-ordinates of the two fixed points are equivalent to five constraining conditions, since the distance of these two points is invariable.*

32. Let us now consider any two positions M_0, M_1 of a rigid body M, given by the positions A_0, B_0, C_0 and A_1, B_1, C_1 of three points A, B, C of the body. The displacement M_0M_1 can be effected in various ways. Thus we might, for instance, begin by giving the whole body a translation equal to A_0A_1 which would bring the point A to its final position while all other points of the body would be displaced by distances parallel and equal to A_0A_1. As the body has now one of its points, A, in its final position, it will (by Art. 28) require only a single rotation about a certain axis passing through this point to bring the whole body into its final position. It thus appears that any displacement of a rigid body can be effected by subjecting that body first to a translation and then to a rotation (or *vice versa*, as is easily seen); and this can be done in an infinite number of ways, as the displacement of *any* point of the body may be selected for the translation.

33. It is to be noticed that for all these different ways of effecting the displacement M_0M_1 the *direction* of the axis of rotation and the *angle* of rotation are the same. To see this more clearly, let the displacement be effected first by the translation A_0A_1 and a rotation of angle α about the axis a_1 passing through A_1; and then let the same displacement be produced by the translation

---

* Interesting remarks on the mechanical means of producing constraints of various degrees will be found in THOMSON and TAIT, *Natural Philosophy*, London, Macmillan, new edition, 1879, Art. 195 sq. (Part I., p. 149).

B_0B_1 of some other point B and a rotation of angle β about an axis b_1 passing through B_1. We wish to show that a_1 and b_1 are parallel and that the angles α and β are equal.

Consider a plane π of the rigid body which in its original position π_0 is perpendicular to the axis a_1. The translation A_0A_1 transfers it into a parallel position and the rotation α about a_1 turns it in itself into its final position π_1; hence π_0 and π_1 are parallel. The translation B_0B_1 likewise moves π into a position parallel to the original one; and as its final position, π_1, is parallel to π_0, the axis of rotation b_1 must necessarily be perpendicular to π_0 and π_1, that is b_1 must be parallel to a_1.

Again, any straight line l in π remains parallel to its original position l_0 after the translations A_0A_1 and B_0B_1. Its change of direction is due to the rotations alone; the angle of rotation must therefore be the same for both rotations, viz. equal to the angle (l_0l_1) formed by the initial and final positions of the line l.

34. Among the different combinations of a translation with a rotation effecting the displacement M_0M_1 there is one of particular importance; it is that for which the axis of rotation is parallel to the translation.

Let us again consider a plane π perpendicular to the common direction of the axes of rotation. To bring any three points of this plane into their final position it is only necessary to give the body a translation at right angles to π such as will bring π into its final position and then to add the necessary rotation for plane motion.

We have therefore the important proposition that *it is always possible to bring a rigid body M from any position M_0 into any other position M_1 by a translation combined with a rotation about an axis parallel to the direction of translation*, and this can be done in only one way. The axis so determined is called the **central axis** of the displacement.

The order of translation and rotation about the central axis is indifferent; indeed, translation and rotation might take place simultaneously.

35. A motion of a rigid body consisting of a rotation about an axis combined with a translation parallel to the axis is called a *screw motion*, or a *twist*. We have proved therefore, in Art. 34, that *the most general displacement of a rigid body can be brought about by a single twist.*

36. To construct the central axis and find the translation and angle of the twist when the displacement is given by the positions A_0, B_0, C_0 and A_1, B_1, C_1 of three points of the body, we first remark that the projection on the central axis of the displacement of any point, say A_0A_1, is equal to the translation of the twist, and hence the projections of the displacements of all points of the body (such as A_0A_1, B_0B_1, C_0C_1) are all equal. If therefore from any point O we draw lines OA, OB, OC equal and parallel to A_0A_1, B_0B_1, C_0C_1, their ends A, B, C will lie in a plane π perpendicular to the central axis, and the perpendicular p dropped from O on this plane π will represent in length and direction the translation of the twist.

The direction of the central axis being thus determined, we find its position in space by projecting the displacements of any two of the three given points, say A_0A_1 and B_0B_1, on the plane π, and finding the intersection of the perpendicular bisectors of these projections. This intersection is evidently a point of the central axis, and a perpendicular through it to the plane π will give the central axis in position.

The angle of the twist is equal to the angle between the projections on π of A_0B_0 and A_1B_1.

37. In the case of continuous motion there exists a central axis for every position of the body; but its position both in space and in the body in general varies in the course of the motion. The central axis at any moment is therefore called in this case the **instantaneous axis.**

The straight lines of space which during the progress of the motion become instantaneous axes for the infinitely small twists of the body form a ruled surface. Similarly, the lines of the moving body which in the course of the motion come to coincide with these axes generate another ruled surface. In any given position of the body these two surfaces are in contact along a line (the instantaneous axis) which is a generator in each of the two surfaces. The body has an infinitely small rotation about this line and at the same time slides along this line through an infinitely small distance.

Thus *the continuous motion of a rigid body in the most general case can be regarded as consisting of the combined rolling and sliding of one ruled surface over another.*

V. *Composition and Resolution of Translations; Vectors.*

38. All the points of a rigid body subjected to a translation describe parallel and equal lines (Art. 4). The change of position due to a translation of the body is therefore fully determined by

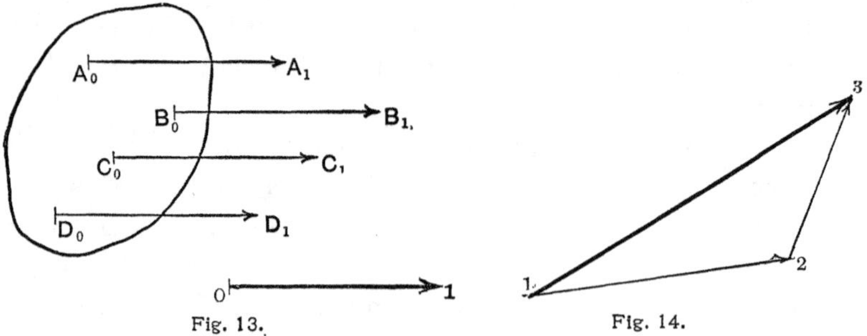

Fig. 13. Fig. 14.

the displacement A_0A_1 of any one point A of the body (Fig. 13), and can be represented geometrically by A_0A_1 or any rectilinear segment equal and parallel to it, like 01.

The *sense* of the segment (see Art. 1) which expresses whether the translation is to take place from 0 to 1 or from 1 to 0, is indicated graphically by an arrow-head, and in naming the segment by the order of the letters, 01 and 10 being segments of opposite sense.

39. Imagine a rigid body subjected to two successive translations. From any point 1 (Fig. 14) draw a segment 12 representing the first translation, and from its end 2 a segment 23 representing the second translation. The segment 13 will then represent a translation that would bring the body directly from its initial to its final position. This segment 13 is called the **geometric sum,** or the **resultant,** of the segments 12 and 23, which are called the **components.** The operation of combining the components into a resultant, or of finding the geometric sum of two segments, is called **geometric addition,** or **composition.**

40. The geometric sum of two segments can also be found by drawing, *from one and the same origin* 1 (Fig. 15), segments 12, 12′, parallel and equal respectively to the given segments, and completing the parallelogram with 12 and 12′ as adjacent sides; the diagonal 13 is the geometric sum, or resultant, of the given segments. Geometric addition is therefore often said to follow the *parallelogram law.*

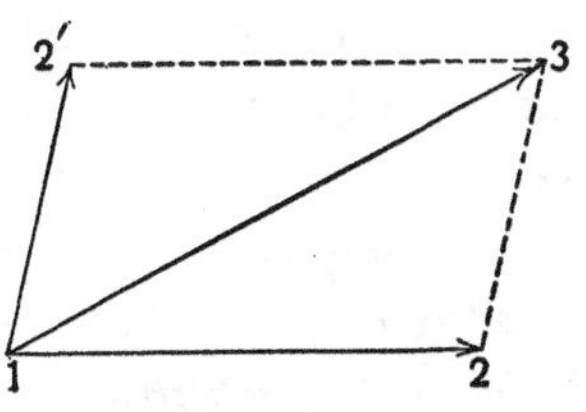

Fig. 15.

A glance at Fig. 15 shows that the order of succession in which the given segments are combined is indifferent for the result; in other words, geometric addition is *commutative,* like algebraic addition.

41. Rectilinear segments, taken as above with a definite *sense* (Art. 38) and subjected to the parallelogram law of combination (Art. 40), are called **vectors.** We can therefore say that a single translation is represented by a vector, and that *the resultant of two consecutive translations is found by adding their vectors geometrically.*

The vector, as the geometric symbol of a translation, has *length, direction,* and *sense;* but it is not restricted to any definite *position,* the same translation being represented by all equal and parallel vectors. We express this by saying that *two vectors are equal if they are of the same length, direction, and sense.*

42. It is easily seen how the operation of geometric or vector addition can be extended to the case of *more than two components*. Having found the resultant of the first and second given vectors, we can geometrically add to this resultant the third vector, and so on. Graphically, this is performed by drawing from any origin O a vector 12 equal to the first given vector, from its end 2 a vector 23 equal to the second given vector, from 3 a vector 34 equal to the third given vector, and so on; the vector drawn from 1 to the end of the last given vector is the geometric or vector sum, or the resultant, of the given vectors.

That here also the order of succession in which the given vectors are combined is indifferent for the result follows by considering that any order of the vectors can be obtained by repeated interchanges of two successive vectors and that for two successive vectors the operation is commutative (Art. 40).

It thus appears that *the succession of any number of translations of a rigid body has for its resultant a single translation whose vector is found by geometrically adding the vectors of the component translations.* (Compare Art. 2.)

43. Translations are not the only magnitudes in mechanics that can be represented by vectors. We shall see later that velocities, accelerations, moments of couples, etc., can all be represented by vectors and are therefore compounded into resultants and resolved into components by geometric addition and subtraction. In this lies the importance of this subject, which in its special application to translations might appear too simple and self-evident to require extended presentation.

The case when the vectors represent concurrent forces is probably known to the student from elementary physics as the "parallelogram, or polygon, of forces."

44. A translation may be *resolved* into two or more translations by resolving its vector into components.

When the resultant translation and one of its components are given by their vectors, the process of finding the other component is called **geometric subtraction.** It is effected, like algebraic subtraction, by reversing the *sense* of the component to be subtracted, and then geometrically adding it to the resultant (Fig. 16). In other words, the *geometric difference* of two vectors AB and CD is found by geometrically adding to AB a vector equal but opposite to CD. Thus, in Fig. 16, 13 is made equal and parallel to AB; 32 is equal and parallel to CD reversed, that is to DC; 12 is the required difference.

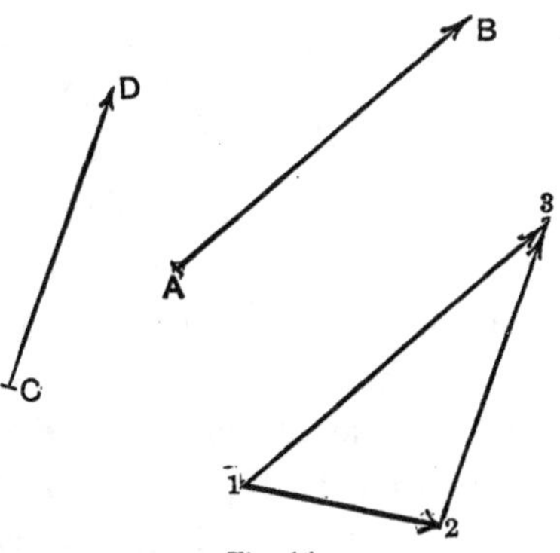

Fig. 16.

45. The composition of translations by geometric addition of their vectors (Art. 42) holds, not for *successive* translations only, but, owing to the commutative law (Art. 40), for *simultaneous* translations as well. This is easily seen by resolving the components into infinitesimal parts.

To obtain a clear idea of two simultaneous translations it is best to imagine the body as having one of these translations with respect to some other body, while the latter itself is subjected to the other translation. A man walking across the deck of a vessel in motion, an object let fall in a moving carriage, a spider running along a branch swayed by the wind, are familiar examples.

46. This leads us to the idea of **relative motion.** Properly speaking, all motion is relative; that is, we can conceive of the motion of a body only with regard to some other body, called the **body of reference.** If the latter be regarded as fixed, the motion of the former is called its **absolute motion.**

Thus in speaking of the motion of a railway train, we usually regard the earth as fixed and can thus call the displacement of the train from one station to another an *absolute displacement.* If, however, the motion of the earth with regard to the sun be taken into account, the displacement of the train from station to station is the relative displacement of the train with respect to the earth; and its absolute displacement would be found by combining this relative displacement with the absolute displacement of the earth (with respect to the sun regarded as fixed).

47. It follows that when the two displacements are translations *the absolute displacement of the body will be found by geometrically adding its relative displacement to the absolute displacement of the body of reference.* And conversely, *the relative displacement of a body is found by geometrically subtracting from its absolute displacement the absolute displacement of the body of reference.*

48. Analytically, the composition and resolution of vectors is merely a problem of trigonometry. Thus, the resultant of two vectors is the diagonal of the parallelogram formed by the two vectors as adjacent sides; the resultant of three vectors is the diagonal of the parallelepiped having the three vectors as concurrent edges.

49. In the case of more than two or three vectors, however, the solution by ordinary trigonometry would become rather tedious, and it is best to proceed as follows:

Assume an origin O and three rectangular axes Ox, Oy, Oz, and project each vector on the three axes; let X, Y, Z be its projections. These projections X, Y, Z are three vectors whose geometrical sum is equal to the vector. If n vectors were originally given, we should now have them replaced by $3n$ components of which n lie in each axis. The components lying in the same axis can be added algebraically; let their respective sums be ΣX, ΣY, ΣZ. The n vectors are therefore equivalent to the three vectors ΣX, ΣY, ΣZ, which form the concurrent edges of a rectangular parallelepiped whose diagonal drawn through the origin O is the resultant vector $OR = R$, *i. e.*

$$R = \sqrt{(\Sigma X)^2 + (\Sigma Y)^2 + (\Sigma Z)^2}.$$

The direction of this vector is given by the equations

$$\cos \alpha = \frac{\Sigma X}{R}, \quad \cos \beta = \frac{\Sigma Y}{R}, \quad \cos \gamma = \frac{\Sigma Z}{R},$$

where α, β, γ are the angles made by OR with the axes Ox, Oy, Oz, respectively.

If all the vectors lie in the same plane, we have simply:

$$R = \sqrt{(\Sigma X)^2 + (\Sigma Y)^2}, \quad \tan\alpha = \frac{\Sigma Y}{\Sigma X}.$$

50. Exercises.

(1) A ship sails first 6 miles N. 60° E., then 16 miles N. 60° W., and finally 18 miles S. 15° E. Find distance and bearing of the point reached: (*a*) graphically, (*b*) analytically.

(2) Is a scale of 10 miles to the inch sufficient to obtain the results of Ex. (1) correctly to whole miles and degrees?

(3) A body is given first a translation of 3 feet along one side of an equilateral triangle, then of 2 feet parallel to the second, then of 1 foot parallel to the third side, the same way around. Find the resulting displacement.

(4) A ship is carried by the current 2 miles due W., and at the same time by the wind 4 miles due N.W., and by her screw 9 miles N.E. Find her resultant displacement.

(5) A ferry-boat crosses a river in a direction inclined at an angle of 60° to the direction of the current. If the width of the river be half a mile, what are the component displacements of the boat along the river and at right angles to it?

(6) Two vectors of equal length are inclined to each other at an angle α. Find the resultant in magnitude and direction.

(7) For what angle α, in Ex. (6), is the resultant equal in magnitude: (*a*) to each component a? (*b*) to $\frac{1}{3} a$?

(8) Resolve a vector a into two components making with the vector angles of 30° and 45° on opposite sides.

(9) Steering his boat directly across a river whose current is due west, a man arrives on the opposite bank at a point from which the starting-point bears S.E.; the width of the river being 1500 feet, how far has he rowed? What is the absolute, and what the relative, displacement of the boat?

(10) Assuming a raindrop to fall 24 feet in a second in a vertical direction, find in what direction it appears to be falling to a man: (*a*) walking at the rate of 4 feet per second, (*b*) driving at the rate of 10

feet per second, (*c*) riding on a bicycle at 18 feet per second, (*d*) in a railroad car running 60 feet per second.

(11) Find in magnitude and direction the resultant of 6 translations of 1, 2, 3, 4, 5, 6 feet, respectively, each component making an angle of 45° with the preceding one: (*a*) graphically, (*b*) analytically.

(12) If a, b, c are three vectors whose geometric sum is 0, prove that $a/\sin(bc) = b/\sin(ca) = c/\sin(ab)$.

(13) Find the resultant of two translations represented in magnitude and direction by two rectangular chords of a circle drawn from a point on its circumference.

(14) From a point C in the plane of a circle whose center is O, draw two lines at right angles to each other and intersecting the circle in A, A' and B, B', respectively. Show that the resultant of the four vectors CA, CA', CB, CB' is equal to twice CO.

(15) Prove that the geometric sum of two vectors P_0P_1, P_0P_2 issuing from the same point P_0 passes through the middle point G of P_1P_2 and has a length $= 2\,P_0G$.

CHAPTER II.

KINEMATICS.

I. *Time.*

51. Before introducing the idea of time into the study of motion, a word must be said on the measurement of time.

It is the province of astronomy to devise methods for measuring time; the usual method consists in transit observations. Thus the fundamental unit of time in astronomy, or the **sidereal day**, is the interval between two successive upper transits of a fixed star over the same meridian, that is, the interval of time in which the earth makes one complete revolution referred to some fixed star.

52. For the purposes of every-day life, it is more convenient to make the measurement of time depend on the apparent revolution of the sun. But the interval between two successive upper transits of the sun over the same meridian, which is the *true*, or *apparent, solar day*, is not constant throughout the year, owing to the inclination of the earth's axis to the plane of its orbit and to the ellipticity of this orbit. The true solar day is thus not well adapted to serve as a unit of time.

Astronomers imagine, therefore, a so-called *mean sun*, which is supposed to travel around the earth at a *uniform* rate, in such a way as to make the *years* of the real and mean suns equal. The interval between two successive upper transits of this mean sun over the same meridian is called the **mean solar day**. This may be regarded as the standard on which all time determinations in mechanics are based.

The mean solar day is subdivided into 24 hours = 1440 minutes = 86 400 seconds. In theoretical mechanics the **second** is generally used as the unit of time.

53. The relation between mean solar time and sidereal time is readily found by considering that the tropical year, *i. e.* the interval between two successive passages of the sun through the mean vernal equinox, has 365.2422 mean solar days, and of course just one more sidereal day. Hence 1 solar day $= 366.2422 \div 365.2422 = 1.002738$ sidereal day; in other words, the sidereal day contains 86 164.1 seconds of mean time, while the solar day contains 86 400 such seconds.*

It will have been noticed that all these methods of measuring time are ultimately based on the assumption that the rotation of the earth on its axis is perfectly uniform. Observation shows this assumption to be true, or at least to have a very high degree of approximation.

II. *Linear Kinematics.*

1. UNIFORM RECTILINEAR MOTION; VELOCITY.

54. Consider a point moving in a straight line. If throughout the whole motion equal spaces are always described in equal times, the motion is said to be **uniform.**

Next consider two points each moving uniformly in a straight line. The motions may still be different; for it is possible that while one of the points moves in a given time t over a space s_1, the other moves during the same time t over a different space s_2. The points are then said to have different velocities, and their velocities are said to be as s_1 is to s_2. The **velocity** v *of uniform motion* is therefore measured by the ratio of the space s described in any time t to this time; that is, $v = s/t$.

55. This equation written in the form

$$s = vt \tag{1}$$

is called the **equation of motion** of the point. It follows from Art. 54 that in uniform motion the velocity v is constant.

With t as abscissa and s as ordinate (or *vice versa*), the equa-

---

* For further particulars see W. CHAUVENET, *Spherical and practical astronomy*, Vol. I., p. 52 sq. and pp. 651–654; also the *American Ephemeris and Nautical Almanac.*

tion of uniform motion (1) represents a straight line; the tangent of the angle made by this line with the axis of t represents the velocity v.

56. Let the point P start at the time $t = 0$ from a point O (Fig. 17); let it reach the point P_0 at the time $t = t_0$ and the point P_1 at the time $t = t$. Then, putting $OP_0 = s_0$, $OP_1 = s$, the space passed over in the time $t - t_0$ is $s - s_0$; hence the velocity

Fig. 17.

$v = (s - s_0)/(t - t_0)$. The equation of uniform motion can therefore be written in the form

$$s - s_0 = v(t - t_0). \qquad (1')$$

If the times be counted from the instant when the moving point is at P_0, we have $t_0 = 0$, and the equation of motion is

$$s = s_0 + vt. \qquad (1'')$$

Finally, if both times and spaces are counted from P_0 as origin, we have $s_0 = 0$, so that (1'') reduces to (1).

57. To measure velocities we must adopt a **unit of velocity.**

In kinematics, the only **fundamental,** *i. e.* independent, units required are those of length and time. All other quantities can be expressed in terms of length and time, and their units are therefore called **derived** units.

Thus, the definition of the velocity of uniform motion as a length divided by a time (Art. 54) can be expressed by the symbolic equation

$$V = \frac{L}{T}, \text{ or } V = LT^{-1},$$

and we say that the **dimensions** of velocity are 1 in length and -1 in time.

When $L = 1$ and $T = 1$, we have $V = 1$. We must therefore select for our unit of velocity that velocity with which unit length is described in unit time.

Hence in the C. G. S. system (see Arts. 8, 9) the unit velocity is a velocity of 1 cm. per second; in the F.P.S. system it is a velocity of 1 ft. per second.

58. In practice other units are often used, and the same concrete velocity can therefore be expressed by different numbers. Thus the same velocity of a railroad train can be described as 30 miles per hour, or 44 feet per second, or (approximately) 13.41 meters per second.

The symbols s, v, t, etc., in the kinematical equations must be understood to represent the numerical **ratios** of the concrete quantities to their respective units. The symbol v, for instance, stands for the ratio V/V_1 of the concrete velocity V to its unit V_1, and we have of course the proportion: 30 miles an hour is to one mile an hour as 44 feet per second is to one foot per second, etc.

59. The full meaning of the equation of dimensions $\mathrm{V} = \mathrm{LT}^{-1}$ is obtained if we substitute V/V_1 for V, L/L_1 for L, T/T_1 for T, where V, L, T are the concrete quantities and V_1, L_1, T_1 their units. We find

$$\frac{V}{V_1} = \frac{L}{L_1} \cdot \frac{T_1}{T};$$

and this equation shows two things which are of frequent application in reductions between different systems of units:

(a) The *numerical value* V/V_1 of a velocity varies directly as the unit of time and inversely as the unit of length;

(b) the *unit* of velocity V_1 varies directly as the unit of length and inversely as the unit of time.*

60. In speaking of velocities, the time unit (usually the second) is frequently understood without being mentioned. This has led to considering velocity as a length (viz. the length passed over in unit time); it can then be represented graphically by a segment of a straight line, and if in addition we combine with the idea of

* See J. D. EVERETT, *C. G. S. system of units*, 1902, p. 3.

velocity that of the *direction* and *sense* of the motion, its geometrical representative will be a vector (see Art. 41). We shall see later that this view is of particular advantage in studying the velocity of curvilinear motion.

Some recent writers on mechanics use the term *velocity* exclusively in this meaning, *i.e.* as denoting a vector, and apply the term *speed* to denote the numerical magnitude of this vector. In linear kinematics the direction is given, and the "speed" alone is the subject of investigation. The + or − sign of the "speed" expresses the *sense* of the motion.

61. Exercises.

(1) A train leaves the station A at 9 h. 5 m., passes (without stopping) B at 9 h. 31 m., C at 9 h. 48 m., and arrives at D at 10 h. 11 m., the distance AD being 46.2 miles. Considering the motion as uniform:

(*a*) What is the velocity?

(*b*) What is the equation of motion?

(*c*) What are the distances BD and CD?

(*d*) If after stopping 10 minutes at D the train goes on with the same velocity as before, when will it reach E, $20\frac{1}{4}$ miles beyond D?

(*e*) Construct a graphical time-table, taking the times as abscissas and the distances as ordinates.

(2) Interpret equations (1′) and (1″) geometrically.

(3) A train leaves Detroit at 8 h. 25 m. A.M. and reaches Chicago at 4 h. 5 m. P.M.; another train leaves Chicago at 10 h. 30 m. A.M. and arrives in Detroit at 5 h. 30 m. P.M. The distance is 284 miles. Regarding the motion as uniform and neglecting the stops, find, both analytically and graphically, when and where the trains will meet.

(4) Four trains leave New York at 8 h., 9 h. 30 m., 10 h., 11 h. 20 m., A.M., respectively, arriving at Boston at 1 h., 2 h. 30 m., 4 h., 5 h., P.M., respectively. Taking the distance as 300 miles, find graphically when and where a train leaving Boston at noon and reaching New York at 5 h. 30 m. will meet the four trains, all velocities being regarded as constant.

(5) Reduce the following velocities to F. P. S. units: (*a*) Walking 4 miles an hour; (*b*) trotting a mile in 2 m. 10 s.; (*c*) railroad

train, from 30 to 50 miles per hour; (d) bicyclist, 2 miles in 4 m. $59\frac{3}{5}$ s.; (e) sound in air, 333 meters per second.

(6) Two men starting (in opposite sense) from the same point walk around a block forming a rectangle of sides a, b; if their constant speeds are v_1, v_2, when and where will they meet?

(7) How is the unit of velocity changed if the minute be adopted as unit of time, the unit of length remaining unchanged?

(8) The mean distance of the sun being $92\frac{1}{3}$ million miles, find the velocity of light if it takes light 16 m. 42 s. to cross the earth's orbit (a) in miles per second, (b) in centimeters per second.

(9) Two trains are running on the same track at the rate of 30 and 24 miles per hour, respectively. If at a certain instant they are 12 miles apart, find both graphically and analytically when they will collide (a) if they are headed the same way; (b) if they run in opposite directions.

(10) In what latitude is a bullet shot west with a velocity of 1320 ft. per second at rest relatively to the earth's axis, the radius being taken as 4000 miles?

(11) Two trains, one 250, the other 420 ft. long, pass each other on parallel tracks in opposite directions with equal velocity. A passenger in the shorter train observes that it takes the longer train just 6 seconds to pass him. What is the velocity?

2. VARIABLE RECTILINEAR MOTION; ACCELERATION.

62. If the motion be not uniform, the definition of velocity given in Art. 54 is not applicable, as it would not give any definite meaning to the term.

We may of course divide the whole space, or any portion of it, by the corresponding time; and the quotient so obtained is called the **mean, or average, velocity for that space or time.** But its value is in general different for different portions of the path. It simply represents that constant velocity with which the space could be described in the same time in which it is actually described.

63. While we cannot speak, generally, of *the velocity* of a variable motion, we attach a perfectly definite meaning to the expression: *the velocity* of the motion *at any particular point or instant.*

To obtain a mathematical expression for this velocity at the point P, or at the time t, let us consider a point moving in a straight line. Let P (Fig. 18) be its position at the time t, P' its position at the time $t + \Delta t$; let the spaces be counted from the point O as origin so that $OP = s$, $OP' = s + \Delta s$. Then, by

Fig. 18.

Art. 62 the quotient $\Delta s / \Delta t$ is the *average* velocity in the interval PP'.

As P' approaches P, *i. e.*, as Δs and Δt approach zero, this average velocity $\Delta s / \Delta t$ will approach a definite limit. This limit is called the **velocity** v *of the point at* P; we have thus the definition

$$v = \lim_{\Delta t = 0} \frac{\Delta s}{\Delta t} = \frac{ds}{dt};$$

i. e., in variable motion in a straight line the velocity at any particular point of the path, or at any instant of time, is the value, at that point or time, of the first derivative of the space with respect to the time; in other words, *velocity is the time-rate of change of space.*

This definition includes uniform motion as a special case; for in this case, v being constant, the equation of uniform motion (1), Art. 55, gives $ds/dt = v$.

64. The definition of velocity given in the last article,

$$v = \frac{ds}{dt}, \tag{2}$$

enables us to find the velocity if the space be given as a function of the time, say $s = f(t)$; and conversely, if the velocity be given as a function of time or space, we find by integrating the differential equation $ds/dt = v$ an integral *equation of motion* $s = f(t)$.

If v be given as function of t, say $v = \phi(t)$, we find from (2) $ds = vdt$, and hence by integration

$$s - s_0 = \int_{t_0}^{t} vdt, \tag{3}$$

where s_0 is the space described during the time t_0.

Similarly, if v be given as a function of s, say $v = \psi(s)$, we have from (2) $dt = ds/v$, and hence

$$t - t_0 = \int_{s_0}^{s} \frac{ds}{v}. \tag{4}$$

65. We have seen (Art. 56, equation (1″)) that in the case of uniform motion the velocity $v = (s - s_0)/t$, *i. e.*, the rate of change of space with time, is constant. The simplest case of variable motion is that in which the velocity varies uniformly. In rectilinear motion, *the rate at which the velocity varies with the time* is called the **acceleration**; we shall denote it by j.

If the velocity vary uniformly, the acceleration is constant, and we have $j = (v - v_0)/t$, where t is the time during which the velocity changes from v_0 to v.

By reasoning analogous to that employed in Art. 63, we find for the acceleration of *any* rectilinear motion at the time t

$$j = \frac{dv}{dt} = \frac{d^2s}{dt^2}; \tag{5}$$

that is, *in rectilinear motion the acceleration at any point or instant is the value, at that point or instant, of the second derivative of the space with respect to the time.*

Negative acceleration will thus indicate a decreasing velocity.

When the acceleration is constant, the motion is said to be **uniformly accelerated.**

66. Conformably to the definition of acceleration, its unit is the "cm. per second per second" in the C. G. S. system, and the "foot per second per second" in the F. P. S. system. As it can rarely be convenient to use two different time units in the unit of acceleration (say, for instance, mile per hour per second), it is customary to mention the time unit but once and to speak of an acceleration of so many feet per second, or cm. per second, it being understood that the other time unit is also the second.

For the *dimensions* of acceleration we have (see Art. 57)

$$J = VT^{-1} = LT^{-2}.$$

Denoting, as in Arts. 58, 59, the concrete value of an acceleration by J, its unit by J_1, and similarly for length and time, we have the equation

$$\frac{J}{J_1} = \frac{L}{L_1} \cdot \frac{T_1^2}{T^2},$$

which shows that (*a*) the numerical value J/J_1 of an acceleration varies directly as the square of the unit of time, and inversely as the unit of length; and (*b*) the unit of acceleration, J_1, varies directly as the unit of length, and inversely as the square of the unit of time.

67. Exercises.

(1) A point moving with constant acceleration gains at the rate of 30 miles an hour in every minute. Express its acceleration in F. P. S. units.

(2) At a place where the acceleration of gravity is $g = 9.810$ meters per second, what is the value of g in feet per second?

(3) A railroad train, 10 minutes after starting, attains a velocity of 45 miles an hour; what was its average acceleration during these 10 minutes?

(4) If the acceleration of gravity, $g = 32$ feet per second, be taken as unit, what is the acceleration of the railroad train in Ex. (3)?

3. APPLICATIONS.

68. Uniformly Accelerated Motion. As in this case the acceleration j is constant (see Art. 65), the equation of motion (5)

$$\frac{d^2s}{dt^2} = j, \quad \text{or} \quad \frac{dv}{dt} = j,$$

can readily be integrated:

$$v = jt + C.$$

To determine the constant of integration C, we must know the value of the velocity at some particular moment of time. Thus, if $v = v_0$ when $t = 0$, we find $v_0 = C$; hence, substituting this value for C,

$$v - v_0 = jt. \tag{6}$$

This equation, which agrees with the definition of j given in Art. 65, gives the velocity at any time t. Substituting ds/dt for v and integrating again, we find $s = v_0 t + \frac{1}{2}jt^2 + C'$, where the constant of integration, C', must again be determined from given "initial conditions." Thus, if we know that $s = s_0$ when $t = 0$, we find $s_0 = C'$; hence

$$s - s_0 = v_0 t + \tfrac{1}{2}jt^2. \tag{7}$$

This equation gives the space or distance passed over in terms of the time.

69. Eliminating j between (6) and (7), we obtain the relation

$$s - s_0 = \tfrac{1}{2}(v_0 + v)t,$$

which shows that in uniformly accelerated motion the space can be found as if it were described uniformly with the mean velocity $\frac{1}{2}(v_0 + v)$.

70. To obtain the velocity in terms of the space, we have only to eliminate t between (6) and (7); we find

$$\tfrac{1}{2}(v^2 - v_0^2) = j(s - s_0). \tag{8}$$

This relation can also be derived by eliminating dt between the differential equations $v = ds/dt$, $dv/dt = j$, which gives $vdv = jds$, and integrating. The same equation (8) is also obtained directly from the fundamental equation of motion $d^2s/dt^2 = j$ by a process very frequently used in mechanics, viz., by multiplying both members of the equation by ds/dt. This makes the left-hand member the exact derivative of $\frac{1}{2}(ds/dt)^2$ or $\frac{1}{2}v^2$, and the integration can therefore be performed.

71. The three equations (6), (7), (8) contain the complete solution of the problem of uniformly accelerated motion. For uniformly retarded motion, taking the direction of motion as positive, we have only to write $-j$ for $+j$.

If the spaces be counted from the position of the moving point at the time $t = 0$, we have $s_0 = 0$, and the equations become

$$v = v_0 + jt, \tag{6'}$$

$$s = v_0 t + \tfrac{1}{2} jt^2, \tag{7'}$$

$$\tfrac{1}{2}(v^2 - v_0^2) = js. \tag{8'}$$

If in addition the initial velocity v_0 be zero, the point starting from rest at the time $t = 0$, the equations reduce to the following:

$$v = jt, \tag{6''}$$

$$s = \tfrac{1}{2} jt^2, \tag{7''}$$

$$\tfrac{1}{2} v^2 = js. \tag{8''}$$

72. The most important example of uniformly accelerated motion is furnished by a body falling in vacuo near the earth's surface. Assuming that the body does not rotate during its fall, its motion relative to the earth is a mere translation, and it is sufficient to consider the motion of any one point of the body. It is known from observation and experiment that under these circumstances the acceleration of a falling body is constant at any given place and equal to about 980 cm., or 32 ft., per second per second; the value of this so-called *acceleration of gravity* is usually denoted by g.

In the exercises on falling bodies (Art. 74) we make throughout the following simplifying assumptions: the falling body does not rotate; the resistance of the air is neglected, or the body falls in vacuo; the space fallen through is so small that g may be regarded as constant; the earth is regarded as fixed.

73. The velocity v acquired by a falling body after falling from rest through a height h is found from (8'') as

$$v = \sqrt{2gh}.$$

This is usually called the **velocity due to the height** (or **head**) h, while

$$h = \frac{v^2}{2g}$$

is called the **height** (or **head**) **due to the velocity** v.

74. Exercises.

(1) A body falls from rest at a place where $g = 32.2$. Find (*a*) the velocity at the end of the fourth second; (*b*) the space fallen through in 4 seconds; (*c*) the space fallen through in the fifth second.

(2) If a railroad train, at the end of 2 m. 40 s. after leaving the station, has acquired a velocity of 30 miles per hour, what was its acceleration (regarded as constant)?

(3) Galileo, who first discovered the laws of falling bodies, expressed them in the following form: (*a*) The velocities acquired at the end of the successive seconds increase as the natural numbers; (*b*) the spaces described during the successive seconds increase as the odd numbers; (*c*) the spaces described from the beginning of the motion to the end of the successive seconds increase as the squares of the natural numbers. Prove these statements.

(4) A stone dropped into the vertical shaft of a mine is heard to strike the bottom after t seconds; find the depth of the shaft, if the velocity of sound be given $= c$. Assume $t = 4$ s., $c = 332$ meters, $g = 980$.

(5) A railroad train in approaching a station makes half a mile in the first, 2000 ft. in the second, minute of its retarded motion. If the motion is *uniformly* retarded: (*a*) When will it stop? (*b*) What is the retardation? (*c*) What was the initial velocity? (*d*) When will its velocity be 4 miles an hour?

(6) Interpret equations (6) and (7) geometrically.

(7) A body being projected vertically upwards with an initial velocity v_0, (*a*) how long and (*b*) to what height will it rise? (*c*) When and (*a*) with what velocity does it reach the starting-point?

(8) A bullet is shot vertically upwards with an initial velocity of 1200 ft. per second. (*a*) How high will it ascend? (*b*) What is its velocity at the height of 16,000 ft.? (*c*) When will it reach the ground

again? (*d*) With what velocity? (*e*) At what time is it 16,000 ft. above the ground? Explain the meaning of the double signs wherever they occur in the answers.

(9) With what velocity must a ball be thrown vertically upwards to reach a height of 100 ft.?

(10) A body is dropped from a point A at a height $AB = h$ above the ground; at the same time another body is thrown vertically upward from the point B, with an initial velocity v_0. (*a*) When and (*b*) where will they collide? (*c*) If they are to meet at the height $\frac{1}{2} h$, what must be the initial velocity?

(11) If a train can attain its regular speed in 3 minutes and can be brought to rest in the same time, how much time is lost by making four stops of 2 minutes each between two stations?

(12) The barrel of a rifle is 30 in. long; the muzzle velocity is 1300 ft./sec.; if the motion in the barrel be uniformly accelerated, what is the acceleration and what the time?

(13) If a stone dropped from a balloon while ascending at the rate of 25 ft./sec. reaches the ground in 6 seconds, what was the height of the balloon when the stone was dropped?

(14) If the speed of a train increases uniformly after starting for 8 minutes while the train travels 2 miles, what is the velocity acquired?

(15) A train running 30 miles an hour is brought to rest, uniformly, in 2 minutes. How far does it travel?

(16) Two particles fall from rest from the same point, at a short interval of time τ; find how far they will be apart when the first particle has fallen through a height h. Take e.g. $h = 900$ ft., $\tau = \frac{1}{400}$ second.

75. The general problem of rectilinear motion requires the integration of the differential equation

$$\frac{d^2s}{dt^2} = j, \qquad (5)$$

where j is a function of s, t, and v, in connection with the equation

$$\frac{ds}{dt} = v. \qquad (2)$$

As these two equations involve four quantities t, s, v, j, a third relation between them, say

$$f(t, s, v, j) = 0, \tag{9}$$

is always necessary in order to express three of these four quantities in terms of the fourth. Next to the case of uniformly accelerated motion where the relation (9) is simply $j =$ const., the most important cases are those when j is given as a function of t, of s, or of v, or of both s and v.

76. Whenever in nature we observe a motion not to remain uniform, we try to account for the change in the character of the motion by imagining a special cause for such change. In rectilinear motion, the only change that can occur in the motion is a change in the velocity, *i.e.* an acceleration (or retardation). It is often convenient to have a special name for this supposed cause producing acceleration or retardation; we call it **force** (attraction, repulsion, pressure, tension, friction, resistance of a medium, elasticity, cohesion, etc.), and assume it to be proportional to the acceleration. A fuller discussion of the nature of force and its relation to mass will be found in Arts. 255–272. The present remark is only intended to make more intelligible the physical meaning and applications of the problems to be discussed in the following articles.

77. Acceleration inversely proportional to the square of the distance, *i. e.* $j = \mu / s^2$ where μ is a constant (viz. the acceleration at the distance $s = 1$) and s is the distance of the moving point from a fixed point in the line of motion.

The differential equation (5) becomes in this case

$$\frac{d^2s}{dt^2} = \frac{\mu}{s^2}; \tag{10}$$

the first integration is readily performed by multiplying both members by ds/dt so as to make the left-hand member the exact derivative of $\frac{1}{2}(ds/dt)^2$ or $\frac{1}{2}v^2$. Thus we find

$$\tfrac{1}{2}v^2 = \mu \int \frac{ds}{s^2} = -\frac{\mu}{s} + C, \tag{11}$$

where the constant of integration, C, must be determined from the so-called initial conditions of the problem. For instance, if $v = v_0$ when $s = s_0$, we have $\frac{1}{2}v_0^2 = -\mu/s_0 + C$; hence, eliminating C between this relation and (11),

$$\tfrac{1}{2}(v^2 - v_0^2) = -\mu\left(\frac{1}{s} - \frac{1}{s_0}\right). \tag{12}$$

To perform the second integration we solve this equation for v and substitute ds/dt for v:

$$\frac{ds}{dt} = \pm\sqrt{v_0^2 + \frac{2\mu}{s_0} - \frac{2\mu}{s}},$$

or putting $v_0^2 + 2\mu/s_0 = 2\mu/\mu'$,

$$\frac{ds}{dt} = \pm\sqrt{\frac{2\mu}{\mu'}} \cdot \sqrt{\frac{s - \mu'}{s}}. \tag{13}$$

Here the variables s and t can be separated, and we find

$$t = \pm\sqrt{\frac{\mu'}{2\mu}}\int\sqrt{\frac{s}{s - \mu'}}\,ds. \tag{14}$$

To integrate put $s - \mu' = x^2$. The result will be different according to the signs of μ, μ', and v, which must be determined from the nature of the particular problem.

It is easily seen that the methods of integration used in this problem apply whenever j is given as a function of s alone.

78. It is an empirical fact that the acceleration of bodies falling in vacuo on the earth's surface is constant only for distances from the surface that are very small in comparison with the radius of the earth. For larger distances the acceleration is found inversely proportional to the square of the distance from the earth's center.

By a bold generalization Newton assumed this law to hold generally between any two particles of matter, and this assumption has been verified by all subsequent observations. It can therefore be regarded as a general law of nature that any particle of matter produces in every other such particle, each particle being regarded as concentrated at a point, an acceleration inversely proportional to the square of the distance between these points. This is known as *Newton's law of universal gravitation*, the acceleration being regarded as caused by a force of attraction inherent in each particle of matter.

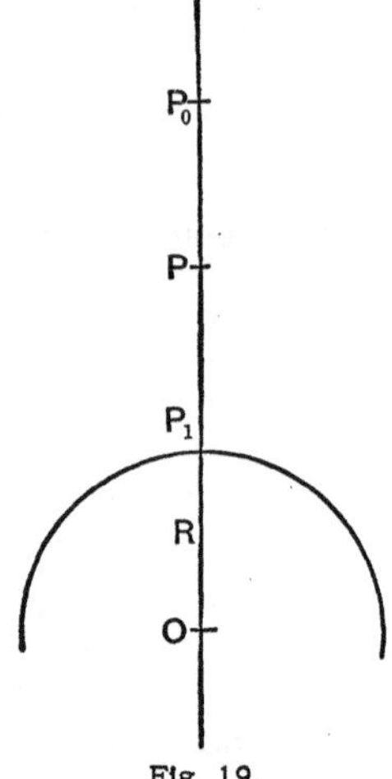

Fig. 19.

It is shown in the theory of attraction that the attraction of a spherical mass, such as the earth, on any particle *outside* the sphere is the same as if the mass of the sphere were concentrated at its center. The acceleration produced by the earth on any particle outside it is therefore inversely proportional to the square of the distance of the particle from the center of the earth.

79. Let us now apply the general equations of Art. 77 to the particular case of a body falling from a great height towards the center of the earth, the resistance of the air being neglected.

Let O be the center of the earth (Fig. 19), P_1 a point on its surface, P_0 the initial position of the moving point at the time $t = 0$, P its position at the time t; let $OP_1 = R$, $OP_0 = s_0$, $OP = s$; and let g be the acceleration at P_1, j the acceleration at P, both in absolute value. Then, according to Newton's law, $j : g = R^2 : s^2$. This relation serves to determine the value of the constant μ in (10); for since the acceleration is to have the value g when $s = R$ we have

$$\left(\frac{d^2s}{dt^2}\right)_{s=R} = \frac{\mu}{R^2} = -g,$$

the minus sign being taken because the acceleration is directed toward the origin O. We have therefore

$$\mu = -gR^2,$$

so that (10) becomes in our case

$$\frac{d^2s}{dt^2} = -\frac{gR^2}{s^2}, \tag{15}$$

the minus sign indicating that the acceleration tends to diminish the distances counted from O as origin.

The integration can now be performed as in Art. 77. Multiplying by ds/dt and integrating, we find $\frac{1}{2}v^2 = gR^2/s + C$. If the initial velocity be zero, we have $v = 0$ for $s = s_0$; hence

$$C = -gR^2/s_0,$$

and

$$v = -R\sqrt{2g}\sqrt{\frac{1}{s} - \frac{1}{s_0}} = -R\sqrt{\frac{2g}{s_0}}\sqrt{\frac{s_0 - s}{s}}. \tag{16}$$

Here again the minus sign before the radical is selected since the velocity v is directed in the sense opposite to that of the distance s.

Substituting ds/dt for v and separating the variables t and s we have

$$dt = -\frac{1}{R}\sqrt{\frac{s_0}{2g}}\sqrt{\frac{s}{s_0 - s}}\,ds;$$

ence, integrating as indicated at the end of Art. 77:

$$t = \frac{1}{R}\sqrt{\frac{s_0}{2g}}\left(\sqrt{s(s_0 - s)} + s_0 \sin^{-1}\sqrt{\frac{s_0 - s}{s_0}}\right),$$

the constant of integration being zero since $s = s_0$ for $t = 0$. The last term can be slightly simplified by observing that

$$\sin^{-1}\sqrt{1 - u^2} = \cos^{-1}u,$$

whence finally:

$$t = \frac{1}{R}\sqrt{\frac{s_0}{2g}}\left(\sqrt{s(s_0 - s)} + s_0 \cos^{-1}\sqrt{\frac{s}{s_0}}\right). \tag{17}$$

80. Exercises.

(1) Find the velocity with which the body arrives at the surface of the earth if it be dropped from a height equal to the earth's radius, and determine the time of falling through this height. Take $R = 4000$ miles, $g = 32$.

(2) Interpret equation (17) geometrically.

(3) Show that formula (16) reduces to $v = \sqrt{2gh}$ (Art. 73) with $s = R$ if $s_0 - s = h$ is small in comparison with R.

(4) Show that when s_0 is large in comparison with R while s differs but slightly from R, the formulæ (16) and (17) reduce approximately to

$$v = -\sqrt{2g}\frac{R}{\sqrt{s}}, \quad t = \frac{\pi}{2\sqrt{2g}}\frac{s_0^{\frac{3}{2}}}{R}.$$

Hence find the final velocity and time of fall of a body falling to the earth's surface (*a*) from an infinite distance; (*b*) from the moon ($s_0 = 60\ R$).

(5) Derive the expressions for v and t corresponding to (16) and (17) when the initial velocity is v_0.

(6) Find the time of fall and final velocity of a meteor if (*a*) $s_0 = 2R$, $v_0 = 1$ mile per second, (*b*) $s_0 = 3R$, $v_0 = 4$ miles per second, (*c*) $s_0 = 4R$, $v_0 = 10$ miles per second.

(7) A particle is projected vertically upwards from the earth's surface with an initial velocity v_0. How far and how long will it rise?

(8) If, in (7), the initial velocity be $v_0 = \sqrt{gR}$, how high and how long will the particle rise? How long will it take the particle to rise and fall back to the earth's surface?

(9) A body is projected vertically upwards. Find the least initial velocity that would prevent it from returning to the earth, taking $g = 32$, $R = 4000$ miles.

81. Acceleration directly proportional to the distance, *i. e.* $j = \kappa s$, where κ is a constant and s is the distance of the moving point from a fixed point in the line of motion.

The equation of motion

$$\frac{d^2s}{dt^2} = \kappa s \tag{18}$$

can be integrated by the method used in Art. 77. The result of the second integration will again be different according to the sign of κ. We shall study here only a special case, reserving the general discussion of this law of acceleration until later (see Arts. 125 sq.).

82. It is shown in the theory of attraction that the attraction of a spherical mass such as the earth on any point *within* the mass produces an acceleration directed to the center of the sphere and proportional to the distance from this center. Thus, if we imagine a particle moving along a diameter of the earth, say in a straight narrow tube passing through the center, we should have a case of the motion represented by equation (18).

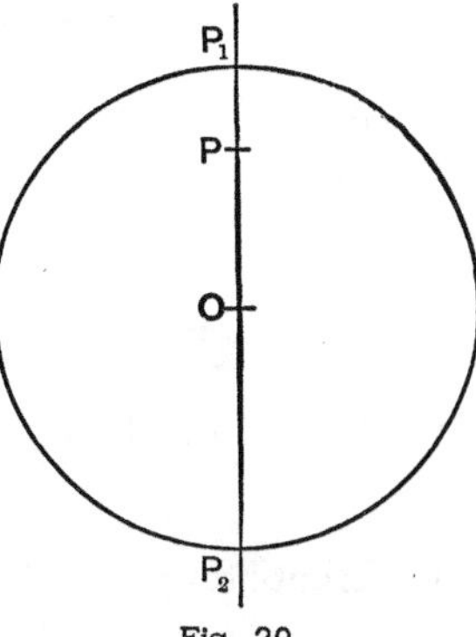

Fig. 20.

To determine the value of κ for our problem we notice that at the earth's surface, that is, at the distance $OP_1 = R$ from the center O (Fig. 20), the acceleration must be g. If, therefore, j denote the numerical value of the acceleration at any distance $OP = s\,(< R)$, we have $j : g = s : R$, or $j = gs/R$. But the acceleration tends to diminish the distance s, hence $\frac{d^2s}{dt^2} = -\frac{g}{R}s$. Denoting the positive constant g/R by μ^2, the equation of motion is

$$\frac{d^2s}{dt^2} = -\mu^2 s, \text{ where } \mu = \sqrt{\frac{g}{R}}. \tag{19}$$

Integrating as in Arts. 77 and 79, we find

$$\tfrac{1}{2}v^2 = -\tfrac{1}{2}\mu^2 s^2 + C.$$

If the particle starts from rest at the surface, we have $v = 0$ when $s = R$; hence $0 = -\frac{1}{2}\mu^2 R^2 + C$; and subtracting this from the preceding equation, we find

$$v = -\mu\sqrt{R^2 - s^2}, \tag{20}$$

where the minus sign of the square root is selected because s and v have opposite sense.

Writing ds/dt for v and separating the variables, we have

$$dt = -\frac{1}{\mu}\frac{ds}{\sqrt{R^2 - s^2}},$$

whence

$$t = \frac{1}{\mu}\cos^{-1}\frac{s}{R} + C'.$$

As $s = R$ when $t = 0$, we have $0 = \frac{1}{\mu}\cos^{-1} 1 + C'$, or $C' = 0$. Solving for s, we find

$$s = R\cos\mu t. \tag{21}$$

Differentiating, we obtain v in terms of t:

$$v = -\mu R\sin\mu t. \tag{22}$$

83. The motion represented by equations (21) and (22) belongs to the important class of *simple harmonic motions* (see Arts. 121 sq.). The particle reaches the center when $s = 0$, *i. e.*, when $\mu t = \pi/2$, or at the time $t = \pi/2\mu$. At this time the velocity has its maximum value. After passing through the center the point moves on to the other end, P_2, of the diameter, reaches this point when $s = -R$, *i. e.* when $\mu t = \pi$, or at the time $t = \pi/\mu$. As the velocity then vanishes, the moving point begins the same motion in the opposite sense.

The time of performing one complete oscillation (back and forth) is called the **period** of the simple harmonic motion; it is evidently

$$T = 4\cdot\frac{\pi}{2\mu} = \frac{2\pi}{\mu}$$

84. Exercises.

(1) Equation (19) is a differential equation whose general integral is known to be of the form

$$s = C_1\sin\mu t + C_2\cos\mu t;$$

determine the constants C_1, C_2 and deduce equations (21) and (22).

(2) Find the velocity at the center and the period, taking $g = 32$ and $R = 4000$ miles.

(3) If the acceleration, instead of being directed toward the center, is directed away from it, the equation of motion would be $d^2s/dt^2 = \mu^2 s$. Investigate this motion, which can be imagined as produced by a force of repulsion emanating from the center.

(4) A point whose acceleration is proportional to its distance from a fixed point O starts at the distance s_0 from O with a velocity v_0 directed away from O; how far will it go before returning?

4. ROTATION; ANGULAR VELOCITY; ANGULAR ACCELERATION.

85. A motion of rotation about a fixed axis can be treated in precisely the same way in which we have treated rectilinear motion in the preceding sections. It is only to be remembered that rotations are measured by angles (see Arts. 8–10), while translations are measured by lengths.

86. The rotation of a rigid body (see Art. 5) about a fixed axis is said to be *uniform* if equal angles are described in equal times; in other words, if the angle of rotation is proportional to the time in which it is described. In this case of uniform rotation, the quotient obtained by dividing the angle of rotation, θ, by the corresponding time, t, is called the **angular velocity**. Denoting it by ω we have $\omega = \theta/t$; and the equation of motion is

$$\theta = \omega t. \tag{1}$$

Thus, expressing the time in seconds and the angle in radians (Art. 10), the angular velocity is equal to the number of radians described per second. (Compare Arts. 54, 55.)

87. If the time of a whole revolution be denoted by T, we have, from (1), $2\pi = \omega T$; hence

$$\omega = \frac{2\pi}{T}. \tag{2}$$

In engineering practice it is customary to take a whole revolution as angular unit and to express the angular velocity of uniform rotation by the number of revolutions made in the unit of time.

Let n, N be the number of revolutions per second and per minute, respectively; then we have evidently

$$n = \frac{\omega}{2\pi}, \quad N = \frac{30\omega}{\pi}. \tag{3}$$

88. When the rotation is *not* uniform, the quotient obtained by dividing the angle of rotation by the time in which it is described, gives the *mean*, or *average, angular velocity* for that time.

The rate of change of the angle of rotation with the time at any particular moment is called the **angular velocity** *at that moment.* By reasoning similar to that in Art. 63 it will be seen that its mathematical expression is

$$\omega = \frac{d\theta}{dt}. \tag{4}$$

The rate at which the angular velocity changes with the time is called the **angular acceleration**; denoting it by α, we have

$$\alpha = \frac{d\omega}{dt} = \frac{d^2\theta}{dt^2}. \tag{5}$$

89. The most important special case of variable angular velocity is that of uniformly accelerated (or retarded) rotation when the angular acceleration is constant. The formulæ for this case have precisely the same form as those given in Arts. 68–71 for uniformly accelerated rectilinear motion. Denoting the constant linear acceleration by j, we have, when the initial velocity is 0,

FOR TRANSLATION:	FOR ROTATION:
$\frac{d^2s}{dt^2} = j$, a constant;	$\frac{d^2\theta}{dt^2} = \alpha$, a constant;
$v = jt$,	$\omega = \alpha t$,
$s = \frac{1}{2}jt^2$,	$\theta = \frac{1}{2}\alpha t^2$,
$\frac{1}{2}v^2 = js$;	$\frac{1}{2}\omega^2 = \alpha\theta$;

(6)

and when the initial velocities are v_0 and ω_0, respectively:

FOR TRANSLATION:	FOR ROTATION:
$v = v_0 + jt$,	$\omega = \omega_0 + \alpha t$,
$s = v_0 t + \frac{1}{2}jt^2$,	$\theta = \omega_0 t + \frac{1}{2}\alpha t^2$,
$\frac{1}{2}v^2 - \frac{1}{2}v_0^2 = js$;	$\frac{1}{2}\omega^2 - \frac{1}{2}\omega_0^2 = \alpha\theta$.

(7)

90. Let a point P, whose perpendicular distance from the axis of rotation is $OP = r$, rotate about the axis with the angular velocity $\omega = d\theta/dt$. In the element of time, dt, it will describe an element of arc $ds = rd\theta = r\omega dt$. Its velocity $v = ds/dt$ (frequently called its **linear** velocity in contradistinction to the angular velocity) is therefore related to the angular velocity of rotation by the equation

$$v = \omega r. \tag{8}$$

91. The radius vector $OP = r$ sweeps over a circular sector which in uniform rotation has an area $S = \frac{1}{2}\theta r^2 = \frac{1}{2}\omega t r^2$, while in variable rotation the infinitesimal sector described during the element of time dt is $dS = \frac{1}{2}r^2 d\theta = \frac{1}{2}\omega r^2 dt$.

The quotient

$$\frac{S}{t} = \frac{1}{2}r^2\frac{\theta}{t} = \frac{1}{2}\omega r^2, \tag{9}$$

for uniform rotation, and the corresponding derivative

$$\frac{dS}{dt} = \frac{1}{2}r^2\frac{d\theta}{dt} = \frac{1}{2}\omega r^2, \tag{10}$$

for variable rotation, represent, therefore, the **sectorial,** or *areal,* **velocity,** *i. e.* the rate of increase of area with the time (see Art. 97).

The rate of change of this velocity with the time,

$$\frac{d^2S}{dt^2} = \frac{1}{2}\frac{d}{dt}\left(r^2\frac{d\theta}{dt}\right), \tag{11}$$

is called the **sectorial,** or *areal,* **acceleration.**

92. Exercises.

(1) If a fly-wheel of 10 ft. diameter makes 30 revolutions per minute, what is its angular velocity, and what is the linear velocity of a point on its rim?

(2) A pulley 5 ft. in diameter is driven by a belt travelling 450 ft. a minute. Neglecting the slipping of the belt, find (a) the angular velocity of the pulley in radians per second, and (b) its number of revolutions per minute.

(3) Find the constant acceleration (such as the retardation caused by a Prony brake) that would bring the fly-wheel in Ex. (1) to rest in ⅓ minute.

(4) How many revolutions does the fly-wheel in Ex. (3) make during its retarded motion before it comes to rest?

(5) A wheel is running at a uniform speed of 32 turns a second when a resistance begins to retard its motion uniformly at a rate of 8 radians per second. (*a*) How many turns will it make before stopping? (*b*) In what time is it brought to rest?

(6) A belt runs over two pulleys turning about parallel axes. Show that the angular velocities of the pulleys are inversely proportional to their diameters. Do the pulleys rotate in the same or opposite sense?

(7) A wheel of 4 ft. diameter rolls on a straight horizontal track at the rate of 15 miles an hour; find the velocities of its foremost and hindmost points.

(8) A wheel of 6 ft. diameter is making 50 rev./min. when thrown out of gear. If it comes to rest in 4 minutes, find (*a*) the angular retardation; (*b*) the linear velocity of a point on the rim at the beginning of the retarded motion; (*c*) the same after two minutes.

III. *Plane Kinematics.*

1. VELOCITY; COMPOSITION OF VELOCITIES; RELATIVE VELOCITY.

93. Consider a point describing a curved path. Let P (Fig. 21) be the position of the point at the time t, P' its position at the time $t + \Delta t$; and let us select on the curve a point P_0 from which to count the distances s along the curve so that arc $P_0P = s$, arc $PP' = \Delta s$. Then, proceeding just as in the case of rectilinear motion (Art. 63) we may define the *speed*, or *magnitude of the velocity*, at the point P (or at the time t) as

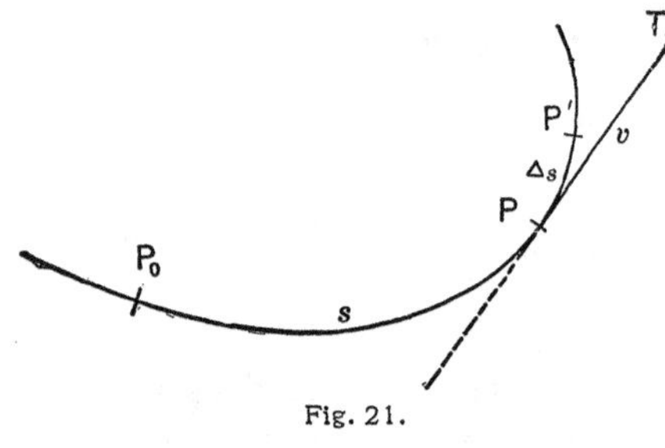

Fig. 21.

$$v = \lim_{\Delta t = 0} \frac{\Delta s}{\Delta t} = \frac{ds}{dt}.$$

It is however found convenient in curvilinear motion to incorporate in the definition of velocity the idea of the varying *direction* of the motion by assigning to the velocity at P as its direction that of the tangent to the curve at the point P. The **velocity** *at the point* P is therefore represented by laying off on the tangent to the curve at P, in the sense of the motion, a segment PT proportional to ds/dt.

94. Velocity is thus defined as having both magnitude and direction and is represented by a rectilinear segment PT. It is assumed, moreover, that these segments representing velocities can be combined and resolved according to the parallelogram law (Arts. 38–45), so that velocity is a **vector** quantity. The results of this assumption are in complete agreement with observed facts; a *direct* experimental verification would generally be difficult.

Thus if a point be subjected to two or more simultaneous velocities, the velocity of the resulting motion will be represented by the vector found by geometrically adding the component velocities. A velocity may be resolved into any number of component velocities whose geometric sum is equal to the given velocity.

We proceed to consider the most important cases of the resolution of a velocity in plane motion.

95. If the curve P_0P (Fig. 22) in which the point moves be re-

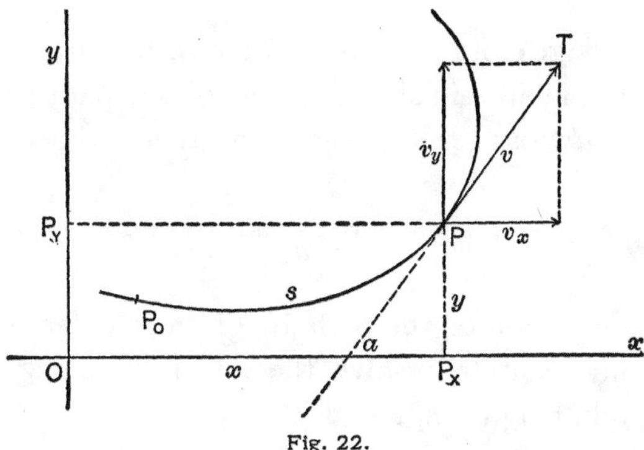

Fig. 22.

ferred to rectangular cartesian co-ordinates x, y, it is convenient to resolve the velocity v along the axes Ox, Oy into v_x, v_y. Denoting by α the angle at which the tangent to the curve at the point P is inclined to the axis Ox we have

$$v_x = v \cos\alpha, \quad v_y = v \sin\alpha.$$

As the point P moves along the curve, its projection P_x on the axis Ox moves along this axis; and since $OP_x = x$, the velocity of the rectilinear motion of the point P_x is dx/dt (Art. 63). Similarly, the projection P_y of P on the axis Oy moves along this axis Oy with the velocity dy/dt. It is easily seen that the components v_x, v_y of v are equal respectively to these velocities of P_x and P_y; that is:

$$v_x = \frac{dx}{dt}, \quad v_y = \frac{dy}{dt}. \tag{1}$$

For we have

$$v_x = v \cos\alpha = \cos\alpha \lim_{\Delta t = 0} \frac{\Delta s}{\Delta t} = \cos\alpha \lim \frac{\Delta s}{\Delta x}\frac{\Delta x}{\Delta t}$$

$$= \cos\alpha \cdot \frac{1}{\cos\alpha} \lim \frac{\Delta x}{\Delta t} = \frac{dx}{dt};$$

and similarly for v_y.

As the components v_x, v_y are rectangular we have

$$v = \sqrt{v_x^2 + v_y^2} = \sqrt{\left(\frac{dx}{dt}\right)^2 + \left(\frac{dy}{dt}\right)^2}. \tag{2}$$

In the language of infinitesimals these results are expressed by saying that the infinitesimal element ds of the path has the components $dx = ds \cos\alpha$, $dy = ds \sin\alpha$, and that division by dt gives

$$\frac{dx}{dt} = \frac{ds}{dt}\cos\alpha = v \cos\alpha = v_x, \quad \frac{dy}{dt} = \frac{ds}{dt}\sin\alpha = v \sin\alpha = v_y.$$

96. If the equation of the path be given in polar co-ordinates, it may be convenient to resolve the velocity v along the radius vector OP and at right angles to it (Fig. 23).

Let r, θ be the polar co-ordinates, ψ the angle between v and r; then $v_r = v\cos\psi$, $v_\theta = v\sin\psi$. The element ds of the curve has in the same directions the components $dr = ds\cos\psi$, $rd\theta = ds\sin\psi$. Hence, dividing by dt, we find

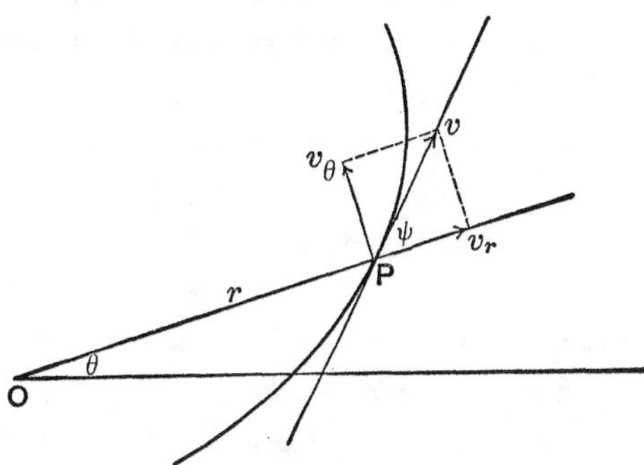

Fig. 23.

$$v_r = \frac{dr}{dt}, \quad v_\theta = r\frac{d\theta}{dt}, \tag{3}$$

and
$$v = \sqrt{v_r^2 + v_\theta^2} = \sqrt{\left(\frac{dr}{dt}\right)^2 + r^2\left(\frac{d\theta}{dt}\right)^2}. \tag{4}$$

97. As the point P moves along the curve its radius vector OP sweeps out the polar area S of the curve, *i. e.*, the area bounded by any two radii vectores and the arc of curve between their ends. If ΔS be the increment of this area in the time Δt, the limit of the ratio $\Delta S/\Delta t$, as Δt approaches zero, is called the **sectorial velocity** dS/dt of the point P (about the origin O):

$$\frac{dS}{dt} = \lim_{\Delta t = 0} \frac{\Delta S}{\Delta t}.$$

It follows from the well-known expression for the element of polar area that in polar co-ordinates

$$\frac{dS}{dt} = \tfrac{1}{2}r^2\frac{d\theta}{dt}, \tag{5}$$

and in rectangular cartesian co-ordinates:

$$\frac{dS}{dt} = \tfrac{1}{2}\left(x\frac{dy}{dt} - y\frac{dx}{dt}\right). \tag{6}$$

98. In the case of **relative motion** we have to distinguish between the *absolute velocity* v of a point, its *relative velocity* v_1, and the *velocity of the body of reference* v_2.

To fix the ideas, imagine a man walking on the deck of a steamboat. His velocity v_1 relative to the boat is his velocity of walking; the velocity of the boat (say with respect to the water regarded as fixed), or more exactly speaking, the velocity of that point of the boat at which the man happens to be at the time, is the velocity v_2 of the body of reference; and the velocity with which the man is moving with respect to the water, is his absolute velocity.

Representing these three velocities by means of their vectors, we evidently find that *the absolute velocity v is the geometric sum of the relative velocity v_1 and the velocity v_2 of the body of reference*, just as in the case of displacements of translation (Art. 47). And conversely, *the relative velocity is found by geometrically subtracting from the absolute velocity the velocity of the body of reference.*

It is often convenient to state the last proposition in a somewhat different form. Imagine that we give the velocity $-v_2$ both to the man and to the boat; then the boat is brought to rest, and the resulting velocity of the man is what was before his relative velocity. Hence *the relative velocity is found as the resultant of the absolute velocity and the velocity of the body of reference reversed.*

99. Exercises.

(1) The components of the velocity of a point are 5 and 3 ft./sec. and enclose an angle of 135°; find the resultant in magnitude and direction.

(2) Find the components of a velocity of 10 ft./sec., along two lines inclined to it at 30° and 90°, (*a*) on opposite sides, (*b*) on the same side.

(3) A man jumps from a car at an angle of 60°, with a velocity of 9 ft./sec. (relatively to the car). If the car is running 10 miles/hour, with what velocity and in what direction does the man strike the ground?

(4) Two men, A and B, walking at the rate of 3 and 4 miles/hour, respectively, cross each other at a rectangular street corner. Find the relative velocity of A with respect to B in magnitude and direction.

(5) How must a man throw a stone from a train running 15 miles an hour to make it move 10 ft. per second at right angles to the track?

(6) The velocity of light being 300,000 km./sec., the velocity of the earth in its orbit 30 km./sec., determine approximately the constant of the aberration of the fixed stars.

(7) A man on a wheel, riding along the railroad track at the rate of 9 miles an hour, observes that a train meeting him takes 3 seconds to pass him, while a train of equal length takes 5 seconds to overtake him. If the trains have the same speed, what is it?

(8) A swimmer starting from a point A on one bank of a river wishes to reach a certain point B on the opposite bank. The velocity v_2 of the current and the angle $\theta(<\frac{1}{2}\pi)$ made by AB with the current being given, determine the least relative velocity v_1 of the swimmer in magnitude and direction.

(9) A wheel of radius a rolls on a straight track with constant velocity (of its center) v_0. Find the velocity v of a point P on the rim. What point of the rim has the velocity v_0?

(10) Show that the tangent to the cycloid described by P, Ex. (9), passes through the highest point of the wheel.

(11) A straight line in a plane turns with constant angular velocity ω about one of its points O, while a point P, starting from O, moves along the line with constant velocity v_0. Determine the absolute path of P and its absolute velocity v.

(12) Show how to construct the tangent and normal to the spiral of Archimedes, $r = a\theta$, where $\theta = \omega t$.

(13) Show that the tangent to the ellipse bisects the angle between the radii vectores r_1, r_2, drawn from any point P on the ellipse to the foci.

(14) Construct the tangent to any conic, a directrix and the corresponding focus being given.

(15) Prove the relations (3), Art. 96, by the method of limits.

2. APPLICATIONS.

100. The motion of the **piston of a steam engine** furnishes interesting illustrations of the application of graphical methods in kinematics.

In Fig. 24, let $OQ = a$ be the crank arm, $PQ = l = ma$ the connecting rod, $P_1P_2 = s$ the "stroke," so that $l = ma = \frac{1}{2}ms$. As $P_1P_2 = A_1A_2 = 2a$, we may regard A_1A_2 as representing the stroke. The position of the piston head P at the time when the crank pin is at Q will then be found as the intersection N of A_1A_2

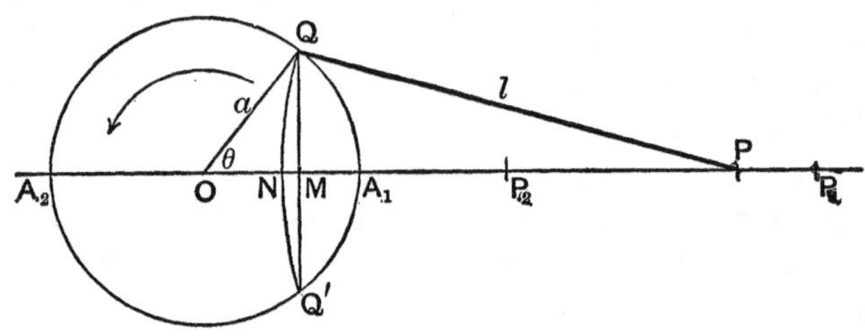

Fig. 24.

with a circle of radius l described about P; in other words, N represents the position of the piston corresponding to the angle $A_1OQ = \theta$ in the forward stroke and to the angle $A_1OQ' = 2\pi - \theta$ in the return stroke.

The crank may generally be assumed to turn uniformly, making n revolutions per second. The linear velocity of the crank pin Q is therefore $u = 2\pi a \cdot n = \pi ns$.

For the piston head P, or for the point N, we must distinguish between its *mean*, or *average*, velocity V, and its variable *instantaneous* velocity v at any particular moment. For each revolution of the crank the piston head completes a double stroke so that its mean speed is $V = 2ns$. Hence we have for the *mean speed* V of the piston:

$$\frac{V}{u} = \frac{2ns}{\pi ns}, \quad \text{or } V = \frac{2}{\pi} u.$$

101. The *instantaneous* velocity v of the piston can be found graphically by constructing, as in Art. 23, the instantaneous center C (Fig. 25) for the motion of the connecting rod. As the instantaneous motion of the rod PQ is a rotation about C, the

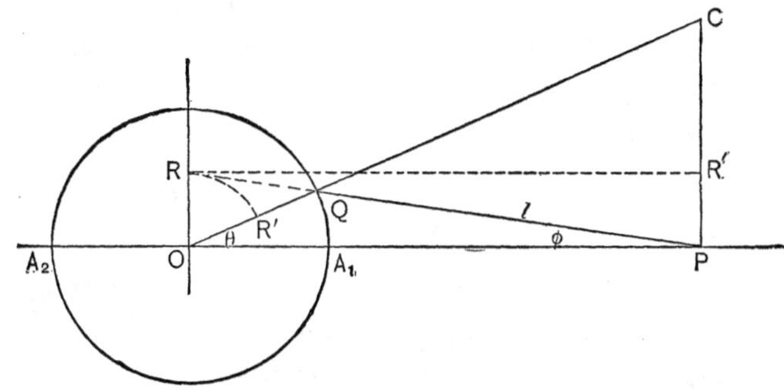

Fig. 25.

linear velocities of P and Q must be as their distances from C, *i. e.*

$$\frac{v}{u} = \frac{CP}{CQ}.$$

Through O draw a perpendicular to OP; if PQ (produced if necessary) meet this perpendicular in R, the similar triangles CPQ and ORQ give:

$$\frac{v}{u} = \frac{CP}{CQ} = \frac{OR}{OQ},$$

whence

$$v = \frac{u}{a} \cdot OR,$$

i. e. v is proportional to OR. If the scale of velocities be selected so that the constant velocity u is represented by the length a of the crank, the instantaneous piston speed v is represented in length by OR.

102. The variation of the piston speed v in the course of the motion can best be exhibited graphically. Thus a *polar curve of piston speed* is obtained by laying off on OQ a length $OR' = OR$, for a number of positions of OQ, and joining the points R' by a continuous curve.

Another convenient method consists in erecting perpendiculars to OP at the various positions of P and laying off, on these perpendiculars, $PR'' = OR = v$.

103. To derive an analytical expression for the piston speed v, let ϕ be the angle OPQ which determines the position of the connecting rod.

The triangle ROQ gives by Art. 101:

$$\frac{v}{u} = \frac{OR}{OQ} = \frac{\sin(\theta + \phi)}{\cos\phi} = \sin\theta + \cos\theta \tan\phi.$$

If, as is usually the case, the connecting rod is much longer than the crank arm, ϕ will be a small angle, and we may substitute $\sin\phi$ for $\tan\phi$. But from the triangle OPQ we have

$$\frac{\sin\phi}{\sin\theta} = \frac{OQ}{PQ} = \frac{a}{l} = \frac{1}{m}.$$

Hence

$$v = u\left(\sin\theta + \cos\theta \cdot \frac{1}{m}\sin\theta\right) = u\left(\sin\theta + \frac{1}{2m}\sin 2\theta\right).$$

104. The motion of the piston head being rectilinear, we find its acceleration j by differentiating with respect to t the expression for v found in the preceding article:

$$j = \frac{dv}{dt} = \left(\sin\theta + \frac{1}{2m}\sin 2\theta\right)\frac{du}{dt} + u\left(\cos\theta + \frac{1}{m}\cos 2\theta\right)\frac{d\theta}{dt},$$

or, since $d\theta / dt = \omega = u/a$,

$$j = \left(\sin\theta + \frac{1}{2m}\sin 2\theta\right)\frac{du}{dt} + \left(\cos\theta + \frac{1}{m}\cos 2\theta\right)\frac{u^2}{a},$$

where $du/dt = 0$ if the crank motion can be regarded as uniform.

105. If the connecting rod were of infinite length so as to make PQ (in Fig. 24) parallel to A_1A_2, the position of the piston corresponding to the position Q of the crank pin would be represented by the projection M of Q on A_1A_2; that is NM would be zero. This length NM is therefore called the *deviation due to the obliquity* of the connecting rod.

If $m = l/a$ is so large that the connecting rod can be regarded as infinite, the cartesian curve of piston speed (Art. 102) becomes a circle. And as $NM = 0$ the expression for the acceleration (Art. 104) reduces to $dv/dt = (u^2/a)\cos\theta$, representing a simple harmonic motion (see Art. 121).

106. The slide valve of a steam engine is generally worked by an eccentric whose radius is set on the shaft at such an angle as to shut off the steam when the crank makes a certain angle θ with the direction of motion of the piston. It follows that the fraction of stroke completed before cut-off takes place is affected by the obliquity of the connecting rod. The rates of cut-off are therefore different in the forward and backward strokes. In the forward stroke, the effect of the obliquity is to put the piston in advance of the position it would have if the connecting rod were of infinite length; in the return stroke, *i. e.* when θ is greater than π, the piston lags behind.

107. An analytical expression for the deviation due to obliquity is readily obtained from Fig. 24. We have

$$MN = PN - PM = l(1 - \cos\phi) = ms\sin^2\tfrac{1}{2}\phi = \tfrac{1}{4}ms(2\sin\tfrac{1}{2}\phi)^2,$$

or approximately, since ϕ is small,

$$MN = \tfrac{1}{4}ms\sin^2\phi.$$

Also, as in Art. 103, $\sin\phi/\sin\theta = 1/m$; hence

$$MN = \frac{s}{4m}\sin^2\theta.$$

The greatest value of MN is thus seen to be $s/4m$; for instance, if the connecting rod be four times the length of the crank, the deviation due to obliquity cannot exceed $\frac{1}{16}$ of the stroke.

108. Exercises.*

(1) Construct a polar diagram exhibiting the *position* of the piston

* Some of these problems are taken with slight modification from COTTERILL'S *Applied mechanics*, 1884, p. 112.

for all angles θ by laying off on the crank arm OQ a length $ON' = ON$ and joining the points N' by a continuous curve.

(2) Construct the curves of piston speed indicated in Art. 102.

(3) Show that for a connecting rod of infinite length the two loops of the curve of Ex. (1) reduce to two equal circles.

(4) The driving wheels of a locomotive are 6 feet in diameter; find the number of revolutions per minute and the angular velocity when running at 45 miles per hour. If the stroke be 2 feet, find the mean speed of the piston.

(5) The pitch of a screw is 24 ft., and the number of revolutions 70 per minute. Find the speed in knots, a knot being a sea mile per hour = 6090 ft./hour. If the stroke is 4 ft., find the speed of piston in feet per minute.

(6) The stroke of a piston is 4 ft., and the connecting rod is 9 ft. long. Find the position of the crank, when the piston has completed the first quarter of the forward and backward strokes respectively. Also find the position of the piston when the crank is upright.

(7) The valve gear is so arranged in the last question as to cut off the steam when the crank is 45° from the dead-points both in the forward and backward strokes. Find the point at which steam will be cut off in the two strokes. Also when the obliquity of the connecting rod is neglected.

(8) If, in Fig. 24, the crank $a = 1$ ft., the connecting rod $l = 4$ ft., the number of revolutions = 90 per minute, what is the velocity of the cross-head P when the crank stands at $\theta = 30°$?

(9) With $m = 4$, $u =$ const., in what position of the piston in the stroke is its velocity greatest?

(10) Determine the velocity and acceleration of a point R on the connecting rod QP if $QR = n \cdot QP$, assuming the rotation of the crank as uniform.

3. ACCELERATION IN CURVILINEAR MOTION.

109. Let the velocity of a moving point be represented by the vector $v = PT$ at the time t, and by the vector $v' = P'T'$ at the time $t + \Delta t$ (Fig. 26). Then, drawing from any point O, OV and OV' respectively equal and parallel to PT and $P'T'$, the vector VV' represents the geometric difference between v' and v; in other words, VV' is the velocity which, geometrically added to v, produces v'. As Δt approaches zero, the vector VV' approaches zero. But in general the quotient $VV'/\Delta t$ will approach a definite finite limit and the direction of VV' will at the same time approach a definite limiting direction. A rectilinear segment of length

$$j = \lim_{\Delta t = 0} \frac{VV'}{\Delta t},$$

laid off in this limiting direction, is called the **acceleration** of the point P at the time t. It can be regarded as the *geometric derivative* with respect to the time of the vector representing the velocity.

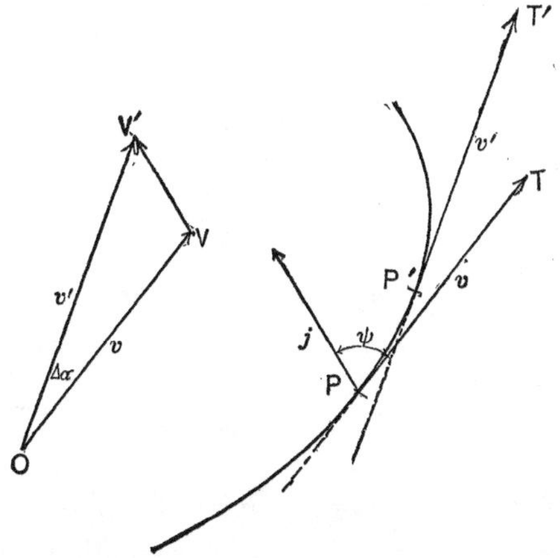

Fig. 26.

The segments representing accelerations are assumed to follow the parallelogram law of composition and resolution (Arts. 38–45), just like the segments representing velocities (Art. 94) and translations. Acceleration is thus defined as a *vector*.

It will be noticed that the sense of the acceleration is towards that side of the tangent of the curve on which the center of curvature is situated.

110. Suppose a point P to move along a curve $P_1P_2P_3 \ldots$ with variable velocity v (Fig. 27). From any fixed origin O draw a vector $OV_1 = v_1$, equal and parallel to the velocity v_1 of P_1, and repeat this construction for every position of the moving point P. The ends V_1, V_2, V_3, ... of all these radii vectores drawn from O will form a continuous curve $V_1V_2V_3 \ldots$ which is called the **hodograph** of the motion of P.

If we imagine a point V describing this curve $V_1V_2V_3 \ldots$ at the same time that P describes the curve $P_1P_2P_3 \ldots$, it is evident that the velocity of V, *i. e.*, $\lim_{\Delta t=0} (VV'/\Delta t)$, laid off on the tangent of the curve $V_1V_2V_3 \ldots$, represents the acceleration of the point P both in magnitude and direction; *i. e.*, *the velocity on the hodograph is the acceleration of the original motion.*

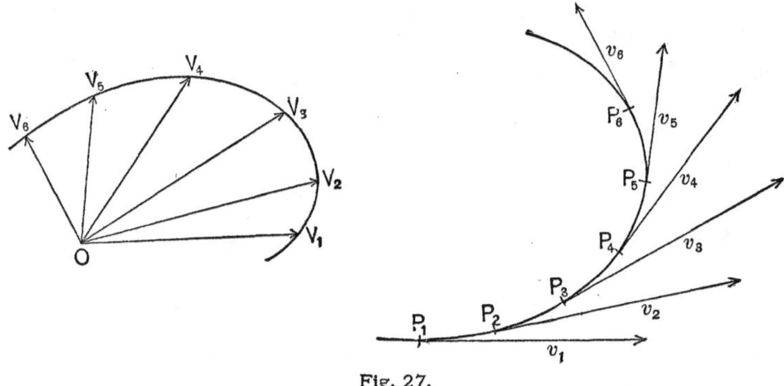

Fig. 27.

111. Acceleration having been defined as a vector, the rules for vector composition and resolution may be applied to acceleration just as they were before applied to displacements and to velocities. Thus, a point subjected to two or more simultaneous accelerations will have a resulting acceleration found by geometrically adding the component accelerations; and conversely, the acceleration of a point may be resolved in various ways.

112. Let the vector j which represents the acceleration of the point P at the time t, make an angle ψ with the vector representing the velocity v at the same time (see Fig. 26). Resolving the vector j along the tangent and normal at P, we obtain the **tangential acceleration** $j_t = j \cos\psi$ and the **normal acceleration** $j_n = j \sin\psi$.

To find expressions for these components, let us put arc $PP' = \Delta s$, $PT = OV = v$, $P'T' = OV' = v' = v + \Delta v$, and $\sphericalangle VOV' = \Delta\alpha$.

As Δv approaches zero, the direction of VV' approaches that of the acceleration j which makes the angle ψ with the velocity v. By resolving the vector VV' along OV and at right angles to it, it appears that

$$\cos\psi = \lim \frac{\Delta v}{VV'}, \quad \sin\psi = \lim \frac{v\Delta\alpha}{VV'}.$$

We find therefore

$$j_t = j \cos\psi = \cos\psi \lim_{\Delta t=0} \frac{VV'}{\Delta t}$$

$$= \cos\psi \lim \frac{VV'}{\Delta v}\frac{\Delta v}{\Delta t} = \lim \frac{\Delta v}{\Delta t} = \frac{dv}{dt},$$

and

$$j_n = j \sin\psi = \sin\psi \lim_{\Delta t=0} \frac{VV'}{\Delta t}$$

$$= \sin\psi \lim \frac{VV'}{v\Delta\alpha}\frac{v\Delta\alpha}{\Delta t} = \lim v\frac{\Delta\alpha}{\Delta t} = v\frac{d\alpha}{dt}.$$

As

$$\frac{d\alpha}{dt} = \frac{d\alpha}{ds}\frac{ds}{dt} = \frac{1}{\rho}\cdot v,$$

where $\rho = ds/d\alpha$ is the radius of curvature at P, we have finally:

$$j_t = \frac{dv}{dt}, \tag{1}$$

$$j_n = \frac{v^2}{\rho}, \tag{2}$$

whence

$$j = \sqrt{j_t^2 + j_n^2} = \sqrt{\left(\frac{dv}{dt}\right)^2 + \frac{v^4}{\rho^2}}. \tag{3}$$

113. When rectangular cartesian co-ordinates are used we may resolve the acceleration j into components $j_x = j \cos\phi, j_y = j \sin\phi$, along the axes Ox, Oy; ϕ being the angle at which the acceleration j is inclined to the axis Ox. We obtain an expression for j_x by projecting the triangle OVV' (Fig. 26) on the axis Ox and denoting the projections of the velocities OV, OV' by v_x, v_x'. This gives

$$VV' \cos\phi = v_x' - v_x = \Delta v_x,$$

whence, dividing by Δt and passing to the limit: $j_x = dv_x/dt$.

Similarly we find $j_y = dv_y/dt$. Hence, by the formulæ (1) of Art. 95:

$$j_x = \frac{dv_x}{dt} = \frac{d^2x}{dt^2}, \quad j_y = \frac{dv_y}{dt} = \frac{d^2y}{dt^2}. \tag{4}$$

These so-called *equations of motion* offer the advantage that the curvilinear motion is replaced by two rectilinear motions, thus avoiding the use of vectors.

By composition, we have of course

$$j = \sqrt{j_x^2 + j_y^2} = \sqrt{\left(\frac{d^2x}{dt^2}\right)^2 + \left(\frac{d^2y}{dt^2}\right)^2}. \tag{5}$$

While the whole present chapter is confined to *plane* motion it may here be stated for the sake of completeness that, if the path is not a plane curve, the acceleration j can be resolved, by the same method, into *three* rectangular components:

$$j_x = \frac{d^2x}{dt^2}, \quad j_y = \frac{d^2y}{dt^2}, \quad j_z = \frac{d^2z}{dt^2}.$$

114. For polar co-ordinates r, θ, we may resolve the acceleration j into a component j_r along the radius vector r and a component j_θ at right angles to r. Expressions for these components are readily found by projecting the components $j_x = d^2x/dt^2$ and

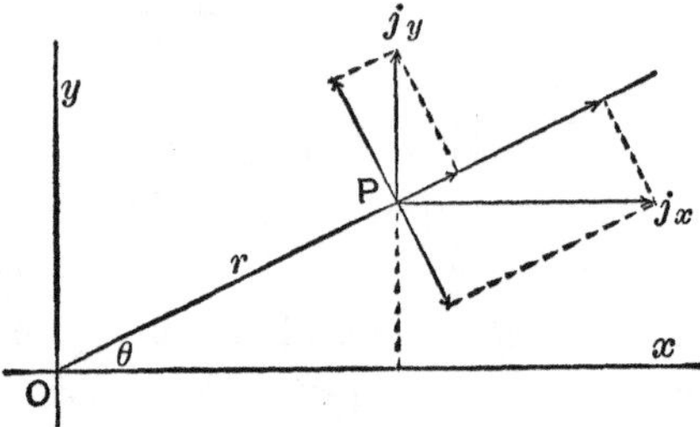

Fig. 28.

$j_y = d^2y/dt^2$ on r and at right angles to r (Fig. 28):

$$j_r = \frac{d^2x}{dt^2}\cos\theta + \frac{d^2y}{dt^2}\sin\theta, \quad j_\theta = -\frac{d^2x}{dt^2}\sin\theta + \frac{d^2y}{dt^2}\cos\theta.$$

Differentiating the relations $x = r\cos\theta$, $y = r\sin\theta$, we find

$$\frac{dx}{dt} = \frac{dr}{dt}\cos\theta - r\sin\theta\frac{d\theta}{dt},\quad \frac{dy}{dt} = \frac{dr}{dt}\sin\theta + r\cos\theta\frac{d\theta}{dt};$$

and differentiating again:

$$\frac{d^2x}{dt^2} = \left[\frac{d^2r}{dt^2} - r\left(\frac{d\theta}{dt}\right)^2\right]\cos\theta - \left(2\frac{dr}{dt}\frac{d\theta}{dt} + r\frac{d^2\theta}{dt^2}\right)\sin\theta,$$

$$\frac{d^2y}{dt^2} = \left[\frac{d^2r}{dt^2} - r\left(\frac{d\theta}{dt}\right)^2\right]\sin\theta + \left(2\frac{dr}{dt}\frac{d\theta}{dt} + r\frac{d^2\theta}{dt^2}\right)\cos\theta.$$

Substituting these expressions for d^2x/dt^2 and d^2y/dt^2 in the above equations for j_r, j_θ, we find:

$$j_r = \frac{d^2r}{dt^2} - r\left(\frac{d\theta}{dt}\right)^2,\quad j_\theta = 2\frac{dr}{dt}\frac{d\theta}{dt} + r\frac{d^2\theta}{dt^2} = \frac{1}{r}\frac{d}{dt}\left(r^2\frac{d\theta}{dt}\right).\quad (6)$$

115. Exercises.

(1) Show that the velocity of a moving point is increasing, constant, or diminishing, according to the value of the angle ψ between v and j (Fig. 26).

(2) Show that the sectorial velocity (Art. 97) is constant whenever $j_\theta = 0$.

(3) Show that the normal component of the acceleration is the product of the radius of curvature into the square of the angular velocity about the center of curvature.

(4) Show that the velocity is the mean proportional between the acceleration and half the chord intercepted by the direction of the acceleration on the osculating circle.

(5) Show that the hodograph of rectilinear motion is a straight line.

(6) If the acceleration of a point P be always directed to a fixed point A, show that the radius vector AP describes equal areas in equal times.

(7) Show that in uniform circular motion the acceleration is directed to the center and proportional to the radius.

(8) Prove Ex. (7) by twice differentiating the equations $x = a\cos\theta$, $y = a\sin\theta$ of the circle, with respect to the time.

(9) A wheel rolls on a straight track; find the acceleration of its lowest and highest points.

4. APPLICATIONS.

116. Inclined Plane. Imagine a body sliding (without rolling or turning) down a smooth rigid plane inclined at an angle θ to the horizon. In addition to the assumptions made in the case of falling bodies (see Art. 72) we assume that the motion takes place along a "line of greatest slope," *i. e.* in a vertical plane at right angles to the intersection of the inclined plane with a horizontal plane. A "smooth" plane means one that offers no frictional resistance. The body is therefore subject only to the acceleration of gravity, g; and it is sufficient to consider the motion of a single point of the body.

Resolving g into two components, $g\cos\theta$ perpendicular to the plane and $g\sin\theta$ along the plane (Fig. 29), it will be seen that the former component, being at right angles to the rigid plane, cannot change the *magnitude* of this velocity. We have therefore simply a rectilinear motion with the constant acceleration $g\sin\theta$, so that all the formulæ of Arts. 68–73 will here apply if for the acceleration j (or g) we substitute $g\sin\theta$.

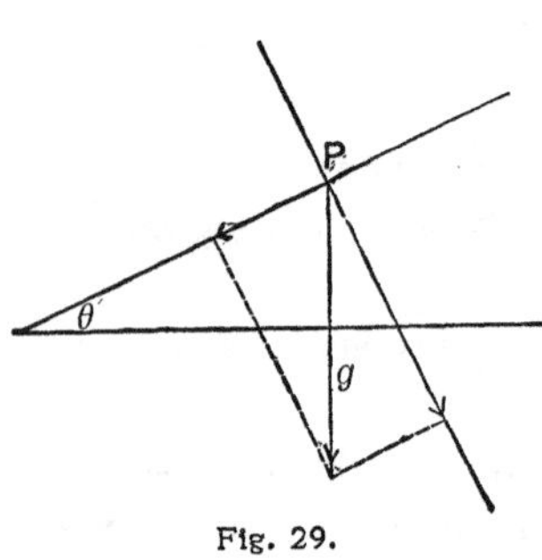

Fig. 29.

Thus, if the initial velocity be 0, the motion is determined by the equations

$$v = g\sin\theta \cdot t, \tag{1}$$

$$s = \tfrac{1}{2}g\sin\theta \cdot t^2, \tag{2}$$

$$\tfrac{1}{2}v^2 = g\sin\theta \cdot s. \tag{3}$$

If there be an initial velocity v_0 parallel to the line of greatest slope of the inclined plane, the equations are

$$v = v_0 + g \sin\theta \cdot t, \tag{1'}$$

$$s = v_0 t + \tfrac{1}{2} g \sin\theta \cdot t^2, \tag{2'}$$

$$\tfrac{1}{2}(v^2 - v_0^2) = g \sin\theta \cdot s, \tag{3'}$$

where v_0 is to be regarded as positive if it is directed down the plane, and negative if directed up the plane.

117. Exercises.

(1) A railroad train is running up a grade of 1 in 200 at the rate of 20 miles an hour when the coupling of the last car breaks. Neglecting friction, (*a*) how far will the car have gone after two minutes from the point where the break occurred? (*b*) When will it begin moving down the grade? (*c*) How far behind the train will it be at that moment? (*d*) If the grade extend 1500 ft. below the point where the break occurred, with what velocity will it arrive at the foot of the grade?

(2) Show that the final velocity is independent of the inclination of the plane; in other words, in sliding down a smooth inclined plane a body acquires the same velocity as in falling vertically through the "height" of the plane.

(3) Show that it takes a body twice as long to slide down a plane of 30° inclination as it would take it to fall through the height of the plane.

(4) At what angle θ should the rafters of a roof of given span $2b$ be inclined to make the water run off in the shortest time?

(5) Prove that the times of sliding from rest down the chords issuing from the highest (or lowest) point of a vertical circle are equal.

(6) Show how to construct geometrically the line of quickest descent from a given point: (*a*) to a given straight line, (*b*) to a given circle, situated in the same vertical plane.

(7) Analytically, the line of quickest or slowest descent from a given point to a curve in the same vertical plane is found by taking the equation of the curve in polar co-ordinates, $r = f(\theta)$, with the given point as origin and the axis horizontal. The time of sliding down the radius vector r is $t = \sqrt{2r/(g \sin\theta)}$. Show that this becomes a maximum or minimum when $\tan\theta = f(\theta)/f'(\theta)$, according as $f(\theta) + f''(\theta)$ is negative or positive.

(8) Show that the line of quickest descent to a parabola from its focus, the axis of the parabola being horizontal, is inclined at an angle of 60° to the horizon.

118. Projectiles. With the same assumptions as in Art. 72, the motion of a projectile reduces to that of a point subject to the constant acceleration of gravity and starting with an initial velocity v_0 inclined to the horizon at any angle ϵ. The angle ϵ between the horizon and the initial velocity is called the **elevation** of the projectile.

Taking the horizontal line through the starting point O as axis of x, the vertical upwards as positive axis of y (Fig. 30), the x-

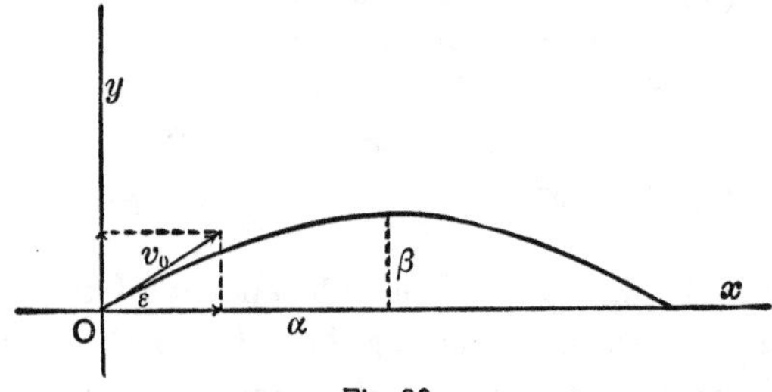

Fig. 30.

component of the acceleration is evidently 0, while the y-component is $-g$; hence, by (4), Art. 113, the equations of motion are

$$\frac{d^2x}{dt^2} = 0, \quad \frac{d^2y}{dt^2} = -g. \tag{4}$$

The first integration gives

$$\frac{dx}{dt} = C_1, \quad \frac{dy}{dt} = -gt + C_2.$$

As $dx/dt = v_x$, $dy/dt = v_y$ are the components of the velocity v at the time t, the constants are determined from the values of v_x, v_y at the time $t = 0$; viz. $v_0 \cos\epsilon = C_1$, $v_0 \sin\epsilon = 0 + C_2$. We have therefore

$$v_x \equiv \frac{dx}{dt} = v_0 \cos\epsilon, \quad v_y \equiv \frac{dy}{dt} = v_0 \sin\epsilon - gt. \tag{5}$$

Integrating again, we find

$$x = v_0 \cos\epsilon \cdot t, \qquad y = v_0 \sin\epsilon \cdot t - \tfrac{1}{2}gt^2, \tag{6}$$

the constants of integration being 0, since for $t = 0$ we have $x = 0$, $y = 0$.

These values of x, y, v_x, v_y show that the motion in the horizontal direction is uniform while in the vertical direction it is uniformly accelerated. This is otherwise directly evident from the nature of the problem.

Eliminating t between the expressions for x and y, we find the *equation of the path*

$$y = \tan\epsilon \cdot x - \frac{g}{2v_0^2 \cos^2\epsilon} \cdot x^2, \tag{7}$$

which represents a parabola passing through the origin. To find its vertex and latus rectum, divide by the coefficient of x^2 and rearrange:

$$x^2 - \frac{2v_0^2}{g} \sin\epsilon \cos\epsilon \cdot x = -\frac{2v_0^2}{g} \cos^2\epsilon \cdot y;$$

completing the square in x, the equation can be written in the form

$$\left(x - \frac{v_0^2}{2g} \sin 2\epsilon\right)^2 = -\frac{2v_0^2}{g} \cos^2\epsilon \left(y - \frac{v_0^2}{2g} \sin^2\epsilon\right).$$

The co-ordinates of the vertex are therefore $\alpha = (v_0^2/2g) \sin 2\epsilon$, $\beta = (v_0^2/2g) \sin^2\epsilon$; the latus rectum $4a = (2v_0^2/g) \cos^2\epsilon$; the axis is vertical, and the directrix is a horizontal line at the distance $a = (v_0^2/2g) \cos^2\epsilon$ above the vertex.

119. Exercises.

(1) Show that the velocity at any time is $v = \sqrt{v_0^2 - 2gy}$.

(2) Prove that the velocity of the projectile is equal in magnitude to the velocity that it would acquire by falling from the directrix: (*a*) at the starting point, (*b*) at any point of the path (see Art. 73).

(3) Show that a body projected vertically upwards with the initial velocity v_0 would just reach the common directrix of all the parabolas

described by bodies projected at different elevations ε with the same initial velocity v_0.

(4) The *range* of a projectile is the distance from the starting point to the point where it strikes the ground. Show that on a horizontal plane the range is $R = 2a = (v_0^2/g) \sin 2\varepsilon$.

(5) The *time of flight* is the whole time from the beginning of the motion to the instant when the projectile strikes the ground. It is best found by considering the horizontal motion of the projectile alone, which is uniform. Show that on a horizontal plane the time of flight is $T = (2v_0/g) \sin\varepsilon$.

(6) Show that the time of flight and the range, on a plane through the starting point inclined at an angle θ to the horizon, are

$$T_\theta = \frac{2v_0}{g}\frac{\sin(\varepsilon - \theta)}{\cos\theta}, \text{ and } R_\theta = \frac{2v_0^2}{g}\frac{\sin(\varepsilon - \theta)\cos\varepsilon}{\cos^2\theta}.$$

(7) What elevation gives the greatest range on a horizontal plane?

(8) Show that on a plane rising at an angle θ to the horizon, to obtain the greatest range, the direction of the initial velocity should bisect the angle between the plane and the vertical.

(9) A stone is dropped from a balloon which, at a height of 625 ft., is carried along by a horizontal air-current at the rate of 15 miles an hour. (*a*) Where, (*b*) when, and (*c*) with what velocity will it reach the ground?

(10) What must be the initial velocity v_0 of a projectile if, with an elevation of 30°, it is to strike an object 100 ft. above the horizontal plane of the starting point at a horizontal distance from the latter of 1200 ft.?

(11) What must be the elevation ε to strike an object 100 ft. above the horizontal plane of the starting point and 5000 ft. distant, if the initial velocity be 1200 ft. per second?

(12) Show that to strike an object situated in the horizontal plane of the starting point at a distance x from the latter, the elevation must be ε or $90° - \varepsilon$, where $\varepsilon = \frac{1}{2}\sin^{-1}(gx/v_0^2)$.

(13) The initial velocity v_0 being given in magnitude and direction, show how to construct the path graphically.

(14) The solution of Ex. (11) shows that a point that can be reached with a given initial velocity can in general be reached by two different elevations. Find the locus of the points that can be reached

by only one elevation, and show that it is the envelope of all the parabolas that can be described with the same initial velocity (in one vertical plane).

(15) If it be known that the path of a point is a parabola and that the acceleration is parallel to its axis, show that the acceleration is constant.

(16) Prove that a projectile whose elevation is 60° rises three times as high as when its elevation is 30°, the magnitude of the initial velocity being the same in each case.

(17) Construct the hodograph for parabolic motion, taking the focus as pole and drawing the radii vectores at right angles to the velocities.

(18) A stone slides down a roof sloping 30° to the horizon, through a distance of 12 ft. If the lower edge of the roof be 50 ft. above the ground, (*a*) when, (*b*) where, (*c*) with what velocity does the stone strike the ground?

(19) If a golf ball be driven from the tee horizontally with initial speed = 300 ft./sec., where and when would it land on ground 16 ft. below the tee if resistance of air and rotation of ball could be neglected?

(20) A man standing 15 ft. from a pole 150 ft. high aims at the top of the pole. If the bullet just misses the top where will it strike the ground if $v_0 = 1000$ ft./sec.?

120. Uniform Circular Motion. Let a point P (Fig. 31) describe a circle of radius a with constant angular velocity ω. Its linear

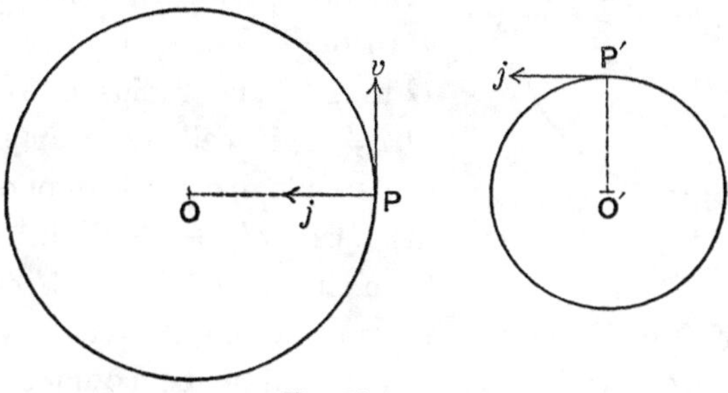

Fig. 31.

velocity $v = \omega a$ is of constant magnitude, but varies in direction. By the formulæ (1), (2) of Art. 112, the tangential acceleration

$j_t = 0$, while the normal acceleration $j_n = v^2/a = \omega^2 a$ represents the total acceleration. Hence, in uniform circular motion, the acceleration is

$$j = \frac{v^2}{a} = \omega^2 a, \tag{8}$$

and is always directed toward the center O of the circle.

This appears also by constructing the hodograph of the motion, which is evidently a circle of radius v (Fig. 31). As the acceleration of P is represented by the velocity of the point P' of the hodograph (see Art. 110), we have only to determine this velocity. Let T be the so-called **period,** or **periodic time,** *i. e.* the time in which the point P makes a whole revolution, so that

$$T = \frac{2\pi a}{v} = \frac{2\pi}{\omega};$$

then, since P' describes the circle of radius v in the same time T, we have for the velocity of P' the expression $2\pi v/T$, or substituting for T its value: v^2/a or $\omega^2 a$ as above.

121. If the positions P of a point moving uniformly in a circle be projected at every instant on any diameter AA' of the circle, the rectilinear motion of the projection P_x along this diameter is called **simple harmonic motion.**

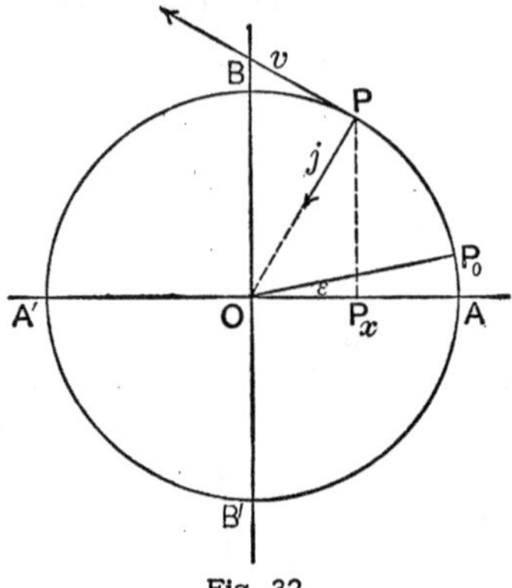

Fig. 32.

While P moves uniformly in the circle (Fig. 32), its projection P_x evidently performs oscillations from A through O to A' and back through O to A. The radius of the circle, $OA = a$, is called the **amplitude** of the simple harmonic motion. The distance $OP_x = x$ of the point P_x from the center O is called the **displacement** of P_x. Denoting the constant angular velocity of P by ω, the angle AOP will be $= \omega t$, if the time be counted from the instant when P is at A. We have therefore

$$x = a \cos\omega t. \tag{9}$$

The projection of the motion of P on the diameter BB' would give similarly

$$y = a \sin\omega t = a \cos(\omega t - \tfrac{1}{2}\pi).$$

122. The time T of completing one whole oscillation forward and backward is called the **period** of the simple harmonic motion; it is obviously equal to the period of the motion of P in the circle, *i. e.*

$$T = \frac{2\pi}{\omega}. \tag{10}$$

The period is therefore independent of the amplitude a. It follows that two simple harmonic motions resulting from two uniform circular motions of the same angular velocity on two concentric circles of different radii have the same period; such motions are called **isochronous.**

If the point P describes the circle n times per second and hence the point P_x on the diameter performs n oscillations per second, the angular velocity is $\omega = 2\pi n$; it follows from (10) that

$$T = \frac{1}{n}, \text{ and } n = \frac{1}{T}; \tag{11}$$

i. e. the number of revolutions of P, or *the number of oscillations* of P_x, the so-called **frequency** of the simple harmonic motion *is the reciprocal of the period.*

123. If the time t be counted, not from A, but from some other point P_0 on the circle for which $\measuredangle AOP_0 = \epsilon$, we have $\measuredangle AOP = \omega t + \epsilon$, and the equation of the simple harmonic motion is

$$x = a \cos(\omega t + \epsilon). \tag{12}$$

The angle $\omega t + \epsilon$ is called the **phase-angle,** while ϵ is the **epoch-angle** of the motion. The names *phase* and *epoch* are sometimes applied to these angles, although, strictly speaking, the phase is the *time* (usually expressed as a fraction of the period T) of passing from the position A of maximum displacement to any position P_x, while the epoch is the phase corresponding to the time $t = 0$.

124. Differentiating equation (12), we find the velocity

$$v_x \equiv \frac{dx}{dt} = -a\omega \sin(\omega t + \epsilon); \tag{13}$$

and differentiating again, we obtain the acceleration

$$j_x \equiv \frac{d^2x}{dt^2} = -a\omega^2 \cos(\omega t + \epsilon) = -\omega^2 x \tag{14}$$

of simple harmonic motion.

The same values can be derived by projecting the velocity and acceleration of the uniform circular motion of P on the diameter AA', as is readily seen from Fig. 32.

125. Equation (14) shows that *in simple harmonic motion the acceleration is directly proportional to the distance from the center.*

Conversely, it can be shown that if the acceleration be proportional to the distance from a fixed point in the direction of the initial velocity, and if it be directed *towards* this point the motion is simply harmonic. For we then have

$$\frac{d^2x}{dt^2} = -\mu^2 x, \tag{15}$$

where μ is constant. The general integral of this differential equation is (compare Arts. 82–84)

$$x = C_1 \sin\mu t + C_2 \cos\mu t.$$

Differentiating, we find for the velocity

$$v = C_1\mu \cos\mu t - C_2\mu \sin\mu t.$$

To determine the constants of integration C_1, C_2, let $x = x_0$ and $v = v_0$ at the time $t = 0$. Substituting these values, we find $x_0 = C_2$ and $v_0 = \mu C_1$; hence

$$x = \frac{v_0}{\mu} \sin\mu t + x_0 \cos\mu t.$$

This expression for x can be simplified by observing that if we construct a right-angled triangle (Fig. 33) with v_0/μ and x_0 as sides (which is always possible) and call a its hypotenuse, ϵ the angle opposite x_0, we have

$$\frac{v_0}{\mu} = a \cos\epsilon, \quad x_0 = a \sin\epsilon.$$

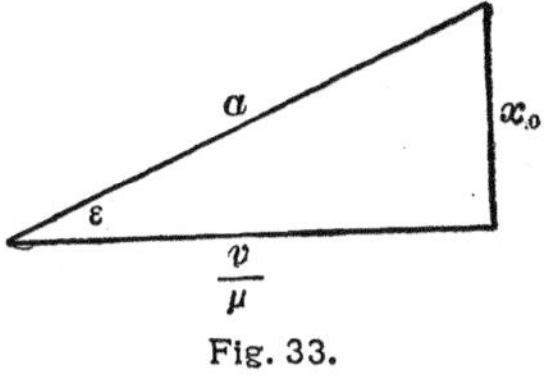

Fig. 33.

Substituting these values we obtain

$$\begin{aligned} x &= a\,(\sin\mu t \cos\epsilon + \cos\mu t \sin\epsilon) \\ &= a \sin(\mu t + \epsilon). \end{aligned} \qquad (16)$$

This represents a simple harmonic motion whose amplitude a and epoch angle ϵ are

$$a = \sqrt{\frac{v_0^2}{\mu^2} + x_0^2}, \quad \epsilon = \tan^{-1}\frac{\mu x_0}{v_0}.$$

As the angular velocity of the corresponding uniform circular motion is μ, the period is $T = 2\pi/\mu$.

126. Simple harmonic motions occur very frequently in applied mechanics and mathematical physics. A particular case has been treated in Arts. 81–84. As another example we may mention the apparent motion of a satellite about its primary as seen from any point in the plane of the motion, provided the satellite be regarded as moving uniformly in a circle relatively to its primary. Thus the moons of Jupiter, as seen from the earth, have approximately a simple harmonic motion.

127. A mechanism for producing simple harmonic motion can readily be constructed as follows. The end A (Fig. 34) of a crank rotating uniformly about the axis O carries a pin running in the slot AB of a T bar ABD whose axis (produced) passes through the center O of the crank circle. The T bar is constrained by guides to move back and forth along the line OD; its motion is evidently simply harmonic, the uniform circular motion of the crank being transformed into rectilinear motion. Compare Art. 105.

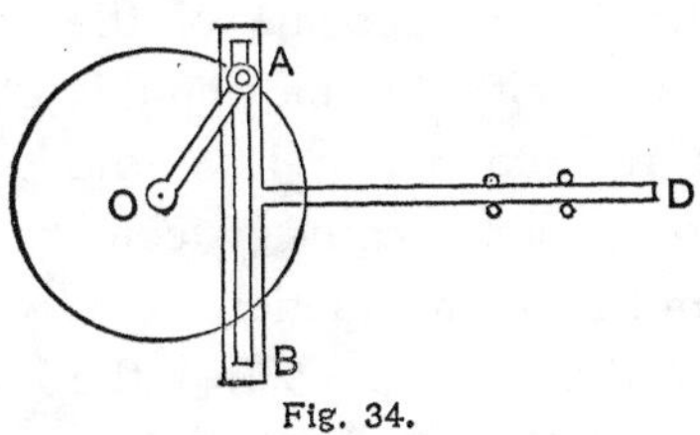

Fig. 34.

128. Exercises.

(1) Show that the maximum acceleration of a simple harmonic motion is numerically equal to the acceleration in the corresponding uniform circular motion.

(2) Find the time of one oscillation from equation (12) without reference to the circular motion.

(3) In the mechanism shown in Fig. 34, if the length of the crank be 2 feet and the number of revolutions 15 per minute, find the velocity and acceleration of the end D of the T bar: (a) when at elongation; (b) when at quarter stroke; (c) when at the middle of the stroke.

(4) Show that the period of a simple harmonic oscillation can be expressed in the form $T = 2\pi\sqrt{-x/j_x}$ where j_x is the acceleration of the oscillating point at the time when its distance from the center, or its displacement, is x.

(5) Show that $v_x = -\omega\sqrt{a^2 - x^2}$.

(6) Show that the equation $x = a \cos\mu t + b \sin\mu t$ represents a simple harmonic motion.

129. Compound Harmonic Motion. We have seen (Art. 125) that the motion of a point, whose acceleration is directly proportional to its distance from a fixed center, and directed towards this center, is simply harmonic, provided the center lies in the line of the initial velocity. Removing this last restriction, we, have the more general case of compound harmonic motion.

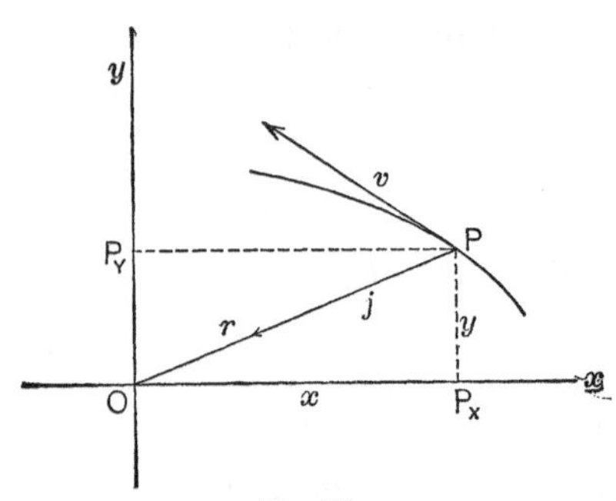

Fig. 35.

Let O (Fig. 35) be the fixed center; P the position, $OP = r$ the distance from the center, v the velocity, of the moving point at any time t. The velocity v and the point O determine a plane to which the motion is confined since the acceleration $j = \mu^2 r$, being directed along PO, lies in this plane.

Taking in this plane the center O as origin and arbitrary rectangular axes Ox, Oy, we have for the direction cosines of OP: x/r, y/r; hence for those of j: $-x/r$, $-y/r$. The components of j are therefore

$$j_x = -\mu^2 x, \quad j_y = -\mu^2 y,$$

and the equations of motion are

$$\frac{d^2x}{dt^2} = -\mu^2 x, \quad \frac{d^2y}{dt^2} = -\mu^2 y.$$

These equations show that the projections P_x, P_y of P on the axes have each a simple harmonic motion, and the motion of P which is in general curvilinear may be regarded as the resultant of these component motions.

We proceed to examine in some detail the most important cases of this **composition of** two or more **simple harmonic motions**, beginning with those cases in which the resultant motion is rectilinear.

As, according to Hooke's law, the particles of elastic bodies, after release from strain within the elastic limits, perform small oscillations for which the acceleration is proportional to the displacement from a middle position, the motions under discussion find a wide application in the theories of elasticity, sound, light, and electricity, and form the basis of the general theory of wave motion in an elastic medium.

130. *Two simple harmonic motions in the same line, of equal period T, but differing in amplitude and phase, compound into a single simple harmonic motion in the same line and of the same period.*

For, by Art. 123, the component displacements can be written

$$x_1 = a_1 \cos(\omega t + \epsilon_1), \quad x_2 = a_2 \cos(\omega t + \epsilon_2),$$

and being in the same line they can be added algebraically, giving the resultant displacement

$$x = x_1 + x_2 = a_1 \cos(\omega t + \epsilon_1) + a_2 \cos(\omega t + \epsilon_2)$$

$$= (a_1 \cos\epsilon_1 + a_2 \cos\epsilon_2) \cos\omega t - (a_1 \sin\epsilon_1 + a_2 \sin\epsilon_2) \sin\omega t.$$

Putting (comp. Art. 125)

$$a_1 \cos\epsilon_1 + a_2 \cos\epsilon_2 = a \cos\epsilon, \quad a_1 \sin\epsilon_1 + a_2 \sin\epsilon_2 = a \sin\epsilon,$$

we have

$$x = a \cos\epsilon \cos\omega t - a \sin\epsilon \sin\omega t = a \cos(\omega t + \epsilon),$$

where

$$a^2 = (a_1 \cos\epsilon_1 + a_2 \cos\epsilon_2)^2 + (a_1 \sin\epsilon_1 + a_2 \sin\epsilon_2)^2$$

$$= a_1^2 + a_2^2 + 2a_1 a_2 \cos(\epsilon_2 - \epsilon_1)$$

and

$$\tan\epsilon = \frac{a_1 \sin\epsilon_1 + a_2 \sin\epsilon_2}{a_1 \cos\epsilon_1 + a_2 \cos\epsilon_2}.$$

131. A geometrical illustration of the preceding proposition is obtained by considering the uniform circular motions corresponding to the two simple harmonic motions (Fig. 36).

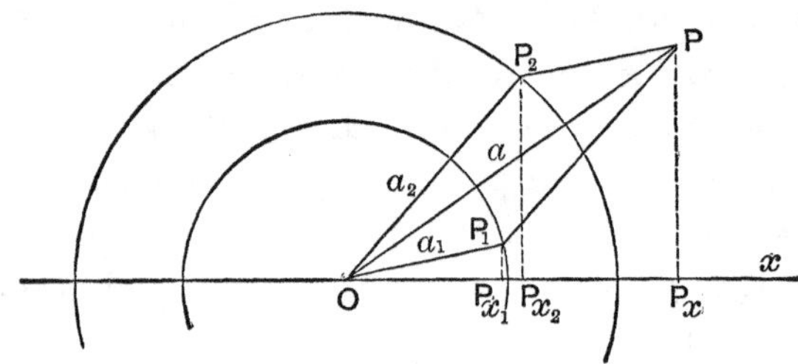

Fig. 36.

Drawing the radii $OP_1 = a_1$, $OP_2 = a_2$ so as to include an angle equal to the difference of phase $\epsilon_2 - \epsilon_1$ and completing the parallelogram OP_1PP_2, it appears from the figure that the diagonal OP of this parallelogram represents the resulting amplitude a.

As P_1P is equal and parallel to OP_2, we have for the projections on any axis Ox the relation $OP_{x_1} + OP_{x_2} = OP_x$, or $x_1 + x_2 = x$. If now the axis Ox be drawn so as to make the angle xOP_1 equal to the epoch-angle ϵ_1, and hence $xOP_2 = \epsilon_2$, the angle xOP represents the epoch ϵ of the resulting motion.

We thus have a simple geometrical construction for the elements a, ϵ of the resulting motion from the elements a_1, ϵ_1 and a_2, ϵ_2 of the component motions. As the period is the same for the two component motions, the points P_1 and P_2 describe their respective circles with equal angular velocity so that the parallelogram OP_1PP_2 does not change its form in the course of the motion.

132. The construction given in the preceding article can be described briefly by saying that two simple harmonic motions of equal period in the same line are compounded by *geometrically adding* their amplitudes, it being understood that the phase-angles determine the directions in which the amplitudes are to be drawn. Analytically, this appears of course directly from the formulæ of Art. 130.

It follows at once that not only two, but *any number of simple harmonic motions, of equal period in the same line, can be compounded by geometric addition of their amplitudes into a single simple harmonic motion in the same line and of the same period.*

Conversely, any given simple harmonic motion can be resolved into two or more components in the same line and of the same period.

133. The kinematical meaning of this composition of simple harmonic motions of equal period in the same line will perhaps be best understood from the mechanism sketched in Fig. 37. A cord runs from the fixed point A over the movable pulleys B, D and the fixed pulleys C, E, and ends in F. Each of the movable pulleys receives a vertical simple harmonic motion from the T bars BG and DH, just as in Fig. 34 (Art. 127). The mechanism is set in motion by imparting uniform rotations to the wheels G, H. If the free end F of the cord be just kept tight, its vertical displacement will be twice the sum of the vertical displacements of B and D, and as these points have

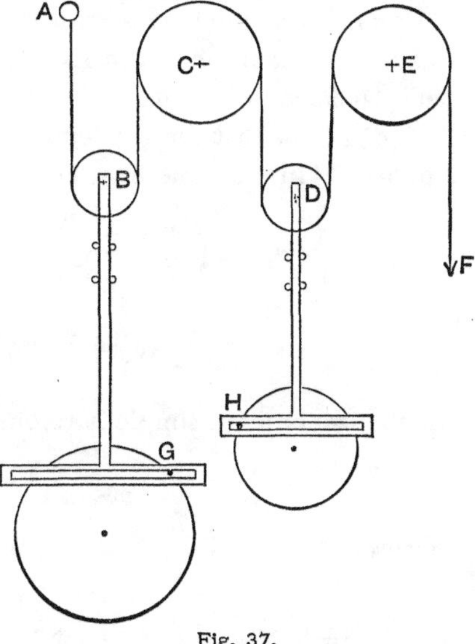

Fig. 37.

simple harmonic motions, the motion of F will be twice the resultant simple harmonic motion.

The idea of this mechanism is due to Lord Kelvin.

134. Exercises.

(1) Find the resultant of three simple harmonic motions in the same line, and all of period $T = 10$ seconds, the amplitudes being 5, 3, and 4 cm., and the phase differences 30° and 60°, respectively, between the first and second, and the first and third motions.

(2) If in the proposition of Art. 130 the amplitudes are equal, $a_1 = a_2 = a$, while the phase-angles differ by $\epsilon_2 - \epsilon_1 = \delta$, show that the resulting motion has the amplitude $2a\cos\frac{1}{2}\delta$ and the phase-angle $\frac{1}{2}\delta$: (*a*) directly, (*b*) from the formulæ of Art. 130, (*c*) by the geometric method of Art. 131.

(3) Find the resultant of two simple harmonic motions in the same line and of equal period when the amplitudes are equal and the phases differ: (*a*) by an even multiple of π, (*b*) by an odd multiple of π.

(4) Resolve $x = 10\cos(\pi t + 45°)$ into two components in the same line with a phase difference of 30°, one of the components having the epoch 0.

(5) Trace the curves representing the component motions as well as the resultant motion in Ex. (1), taking the time as abscissa and the displacement as ordinate.

(6) Show that the resultant of n simple harmonic motions of equal period T in the same line, viz.:

$$x_1 = a_1 \cos\left(\frac{2\pi}{T}t + \epsilon_1\right), \quad x_2 = a_2 \cos\left(\frac{2\pi}{T}t + \epsilon_2\right), \quad \cdots$$

$$x_n = a_n \cos\left(\frac{2\pi}{T}t + \epsilon_n\right),$$

is the isochronous simple harmonic motion

$$x = a \cos\left(\frac{2\pi}{T}t + \epsilon\right),$$

where

$$a^2 = \left(\sum_1^n a_i \cos\epsilon_i\right)^2 + \left(\sum_1^n a_i \sin\epsilon_i\right)^2, \quad \tan\epsilon = \frac{\sum_1^n a_i \sin\epsilon_i}{\sum_1^n a_i \cos\epsilon_i}.$$

135. The composition of two or more simple harmonic motions in the same line can readily be effected, even *when the components differ in period. But the resultant motion is in general not simply harmonic.*

Thus, with two components

$$x_1 = a_1 \cos(\omega_1 t + \epsilon_1), \quad x_2 = a_2 \cos(\omega_2 t + \epsilon_2),$$

putting $\omega_2 t + \epsilon_2 = \omega_1 t + (\omega_2 - \omega_1)t + \epsilon_2 = \omega_1 t + \epsilon_1 + \delta$, say, where $\delta = (\omega_2 - \omega_1)t + \epsilon_2 - \epsilon_1$ is the difference of phase at the time t, we have for the resulting motion

$$x = x_1 + x_2 = a_1 \cos(\omega_1 t + \epsilon_1) + a_2 \cos(\omega_1 t + \epsilon_1 + \delta);$$

$$= (a_1 + a_2 \cos\delta) \cos(\omega_1 t + \epsilon_1) - a_2 \sin\delta \sin(\omega_1 t + \epsilon_1),$$

or putting $a_1 + a_2 \cos\delta = a \cos\epsilon$, $\quad a_2 \sin\delta = a \sin\epsilon$:

$$x = a \cos(\omega_1 t + \epsilon_1 + \epsilon),$$

where

$$a^2 = a_1^2 + a_2^2 + 2a_1 a_2 \cos\delta, \quad \tan\epsilon = \frac{a_2 \sin\delta}{a_1 + a_2 \cos\delta},$$

$$\delta = (\omega_2 - \omega_1)t + \epsilon_2 - \epsilon_1.$$

It can be shown that this represents a simple harmonic motion only when $\omega_2 = \pm \omega_1$.

The formulæ can be interpreted geometrically by Fig. 36, as in Art. 131. But as in the present case the angle δ, and consequently the quantities a and ϵ in the expression for x, vary with the time, the parallelogram OP_1PP_2 while having constant sides has variable angles and changes its form in the course of the motion.

A mechanism similar to that of Fig. 37 (Art. 133) can be used to effect mechanically the composition of simple harmonic motions in the same line whether the periods be equal or not. This is the principle of the tide-predicting machine devised by Lord Kelvin.*

136. To show the connection of the present subject with the theory of **wave motion,** imagine a flexible cord AB of which one

* See THOMSON and TAIT, *Natural philosophy*, Vol. I., Part I., new edition, 1879, p. 43 sq. and p. 479 sq., and J. D. EVERETT, *Vibratory motion and sound*, 1882.

end B is fixed, while the other A is given a sudden jerk or transverse motion from A to C and back through A to D, etc. (Fig. 38). The displacement given to A will, so to speak, run along the cord, travelling from A to B and producing a wave, while any particular point of the cord has approximately a rectilinear motion at right angles to AB. The figure exhibits the successive stages of the

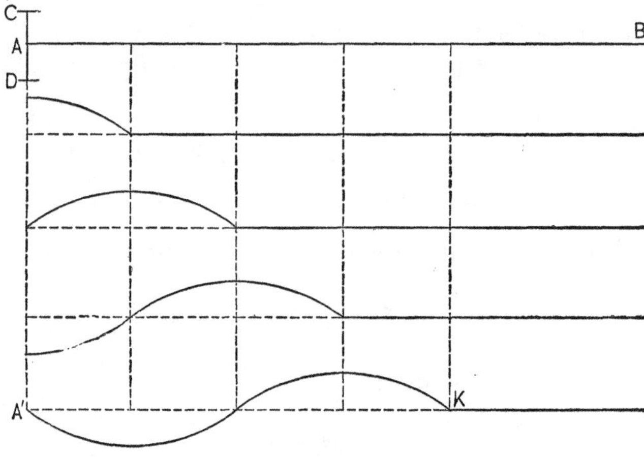

Fig. 38.

motion up to the time when a complete wave $A'K$ has been produced.

The distance $A'K = \lambda$ is called the **length of the wave.** Let T be the time in which the motion spreads from A' to K, that is, the time of a complete vibration of the point A, from A to C, back to D, and back again to A; then

$$\frac{\lambda}{T} = V \tag{17}$$

is called the **velocity of propagation** of the wave.

137. Suppose now that the vibration of A is a simple harmonic motion, say $y = a \sin \omega t$. As the time of vibration of A is T we must have $\omega = 2\pi / T$, and hence, by (17),

$$\omega = \frac{2\pi}{\lambda} V.$$

If we assume that the vibrations of the successive points of the cord differ from the motion of A only in phase, the displacements of all points of the cord at any time t can be represented by

$$y = a \sin(\omega t - \epsilon),$$

where ϵ varies from 0 to 2π as we pass from A' to K.

If we further assume that the phase-angle ϵ of any point of the cord is proportional to the distance x of the point from A' we have $\epsilon = kx$, or since $\epsilon = 2\pi$ for $x = \lambda$:

$$\epsilon = \frac{2\pi}{\lambda} x.$$

Substituting the values of ω and ϵ we find

$$y = a \sin\left[\frac{2\pi}{\lambda}(Vt - x)\right] \qquad (18)$$

The assumptions here made can be regarded as roughly suggested by the experiment of Art. 136 or similar observations. The motion represented by the final equation (18) may be called **simple harmonic wave motion.**

138. To understand the full meaning of the equation (18) it may be observed that, as (in accordance with the assumptions of Art. 137) the quantities a, λ, V are regarded as constant, the displacement y is a function of the two variables t and x.

If t be given a particular value t_1, equation (18) represents the displacements of all points of the cord at the time t_1. The substitution for x of $x + n\lambda$, where n is any positive or negative integer, changes the angle $(2\pi/\lambda)(Vt - x)$ by $2\pi n$ and hence leaves y unchanged. This means that the displacements of all points whose distances from A differ by whole wave-lengths are the same; in other words, the state of motion at any instant is given by a series of equal waves.

If, on the other hand, we assign a particular value x_1 to x and let t vary, the equation represents the rectilinear vibration of the point whose abscissa is x_1. Substituting for t the value $t + nT = t + n\lambda/V$, the angle $(2\pi/\lambda)(Vt - x)$ is again changed by $2\pi n$, so that y remains unchanged. This shows the periodicity of the motion of any point.

139. It may be well to state once more, and as briefly as possible, the fundamental assumptions that underlie the important formula (18).

The idea of simple harmonic wave motion implies that the displacement y should be a periodic function of x and t such as to fulfil the following conditions: y must assume the same value (a) when x is changed into $n\lambda$, (b) when t is changed into $t+T$, (c) when both changes are made simultaneously; the constants λ and T being connected by the relation $\lambda = VT$.

The condition (c) requires y to be of the form $y = f(Vt - x)$; for $Vt - x$ remains unchanged when x is replaced by $x + VT$ and at the same time t by $t + T$.

A particular case of such a function is $y = a \sin c(Vt - x)$. As y should remain unchanged when t is replaced by $t + T$, we must have $c = 2\pi / VT = 2\pi / \lambda$. Thus the function

$$y = a \sin \frac{2\pi}{\lambda}(Vt - x)$$

fulfils the three conditions (a), (b), (c).

Putting $2\pi x / \lambda = -\epsilon$ we have

$$y = a \sin\left(\frac{2\pi}{T} t + \epsilon\right). \qquad (19)$$

The importance of this particular solution of our problem lies in the fact that, according to *Fourier's theorem*,* any single-valued periodic function of period T can be expanded, between definite limits of the variable, in a series of the form:

* For a discussion of Fourier's theorem and its applications the student is referred to THOMSON and TAIT, Natural philosophy, I., 1, London, Macmillan, 1879; G. M. MINCHIN, Uniplanar Kinematics, Oxford, Clarendon Press, 1882, pp. 13 sq.; W. E. BYERLY, An elementary treatise on Fourier's series and spherical, cylindrical, and ellipsoidal harmonics, with applications to problems in mathematical physics, Boston, Ginn, 1893; and H. WEBER, Die partiellen Differentialgleichungen der mathematischen Physik, nach Riemann's Vorlesungen, I., Braunschweig, Vieweg, 1900, pp. 32–80.

$$f(t) = a_0 + a_1 \sin\left(\frac{2\pi}{T}\cdot t + \epsilon_1\right) + a_2 \sin\left(\frac{2\pi}{T}\cdot 2t + \epsilon_2\right)$$
$$+ a_3 \sin\left(\frac{2\pi}{T}\cdot 3t + \epsilon_3\right) + \cdots. \qquad (20)$$

As applied to the theory of wave motion this means that any wave motion, however complex, can be regarded as made up of a series of superposed simple harmonic wave motions of periods T, $\frac{1}{2}T$, $\frac{1}{3}T$, $\cdots$, or since $T = \lambda / V$, of wave-lengths λ, $\frac{1}{2}\lambda$, $\frac{1}{3}\lambda$, $\cdots$. For, if the point A (Fig. 38) be subjected simultaneously to more than one simple harmonic motion, the displacements resulting from each can be added algebraically, thus forming a compound wave which can readily be traced by first tracing the component waves and then adding their ordinates.

The motion due to the superposition of two or more simple harmonic waves may be called *compound harmonic wave motion.*

140. Exercises.

(1) Trace the wave produced by the superposition of two simple harmonic wave motions in the same line of equal amplitudes, the periods being as 2 : 1, (*a*) when they do not differ in phase, (*b*) when their epochs differ by 7/16 of the period.

(2) In the problem of Art. 135, determine the maximum and minimum of the resulting amplitude a and show that the number of maxima per second is equal to the difference of the number of vibrations per second.

141. We proceed to the *composition of simple harmonic motions not in the same line.* We shall, however, assume that all the component motions lie in the same plane.

It is evident that the projection of a simple harmonic motion on any line is again a simple harmonic motion of the same period and phase and with an amplitude equal to the projection of the original amplitude.

Hence, to compound any number of simple harmonic motions along lines lying in the same plane, we may project all these motions on any two rectangular axes Ox, Oy taken in this plane, and compound, by Art. 130 or 135, the components lying on the same axis. It then only remains to compound the two motions, one along Ox, the other along Oy, into a single motion.

142. Just as in Arts. 130, 135, we must distinguish two cases: (a) When the given motions have all the same period, and (b) when they have not.

In the former case, by Art. 130, the two components along Ox and Oy will have equal periods, *i. e.* they will be of the form

$$x = a\cos\omega t, \quad y = b\cos(\omega t + \delta). \tag{21}$$

The path of the resulting motion is obtained by eliminating t between these equations. We have

$$\frac{y}{b} = \cos\omega t\cos\delta - \sin\omega t\sin\delta$$

$$= \frac{x}{a}\cos\delta - \sqrt{1 - \frac{x^2}{a^2}}\sin\delta.$$

Writing this equation in the form

$$\left(\frac{y}{b} - \frac{x}{a}\cos\delta\right)^2 = \left(1 - \frac{x^2}{a^2}\right)\sin^2\delta,$$

or

$$\frac{x^2}{a^2} - \frac{2xy}{ab}\cos\delta + \frac{y^2}{b^2} = \sin^2\delta, \tag{22}$$

we see that it represents an ellipse (since $\frac{1}{a^2}\cdot\frac{1}{b^2} - \frac{\cos^2\delta}{a^2b^2} = \left(\frac{\sin\delta}{ab}\right)^2$ is positive) whose center is at the origin. The resultant motion is therefore called **elliptic harmonic motion.**

We have thus the general result that *any number of simple harmonic motions of the same period and in the same plane, whatever may be their directions, amplitudes, and phases, compound into a single elliptic harmonic motion.*

143. A few particular cases may be noticed. The equation (22) will represent a (double) straight line, and hence the elliptic vibration will degenerate into a simple harmonic vibration, whenever $\sin^2\delta = 0$, *i.e.* when $\delta = n\pi$, where n is a positive or negative integer. In this case $\cos\delta$ is $+1$ or -1, and (22) reduces to

$$\frac{x}{a} - \frac{y}{b} = 0, \text{ if } \delta = 2m\pi,$$

and to

$$\frac{x}{a} + \frac{y}{b} = 0, \text{ if } \delta = (2m+1)\pi.$$

Thus two rectangular vibrations of the same period compound into a simple harmonic vibration when they differ in phase by an integral multiple of π, that is when one lags behind the other by half a wave-length.

Again, the ellipse (22) reduces to a circle only when $\cos\delta = 0$, *i. e.* $\delta = (2m+1)\pi/2$, and in addition $a = b$, the co-ordinates being assumed orthogonal.

Thus two rectangular vibrations of equal period and amplitude compound into a circular vibration if they differ in phase by $\pi/2$, *i. e.* if one lags behind the other by a quarter of a wave-length.

This circular harmonic motion is evidently nothing but uniform motion in a circle; and we have seen in Art. 121 that, conversely, uniform circular motion can be resolved into two rectangular simple harmonic vibrations of equal period and amplitude, but differing in phase by $\pi/2$.

The results of Arts. 141–143 can also be established by purely geometrical methods of an elementary character.*

144. It remains to consider the case when the given simple harmonic motions do not all have the same period. It follows from Art. 135 that in this case, if we again project the given motions on two rectangular axes Ox, Oy, the resulting motions along Ox, Oy are in general not simply harmonic.

The elimination of t between the expressions for x and y may present difficulties. But, of course, the curve can always be traced by points, graphically.

We shall here consider only the case when the motions along Ox and Oy are simply harmonic.

---

* See, for instance, J. G. MACGREGOR, *An elementary treatise on kinematics and dynamics*, London, Macmillan, 1887, pp. 115 sq.

145. *If two simple harmonic motions along the rectangular directions Ox, Oy, viz.:*

$$x = a_1 \cos(\omega_1 t + \epsilon_1), \quad y = a_2 \cos(\omega_2 t + \epsilon_2),$$

of different amplitudes, phases, and *periods* are to be compounded, the resulting motion will be confined within a rectangle whose sides are $2a_1$, $2a_2$, since these are the maximum values of $2x$ and $2y$.

The path of the moving point will be a *closed* curve only when the quotient $\omega_2/\omega_1 = T_1/T_2$ is a rational number, say $= m/n$, where m is prime to n. The x co-ordinate of the curve will have m maxima, the y co-ordinate n, and the whole curve will be traversed after m vibrations along Ox and n along Oy.

The formation of the resulting curve will best be understood from the following example.

146. Let $a_1 = a_2 = a$, $\epsilon_1 = 0$, $\epsilon_2 = \delta$, and let the ratio of the periods be $T_1/T_2 = 2/1$. The equations of the component simple harmonic vibrations are

$$x = a \cos\omega t, \quad y = a \cos(2\omega t + \delta).$$

Here it is easy to eliminate t. We have

$$y = a \cos 2\omega t \cos\delta - a \sin 2\omega t \sin\delta$$

$$= a\left(2\frac{x^2}{a^2} - 1\right) \cos\delta - 2a\frac{x}{a}\sqrt{1 - \frac{x^2}{a^2}} \sin\delta.$$

Hence the equation of the path is:

$$ay = (2x^2 - a^2) \cos\delta - 2x\sqrt{a^2 - x^2} \sin\delta.$$

If there be no difference of phase between the components, *i. e.* if $\delta = 0$, this reduces to the equation of a parabola:

$$x^2 = \tfrac{1}{2}a(y + a).$$

For $\delta = \pi/2$, the equation also assumes a simple form.

$$a^2y^2 = 4x^2(a^2 - x^2).$$

147. It is instructive to trace the resulting curves for a given ratio of periods and for a series of successive differences of phase (*Lissajous's Curves*).

Thus in Fig. 39, the curve for $T_1/T_2 = \frac{3}{4}$, and for a phase difference $\delta = 0$ is the heavily drawn curve, while the dotted curve represents the path for the same ratio of the periods when the phase difference is one-twelfth of the smaller period. The equations of the components are for the heavy curve

$$x = 6 \cos \frac{2\pi}{3} t, \quad y = 5 \cos \frac{2\pi}{4} t,$$

and for the dotted curve

$$x = 6 \cos \left(\frac{2\pi}{3} t + \frac{2\pi}{12} \right), \quad y = 5 \cos \frac{2\pi}{4} t.$$

In tracing these curves, imagine the simple harmonic motions replaced by the corresponding uniform circular motions (Fig. 39).

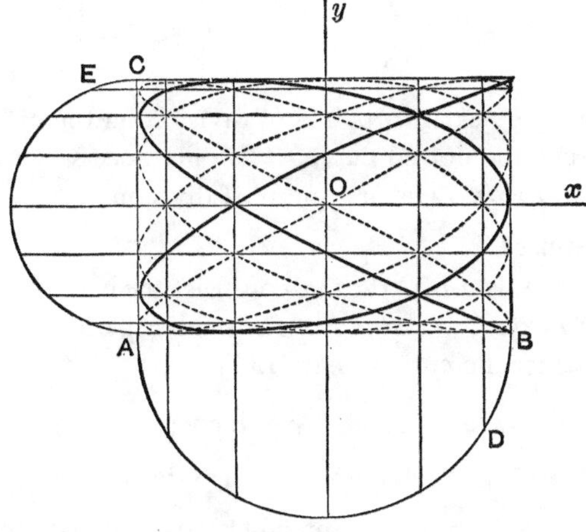

Fig. 39.

With the amplitudes 6, 5, as radii, describe the semi-circles ADB, AEC, so that BC is the rectangle within which the curves are confined; the intersection of the diagonals of this rectangle is the origin O, AB is parallel to the axis of x, AC to the axis of y. Next divide the circles on AB, AC into parts corresponding to equal intervals of time. In the present case, the periods for AB, AC being as 3 to 4, the circle on AB must be divided into $3n$ equal parts, that on AC into $4n$. In the figure, n is taken as 4, the circles being divided into 12 and 16 equal parts, respectively.

The first point of the heavily drawn curve corresponds to $t = 0$, that is $x = 6$, $y = 5$; this gives the upper right-hand corner of the rectangle. The next point is the intersection of the vertical line through D and the horizontal line through E, the arcs $BD = 1/12$ of the circle over AB, and $CE = 1/16$ of that over AC being described in the same time, so that the co-ordinates of the corresponding point are

$$x = 6 \cos\left(\frac{2\pi}{3} \cdot \frac{3}{12}\right) = 6 \cos\left(2\pi \cdot \frac{1}{12}\right),$$

$$y = 5 \cos\left(\frac{2\pi}{4} \cdot \frac{4}{16}\right) = 5 \cos\left(2\pi \cdot \frac{1}{16}\right).$$

Similarly the next point

$$x = 6 \cos\left(2\pi \cdot \frac{2}{12}\right), \quad y = 5 \cos\left(2\pi \cdot \frac{2}{16}\right)$$

is found from the next two points of division on the circles, etc.

To construct the dotted curve, it is only necessary to begin on the circle over AB with D as first point of division.

148. Exercises.

(1) With the data of Art. 147 construct the curves for phase differences of $2/12$, $3/12$, $\cdots$ $11/12$ of the smaller period.

(2) Construct the curves (Art. 146)

$$x = a \cos\omega t, \quad y = a \cos(2\omega t + \delta)$$

for $\delta = 0, \pi/4, \pi/2, 3\pi/4, \pi, 5\pi/4, 3\pi/2, 7\pi/4, 2\pi$.

(3) Trace the path of a point subjected to two circular vibrations of the same amplitude, but differing in period: (a) when the sense is the same; (b) when it is opposite.

149. The **mathematical pendulum** is a point compelled to move in a vertical circle under the acceleration of gravity.

Let O be the center (Fig. 40), A the lowest, and B the highest point of the circle. The radius $OA = l$ of the circle is called the length of the pendulum. Any position P of the moving point is determined by the angle $AOP = \theta$ counted from the vertical radius OA in the positive (counterclockwise) sense of rotation.

If P_0 be the initial position of the moving point at the time $t = 0$, and $\measuredangle AOP_0 = \theta_0$, then the arc $P_0P = s$ described in the time t is $s = l(\theta_0 - \theta)$; hence $v = ds/dt = -ld\theta/dt$, and $dv/dt = -ld^2\theta/dt^2$, the negative sign indicating that θ diminishes as s and t increase.

Resolving the acceleration of gravity, g, into its normal and tangential components $g\cos\theta$, $g\sin\theta$, and considering that the former is without effect owing to the condition that the point is

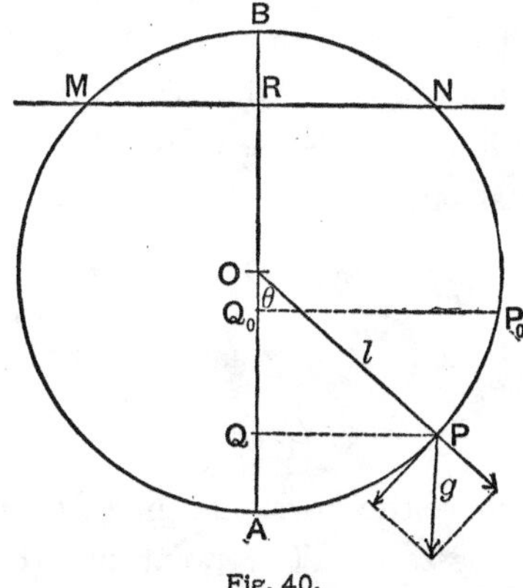

Fig. 40.

constrained to move in a circle, we obtain the equation of motion in the form $dv/dt = g\sin\theta$, or

$$l\frac{d^2\theta}{dt^2} + g\sin\theta = 0. \tag{23}$$

150. The first integration is readily performed by multiplying the equation by $d\theta/dt$ which makes the left-hand member an exact derivative,

$$\frac{d}{dt}\left[\frac{l}{2}\left(\frac{d\theta}{dt}\right)^2 - g\cos\theta\right];$$

hence integrating, we obtain

$$\tfrac{1}{2}l\left(\frac{d\theta}{dt}\right)^2 - g\cos\theta = C,$$

or considering that $v = -ld\theta/dt$,

$$\tfrac{1}{2}v^2 - gl\cos\theta = Cl.$$

To determine the constant C, the initial velocity v_0 at the time $t = 0$ must be given. We then have $\frac{1}{2}v_0^2 - gl\cos\theta_0 = Cl$; hence

$$\begin{aligned}\tfrac{1}{2}v^2 &= \tfrac{1}{2}v_0^2 - gl\cos\theta_0 + gl\cos\theta \\ &= g\left(\frac{v_0^2}{2g} - l\cos\theta_0 + l\cos\theta\right).\end{aligned} \tag{24}$$

The right-hand member can readily be interpreted geometrically; $v_0^2/2g$ is the height by falling through which the point would acquire the initial velocity v_0 (see Art. 73); $l\cos\theta - l\cos\theta_0 = OQ - OQ_0 = Q_0Q$, if Q, Q_0 are the projections of P, P_0 on the vertical AB. If we draw a horizontal line MN at the height $v_0^2/2g$ above P_0 and if this line intersect the vertical AB at R, we have for the velocity v the expression:

$$\tfrac{1}{2}v^2 = g \cdot RQ.$$

If the initial velocity be zero, the equation would be

$$\tfrac{1}{2}v^2 = g \cdot Q_0Q.$$

At the points M, N where the horizontal line MN intersects the circle the velocity becomes zero. The point can therefore never rise above these points.

Now, according to the value of the initial velocity v_0, the line MN may intersect the circle in two real points M, N, or touch it at B, or not meet it at all. In the first case the point P performs oscillations, passing from its initial position P_0 through A up to M, then falling back to A and rising to N, etc. In the third case P makes complete revolutions.

151. The second integration of the equation of motion cannot be effected in finite terms, without introducing elliptic functions. But for the case of most practical importance, viz. for very small values of θ, it is easy to obtain an approximate solution. In this case θ can be substituted for $\sin\theta$, and the equation becomes:

$$\frac{d^2\theta}{dt^2} + \frac{g}{l}\theta = 0,$$

or, putting $g/l = \mu^2$:

$$\frac{d^2\theta}{dt^2} = -\mu^2\theta. \qquad (25)$$

This is a well known differential equation (compare Art. 82, eq. (19), and Art. 125), whose general integral is

$$\theta = C_1 \cos\mu t + C_2 \sin\mu t.$$

The constants C_1, C_2 can be determined from the initial conditions for which we shall now take $\theta = \theta_0$ and $v = 0$ when $t = 0$; this gives $C_1 = \theta_0$, $C_2 = 0$; hence

$$\theta = \theta_0 \cos\mu t, \quad t = \frac{1}{\mu}\cos^{-1}\frac{\theta}{\theta_0}.$$

The last equation gives with $\theta = -\theta_0$ the time t_1 of one **swing** or *beat*, that is, half the period:

$$t_1 = \frac{\pi}{\mu} = \pi\sqrt{\frac{l}{g}}. \qquad (26)$$

The time of a small oscillation or swing is thus seen to be independent of the arc through which the pendulum swings; in other words, for all small arcs the times of swing of the same pendulum are very nearly the same; such oscillations are therefore called *isochronous.*

152. The formula (26) shows that for a pendulum of given length l_1 the time of one swing t_1 varies for different places owing to the variation of g. As l_1 and t_1 can be measured very accurately, the pendulum can be used to determine g, the acceleration of gravity at any place; (26) gives:

$$g = \frac{\pi^2 l_1}{t_1^2}. \qquad (27)$$

Now let l_0 be the length of a pendulum which *beats seconds, i. e.* makes just one swing per second; by (26) and (27) we find for the length l_0 of such a *seconds pendulum:*

$$l_0 = \frac{g}{\pi^2} = \frac{l_1}{t_1^2}. \qquad (28)$$

The length l_0 of the seconds pendulum is therefore found by measuring the length l_1 and the time of swing t_1 of any pendulum. This length l_0 is very nearly a meter; it varies slightly with g; thus, for points at the sea level it varies from $l_0 = 99.103$ cm. at the equator to $l_0 = 99.610$ at the poles.

If g_0 be the value of g at sea level, *i. e.* at the distance R from the center of the earth, g_1 the value of g at an elevation h above sea level in the same latitude, it is known that

$$\frac{g_0}{g_1} = \frac{(R + h)^2}{R^2}.$$

Hence, if g_0 be known, pendulum experiments might serve to find the altitude of a place above sea level; but the observations would have to be of very great accuracy.

153. Let n be the number of swings made by a pendulum of length l in any time T so that $t_1 = T/n$. Then, by (26),

$$\frac{T}{n} = \pi \sqrt{\frac{l}{g}}. \tag{29}$$

If T and one of the three quantities n, l, g in this equation be regarded as constant, the small variations of the two others can be found approximately by differentiation. For instance, if the daily number of oscillations of a pendulum of constant length be observed at two different places, T and l keep the same values while n and g vary by small amounts, say Δn and Δg. Now the differentiation of (29) gives

$$-\frac{T}{n^2}\,dn = -\frac{\pi\sqrt{l}}{2}\,\frac{dg}{g^{\frac{3}{2}}},$$

or, dividing by (29):

$$\frac{dn}{n} = \frac{1}{2}\,\frac{dg}{g}.$$

We have therefore approximately, for small variations Δn, Δg:

$$\frac{\Delta n}{n} = \frac{1}{2}\,\frac{\Delta g}{g}. \tag{30}$$

154. Exercises.

(1) Find the number of swings made in a second and in a day by a pendulum 1 meter long, at a place where $g = 980.5$.

(2) Find the length of the seconds pendulum at a place where $g = 32.17$.

(3) Find the value of g at a place where a pendulum of length 3.249 ft. is found to make 86522 swings in 24 hours.

(4) A chandelier suspended from the ceiling is seen to make 20 swings a minute; find its distance from the ceiling.

(5) A pendulum of length 1 meter is carried from the equator where $g = 978.1$ to another latitude; if it gains 100 swings a day find the value of g there.

(6) Investigate whether the approximate formula (30) is sufficiently accurate for Ex. (5).

(7) If the length of a pendulum be increased by a small amount $\varDelta l$, show that the daily number of swings, n, will be diminished by $\varDelta n$ so that approximately

$$\frac{\varDelta n}{n} = -\tfrac{1}{2}\frac{\varDelta l}{l}.$$

(8) A clock beating seconds is gaining 5 minutes a day; how much should the pendulum bob be screwed up or down?

(9) A clock beating seconds at a place where $g = 32.20$ is carried to a place where $g = 32.15$; how much will it gain or lose per day if the length of the pendulum be not changed?

(10) A pendulum of length 100.18 cm. is found to beat 3585 times per hour; find the elevation of the place if in the same latitude $g = 981.02$ at sea level.

(11) Show that for small oscillations the motion of the pendulum bob is nearly a simple harmonic motion, and deduce from this fact the relation (26), Art. 151.

155. When the oscillations of a pendulum are not so small that the angle can be substituted for its sine as was done in Art. 151, an expression for the time t_1 of one swing can be obtained as follows.

We have by (24), Art. 150:

$$\tfrac{1}{2}v^2 - \tfrac{1}{2}v_0^2 = gl(\cos\theta - \cos\theta_0). \qquad (24)$$

Let the time be counted from the instant when the moving point has its highest position (N in Fig. 40), so that $v_0 = 0$. Substituting $v = -l d\theta/dt$ and applying the formula $\cos\theta = 1 - 2\sin^2\frac{1}{2}\theta$ we find:

$$\tfrac{1}{2}l\left(\frac{d\theta}{dt}\right)^2 = 2g(\sin^2\tfrac{1}{2}\theta_0 - \sin^2\tfrac{1}{2}\theta),$$

whence

$$dt = \tfrac{1}{2}\sqrt{\frac{l}{g}}\frac{d\theta}{\sqrt{\sin^2\frac{1}{2}\theta_0 - \sin^2\frac{1}{2}\theta}}.$$

Integrating from $\theta = 0$ to $\theta = \theta_0$ and multiplying by 2 we find for the time t_1 of one swing:

$$t_1 = \sqrt{\frac{l}{g}}\int_0^{\theta_0}\frac{d\theta}{\sqrt{\sin^2\frac{1}{2}\theta_0 - \sin^2\frac{1}{2}\theta}}. \qquad (31)$$

As θ cannot become greater than θ_0 we may put $\sin\frac{1}{2}\theta = \sin\frac{1}{2}\theta_0 \sin\phi$, thus introducing a new variable ϕ for which the limits are 0 and $\pi/2$. Differentiating the equation of substitution, we have

$$\tfrac{1}{2}\cos\tfrac{1}{2}\theta\, d\theta = \sin\tfrac{1}{2}\theta_0 \cos\phi\, d\phi,$$

or, as $\cos\frac{1}{2}\theta = \sqrt{1 - \sin^2\frac{1}{2}\theta_0 \sin^2\phi}$,

$$d\theta = \frac{2\sin\frac{1}{2}\theta_0 \cos\phi\, d\phi}{\sqrt{1 - \sin^2\frac{1}{2}\theta_0 \sin^2\phi}}.$$

Substituting these values and putting for the sake of brevity

$$\sin\tfrac{1}{2}\theta_0 = \kappa, \qquad (32)$$

we find for the time t_1 of one swing:

$$t_1 = 2\sqrt{\frac{l}{g}}\int_0^{\pi/2}\frac{d\phi}{\sqrt{1 - \kappa^2\sin^2\phi}}. \qquad (33)$$

The integral in this expression is called the complete elliptic integral of the first species, and is usually denoted by K. Its value can be found from tables of elliptic integrals or by expanding the argument into an infinite series by the binomial theorem (since κ sinϕ is less than 1), and then performing the integration. We have

$$(1 - \kappa^2 \sin^2\phi)^{-\frac{1}{2}} = 1 + \tfrac{1}{2}\kappa^2 \sin^2\phi + \frac{1 \cdot 3}{2 \cdot 4}\kappa^4 \sin^4\phi + \cdots;$$

hence

$$t_1 = \pi\sqrt{\frac{l}{g}}\left[1 + \left(\frac{1}{2}\right)^2\kappa^2 + \left(\frac{1 \cdot 3}{2 \cdot 4}\right)^2\kappa^4 + \cdots\right]. \qquad (34)$$

If H be the height of the initial point $N(\theta = \theta_0)$ above the lowest point A of the circle, we have by (32)

$$\kappa^2 = \sin^2\tfrac{1}{2}\theta_0 = \frac{1 - \cos\theta_0}{2} = \frac{H}{2l},$$

so that (34) can be written in the form

$$t_1 = \pi\sqrt{\frac{l}{g}}\left[1 + \left(\frac{1}{2}\right)^2\frac{H}{2l} + \left(\frac{1 \cdot 3}{2 \cdot 4}\right)\left(\frac{H}{2l}\right)^2 + \cdots\right].$$

156. Exercises.

(1) Show that $t_1 = \pi\sqrt{\frac{l}{g}}\left(1 + \frac{1}{16} + \frac{9}{1024} + \frac{25}{16384} + \cdots\right)$ if the angle $2\theta_0$ of the swing is 120°.

(2) Show that as second approximation to the time of a small swing we have $t_1 = \pi\sqrt{l/g}\,(1 + \frac{1}{16}\theta_0^2)$.

(3) Find the time of oscillation of a pendulum whose length is 1 meter at a place where $g = 980.8$, to four decimal places, the amplitude θ_0 of the swing being 6°.

(4) Denoting by t_0 the first approximation, $\pi\sqrt{l/g}$, to the time t_1 of one swing, the quotient $(t_1 - t_0)/t_0$ is called the *correction for amplitude*. Show that its value is 0.0005 for $\theta_0 = 5°$.

(5) A pendulum hanging at rest is given an initial velocity v_1. Find to what height h_1 it will rise.

(6) Discuss the pendulum problem in the particular case when MN (Fig. 40) touches the circle at B, that is when the initial velocity is due to falling from the highest point of the circle.

157. Central Motion. The motion of a point P is called **central** if the *direction* of the acceleration constantly passes through a fixed point O. The most important case is that when, moreover, the *magnitude* of the acceleration is a function of the distance $OP = r$ alone, say

$$j = f(r).$$

The fixed point O is in this case usually regarded as the seat of an attractive or repulsive force producing the acceleration, and is therefore called the *center of force.*

Harmonic motion as discussed in Arts. 121–148 is a special case of central motion, viz. the case in which the acceleration j is directly proportional to the distance from the fixed center O, *i. e.* $f(r) = \mu r$.

Another very important particular case is that of Newton's law, *i. e.* $f(r) = \mu/r^2$; this will be discussed below, Arts. 170–173.

158. Any central motion is fully determined if in addition to the form of the function $f(r)$ we know the "initial conditions," say the initial distance $OP_0 = r_0$ (Fig. 41) and the initial velocity v_0 of the point at the time $t = 0$. As v_0 must be given both in magnitude and direction, the angle ψ_0 between r_0 and v_0 must be known.

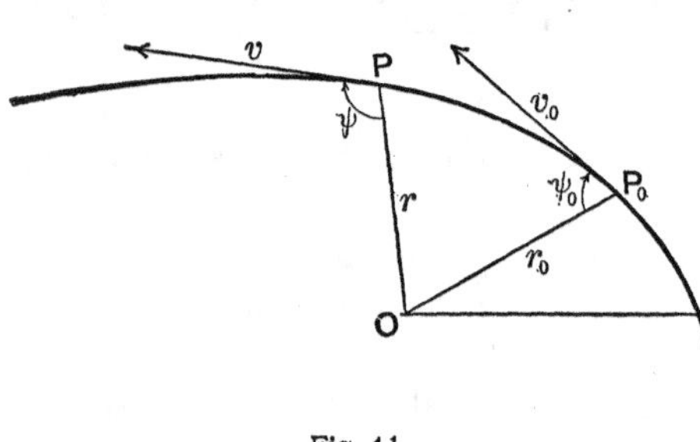

Fig. 41.

It is evident, geometrically, that the motion is confined to the plane determined by O and v_0 since the acceleration always lies in this plane. Hence, *any central motion*, whatever may be the law of acceleration, *is a plane motion.*

159. Another fundamental property is that *in any central motion*, whatever the law of acceleration, *the sectorial velocity is constant.* This is most readily proved by taking the center O as origin for polar co-ordinates r, θ. As by the definition of central motion (Art. 157) the acceleration j is directed along the radius vector

$OP = r$ drawn from the center O to the moving point P, the component j_θ of the acceleration, at right angles to the radius vector, is always zero. We have therefore by the last of the equations (6) of Art. 114:

$$j_\theta = \frac{1}{r}\frac{d}{dt}\left(r^2 \frac{d\theta}{dt}\right) = 0,$$

whence

$$r^2 \frac{d\theta}{dt} = c, \tag{35}$$

where c is the constant of integration. By Art. 97 this equation means that the sectorial velocity is constant and equal to $\frac{1}{2}c$.

160. Let S be the sector P_0OP described by the radius vector r in the time t, so that $dS = \frac{1}{2}r^2 d\theta$ is the elementary sector described in the element of time dt. Then (35) can be written

$$\frac{dS}{dt} = \tfrac{1}{2}c,$$

whence integrating, since $S = 0$ for $t = 0$:

$$S = \tfrac{1}{2}ct.$$

This shows that *the sector is proportional to the time in which it is described*, which is merely another way of stating that the sectorial velocity is constant.

It can be shown conversely, by reversing the steps of the above argument, that if in a plane motion the areas swept out by the radius vector drawn from a fixed point of the plane are proportional to the time, the acceleration must constantly pass through that point.

It is well known that Kepler had found by a careful examination of the observations available to him that *the orbits described by the planets are plane curves, and the sector described by the radius vector drawn from the sun to any planet is proportional to the time in which it is described.* This constitutes **Kepler's first law** of planetary motion.

He concluded from it that the acceleration must constantly pass through the sun.

161. To express the value of the constant of integration c in terms of the given initial conditions (Art. 158), *i. e.* by means of r_0, v_0, ψ_0, observe that at any time t

$$c = r^2 \frac{d\theta}{dt} = r \cdot \frac{rd\theta}{ds} \cdot \frac{ds}{dt} = r \sin\psi \cdot v;$$

hence at the time $t = 0$ we have

$$c = v_0 r_0 \sin\psi_0.$$

Denoting by p_0 and p the perpendiculars let fall from O on v_0 and v we have $r_0 \sin\psi_0 = p_0$, $r \sin\psi = p$; hence

$$c = p_0 v_0 = pv,$$

i. e. the velocity at any time is inversely proportional to its distance from the center.

162. Let us now assume that the acceleration j of a central motion is a given function, $f(r)$, of the radius vector $OP = r$ drawn from the center O to the moving point P. With O as origin, let x, y be the rectangular cartesian co-ordinates of the moving point P, and r, θ its polar co-ordinates, at any time t. Then $\cos\theta = x/r$, $\sin\theta = y/r$ are the direction cosines of $OP = r$, and, therefore, those of the acceleration j, provided the sense of j be away from the center, *i. e.* provided the force causing the acceleration be *repulsive*. In the case of *attraction*, the direction cosines of j are of course $-x/r$, $-y/r$.

Thus the *equations of motion* are in the case of attraction:

$$j_x \equiv \frac{d^2x}{dt^2} = -f(r)\frac{x}{r}, \quad j_y \equiv \frac{d^2y}{dt^2} = -f(r)\frac{y}{r}. \qquad (36)$$

For repulsion, it would only be necessary to change the sign of $f(r)$.

To integrate the equations (36) we cannot, in general, treat each equation by itself; for, as $r = \sqrt{x^2 + y^2}$, each equation contains three variables x, y, t. We must therefore try to combine the equations so as to form integrable combinations.

163. Let us first multiply the equations (36) by y, x and subtract; the right-hand member of the resulting equation is zero while the left-hand member is an exact derivative:

$$x\frac{d^2y}{dt^2} - y\frac{d^2x}{dt^2} \equiv \frac{d}{dt}\left(x\frac{dy}{dt} - y\frac{dx}{dt}\right) = 0.$$

Integrating we find

$$x\frac{dy}{dt} - y\frac{dx}{dt} = c,$$

or, introducing polar co-ordinates:

$$r^2\frac{d\theta}{dt} = c, \qquad (35)$$

which is the equation (35), Art. 159; comp. Art. 97.

164. Next multiply the equations (36) by dx/dt, dy/dt and add. The left-hand member of the resulting equation is

$$\frac{dx}{dt}\frac{d^2x}{dt^2} + \frac{dy}{dt}\frac{d^2y}{dt^2} = \frac{d}{dt}\left[\tfrac{1}{2}\left(\frac{dx}{dt}\right)^2 + \tfrac{1}{2}\left(\frac{dy}{dt}\right)^2\right] = \frac{d}{dt}(\tfrac{1}{2}v^2);$$

the right-hand member becomes

$$-\frac{f(r)}{r}\left(x\frac{dx}{dt} + y\frac{dy}{dt}\right) = -\frac{f(r)}{r}\frac{d}{dt}\tfrac{1}{2}(x^2 + y^2) = -\frac{f(r)}{r}\frac{d}{dt}(\tfrac{1}{2}r^2)$$

$$= -f(r)\frac{dr}{dt}.$$

The resulting equation

$$d(\tfrac{1}{2}v^2) = -f(r)\,dr$$

can be integrated and gives

$$\tfrac{1}{2}v^2 - \tfrac{1}{2}v_0^2 = -\int_{r_0}^{r} f(r)dr; \qquad (37)$$

i. e. it determines the velocity v as a function of r.

165. The two methods of combining the differential equations of motion (36) used in Arts. 163 and 164 are known, respectively, as the *principle of areas* and the *principle of kinetic energy and work.* The former name explains itself (see Arts. 159, 160). The latter is due to the fact (to be more fully explained in kinetics) that if equation (37) be multiplied by the mass of the moving body, the left-hand member will represent the increase in kinetic energy while the right-hand member is the work of the central force.

Each of these methods of preparing the equations of motion for integration consists merely in combining the equations so as to obtain an exact derivative in the left-hand member of the resulting equation. If by this combination the right-hand member happens to vanish or to become likewise an exact derivative, an integration can at once be performed. This is the case in our problem.

166. The two equations (35) and (37), each of which was found by a first integration, are called *first integrals* of the equations of motion. By combining them and integrating again, the equation of the path is found.

We have, by (4), Art. 96, for *any* curvilinear motion

$$v^2=\left(\frac{dr}{dt}\right)^2+r^2\left(\frac{d\theta}{dt}\right)^2=\left(\frac{d\theta}{dt}\right)^2\left[\left(\frac{dr}{d\theta}\right)^2+r^2\right];$$

eliminating $d\theta/dt$ by means of (35) we find for any central motion:

$$v^2=\frac{c^2}{r^4}\left[\left(\frac{dr}{d\theta}\right)^2+r^2\right]=c^2\left[\left(\frac{d}{d\theta}\frac{1}{r}\right)^2+\left(\frac{1}{r}\right)^2\right]. \qquad (38)$$

Substituting this expression of v^2 in (37) we have the differential equation of the path in which the variables are separable. Shorter methods may occasionally suggest themselves in particular cases; see, for instance, Art. 171.

167. To solve the converse problem, viz. to find the law of acceleration when the path is known, we have only to substitute the expression (38) of v^2 in the equation preceding (37); this gives:

$$f(r) = -\frac{d}{dr}\tfrac{1}{2}v^2 = -\frac{c^2}{2}\frac{d}{dr}\left[\left(\frac{d}{d\theta}\frac{1}{r}\right)^2 + \left(\frac{1}{r}\right)^2\right]$$

$$= -\frac{c^2}{2}\frac{d}{d\theta}\left[\left(\frac{d}{d\theta}\frac{1}{r}\right)^2 + \left(\frac{1}{r}\right)^2\right]\frac{d\theta}{dr}$$

$$= -c^2\left(\frac{d^2}{d\theta^2}\frac{1}{r}\cdot\frac{d}{d\theta}\frac{1}{r} + \frac{1}{r}\frac{d}{d\theta}\frac{1}{r}\right)\frac{d\theta}{dr}$$

$$= -c^2\left(\frac{d^2}{d\theta^2}\frac{1}{r} + \frac{1}{r}\right)\frac{d}{dr}\frac{1}{r};$$

hence

$$f(r) = \frac{c^2}{r^2}\left(\frac{d^2}{d\theta^2}\frac{1}{r} + \frac{1}{r}\right). \tag{39}$$

168. Kepler in his **second law** had established the empirical fact that *the orbits of the planets are ellipses, with the sun at one of the foci.*

From this Newton concluded that the law of acceleration must be that of the inverse square of the distance from the sun.

Our equation (39) enables us to draw this conclusion. The polar equation of an ellipse referred to focus and major axis is

$$r = \frac{l}{1 + e\cos\theta},$$

where $l = b^2/a = a(1 - e^2)$; a, b being the semi-axes, l the semi-latus rectum, and e the eccentricity. Hence

$$\frac{1}{r} = \frac{1}{l} + \frac{e}{l}\cos\theta, \quad \frac{d^2}{d\theta^2}\frac{1}{r} = -\frac{e}{l}\cos\theta,$$

so that (39) becomes

$$f(r) = \frac{c^2}{l}\cdot\frac{1}{r^2} = \frac{c^2}{a(1 - e^2)}\cdot\frac{1}{r^2}. \tag{40}$$

169. The **third law of Kepler,** found by him likewise as an empirical fact, asserts that *the squares of the periodic times of different planets are as the cubes of the major axes of their orbits.*

From this fact Newton drew the conclusion that in the law of acceleration,

$$j \equiv f(r) = \frac{\mu}{r^2},$$

the constant μ has the same value for all the planets.

Our formulæ show this as follows. Let T be the *periodic time* of any planet, *i. e.* the time of describing an ellipse whose semi-axes are a, b. Then, since the sector described in the time T is the area πab of the whole ellipse, we have by Art. 160

$$\pi ab = \tfrac{1}{2} cT.$$

Substituting in (40) the value of c found from this equation we have

$$f(r) = \frac{4\pi^2 a^2 b^2}{lT^2} \cdot \frac{1}{r^2} = \frac{4\pi^2 a^3}{T^2} \cdot \frac{1}{r^2}.$$

Hence

$$\mu = \frac{4\pi^2 a^3}{T^2}$$

is constant by Kepler's third law.

170. Planetary motion in its simplest form is that particular case of central motion in which the acceleration is inversely proportional to the square of the distance from the center O so that

$$j \equiv f(r) = \frac{\mu}{r^2},$$

where μ is a constant, viz., the acceleration at the distance $r = 1$ from O.

The equations of motion (36) are in this case, with O as origin,

$$\frac{d^2x}{dt^2} = -\mu\frac{x}{r^3}, \quad \frac{d^2y}{dt^2} = -\mu\frac{y}{r^3}. \tag{41}$$

Combining these by the principle of energy (Arts. 164, 165), we find

$$\frac{d(\frac{1}{2}v^2)}{dt} = -\frac{\mu}{r^3}\left(x\frac{dx}{dt} + y\frac{dy}{dt}\right) = -\frac{\mu}{r^3}\frac{d}{dt}\left(\frac{x^2+y^2}{2}\right)$$

$$= -\frac{\mu}{r^2}\frac{dr}{dt} = \mu\frac{d}{dt}\frac{1}{r};$$

hence integrating

$$\tfrac{1}{2}v^2 - \tfrac{1}{2}v_0^{\ 2} = \frac{\mu}{r} - \frac{\mu}{r_0}. \tag{42}$$

171. To find the equation of the path, or *orbit*, write the equations (41) in the form

$$\frac{d^2x}{dt^2} = -\frac{\mu}{r^2}\cos\theta, \quad \frac{d^2y}{dt^2} = -\frac{\mu}{r^2}\sin\theta,$$

and eliminate r^2 by means of (35);

$$\frac{d^2x}{dt^2} = -\frac{\mu}{c}\cos\theta\frac{d\theta}{dt}, \quad \frac{d^2y}{dt^2} = -\frac{\mu}{c}\sin\theta\frac{d\theta}{dt}.$$

Each of these equations can be integrated by itself:

$$\frac{dx}{dt} - v_1 = -\frac{\mu}{c}\sin\theta, \quad \frac{dy}{dt} - v_2 = \frac{\mu}{c}(\cos\theta - 1), \tag{43}$$

where v_1, v_2 are the components of the velocity when $\theta = 0$.

Multiplying by y, x and subtracting we find, by Art. 163:

$$\left(\frac{\mu}{c} - v_2\right)x + v_1 y + c = \frac{\mu}{c}(x\cos\theta + y\sin\theta) = \frac{\mu}{c}\sqrt{x^2+y^2}. \tag{44}$$

172. The geometrical meaning of this equation is that the radius vector $r = \sqrt{x^2+y^2}$ drawn from the fixed point O to the moving point P is proportional to the distance of P from the fixed straight line

$$\left(\frac{\mu}{c} - v_2\right)x + v_1 y + c = 0. \tag{45}$$

It represents, therefore, a conic section having O for a focus and the line (45) for the corresponding directrix.

The character of the conic depends on the absolute value of the ratio of the radius vector to the distance from the directrix; according as this ratio,

$$\frac{c}{\mu}\sqrt{\left(\frac{\mu}{c}-v_2\right)^2+v_1^2},$$

is <1, $=1$, or >1, the conic will be an ellipse, a parabola, or a hyperbola. This criterion can be simplified. Multiplying by μ/c and squaring, we have

$$-\frac{2\mu v_2}{c}+v_2^2+v_1^2 \lesseqgtr 0,$$

or since $v_1^2+v_2^2=v_0^2$ and $c=r_0v_0\sin\psi_0=r_0v_2$:

$$v_0^2 \lesseqgtr \frac{2\mu}{r_0}. \tag{46}$$

173. If polar co-ordinates be introduced in (44), the equation of the orbit assumes the form

$$\frac{1}{r}=\frac{\mu}{c^2}+\left(\frac{v_2}{c}-\frac{\mu}{c^2}\right)\cos\theta-\frac{v_1}{c}\sin\theta,$$

or putting $(cv_2-\mu)/c^2=C\cos\alpha$, $v_1/c=C\sin\alpha$,

$$\frac{1}{r}=\frac{\mu}{c^2}+C\cos(\theta+\alpha). \tag{47}$$

This equation might have been obtained directly by integrating (39), which in our case, with $f(r)=\mu/r^2$, reduces to

$$\frac{d^2}{d\theta^2}\frac{1}{r}+\frac{1}{r}=\frac{\mu}{c^2};$$

the general integral of this differential equation is of the form (47), C and α being the constants of integration.

Equation (47) represents a conic section referred to the focus as origin and a line making an angle α with the focal axis as polar axis.

174. Exercises.

(1) If $2k$ be the chord intercepted by the osculating circle on the radius vector drawn from the fixed center, show that $v^2 = k \cdot f(r)$.

(2) A point moves in a circle; if the acceleration be constant in direction, what is its magnitude?

(3) A point describes a circle; if the acceleration be constantly directed towards the center, what is its magnitude?

(4) A point has a central acceleration proportional to the distance from the center and directed away from the center; find the equation of the path.

(5) A point P is subject to two accelerations, $\mu^2 \cdot O_1P$ directed toward the fixed point O_1, and $\mu^2 \cdot O_2P$ directed away from the fixed point O_2. Show that its path is a parabola.

(6) A point P describes an ellipse owing to a central acceleration $f(r) = \mu/r^2$ directed toward the focus S. Its initial velocity v_0 makes an angle ψ_0 with the initial radius vector r_0. Determine the semi-axes a, b of the ellipse in magnitude and position.

(7) The rectangular components of the acceleration of a point are both constant; the initial velocity v_0 is parallel to one of these components; find the path.

5. VELOCITIES IN THE RIGID BODY.

175. A rigid body is said to have plane motion when all its points move in parallel planes. Its motion is then fully determined by the motion of any plane section of the body in its plane.

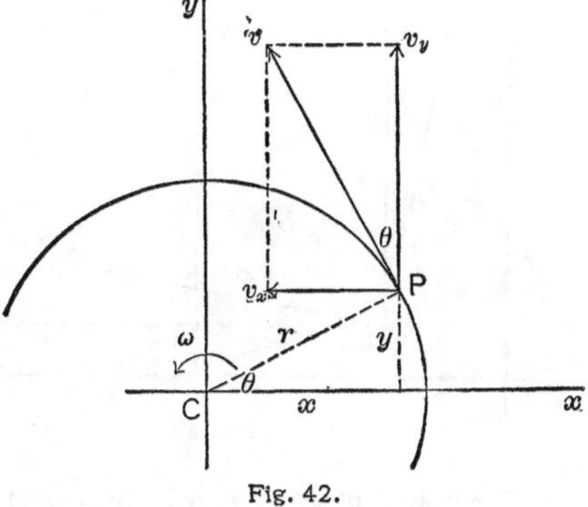

Fig. 42.

It has been shown in Arts. 12–18 that the continuous motion of an invariable plane figure in its plane can be regarded as the limit of a series of instantaneous rotations about the successive instantaneous centers, *i. e.* about the points of the fixed centrode.

If at any instant the center of rotation C and the angular velocity ω about it be known, *the linear velocity of any point P of the plane figure at the distance $CP = r$ from C can be found, being $v = \omega r$, at right angles to r.*

The components v_x, v_y of this velocity along rectangular axes through C are evidently (Fig. 42)

$$v_x = \omega r \cdot \left(-\frac{y}{r}\right) = -\omega y, \quad v_y = \omega r \cdot \left(\frac{x}{r}\right) = \omega x. \qquad (1)$$

These results can also be obtained by differentiating, with respect to the time t, the relations

$$x = r\cos\theta, \quad y = r\sin\theta$$

between the cartesian and polar co-ordinates of the point P and observing that $d\theta/dt$ is the angular velocity ω about the instantaneous center C. For we thus find:

$$v_x = \frac{dx}{dt} = -r\sin\theta\frac{d\theta}{dt} = -\omega y,$$

$$v_y = \frac{dy}{dt} = r\cos\theta\frac{d\theta}{dt} = \omega x.$$

176. In studying the motion of an invariable plane figure in its plane it is generally convenient to use two sets of rectangular co-ordinate axes; one set Ox, Oy, fixed in the plane, called the space axes or *fixed axes*, the other $O'\xi$, $O'\eta$, fixed in the figure and moving with it, called the *moving axes* (Fig. 43).

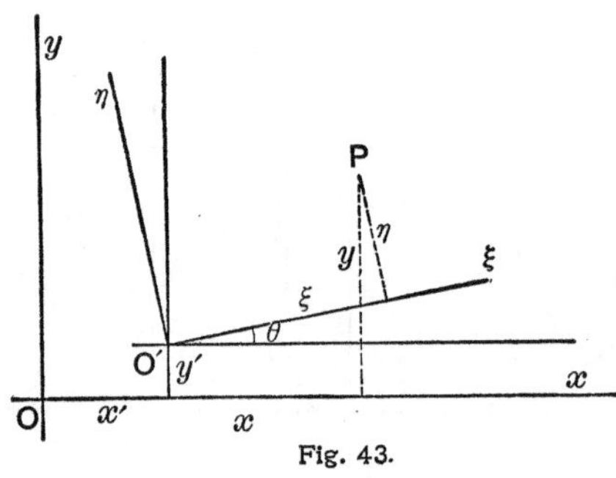

Fig. 43.

At any instant t, let the moving origin O' have the co-ordinates x', y' and let the moving axis $O'\xi$ make an angle θ with the fixed axis Ox; let x, y be the co-ordinates of any point P of the moving figure with respect to the fixed axes; ξ, η the co-ordinates of the same point P with respect to the moving axes. Then we have the obvious relations

$$x = x' + \xi \cos\theta - \eta \sin\theta, \quad y = y' + \xi \sin\theta + \eta \cos\theta,$$

in which ξ, η are constant while x, y, x', y', θ are functions of the time.

Differentiating we find for the component velocities of P parallel to the fixed axes Ox, Oy:

$$\begin{aligned} v_x &= \frac{dx'}{dt} - (\xi \sin\theta + \eta \cos\theta)\frac{d\theta}{dt}, \\ v_y &= \frac{dy'}{dt} + (\xi \cos\theta - \eta \sin\theta)\frac{d\theta}{dt}. \end{aligned} \tag{2}$$

Now, $d\theta/dt$ is the angular velocity ω about the point O' while dx'/dt, dy'/dt are the velocities of O' parallel to the fixed axes, say $v_{x'}, v_{y'}$. Considering moreover that $\xi \sin\theta + \eta \cos\theta = y - y'$, $\xi \cos\theta - \eta \sin\theta = x - x'$, we have

$$v_x = v_{x'} - (y - y')\omega, \quad v_y = v_{y'} + (x - x')\omega. \tag{3}$$

The velocity of P with respect to the fixed axes Ox, Oy consists, therefore, *of two parts, a velocity of translation equal to that of O' and a velocity of rotation about O equal to that of P about O'.*

177. The instantaneous center being the point whose velocity is zero at the given instant, we find its co-ordinates x_0, y_0 from the equations

$$0 = v_{x'} - (y_0 - y')\omega, \quad 0 = v_{y'} + (x_0 - x')\omega,$$

whence

$$x_0 = x' - \frac{v_{y'}}{\omega}, \quad y_0 = y' + \frac{v_{x'}}{\omega}. \tag{4}$$

When the angular velocity ω, the co-ordinates x', y', and the velocity $(v_{x'}, v_{y'})$ of any point O' are known as functions of the time, the elimination of t between the equations (4) gives the equation of the fixed centrode.

The co-ordinates ξ_0, η_0 of the instantaneous center referred to the moving axes are found in a similar way from the equations (2):

$$\xi_0 = \frac{1}{\omega}(v_{x'} \sin\theta - v_{y'} \cos\theta), \quad \eta_0 = \frac{1}{\omega}(v_{x'} \cos\theta + v_{y'} \sin\theta), \qquad (5)$$

from which the body centrode can be found by eliminating t.

178. Exercises.

(1) When a straight line moves in a plane show that the velocities of all its points have equal projections on the line. Hence show how to construct the velocity v of the end P of the connecting rod (Fig. 25, p. 59) when the velocity u of the other end Q is known.

(2) Two points A, A' of a plane figure move on two fixed circles described with radii a, a' about O, O'; show that the angular velocities ω, ω' of OA, $O'A'$ about O, O' are inversely proportional to OM, $O'M$, M being the point of intersection of OO' with AA'.

(3) Given the magnitudes v, v' of the velocities of two points A, A' of an invariable plane figure and the angle (v, v') formed by their directions; find the instantaneous center C and the angular velocity ω about C.

(4) Show that in the "elliptic motion" of a plane figure (Arts. 20–22) the velocity of any point (x', y') is

$$v = [a^2 + x'^2 + y'^2 - 2a(x' \cos 2\varphi + y' \sin 2\varphi)]^{\frac{1}{2}} \frac{d\varphi}{dt}.$$

(5) In the same motion find the velocities of B and O' (Fig. 7, p. 12) when A moves uniformly along the axis of x.

(6) A straight rod AB, 4 ft. long, moves in a plane; the velocity of one end A is 20 ft./sec., along AB, that of B is inclined at 45° to AB. Find the velocity of B and that of the middle point of AB.

179. The continuous motion of a rigid body is called a **translation** when the velocities of all its points are equal and parallel at every moment (Art. 4). All points describe therefore equal and parallel curves, and every line of the body remains parallel to itself. The velocity $v = ds/dt$ of any point is called the velocity of translation of the body.

We can imagine a rigid body subjected to several velocities of translation simultaneously; the resulting motion is a translation whose velocity is found by geometrically adding the component velocities.

Conversely, the velocity of translation of a rigid body can be resolved into components in given directions.

180. The continuous motion of a rigid body is called **rotation** when two points of the body are fixed; the line joining these points is the axis of rotation. All points excepting those on the axis describe arcs of circles whose centers lie on the axis and whose planes are perpendicular to the axis.

The velocity of any point P of the body at the distance $OP = r$ from the axis is $v = \omega r$, if ω is the angular velocity of the rigid body; its direction is at right angles to the plane through P and the axis. The velocities of the different points of the body at any given moment are therefore directly proportional to their distances from the axis, and the velocities of all points at this moment are known if the axis and the instantaneous angular velocity ω are given. It is therefore convenient to imagine this angular velocity represented by a vector ω laid off on the axis of rotation.

The counter-clockwise sense of rotation being adopted as positive, the vector $01 = \omega$ (Fig. 44) should be placed on the axis so that the rotation appears counter-clockwise to a person looking from the arrow-head 1, or end-point, of ω on the plane through the initial point 0 at right angles to the axis.

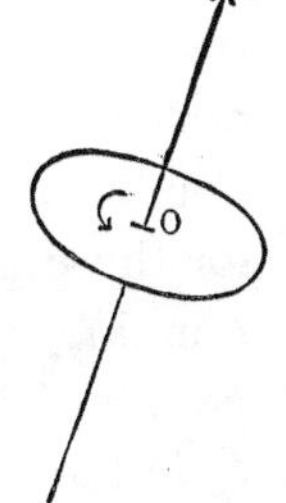

Fig. 44.

181. A special name, **rotor,** is sometimes used for this kind of vector because it is *localized* or attached to a definite line, viz., the axis of rotation. Thus, while any two parallel and equal vectors of the same sense represent the same translation of a rigid body, two parallel and equal rotors of the same sense evidently represent different rotations, viz., rotations about different axes. Two rotors are therefore regarded as equal only when they are of equal magnitude and sense and situated on the same straight line.

182. A body may have several simultaneous rotations, just as it may have several simultaneous translations. Imagine, for instance, a wheel or spindle turning about its axis while the bearings in which its axle runs are attached to a piece of machinery that has itself a motion of rotation. The resulting motion can be found from the rotors of the component rotations according to definite rules. We shall, however, here consider only two particular cases, the composition of parallel and of intersecting rotors.

183. Composition of Parallel Rotors. In the case of *plane* motion of a rigid body the axes of rotation are perpendicular to the plane and hence parallel.

Consider a body turning with angular velocity ω_1 about an axis l_1 (passing through the point L_1, Fig. 45, at right angles to the plane of this figure) and at the same time with angular velocity ω_2 about an axis l_2 (through L_2) parallel to l_1.

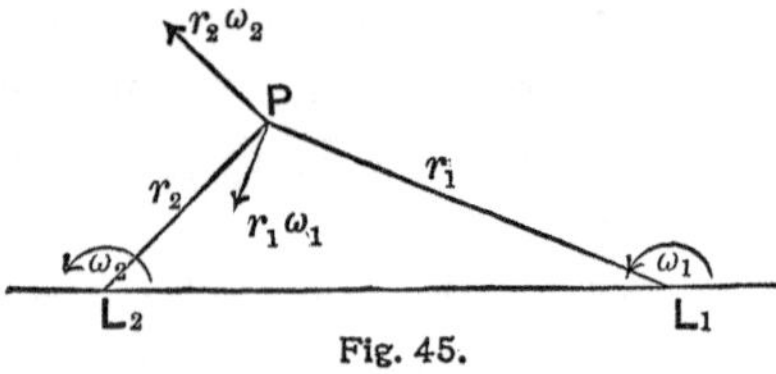

Fig. 45.

Any point P of the body receives from ω_1 a linear velocity $\omega_1 r_1$ perpendicular to L_1P and from ω_2 a linear velocity $\omega_2 r_2$ perpendicular to L_2P; the resultant of these two is the total velocity of P. The two components $\omega_1 r_1$ and $\omega_2 r_2$ fall into the same straight line only for points in the plane $(l_1 l_2)$, and their resultant will be zero only for those points of this plane which divide the distance between l_1 and l_2 in the inverse ratio of ω_1 and ω_2. In other words, the points of zero velocity lie on a straight line l, parallel to l_1 and l_2, in the plane $(l_1 l_2)$, so situated that if L be its intersection with L_1L_2, we have

$$\omega_1 \cdot L_1L = \omega_2 \cdot LL_2.$$

As this line l is instantaneously fixed, the resulting motion of the body is a rotation about l. To find the angular velocity ω of this rotation consider a particular point, for instance L_2; its linear velocity being due entirely to ω_1 about l_1 is $= \omega_1 \cdot L_1L_2$, but it can also be regarded as due to ω about l; hence

$$\omega_1 \cdot L_1L_2 = \omega \cdot LL_2.$$

These two relations give

$$\frac{L_1L}{\omega_2} = \frac{LL_2}{\omega_1} = \frac{L_1L_2}{\omega},$$

and as $L_1L + LL_2 = L_1L_2$, we also have

$$\omega = \omega_1 + \omega_2.$$

Thus, *the resultant of two angular velocities* ω_1, ω_2 *about parallel axes* l_1, l_2 *is an angular velocity* ω *equal to their algebraic sum,* $\omega = \omega_1 + \omega_2$, *about a parallel axis* l *that divides the distance between* l_1, l_2 *in the inverse ratio of* ω_1 *and* ω_2.

Conversely, an angular velocity ω about an axis l can always be replaced by two angular velocities ω_1, ω_2 whose sum is equal to ω and whose axes l_1, l_2 are parallel to l and so selected that l divides the distance between l_1, l_2 inversely as ω_1 is to ω_2.

184. The resulting axis lies between L_1 and L_2 when the components ω_1, ω_2 have the same sense; when they are of opposite sense, it lies without, on the side of the greater one of these components.

If ω_1 and ω_2 are equal and opposite, say $\omega_1 = \omega$, $\omega_2 = -\omega$, the resulting axis lies at infinity. Two such equal and opposite angular velocities about parallel axes are said to form a **rotor-couple**; its effect on the rigid body is that of a velocity of translation $v = L_1L_2 \cdot \omega = p \cdot \omega$ at right angles to the plane of the axes. The distance of the rotors, $L_1L_2 = p$, is called the *arm* of the couple, and the product $p\omega = v$ its *moment*.

A velocity of translation v can therefore always be replaced by a rotor-couple $p\omega = v$, whose axes have the distance p and lie in a plane at right angles to v.

Again, an angular velocity ω about an axis l can be replaced by an equal angular velocity ω about a parallel axis l' at the distance p from l, in combination with a velocity of translation $v = \omega p$ at right angles to the plane determined by l and l'.

It easily follows from these propositions that *the resultant of any number of velocities of translation $v, v', \cdots$, parallel to the same plane, and any number of angular velocities $\omega, \omega', \cdots$ about axes perpendicular to this plane is always a single angular velocity about an axis perpendicular to the plane or a single velocity of translation parallel to the plane.*

185. Composition of Intersecting Rotors. Although not belonging to the theory of plane motion we here subjoin the proof of the important proposition that *angular velocities about intersecting axes combine by the parallelogram law; i. e.* if a rigid body has an angular velocity ω_1 about an axis l_1 and at the same time an angular velocity ω_2 about an axis l_2, intersecting l_1 at a point O (Fig. 46), its motion is a rotation of angular velocity ω about an axis l, through O, such that the rotor ω is the geometric sum of the rotors ω_1 and ω_2. This means analytically that

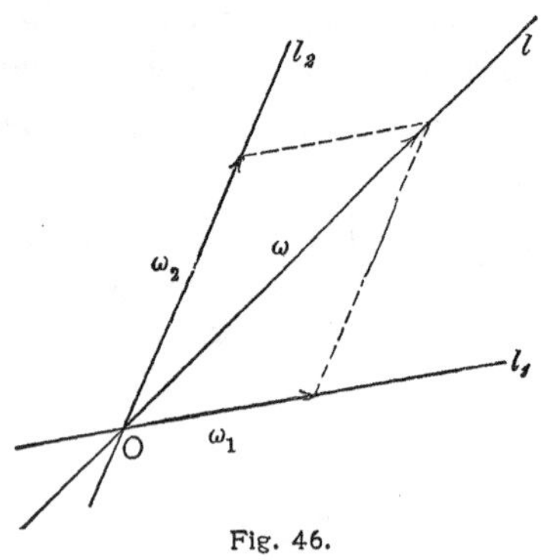

Fig. 46.

$$\omega^2 = \omega_1{}^2 + \omega_2{}^2 + 2\omega_1\omega_2 \cos l_1 l_2,$$

$$\frac{\sin l_1 l}{\omega_2} = \frac{\sin l l_2}{\omega_1} = \frac{\sin l_1 l_2}{\omega}.$$

The proposition is known as the **parallelogram of angular velocities.** Its proof is similar to that of the proposition in Art. 183. The linear velocity of any point P of the body has two components, $\omega_1 r_1$ and $\omega_2 r_2$, where r_1, r_2 are the perpendiculars let fall from P on the axes l_1, l_2. These components lie in the same line only for the points of the plane $(l_1 l_2)$; and they are equal and opposite only for the points on the diagonal of the parallelogram constructed on ω_1, ω_2 as sides. All the points of this diagonal having the velocity zero, this line is the axis of rotation. The above equations follow at once from the parallelogram construction.

Conversely, the angular velocity ω of a rigid body about an axis l can be replaced by the three angular velocities ω_x, ω_y, ω_z which are the components of the vector ω along any three rectangular axes meeting at any point O of l. It is easily proved that, if x, y, z are the co-ordinates of any point P of the body with respect to these axes, the components of the linear velocity v of $P(x, y, z)$ are

$$v_x \equiv \frac{dx}{dt} = \omega_y z - \omega_z y, \quad v_y \equiv \frac{dy}{dt} = \omega_z x - \omega_x z,$$

$$v_z \equiv \frac{dz}{dt} = \omega_x y - \omega_y x.$$

For, the component ω_x produces at P a velocity whose components, by Art. 175, are 0, $-\omega_x z$, $\omega_x y$; similarly, ω_y gives the components $\omega_y z$, 0, $-\omega_y x$; and ω_z gives $-\omega_z y$, $\omega_z x$, 0. By combining the velocities having the same direction the above equations are obtained.

6. APPLICATIONS.

186. Kinematics of Machinery. A large majority of the cases of motion that are of importance in mechanical engineering can be reduced to plane motion.

At first glance the application of theoretical kinematics to machines might seem to lead to rather complicated problems owing to the fact that a machine is never formed by a single rigid body, but always consists of an assemblage of several bodies some of which may even be not rigid (belting, springs, water, steam). The problem is, however, very much simplified by a characteristic of all machines, properly so called, that was first pointed out and insisted upon by recent writers on applied kinematics, in particular by Reuleaux. This characteristic is the *complete constraint* of the motions of the parts of a machine.

Thus Professor Kennedy defines a machine as "a combination of resistant bodies whose relative motions are completely constrained, and by means of which the natural energies at our disposal may be transformed into any special form of work."

With the latter clause of this definition we are not at present concerned; it will be considered in kinetics. To explain the former in detail would lead us too far into the domain of applied mechanics. A brief indication of the fundamental ideas must be sufficient.

187. By considering machines of various types it appears that the bodies, or *elements*, composing a machine always occur in **pairs.** Thus a single rigid bar will form a lever only when taken in connection with a support, or fulcrum; a shaft to be used in a machine must rest in bearings; a screw must turn in a nut. To take a more complex illustration, consider the mechanism formed by the crank and connecting rod of a steam-engine (Fig. 47). It may be regarded as composed of four pairs, three so-called turning pairs at O, A, B, and a sliding pair at B; and these are connected by two rigid bars, called **links,** OA, AB, and a fixed link OB of variable length.

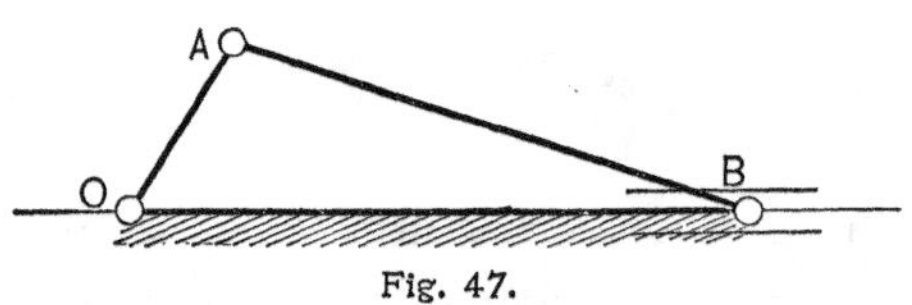

Fig. 47.

188. A **sliding pair** is formed by two bodies so connected that one is constrained to have a motion of translation relatively to the other. A pin moving in a groove or slot, a sleeve sliding along a shaft, are familiar examples.

A **turning pair** constrains one body to rotate about a fixed axis in another, as in the case of a shaft turning in its bearings.

A **twisting pair** makes one body have a screw motion about an axis fixed in the other.

These three pairs are the only so-called **lower pairs.** They are characterized as such by the fact that their elements have surface contact.

189. All other pairs are called **higher pairs.** The contact in such pairs is usually line contact.

Higher pairs are of far less frequent occurrence in ordinary machines than lower pairs. The only very common example of higher pairing is found in toothed wheel gearing.

In any pair, whether higher or lower, the relative motion of either element with respect to the other is *motion with one degree of freedom* (Art. 31).

190. For the purposes of kinematics a machine may be regarded as consisting of a number of bodies (*links*) connected by pairs in such a way that when one of the links is fixed all other links are constrained in their motion. In most cases this constraint is such as to leave but one degree of freedom to every link.

A system of links of this kind forming, so to speak, a skeleton of the machine is called a **kinematic chain** (Reuleaux). When one link of such a chain is fixed, the chain becomes a **mechanism.** As a typical example we may take the "slider crank" in Fig. 47.

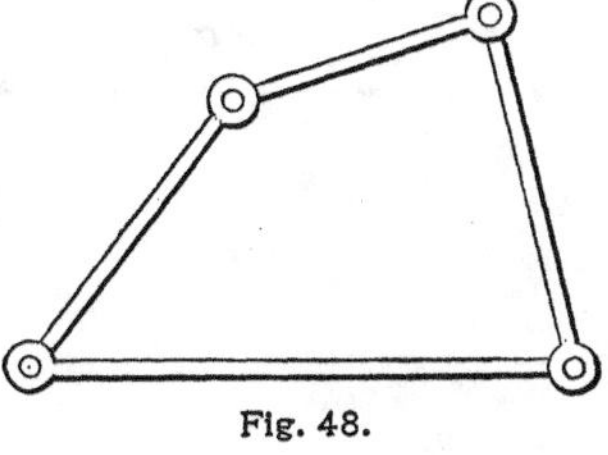
Fig. 48.

If the pairs are all turning pairs with parallel axes, the chain is called a **linkage** (Sylvester). A typical example is the four bar linkage in Fig. 48. A linkage with one link fixed has been called a **linkwork** (Sylvester). The four bar linkwork in Fig. 48 is also called a "lever crank" (Kennedy).

191. The Four Bar Linkage 1 2 3 4 (Fig. 49). Whatever may be its motion, each link considered separately moves as an invariable plane figure and has therefore at any moment an instantaneous center C and an angular velocity ω about this center.

The center C_{12} of 1 2 and the center C_{23} of 2 3 must always lie on a line passing through 2 since the velocity of 2 is perpendicular to both C_{12} 2 and C_{23} 2.

Similarly, 3 must lie on the line joining the centers C_{23} and C_{34}; and so on.

The quadrilateral 1 2 3 4 is therefore, and always remains, inscribed in the quadrangle $C_{12}C_{23}C_{34}C_{41}$. This can be shown to hold even for the *complete* quadrilateral and quadrangle. The complete quadrilateral, or four-side, 1 2 3 4 has six vertices, viz. the six intersections 1, 2, 3, 4, 5, 6 of its four sides, the complete quadrangle, or four-point, $C_{12}C_{23}C_{34}C_{41}$ has six sides, viz. the six lines $C_{41}C_{12}$, $C_{12}C_{23}$, $C_{23}C_{34}$, $C_{34}C_{41}$, $C_{12}C_{34}$, $C_{23}C_{41}$ joining its four vertices; and these six sides of the quadrangle pass through the six vertices of the quadrilateral, respectively.

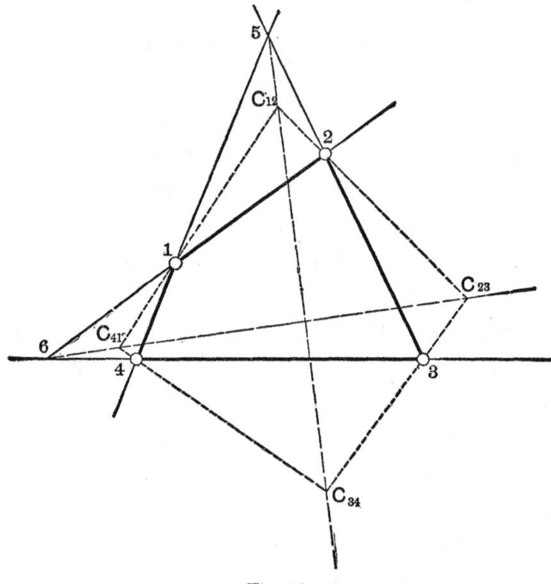

Fig. 49.

It remains to prove that $C_{12}C_{34}$ passes through 5 and that $C_{23}C_{41}$ passes through 6.

Now the velocity of 2 can be expressed by $\omega_1 \cdot C_{12}2$ and also by $\omega_2 \cdot C_{23}2$; hence $C_{12}2/C_{23}2 = \omega_2/\omega_1$; similarly $C_{23}3/C_{34}3 = \omega_3/\omega_2$. We have therefore, by the proposition of Menelaus,* for the intersection of 2 3 with $C_{12}C_{34}$:

* If the sides of a triangle ABC be cut by any transversal, in the points A', B', C', then $\frac{BA'}{A'C} \cdot \frac{CB'}{B'A} \cdot \frac{AC'}{C'B} = -1$. See for instance BEMAN and SMITH, New plane and solid geometry, Boston, Ginn, 1899, p. 240.

$$\frac{5\,C_{12}}{5\,C_{34}} = \frac{\omega_3}{\omega_1}.$$

The same value is obtained by determining the intersection of 1 4 with $C_{12}C_{34}$; the two intersections must therefore coincide.

The proof for the point 6 is analogous.

A corresponding proposition holds of course for four bar linkages with crossed bars 1 2 4 3, or 1 3 2 4.

192. Lever-crank. The linkage considered in the preceding article becomes a mechanism, or linkwork, as soon as one of its four links is fixed. It occurs in machines under a variety of forms some of which are referred to below.

Let the link 3 4 be fixed; then the center C_{34} (Fig. 49) disappears; C_{41} falls into 4, C_{23} into 3, and C_{12} becomes the intersection 5 of 4 1 and 3 2. If 1 2 were fixed instead of 3 4, 3 4 would have its center at 5. Similarly, if either 4 1 or 2 3 be fixed, the center of the other is 6. Hence whichever of the four links be fixed, the centers of all the links lie at some of the six vertices of the complete quadrilateral 1 2 3 4.

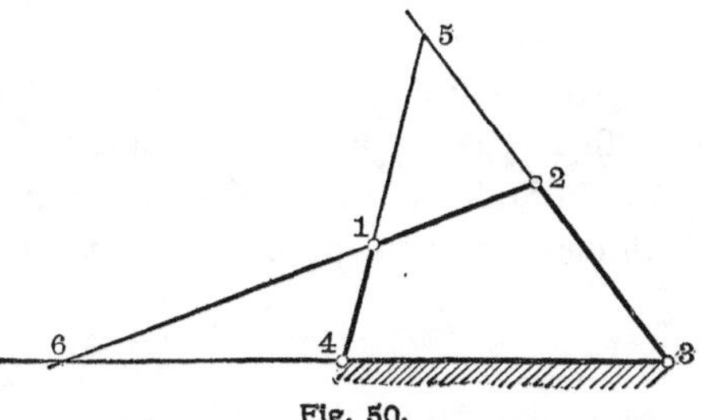

Fig. 50.

If 3 4 be the fixed link (Fig. 50), the ratio of the angular velocities ω_1 of 4 1 and ω_2 of 3 2 can be found. For if ω denote the angular velocity of 1 2 about 5, we have

$$41 \cdot \omega_1 = 51 \cdot \omega, \quad 32 \cdot \omega_2 = 52 \cdot \omega;$$

hence

$$\frac{\omega_2}{\omega_1} = \frac{41}{32} \cdot \frac{52}{51} = \frac{52}{32} \Big/ \frac{51}{41};$$

or, by the proposition of Menelaus:

$$\frac{\omega_2}{\omega_1} = \frac{46}{36}.$$

193. Parallelogram: $41 = 32 = a$, $43 = 12 = b$ (Fig. 51). The link 1 2 has evidently a motion of translation, its instantaneous center lying at the intersection of the parallel lines 4 1, 3 2.

The fixed centrode is the line at infinity; the body centrode may be

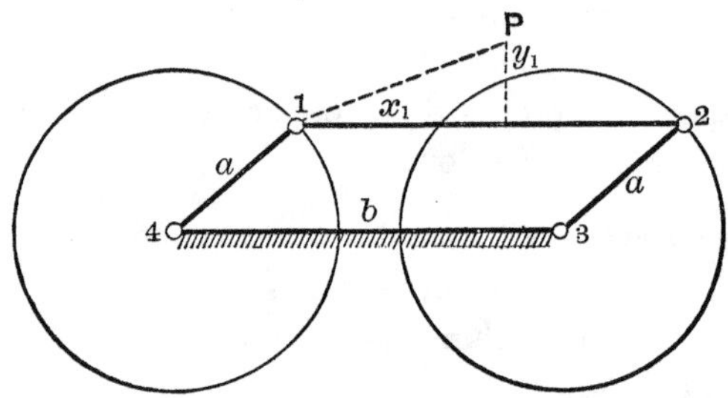

Fig. 51.

regarded as a circle of infinite radius described about the midpoint of 3 4 as center.

To find the equation of the path of any point P rigidly connected with 1 2, let x, y be the rectangular co-ordinates, with respect to 4 as origin and 4 3 as axis of x, and x_1, y_1 its co-ordinates for parallel axes through 1; then, putting $\measuredangle\, 341 = \theta$, we have

$$x = a \cos\theta + x_1, \quad y = a \sin\theta + y_1;$$

hence, eliminating θ,

$$(x - x_1)^2 + (y - y_1)^2 = a^2,$$

which represents a circle of radius a whose center has the fixed co-ordinates x_1, y_1 (comp. Art. 26, Ex. (9)).

For the velocity of P we have $dx/dt = -a\omega \sin\theta$, $dy/dt = a\omega \cos\theta$; hence $v = a\omega$, as is otherwise apparent.

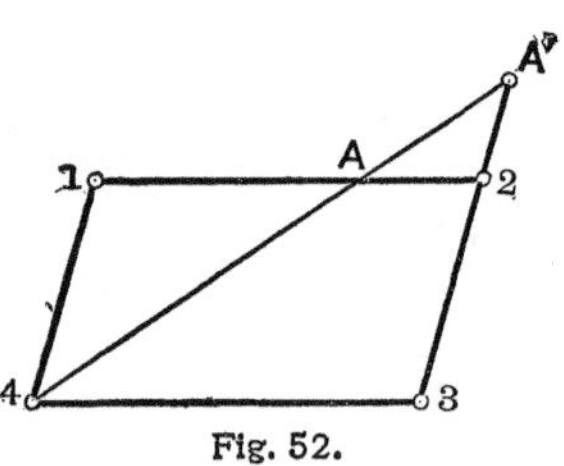

Fig. 52.

194. If in the parallelogram 1 2 3 4 the point 4 alone be fixed, we have a linkage called the **pantograph.**

It can serve to trace a curve similar to a given curve. Indeed, any line through 4 (Fig. 52) cuts the opposite links 1 2, 2 3 (produced if necessary) in points A, A' whose paths are homothetic (similar and similarly situated) curves. For the points 4, A, A' remain always in line and the ratio $4A/4A'$ remains constant. Hence if a pencil be attached to A' and A be made to trace a given curve, A' will trace a similar curve.

Instead of fixing 4, the point A' might be fixed; then 4 and A will describe similar curves. This property is utilized in Watt's parallel motion (see Art. 199).

The parallelogram linkage furnishes also a simple instrument for describing ellipses. Let the sides of the parallelogram be $2\,3 = 4\,1 = a$, $1\,2 = 3\,4 = b$; and let a point A' on 2 3 produced, at the distance b from 2, be fixed (Fig. 53). Then, if 1 be made to describe a

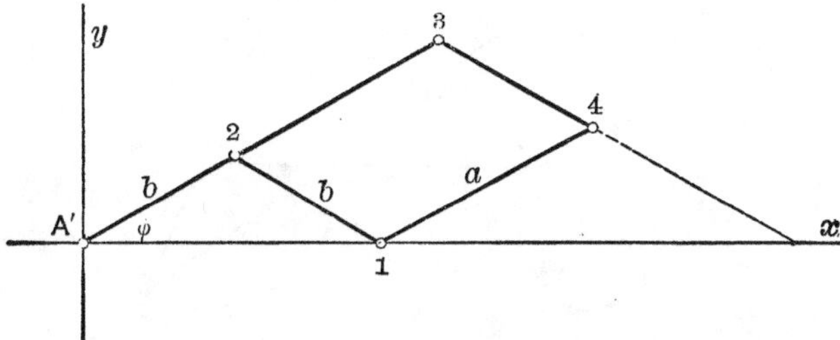

Fig. 53.

straight line passing through A', 4 will describe an ellipse. For, taking A' as origin and A'. 1 as axis of x, we have for the co-ordinates of 4: $x = (a + 2b) \cos\varphi$, $y = a \sin\varphi$, whence

$$\frac{x^2}{(a + 2b)^2} + \frac{y^2}{a^2} = 1.$$

195. In the parallelogram 1 2 3 4, let the link 1 2 be turned so as to coincide in direction with 4 3, and then give the links 4 1 and 3 2 rotations of opposite sense. We thus obtain a linkage with equal, but intersecting, opposite sides, a so-called **anti-parallelogram** (Fig. 54). If 3 4 be fixed, the instantaneous center of 1 2 is the intersection 5 of 4 1 and 2 3.

To obtain the centrodes in this case, notice that as the triangles 1 5 2 and 5 3 4 are equal, the triangle 5 4 2 is isosceles; hence $5\,1 = 5\,3$, and $4\,5 - 3\,5 = 4\,1 = a$. The difference of the radii vectores of

5 drawn from 4 and 3 being thus constant, it follows that the fixed centrode is a hyperbola whose foci are 4, 3, and whose real axis $= a$. As 4 3 = 1 2 = b, the equation of this hyperbola is

$$\frac{x^2}{\left(\frac{a}{2}\right)^2} - \frac{y^2}{\frac{b^2 - a^2}{4}} = 1,$$

for 4 3 as axis of x and the midpoint of 4 3 as origin.

It is easy to see that the fixed centrode becomes an ellipse when $b < a$.

As the triangles 1 5 2 and 3 5 4 are equal the body centrode is an

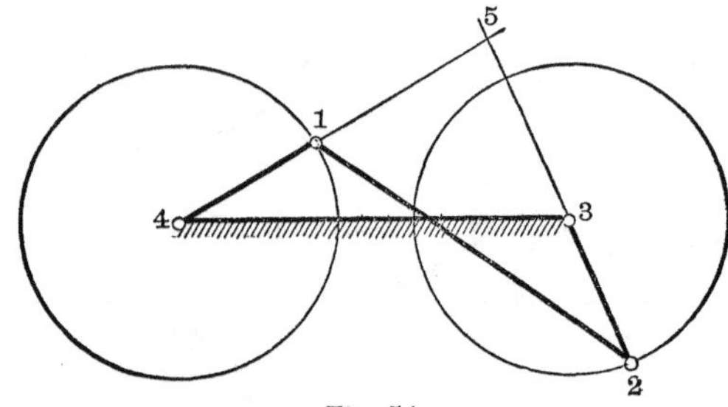

Fig. 54.

equal hyperbola or ellipse. The two centrodes lie symmetrically with respect to their common tangent at 5.

For a given anti-parallelogram the centrodes are hyperbolas when one of the larger links is fixed; they are ellipses when one of the shorter links is fixed.

196. If in the anti-parallelogram only one point, say 4, be fixed, it can be used as an **inversor**, *i. e.* as an instrument for describing the inverse of a given curve.

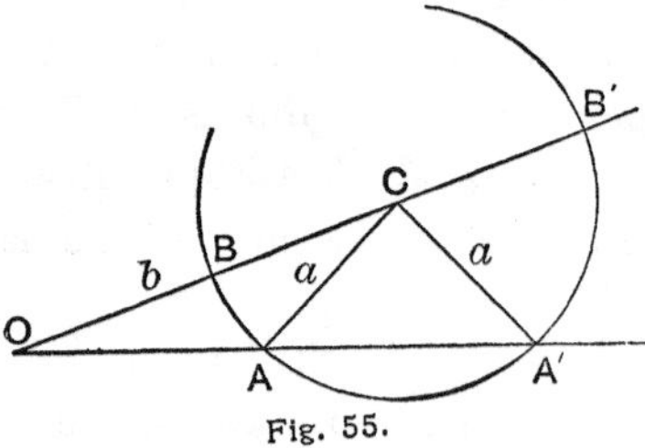

Fig. 55.

Let $r = OP$ be the radius vector drawn from an arbitrary fixed origin, or pole, O to a given curve; on OP lay off a length $OP' \equiv r' = \kappa^2/r$, where κ is a constant; then P' is said to describe the *inverse* of the given curve.

The theory of inversors is based on the following geometrical proposition: If three lines $CA = a$, $CA' = a$, $CO = b$ (Fig. 55) turn about C so that O, A, A' are always in line, the product $OA \cdot OA'$ remains constant, viz. $OA \cdot OA' = b^2 - a^2$. For if the circle of radius a described about C intersect the line OC in B and B', we have $OA \cdot OA' = OB \cdot OB' = (b - a)(b + a)$.

This proposition shows that in the anti-parallelogram 1 2 3 4 (Fig. 56), with the vertex 4 fixed, the line joining the vertices 4 and 2 in-

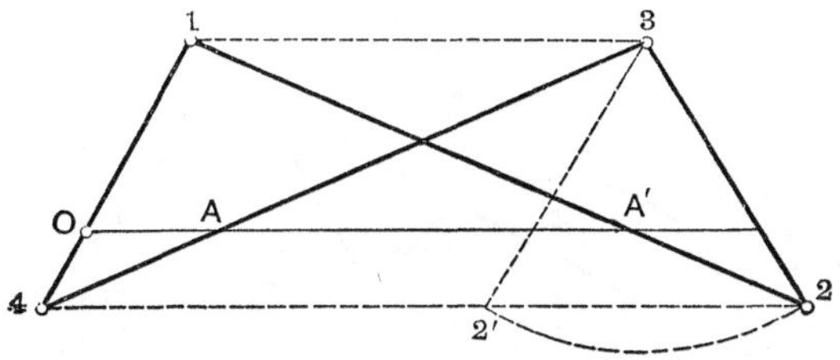

Fig. 56.

tersects the circle described about 3 with radius 3 2 in a point 2′ such that 2 and 2′ describe inverse curves with respect to 4 as pole. For we have $4\,2' \cdot 4\,2 = 4\,3^2 - 2\,3^2 = b^2 - a^2$.

Moreover, any parallel to 4 2 will intersect the links 4 1, 4 3, 2 1 in points O, A, A' dividing the three lines in the same ratio; hence

$$\frac{4\,2'(=1\,3)}{OA} = \frac{OA'}{4\,2},$$

i. e. $OA \cdot OA' = 4\,2' \cdot 4\,2 = b^2 - a^2$, so that if O be fixed, A and A' will describe inverse curves for O as pole. This is the principle on which **Hart's inversor** is based.

197. Peaucellier's cell is another inversor (Fig. 57). It consists of the linked rhombus $A\,B\,A'\,B'$ whose side we denote by a, and the two equal links $O\,B$, $O\,B'$ of length b. If O be fixed, A and A' evidently describe inverse curves for O as pole.

The figure shows that with $\measuredangle A\,O\,B = \chi$, $\measuredangle A'\,A\,B = \psi$ we have $OA = b \cos\chi - a \cos\psi$, $OA' = b \cos\chi + a \cos\psi$, whence

$$OA \cdot OA' = b^2 \cos^2\chi - a^2 \cos^2\psi,$$

and as, moreover, $b \sin\chi = a \sin\psi$, we find by adding to the preceding equation the relation $0 = b^2 \sin^2\chi - a^2 \sin^2\psi$:

$$OA \cdot OA' = b^2 - a^2.$$

The practical application of inversors is based on the property that they enable us to transform circular motion into rectilinear motion (see Art. 199).

The inverse of a circle $r = 2c \cos\theta$ passing through the pole is a straight line; for we have for the radius vector r' of the inverse curve

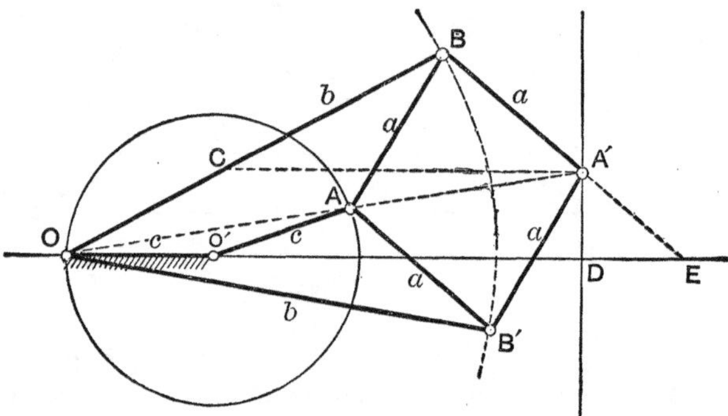

Fig. 57.

$r' = \kappa^2/r = \kappa^2/2c \cos\theta$; hence $r' \cos\theta = \kappa^2/2c$ which is the equation of a straight line at right angles to the polar axis, at the distance $\kappa^2/2c$ from the pole.

If therefore the point A of an inversor be made to describe an arc of a circle passing through O, the point A' will describe a segment of a straight line. The vertex A (Fig. 57) can be compelled to describe a circle by inserting the additional link $O'A$ turning about the fixed point O'. If O' be selected so as to make $O'O = O'A$, say $= c$, the circle described by A will pass through O; and the motion of A' will be confined to the straight line $A'D$ perpendicular to OO', at the distance $OD = (b^2 - a^2)/2c$ from O.

The linkage has thus become a linkwork, OO' being the fixed link.

198. To determine the linear velocity v of A' along DA' when the angular velocity ω of the link OB is given, we notice that the instantaneous center C of the link BA' lies at the intersection of OB with the line drawn through A' parallel to OO'. Let ω' be the angular velocity of BA' about C. Then $v = \omega' \cdot CA'$; also since the point B describes a circle about O, $\omega b = \omega' \cdot CB$; hence

$$v = \omega \cdot \frac{CA'}{CB} b.$$

If BA' intersect OO' in E, we have from similar triangles $CA' : CB = OE : OB$; hence

$$v = \omega \cdot OE.$$

The variable length OE depends on the angles $EOB = \theta$ and $BEO = \varphi$ which are connected by the relation (Art. 197)

$$a \cos\varphi + b \cos\theta = OD = \frac{b^2 - a^2}{2c}.$$

The figure gives $OE = b \cos\theta + b \sin\theta \cot\varphi$; hence, finally,

$$v = \omega b \sin\theta (\cot\theta + \cot\varphi).$$

199. In the steam engine and other machines mechanisms are required for transforming the alternating rectilinear motion of the piston

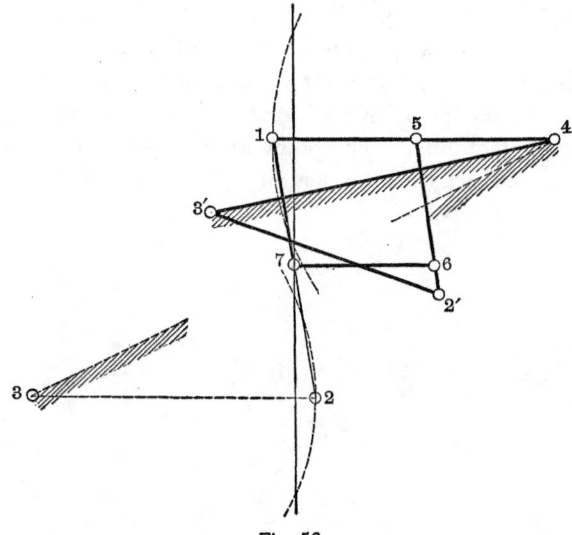

Fig. 58.

into the reciprocating circular motion of a crank, eccentric, or beam; a mechanism of this kind is called, rather inappropriately, a **parallel**

motion. The problem of effecting this transformation has been solved in various ways. Peaucellier's inversor (1864) was the first *accurate* solution. Generally, an approximate solution is sufficient for practical purposes. The most common of such approximations is **Watt's parallel motion.** This mechanism is a combination of a linked parallelogram with a four bar linkwork with crossed links.

To fix the ideas, let 4 1 (Fig. 58) be the horizontal middle position of the beam of a beam engine; 4 is fixed and 1 describes an arc of a circle of radius 4 1 = a. We might place a counter-beam 3 2 of equal length turning about the fixed end 3 so as to be in its middle position parallel to 4 1 and so as to make the connecting link 1 2 nearly vertical. The middle point of 1 2 would then describe a looped curve whose central portion does not differ very much from a straight line; connecting this middle point with the piston rod, the problem would be solved.

But the introduction of the large counter-beam 3 2 in the position indicated above would be very inconvenient. To reduce the size of the mechanism the counter-beam 3 2 is placed nearer to 4 1, in the position 3′ 2′, and the parallelogram 1 5 6 7 is introduced, the piston rod being attached at 7. Owing to the property of the linked parallelogram (Art. 194), the point 7 has a motion similar to that of the point of intersection of 4 7 with 5 6; it describes therefore approximately a straight line. The point of intersection of 4 7 with 5 6 can be used to connect with the pump rods of the engine.

PART II:
INTRODUCTION TO DYNAMICS; STATICS.

CHAPTER III.

INTRODUCTION TO DYNAMICS.

I. *Mass; Moments of Mass; Centroids.*

1. MASS; DENSITY.

200. In the first part of this work only the geometrical and kinematical properties of motion have been considered, the moving object being regarded as a mere point or as a geometrical configuration. It is, however, known, from observation and experiment, that the motions of actual physical bodies are not fully described and determined by those properties alone.

Physical bodies are distinguished from geometrical configurations by being possessed of **mass**; and this property as affecting their motion must be taken into account in dynamics.

201. In physics the mass of a body is usually said to be *the quantity of matter contained in the body.* Postponing for the present the full discussion of the idea of mass in its relation to acceleration and force, and of the methods for comparing and measuring masses, it will suffice for our present purpose to think of the mass of a body as a certain constant quantity, independent of the body's position or motion with respect to the earth or other bodies, as an indestructible something underlying every physical body.

The student must be warned not to confound mass with weight. The weight of a body, as we shall see later, is the force with which the body is attracted by the earth; it varies, therefore, with the distance of the body from the earth's center, and would vanish completely if the earth were suddenly annihilated; while the *indestructibility of mass* is the first fundamental principle of chemistry and physics.

202. To compare the masses of different bodies, we may adopt any given body as a standard.

Thus in the F.P.S. system, the *standard mass* is a certain bar of platinum marked "P. S., 1844, 1 lb.," and preserved at the Office of the Exchequ er, London, England. This is called the "imperial standard pound avoirdupois"; any mass equal to it is a *unit of mass* in this system.

In the C.G.S. system, the standard of mass is the "kilogramme des archives," a bar of platinum kept in the Palais des archives, in Paris, France. A mass equal to one thousandth of this standard is the unit of mass in this system; this unit is called the *gram*.

The numerical relation between the British and metric units of mass is as follows:

$$1 \text{ lb.} = 453.592\,65 \text{ gm.}$$

$$1 \text{ gm.} = 0.002\,204\,621\,2 \text{ lb.} = 15.432 \text{ grains.}$$

203. The three units of *space*, *time*, and *mass* are called *the fundamental units of mechanics*, because with the aid of these three, the units of all other quantities occurring in mechanics can be expressed. Thus we have seen how the units of velocity and acceleration are based on those of space and time, and we shall have many more illustrations in what follows. Any unit that can be expressed mathematically by means of one or more of the fundamental units is called a *derived unit*.

204. From the mathematical point of view, mass appears in our dynamical equations as a coefficient, generally to be regarded as an absolute positive constant. It serves to give different values (different valency, or "weight" in the meaning of the theory of least squares) to the moving points, lines, areas, volumes, apart from their geometrical extension.

205. Thus, a geometrical point endowed with mass is called a *material particle*. We may regard such a mass-point, or particle, as the limit to which a physical body approaches if its volume be imagined to decrease indefinitely, approaching the limit zero, while its mass remains constant. From the physical point of view a particle must be regarded as as much of an abstraction as a geometrical point, since every finite physical mass occupies a finite space and cannot be identified with a point. We shall see, however, that in dynamics this idea of the mass-point, or particle, is of the greatest importance not only because physical matter is usually considered as made up of an aggregation of such points or centers possessing mass (molecules, atoms), but principally because in many cases the motion of a solid body can be fully represented by the motion of a certain point in it, called its *center of mass* or **centroid**, the whole mass being regarded as concentrated at this point.

206. It is also customary in dynamics to speak of *material lines* and *material surfaces*, which may be regarded as the limits of physical bodies obtained by letting two dimensions or one dimension approach zero. Thus a material line represents the limit of a wire, chain, or bar, in which two dimensions are neglected; a material surface can be imagined as the limit of a thin shell, or lamina, with one dimension reduced to zero.

207. A continuous mass of one, two, or three dimensions is said to be *homogeneous* if the masses contained in *any* two equal lengths, areas, or volumes (as the case may be) are equal. The mass is then said to be distributed *uniformly*. In all other cases the mass is said to be *heterogeneous*.

208. The whole mass M of a homogeneous body divided by the space V it fills is called the **density** of the mass or body; denoting density by ρ we have therefore

$$\rho = \frac{M}{V},$$

for homogeneous bodies. It follows from the definition of homogeneity that the density of a homogeneous mass can also be found by dividing any portion ΔM of the whole mass M by the space ΔV occupied by ΔM.

In a heterogeneous body, the quotient $\Delta M / \Delta V$ is called the *average*, or *mean*, density of the portion ΔM. The limit of this average density as the space ΔV approaches zero while always containing a certain point P is called *the density of the mass M at the point P:*

$$\rho = \lim_{\Delta V = 0} \frac{\Delta M}{\Delta V} = \frac{dM}{dV}.$$

209. The *unit of density* is the density of a substance such that the unit of volume contains the unit of mass. If the units of volume and mass are selected arbitrarily, there need not of course necessarily exist any physical substance having unit density exactly. Thus in the F.P.S. system, unit density is the density of an ideal substance one pound of which would just fill a cubic foot. As a cubic foot of water has a mass of about $62\frac{1}{2}$ pounds, or 1,000 ounces, the density of water is about $62\frac{1}{2}$ times the unit density.

The *specific density*, or *specific gravity*, of a substance, is the ratio of its density to that of water at 4° C. Let ρ be the density, ρ_1 the specific density, M the mass, V the volume of a homogeneous mass, then in British units

$$M = \rho V = 62.5 \rho_1 V.$$

In the C.G.S. system, the unit of mass has been so selected as to make the density of water equal to 1 very nearly; in other words, the unit mass (1 gram) of water, at the temperature of 4° C., fills one cubic centimeter.

In the metric system, then, there is no difference between density and specific density or specific gravity.

2. MOMENTS AND CENTERS OF MASS.

210. The product of a mass m, concentrated at a point P, into the distance of the point P from any given point, line, or plane is called the **moment** of this mass with respect to the point, line, or plane.

Thus, denoting by r, q, p, the distance of the point P from the point O, the line l, and the plane π, respectively, we have for the moments of m with respect to O, l, π, the expressions mr, mq, mp.

211. Let a system of n points, or particles, $P_1, P_2, \cdots P_n$ be given; let $m_1, m_2, \cdots m_n$ be their masses, and $p_1, p_2, \cdots p_n$ their distances from a given plane π. Then we call **moment of the system** with respect to the plane π the algebraic sum

$$m_1 p_1 + m_2 p_2 + \cdots + m_n p_n = \Sigma mp,$$

the distances $p_1, p_2, \cdots p_n$ being taken with the same sign or opposite signs according as they lie on the same side or on opposite sides of the plane π.

It is always possible to determine one and only one distance $\bar{p}$ such that $\Sigma mp = M\bar{p}$, where $M = \Sigma m = m_1 + m_2 + \cdots + m_n$ is the total mass of the system. If this distance $\bar{p}$ should happen to be equal to zero, the moment of the system would evidently vanish with respect to the plane π.

212. Let us now refer the points P to a rectangular set of axes and let x, y, z be their co-ordinates. Then we have for the moments of the system with respect to the co-ordinate planes yz, zx, xy, respectively

$$m_1 x_1 + m_2 x_2 + \cdots + m_n x_n = \Sigma mx = M\bar{x},$$

$$m_1 y_1 + m_2 y_2 + \cdots + m_n y_n = \Sigma my = M\bar{y},$$

$$m_1 z_1 + m_2 z_2 + \cdots + m_n z_n = \Sigma mz = M\bar{z},$$

The point G whose co-ordinates are

$$\bar{x} = \frac{\Sigma mx}{M}, \quad \bar{y} = \frac{\Sigma my}{M}, \quad \bar{z} = \frac{\Sigma mz}{M}, \tag{1}$$

is called the *center of mass*, or the **centroid**, of the system.

The centroid is, therefore, defined as *a point such that if the whole mass M of the system be concentrated at this point, its moment with respect to any one of the co-ordinate planes is equal to the moment of the system.*

213. It is easy to see that this holds not only for the co-ordinate planes but for any plane whatever. Let

$$\alpha x + \beta y + \gamma z - p_0 = 0$$

be the equation of any plane in the normal form; $\bar{p}, p_1, p_2, \dots p_n$, the distances of the points $G, P_1, P_2, \dots P_n$ from this plane. Then we wish to prove that $\Sigma mp = M\bar{p}$.

Now

$$\bar{p} = \alpha\bar{x} + \beta\bar{y} + \gamma\bar{z} - p_0, \quad p_1 = \alpha x_1 + \beta y_1 + \gamma z_1 - p_0, \dots;$$

hence

$$\begin{aligned} \Sigma mp &= \alpha\Sigma mx + \beta\Sigma my + \gamma\Sigma mz - p_0\Sigma m \\ &= M(\alpha\bar{x} + \beta\bar{y} + \gamma\bar{z} - p_0) \\ &= M\bar{p}. \end{aligned}$$

The centroid can therefore be defined as *a point such that its moment with respect to any plane is equal to that of the whole system, with respect to the same plane.*

It follows that *the moment of the system vanishes for any plane passing through the centroid.*

214. In the case of a continuous mass, whether it be of one, two, or three dimensions, the same reasoning will apply if we imagine the mass divided up into elements dM of one, two, or three infinitesimal dimensions, respectively. The summations indicated above by Σ will then become integrations, so that the centroid of a continuous mass has the co-ordinates

$$\bar{x} = \frac{\int x dM}{\int dM}, \quad \bar{y} = \frac{\int y dM}{\int dM}, \quad \bar{z} = \frac{\int z dM}{\int dM}. \tag{2}$$

According as the mass is of one, two, or three dimensions, a single, double, or triple integration over the whole mass will in general be required for the determination of the moments $\int x dM$, $\int y dM$, $\int z dM$ of the mass with respect to the co-ordinate planes, as well as of the total mass $\int dM = M$.

Thus, for a mass distributed along a line or a curve we have, if ds be the line-element,

$$dM = \rho'' ds,$$

where ρ'' is called the *linear density*; for a mass distributed over a surface-area we have, with dS as a surface-element,

$$dM = \rho' dS,$$

where ρ' is the *surface* (or *areal*) *density*; finally, for a mass distributed throughout a volume whose element is dV,

$$dM = \rho\, dV,$$

where ρ is the *volume density*.

If the mass be distributed along a straight line, the centroid lies of course on this line, and one co-ordinate is sufficient to determine the position of the centroid. In the case of a plane area, the centroid lies in the plane and two co-ordinates determine its position; we then speak of moments with respect to lines, instead of planes.

215. If the mass be homogeneous (Art. 207), *i. e.*, if the density ρ be constant, it will be noticed that ρ cancels from the numerator and denominator in the equations (2), and does not enter into the problem. Instead of speaking of a center of mass, we may then speak of a center of arc, of area, of volume. The term *centroid* is, however, to be preferred to *center*, the latter term having a recognized geometrical meaning different from that of the former.

The geometrical center of a curve or surface is a point such that any chord through it is bisected by the point; there are but few curves and surfaces possessing a center.

The centroid (Art. 213) is a point such that, for any plane passing through it, the moment of the system is equal to zero. Such a point exists for every mass, volume, area, or arc. The centroid coincides, of course, with the center, when such a center exists and the distribution of mass is uniform.

216. As soon as ρ is given either as a constant or as a function of the co-ordinates, the problem of determining the centroid of a continuous mass is merely a problem in integration. To simplify the integrations, it is of importance to select the element in a convenient way conformably to the nature of the particular problem.

Considerations of symmetry and other geometrical properties will frequently make it possible to determine the centroid without resorting to integration. Thus, in a homogeneous mass, any plane of symmetry, or any axis of symmetry, must contain the centroid, since for such a plane or line the sum of the moments is evidently zero (see Art. 244).

It is to be observed that the whole discussion is entirely independent of the physical nature of the masses m which appear here simply as numerical coefficients, or "weights," attached to the points (comp. Art. 204). Some of the masses might even be negative.

It will be shown later that the *center of gravity*, as well as the *center of inertia*, of a body coincides with its centroid.

217. In determining the centroid of a given system it will often be found convenient to break the system up into a number of partial systems whose centroids are either known or can be found more readily. *The moment of the whole system is* obviously *equal to the sum of the moments of the partial systems.*

Thus let the given mass M be divided into k partial masses $M_1, M_2, \ldots M_k$, so that $M = M_1 + M_2 + \ldots + M_k$; let $G, G_1, G_2, \ldots G_k$ be the centroids of $M, M_1, M_2, \ldots M_k$, and $\bar{p}, \bar{p}_1, \bar{p}_2, \ldots \bar{p}_k$ their distances from some fixed plane. Then we have

$$M\bar{p} = M_1 p_1 + M_2 p_2 + \ldots + M_k p_k.$$

218. The particular case of *two* partial systems occurs most frequently. We then have with reference to any plane

$$M\bar{p} = M_1 p_1 + M_2 p_2, \quad M = M_1 + M_2.$$

Letting the plane coincide successively with the three co-ordinate planes, it will be seen that G must lie on the line joining G_1, G_2. Now taking the plane at right angles to G_1G_2 through G_1, we have

$$M \cdot G_1G = M_2 \cdot G_1G_2;$$

similarly for a plane through G_2,

$$M \cdot GG_2 = M_1 \cdot G_1G_2;$$

whence

$$\frac{G_1G}{M_2} = \frac{GG_2}{M_1} = \frac{G_1G_2}{M};$$

i. e. the centroid of the whole system divides the distance of the centroids of the two partial systems in the inverse ratio of their masses.

3. CENTROIDS OF PARTICLES AND LINES.

219. Two Particles. The centroid G of two particles of masses m_1, m_2 concentrated at two points P_1, P_2 lies on the line P_1P_2 and divides the distance P_1P_2 in the inverse ratio of their masses, *i. e.* so that

$$\frac{P_1G}{m_2} = \frac{GP_2}{m_1} = \frac{P_1P_2}{m_1 + m_2}.$$

(See Art. 218.) These formulæ hold even when one of the masses is positive and the other negative, in which case the sense of the segments must be taken into account.

220. Three Particles. We find first the centroid P' of m_2 at P_2 and m_3 at P_3 (Fig. 59) by Art. 219; then, by the same rule, the centroid G of $m_2 + m_3$ at P' and m_1 at P_1. We might have begun with P_3 and P_1, finding P''; or with P_1 and P_2, finding P'''. G lies at the intersection of the three lines P_1P', P_2P'', P_3P''', and can therefore be constructed graphically.

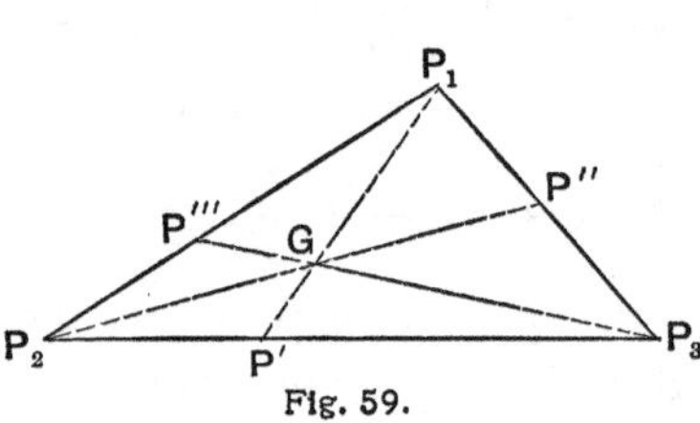

Fig. 59.

221. Four Particles. Find the centroid P' of m_1 at P_1 and m_2 at P_2; also the centroid P'' of m_3 at P_3 and m_4 at P_4; then the centroid G of $m_1 + m_2$ at P' and $m_3 + m_4$ at P''.

The four particles can be arranged in groups of two in three different ways. There are therefore three lines, like $P'P''$, on each of which G lies. Any two of these are sufficient to construct G geometrically.

222. The centroid of a **homogeneous rectilinear segment** (thin rod or wire of constant cross-section) is evidently at its middle point.

If the density of a rectilinear segment be proportional to the nth power of the distance from one end, say $\rho'' = kx^n$, we have

$$\bar{x} = \frac{\int_0^l \rho'' x dx}{\int_0^l \rho'' dx} = \frac{k\int_0^l x^{n+1} dx}{k\int_0^l x^n dx} = \frac{n+1}{n+2} l,$$

where l is the length of the segment.

(a) For $n = 0$, this gives $\bar{x} = \frac{1}{2}l$ which determines the centroi of a *homogeneous straight segment* (see above).

(b) For $n = 1$, we have $\bar{x} = \frac{2}{3}l$. This determines the distance from the vertex, of the centroid of a *homogeneous triangular area* For such an area can be resolved (Fig. 60) by parallels to th base into elements each of which may be regarded as homogeneous segment PQ. If we imagine the mass of every such element concentrated at its middle point, the homogeneous triangle is replaced by its median CC' in which the density is proportional to the distance from the vertex C.

The centroid of a homogeneous triangular area lies therefore on the median at two thirds of its length from the vertex; as this holds for each median, the intersection of the three medians is the centroid (comp. Art. 227).

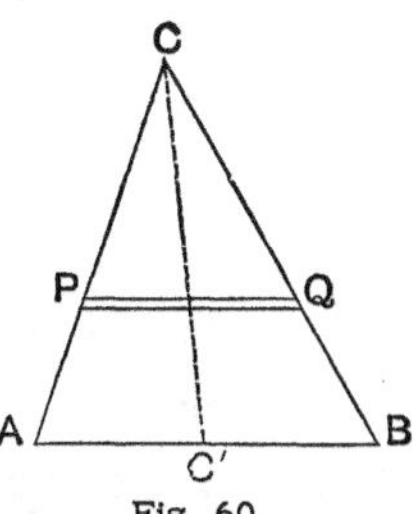

Fig. 60.

(*c*) For $n = 2$, we have $\bar{x} = \frac{3}{4}l$. This gives the position of the centroid of a *homogeneous pyramid* or *cone*, by reasoning similar to that used in (*b*).

Thus, to find the centroid of any homogeneous pyramid or cone, join the vertex to the centroid of the area of the base; the required centroid lies on this line at a distance equal to $\frac{3}{4}$ of its length from the vertex.

223. Homogeneous Circular Arc (Fig. 61). Let O be the center, r the radius of the circle; $ACB = s$ the arc, C its middle point. The centroid G must lie on the bisecting radius OC, since this being a line of symmetry, the sum of the moments of the elements of the arc is zero with respect to this line (Art. 216). To find the distance $\bar{x} = OG$, we take moments with respect to the diameter perpendicular to OC. With OC as axis of x, we have

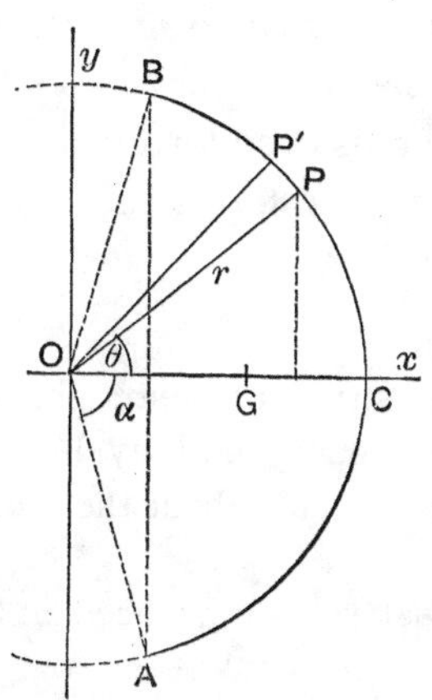

Fig. 61.

$$s \cdot \bar{x} = \int x ds = r \int ds \cos COP = r \int dy.$$

Hence, $s \cdot \bar{x} = r \cdot c$, if c be the length of the chord AB.

If the angle $AOB = 2\alpha$ of the arc AB were given, we might obtain the result by taking the angle $COP = \theta$ as independent variable. We have then

$$s \cdot \bar{x} = \int_{-\alpha}^{+\alpha} r \cos\theta \cdot r d\theta = 2r^2 \sin\alpha,$$

whence

$$\bar{x} = r \cdot \frac{\sin\alpha}{\alpha}.$$

This can be written

$$\bar{x} = r \cdot \frac{2r \sin\alpha}{2r\alpha} = r \cdot \frac{c}{s},$$

which agrees with the expression found above.

224. The First Proposition of Pappus and Guldinus. If an arc of a plane curve be made to rotate about an axis situated in its plane, it generates a surface of revolution whose surface-area is $S = 2\pi \int y ds$, where ds is the element of the curve and the axis of rotation is taken as axis of x. On the other hand we have, if s be the length of the generating arc and $\bar{y}$ the ordinate of its centroid, $s \cdot \bar{y} = \int y ds$; hence

$$S = 2\pi \cdot s\bar{y} = 2\pi\bar{y} \cdot s,$$

i. e., the surface-area of a solid of revolution is obtained by multiplying the generating arc into the path described by its centroid.

It is easy to see that this proposition holds even for incomplete revolutions. When the generating arc cuts the axis, proper regard must be had for signs and sense of rotation.

225. Exercises.

(1) Three beads of masses 3, 5, 12, are strung on a straight wire whose mass is neglected, the bead of mass 5 being midway between the other two. Find the centroid. (Take moments about the middle point.)

(2) Show that the centroid of three equal particles placed at the vertices of a triangle is at the intersection of the medians of the triangle.

(3) Show that the centroid of three masses m_1, m_2, m_3, situated at the vertices of a triangle and proportional to the opposite sides, is at the center of the inscribed circle.

(4) Equal particles are placed at five of the six vertices of a regular hexagon. Find the distance of the centroid from the center of figure.

(5) Find the centroid of a homogeneous triangular frame.

(6) Find the centroid of earth and moon, the moon's mass being $\frac{1}{80}$ of that of the earth, and the distance of their centers 240,000 miles.

(7) Show that the centroid of a homogeneous wire bent into a semicircle lies at the distance $(2/\pi)r$ from the center, r being the radius.

(8) Find the co-ordinates of the centroid of the arc of a quadrant of a circle by using the first proposition of Pappus (Art. 224).

(9) Find the centroid of a circular arc AB of angle $AOB = \alpha$, whose density varies as the length of the arc measured from A.

Find the centroids of the following homogeneous arcs of curves:

(10) Parabola $y^2 = 4ax$ from the vertex to the end of the latus rectum.

(11) Cycloid $x = a(\theta - \sin\theta), y = a(1 - \cos\theta)$, from cusp to cusp.

(12) Half the cardioid $r = a(1 + \cos\theta)$.

(13) Catenary $y = \frac{1}{2}c(e^{\frac{x}{c}} + e^{-\frac{x}{c}})$ between two points equally distant from the axis of x. Show also that the centroid of any arc has the same abscissa as the intersection of the tangents at its ends; and that the ordinate of the centroid of any arc beginning at the vertex is equal to one-half the intercept made on the axis of y by the normal at the end of the arc.

(14) Common helix: $x = r\cos\theta, y = r\sin\theta, z = kr\theta$, from $\theta = 0$ to $\theta = \theta$.

4. CENTROIDS OF AREAS.

226. It follows from symmetry that the centroid of a **homogeneous circular** or **elliptic area** (plate, lamina) is at the geometrical center of figure. Similarly, the centroid of a **homogeneous parallelogram** is at the intersection of its diagonals.

In general, if a homogeneous plane figure have two axes of symmetry, the centroid must be at the intersection of these lines since the sum of the moments is zero for each of these lines.

227. It has been shown in Art. 222 (*b*) how the centroid of a **homogeneous triangular area** ABC can be found.

Dividing the area into linear elements by drawing lines parallel to one of the sides, say AB (Fig. 60, p. 139), it appears that the centroid of each element, such as PQ, lies at its middle point.

The locus of these middle points is the median CC' of the triangle; on this line, then, the centroid G of the triangle must be situated. Resolving the triangle into linear elements parallel to the side BC, or to CA, it follows in the same way that G must lie on each of the other two medians of the triangle. The intersection of these medians is therefore the centroid G.

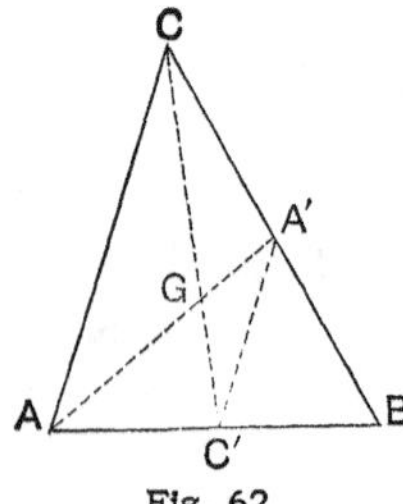

Fig. 62.

The point G trisects each median so that $CG/GC' = 2$. For if AA' (Fig. 62) is another median, the triangles AGC and $A'GC'$ are similar, and $A'C' = \frac{1}{2}AC$; hence $C'G = \frac{1}{2}CG$.

It follows from Art. 220, that the centroid of the homogeneous triangular area coincides with that of three particles of equal mass placed at the vertices.

228. Homogeneous Quadrilateral. The centroid is found graphically by resolving the quadrilateral into triangles, finding their centroids, and deducing from them the centroid of the quadrilateral. This process applies generally to *any polygon* and can be carried out in various ways.

Thus for the quadrilateral $ABCD$ (Fig. 63) drawing the diagonal AC and determining the centroids of the triangles ABC and ADC, we obtain by joining these centroids one line on which the required centroid of the quadrilateral must lie. Repeating the same construction for the triangles obtained by drawing the other diagonal BD, we find a second line on which the centroid must lie. The intersection of these lines gives the centroid of the quadrilateral.

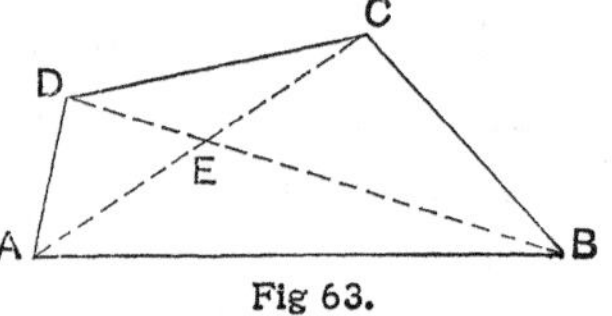

Fig 63.

229. For some purposes it is convenient to find a system of particles whose centroid shall be the same as that of a quadrilateral. The problem is of course indeterminate and may be solved in various ways.

Let m be the mass of the quadrilateral $ABCD$; m_1, m_2 the masses of the triangles ABC, ADC. By Art. 227, each of these triangles can be replaced by three equal particles $\frac{1}{3}m_1$, $\frac{1}{3}m_2$, placed at the vertices. We thus have at A, as well as at C, a mass $\frac{1}{3}(m_1 + m_2) = \frac{1}{3}m$.

The masses $\frac{1}{3}m_1$ at B and $\frac{1}{3}m_2$ at D, whose sum is also $= \frac{1}{3}m$, are proportional to the areas of the triangles ABC, ADC, or to the lengths EB, ED, if E be the intersection of the diagonals. Now these two different masses at B and D can be replaced by a system of three masses, $\frac{1}{3}m$ at B, $\frac{1}{3}m$ at D, and $-\frac{1}{3}m$ at E. For (1) the total mass evidently remains the same, and (2) the centroids of the two systems coincide as is easily seen by taking moments with respect to E.

Indeed, the centroid G' of $\frac{1}{3}m_1$ at B and $\frac{1}{3}m_2$ at D is determined by the equation

$$(m_1 + m_2) \cdot EG' = m_1 \cdot EB - m_2 \cdot ED;$$

substituting for m_1, m_2 their values as found from the relations $m_1 + m_2 = m$, $m_1/m_2 = EB/ED$, this reduces to

$$m \cdot EG' = m \cdot (EB - ED).$$

The centroid G'' of $\frac{1}{3}m$ at B, $\frac{1}{3}m$ at D, and $-\frac{1}{3}m$ at E is given by

$$m \cdot EG'' = m \cdot EB - m \cdot ED - m \cdot 0.$$

Hence G' and G'' coincide.

The centroid of the area of a homogeneous quadrilateral is therefore the same as that of four equal particles placed at its vertices together with a fifth particle of equal but negative mass, placed at the intersection of the diagonals.

230. In the particular case of a **homogeneous trapezoid** (Fig. 64), it may be noticed that the figure can be divided into rectilinear elements by lines drawn parallel to the parallel sides of the trapezoid. Every such element has its centroid at its middle point; the locus of all these points is the so-called median; and the centroid G of the trapezoid must lie on this median, *i. e.* on the line joining the middle points E, F of the parallel sides.

To find the ratio in which G divides the length EF, we use again the method of taking moments. We divide the trapezoid into two triangles by the diagonal BC and remember that the distance of the centroid of a triangle from its base is equal to one-third of its height; then taking moments with respect to the two parallel sides $AB = a$, $CD = b$, denoting the height of the trapezoid by h, and the distances of G from a and b by $\bar{y}$ and $\bar{y}'$, we obtain

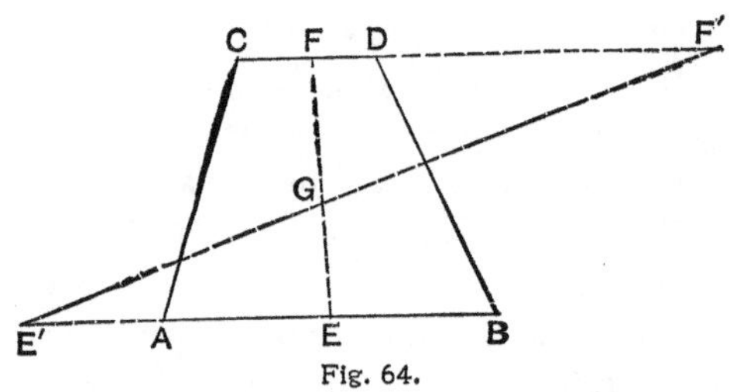

Fig. 64.

$$\tfrac{1}{2}(a+b)h\cdot\bar{y} = \tfrac{1}{2}ah\cdot\tfrac{1}{3}h + \tfrac{1}{2}bh\cdot\tfrac{2}{3}h,$$

$$\tfrac{1}{2}(a+b)h\cdot\bar{y}' = \tfrac{1}{2}ah\cdot\tfrac{2}{3}h + \tfrac{1}{2}bh\cdot\tfrac{1}{3}h.$$

Dividing, we find

$$\frac{\bar{y}}{\bar{y}'} = \frac{EG}{GF} = \frac{a+2b}{2a+b} = \frac{\tfrac{1}{2}a+b}{a+\tfrac{1}{2}b}.$$

This gives the following construction: Make $AE' = b$ on the prolongation of a, and $DF' = a$ on the prolongation of b, in the opposite sense; then $E'F'$ will intersect EF in G.

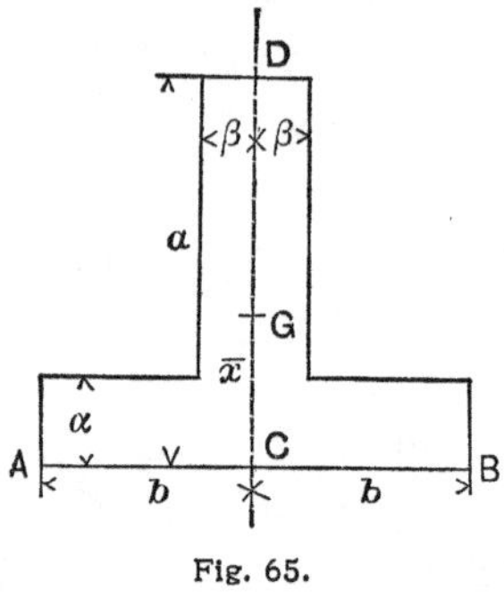

Fig. 65.

231. To find the centroid of the **cross-section of a T-iron** (Fig. 65), it is only necessary to find its distance $\bar{x}$ from the lower side AB; for it must lie on the axis of symmetry CD. Taking moments with respect to AB we obtain with the notation indicated in the figure:

$$[2a\beta + 2(b-\beta)\alpha]\cdot\bar{x} = 2a\beta\cdot\frac{a}{2} + 2(b-\beta)\alpha\cdot\frac{\alpha}{2},$$

hence

$$\bar{x} = \tfrac{1}{2}\,\frac{a^2\beta + b\alpha^2 - \alpha^2\beta}{a\beta + b\alpha - \alpha\beta}.$$

If α, β are nearly equal and very small in comparison with a, b, we have approximately

$$\bar{x} = \frac{1}{2} \frac{a^2 + b\alpha}{a + b - \alpha},$$

or still more roughly

$$\bar{x} = \frac{1}{2} \frac{a^2}{a + b}.$$

232. The area of a **homogeneous circular sector** (Fig. 61, p. 139), of radius r and angle $AOB = 2\alpha$ can be resolved into triangular elements $POP' = \frac{1}{2}r^2 d\theta$, the bisecting radius OC being taken as polar axis. The centroid of such an element lies, by Art. 227, at the distance $\frac{2}{3}r$ from the center O. Regarding the mass, $\rho' \cdot \frac{1}{2}r^2 d\theta$, of each element as concentrated at its centroid, the sector is replaced by a homogeneous circular arc of radius $\frac{2}{3}r$ and density $\frac{1}{2}\rho' r^2 d\theta$. By Art. 223, the centroid of such an arc, which is the required centroid of the sector, lies on the bisecting radius OC at the distance $\frac{2}{3}r \cdot \sin\alpha / \alpha$ from the center O. Hence

$$\bar{x} = \frac{2}{3} r \frac{\sin\alpha}{\alpha}.$$

233. In general, for **areas bounded by curves** we must resort to integration, using the general formulæ of Art. 214.

If the area S be plane, we have in rectangular co-ordinates

$$M = \int_{x_1}^{x_2} \int_{y_1}^{y_2} \rho' dx dy,$$

$$M \cdot \bar{x} = \int_{x_1}^{x_2} \int_{y_1}^{y_2} \rho' x dx dy, \quad M \cdot \bar{y} = \int_{x_1}^{x_2} \int_{y_1}^{y_2} \rho' y dx dy;$$

and if the mass be homogeneous, *i. e.* $\rho' =$ const., since then the first integration can at once be effected:

$$S = \int_{x_1}^{x_2} (y_2 - y_1) dx,$$

$$S \cdot \bar{x} = \int_{x_1}^{x_2} x(y_2 - y_1) dx, \quad S \cdot \bar{y} = \frac{1}{2} \int_{x_1}^{x_2} (y_2^2 - y_1^2) dx,$$

or similar expressions for y as independent variable.

In polar co-ordinates, the element of area is $rdrd\theta$, and we have $x = r\cos\theta$, $y = r\sin\theta$; hence

$$S = \int\int rdrd\theta,$$

$$S\cdot\bar{x} = \int\int r^2\cos\theta drd\theta,\quad S\cdot\bar{y} = \int\int r^2\sin\theta drd\theta;$$

or, performing the first integration,

$$S = \tfrac{1}{2}\int_{\theta_1}^{\theta_2} r^2 d\theta,$$

$$S\cdot\bar{x} = \tfrac{1}{3}\int_{\theta_1}^{\theta_2} r^3\cos\theta d\theta,\quad S\cdot\bar{y} = \tfrac{1}{3}\int_{\theta_1}^{\theta_2} r^3\sin\theta d\theta.$$

It will be noticed that these last formulæ express also that the infinitesimal sector $\frac{1}{2}r^2 d\theta$ is taken as element, the centroid of this element having the co-ordinates $\frac{2}{3}r\cos\theta$, $\frac{2}{3}r\sin\theta$.

234. As a somewhat more complicated example let us consider a circular disk of radius a, in which the density varies directly as the distance from the center. Let a circle described upon a radius as diameter be cut out of this disk; it is required to find the centroid of the remainder.

Let O be the center of the disk of radius a, C that of the disk of radius $\frac{1}{2}a$; G_1 the centroid of the latter, G the required centroid; and put $OG_1 = \bar{x}_1$, $OG = \bar{x}$. Then if M_1 be the mass of the smaller disk, M_2 that of the larger, we must have $(M_2 - M_1)\cdot\bar{x} = M_1\bar{x}_1$.

The equation of the smaller circle is $r = a\cos\theta$. Taking as element of the mass of the smaller disk the mass contained between two arcs of radii r and $r + dr$, we have for this element:

$$dM_1 = \rho'\cdot 2\theta rdr,$$

or since $\rho' = kr,\ r = a\cos\theta,$

$$dM_1 = 2ka^3\theta\cos^2\theta d(\cos\theta).$$

Hence

$$M_1 = \tfrac{2}{3}ka^3\int_{\pi/2}^{0}\theta d(\cos^3\theta)$$

$$= \tfrac{2}{3}ka^3\left(\theta\cos^3\theta - \int\cos^3\theta d\theta\right)_{\pi/2}^{0}$$

$$= \tfrac{2}{3}ka^3\int_0^{\pi/2}\cos^3\theta d\theta = \tfrac{2}{3}ka^3\cdot\tfrac{2}{3} = \tfrac{4}{9}ka^3.$$

The centroid of the element dM_1 lies, according to Art. 223, at the distance $r\sin\theta/\theta$ from O. We have therefore

$$M_1\bar{x}_1 = -2ka^3\int_{\pi/2}^{0}\theta\sin\theta\cos^2\theta d\theta\cdot r\frac{\sin\theta}{\theta}$$

$$= 2ka^4\int_0^{\pi/2}\sin^2\theta\cos^3\theta d\theta = \tfrac{4}{15}ka^4.$$

The mass of the larger disk is

$$M_2 = \int_0^a kr\cdot 2\pi r\cdot dr = 2\pi k\int_0^a r^2 dr = \tfrac{2}{3}\pi ka^3.$$

Substituting these values in the equation of moments we find:

$$\bar{x} = \frac{M_1\bar{x}_1}{M_2 - M_1} = \frac{6}{5(3\pi - 2)}a = 0.1616\,a.$$

235. Proceeding to the determination of the centroids of *curved surface-areas*, we begin with the special case of the homogeneous area of a **surface of revolution,** bounded by two planes at right angles to the axis. The centroid evidently lies on the axis of revolution, which we take as axis of x; it is therefore sufficient to take moments with respect to the yz-plane. As element we take the strip contained between two planes, parallel to the yz-plane, at the distances x and $x + dx$ from it; if the equation of the meridian section be $\eta = f(x)$, where η is the distance of any point of the surface from the axis of revolution, we have for this element:

$$dS = 2\pi\eta ds = 2\pi\eta\sqrt{dx^2 + dy^2} = 2\pi f(x)\sqrt{1 + f'^2}dx.$$

Hence, if the bounding planes have the distances x_1, x_2 from the yz-plane:

$$S = 2\pi\int_{x_1}^{x_2} f(x)\sqrt{1 + f'^2}dx,\quad S\cdot\bar{x} = 2\pi\int_{x_1}^{x_2} xf(x)\sqrt{1 + f'^2}dx.$$

236. When the area is bounded by two planes perpendicular to the axis of revolution and any two meridian planes, inclined to each other at an angle ϕ, we may take one of these meridian planes as xy-plane and find, with

$$dS = \eta d\phi \ ds = \eta \sqrt{dx^2 + dy^2} d\phi = f(x) \sqrt{1 + f'^2} dx d\phi :$$

$$S = \int_{x_1}^{x_2} \int_0^{\phi} f(x) \sqrt{1 + f'^2} dx d\phi = \phi \int_{x_1}^{x_2} f(x) \sqrt{1 + f'^2} dx,$$

$$S \cdot \bar{x} = \int_{x_1}^{x_2} \int_0^{\phi} x f(x) \sqrt{1 + f'^2} dx d\phi = \phi \int_{x_1}^{x_2} x f(x) \sqrt{1 + f'^2} dx,$$

$$S \cdot \bar{y} = \int_{x_1}^{x_2} \int_0^{\phi} f^2 \cos\phi \sqrt{1 + f'^2} dx d\phi = \sin\phi \int_{x_1}^{x_2} f^2 \sqrt{1 + f'^2} dx,$$

$$S \cdot \bar{z} = \int_{x_1}^{x_2} \int_0^{\phi} f^2 \sin\phi \sqrt{1 + f'^2} dx d\phi$$

$$= (1 - \cos\phi) \int_{x_1}^{x_2} f^2 \sqrt{1 + f'^2} dx,$$

where f and f' are known functions of x.

Instead of x, η might be taken as independent variable.

237. In the case of *spherical surfaces*, although the preceding formulæ can of course be used, it is often more convenient to make use of the geometrical property of the sphere that any spherical area is equal to the area of its projection on a cylinder circumscribed about the sphere.

Thus the *area on the sphere contained between two parallel planes* is equal to the area cut out by the same two planes from the circumscribed cylinder whose axis is perpendicular to the planes. The centroid of such a spherical area is therefore on the radius at right angles to the bounding planes midway between these planes.

238. The Second Proposition of Pappus and Guldinus (compare Art. 224).

A plane area S (Fig. 66) rotating about any axis situated in its plane generates a solid of revolution whose volume is $V = \pi \int (y_2^2 - y_1^2) dx$, if the axis of revolution is taken as axis of x and y_1, y_2 are the two ordinates of the curve bounding the area. On the other hand, if $\bar{y}$ be the distance of the centroid G of the plane area from the axis, we have

$$S\cdot\bar{y} = \tfrac{1}{2}\int(y_2^2 - y_1^2)dx,$$

by Art. 233. Combining these two results, we find

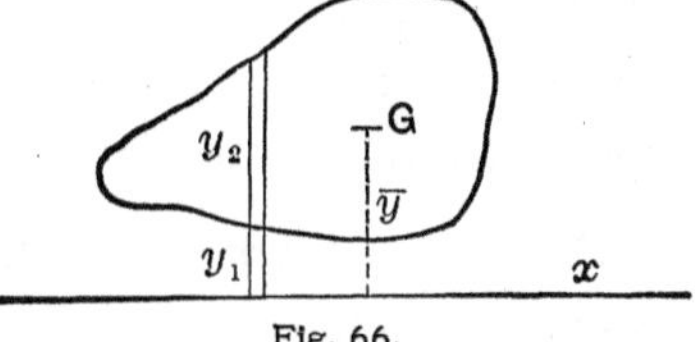

Fig. 66.

$$V = 2\pi\bar{y}\cdot S,$$

i. e., the volume of a solid of revolution is obtained by multiplying the generating area into the path described by its centroid.

The proposition evidently holds even for a partial revolution.

239. To find the centroid of a **portion of any curved surface** $F(x, y, z) = 0$, we have only to substitute $dM = \rho' dS$ in the general formulæ of Art. 214, and then express dS by the ordinary methods of analytic geometry.

Denoting by l, m, n the direction cosines of the normal to the surface at the point (x, y, z), and putting for shortness $\partial F/\partial x = F_x$, $\partial F/\partial y = F_y$, $\partial F/\partial z = F_z$, we have

$$dS = \frac{dydz}{l} = \frac{dzdx}{m} = \frac{dxdy}{n},$$

$$\frac{l}{F_x} = \frac{m}{F_y} = \frac{n}{F_z} = \frac{1}{\sqrt{F_x^2 + F_y^2 + F_z^2}}.$$

Hence, substituting

$$dS = dxdy\cdot\frac{\sqrt{F_x^2 + F_y^2 + F_z^2}}{F_z}$$

in the formulæ of Art. 214, we find

$$M = \int_{y_1}^{y_2}\int_{x_1}^{x_2}\rho' dxdy\cdot\frac{\sqrt{F_x^2 + F_y^2 + F_z^2}}{F_z},$$

where the integration is to be extended over the projection of the portion of surface under consideration on the plane xy. The equation of the curve bounding this projection must be given;

it determines the limits of integration. It is obvious how the formula has to be modified when the projection of the area on either of the other co-ordinate planes is given.

The expressions for $M\cdot\bar{x}$, $M\cdot\bar{y}$, $M\cdot\bar{z}$ differ from the above expression for M only in containing the additional factor x, y, z, respectively, under the integral sign.

240. If the equation of the surface be given in the form $z=f(x, y)$, as is frequently the case, we have

$$F(x, y, z) \equiv z - f(x, y);$$

hence with the usual Gaussian notation $\partial z/\partial x \equiv \partial f/\partial x \equiv p$, $\partial z/\partial y \equiv \partial f/\partial y \equiv q$,

$$F_x = -p, \quad F_y = -q, \quad F_z = 1,$$

which gives

$$M = \int_{y_1}^{y_2}\int_{x_1}^{x_2} \rho' \sqrt{1 + p^2 + q^2}\, dxdy,$$

while $M\cdot\bar{x}$, $M\cdot\bar{y}$, $M\cdot\bar{z}$ contain the additional factor x, y, z, respectively, under the integral sign.

In the case of a homogeneous *spherical surface* $x^2 + y^2 + z^2 = a^2$, we have $p = \partial z/\partial x = -x/z$, $q = \partial z/\partial y = -y/z$; hence $z\sqrt{1 + p^2 + q^2} = a$, so that

$$S\cdot\bar{z} = a\int\int dxdy = a\cdot S_z,$$

where S is the area of the surface and S_z the area of its projection on the plane xy. The formula shows that the distance $\bar{z}$ of the centroid of any spherical area S from a plane passing through the center is equal to the radius a multiplied by the ratio of the projection S_z of the area on the plane to the area itself.

241. Exercises.

(1) The sides of a right-angled triangle are a and b. Find the distances of the centroid of the triangular area from the vertices.

(2) From a square $ABCD$ one corner EAF is cut off so that $AE = \frac{3}{4}a$, $AF = \frac{1}{4}a$, a being the side of the square. Find the centroid of the remaining area.

(3) An isosceles right-angled triangle of sides a being cut out of the area of its circumscribed circle, find the centroid of the remaining area.

(4) If one-fourth be cut away from a triangle by a parallel to the base, show that in the remaining trapezoid the centroid divides the median in the ratio 4 : 5.

(5) Prove that the centroid of any plane quadrilateral $ABCD$ coincides with that of the triangle ACF, if the point F be constructed by laying off $BF = DE$ on the diagonal BD, E being the intersection of the diagonals.

(6) Find the centroid of the cross-section of a retaining wall in the form of a trapezoid with two perpendicular sides, the lower base being a, the upper b, the height h.

(7) Find the centroid of the cross-section of an angle-iron, or L, the sides being a, b, the thickness of each flange δ (Fig. 67).

(8) Find the centroid of the cross-section of a bar formed by placing four angle-irons with their edges together, two of the irons having the

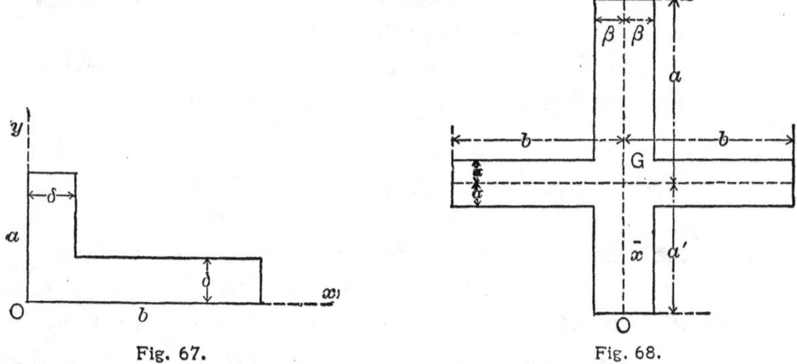

Fig. 67. Fig. 68.

dimensions a, b, α, β, while the other two have the dimension a different, say a' (Fig. 68).

(9) Find the centroid of the cross-section of a U-iron, the length of the flanges being $a = 12$ in., that of the web $2b = 8$ in., and the thickness $\delta = 1$ in. Deduce the general formula for $\bar{x}$, and an approximate formula for a small δ, and compare the numerical results.

(10) In the cross-section of an unsymmetrical double T (Fig. 69), the flanges are $2b = 16$ in., $2b' = 10$ in.; the web is $a = 10$ in.; and the thickness of each of the two channel-irons forming the bar is $\delta = 1$ in. throughout; find the centroid.

(11) In a T-iron (Fig. 70) the width of the flange is b, its thickness α; the depth of the web is a, its thickness β. Find the distance of the centroid from the outer side of the flange; give an approximate expression and investigate it for $a = b$, $\alpha = \beta = \frac{1}{5}a$.

(12) Find the centroid of a circular sector (comp. Art. 232) by integration, using the formulæ of Art. 233, and deduce from the result

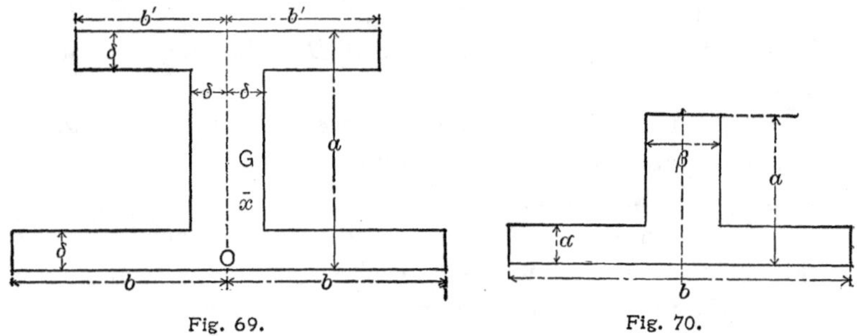

Fig. 69. Fig. 70.

that the centroid of a homogeneous semicircular area of radius r lies at the distance $\bar{x} = (4/3\pi)r$ from the center.

(13) The centroid of the area of a homogeneous circular segment of radius r subtending at the center an angle 2α is at the distance

$$\bar{x} = \frac{2}{3}r \cdot \frac{\sin^3\alpha}{\alpha - \sin\alpha \cos\alpha} = \frac{1}{6}\frac{c^3}{rs - ch},$$

if c is the chord, h its distance from the center, and s the arc.

(14) A painter's palette is formed by cutting a small circle of radius b out of a circular disk of radius a, the distance between the centers being c. It is required to find the distance of the centroid of the remainder from the center of the larger circle. (Routh.)

(15) The arch constructed of brick over a door is in the form of a quadrant of a circular ring. The door is 5 ft. wide; $1\frac{1}{2}$ lengths of brick are used (say 12 in.). Find the centroid of the arch.

Find the co-ordinates of the centroid for the following plane areas:

(16) Area bounded by the parabola $y^2 = 4ax$, the axis of x, and the ordinate y.

(17) Area bounded by the curve $y = \sin x$ from $x = 0$ to $x = \pi$ and the axis of x.

(18) Quadrant of an ellipse.

(19) Elliptic segment bounded by the chord joining the ends of the major and minor axes.

(20) Show, by Art. 222, that the centroid of the surface of a right circular cone lies at a distance from the base equal to one third of the height.

(21) Find the centroid of the portion of the surface of a right circular cone cut out by two planes through the axis inclined at an angle φ.

(22) Find the centroid of the area of the earth's surface contained between the tropic of Cancer (latitude $= 23° 27'$) and the arctic circle (polar distance $= 23° 27'$).

(23) A bowl in the form of a hemisphere is closed by a circular lid of a material whose density is three times that of the bowl. Find the centroid.

(24) The cissoid $(2a - x)y^2 = x^3$ can be represented by the equations $x = 2a \sin^2\theta$, $y = 2a \sin^3\theta/\cos\theta$, where θ is the polar angle, $2a$ the distance from cusp to asymptote. Show that the centroid of the area between the curve and its asymptote divides the distance between cusp and asymptote in the ratio $5 : 1$.

5. CENTROIDS OF VOLUMES.

242. We proceed to the methods of finding the centroids of volumes or solids.

Considerations of symmetry make it clear that the centroid of a **homogeneous parallelepiped** lies at the intersection of its diagonals; similarly, that of a **homogeneous prism** or **cylinder** coincides with the centroid of the area of its middle section (*i. e.* a plane section parallel to, and equally distant from, the bases).

243. For a **homogeneous pyramid** or **cone**, we have found in Art. 222 (*c*) that the centroid lies on the line joining the vertex to the centroid of the area of the base, at a distance from the base equal to $\frac{1}{4}$ of this line. This is, of course, easily shown directly by resolving the pyramid or cone into plane elements parallel to the base, in a manner analogous to that used for the triangular area in Art. 227.

244. It may, perhaps, be well to state formally the principal **laws of symmetry** for homogeneous solids, although they present themselves so naturally that they are used almost instinctively. For however simple and obvious these propositions may appear, the beginner may be led into error if he does not use them cautiously. The proof rests on the fundamental definition of the centroid as a point such that for any plane through it the sum of the moments is zero.

(*a*) *If the surface of the solid have a plane of symmetry*, *i. e.* a plane such that every line perpendicular to it intersects the surface in two points equidistant from the plane, *the centroid lies in this plane.* Hence, the centroid of a homogeneous solid is at once known if its surface possesses three planes of symmetry (not passing through the same straight line). If the surface has two planes of symmetry, the centroid lies on their line of intersection.

(*b*) *If the surface have an axis of symmetry*, *i. e.* a line such that every line perpendicular to it intersects the surface in two points equidistant from the line, *the centroid must lie on this axis.* Two axes of symmetry in the same homogeneous solid determine its centroid by their intersection.

(*c*) *If the surface have a center*, *i. e.* a point such that every line through it intersects the surface in two points equidistant from it, *the centroid coincides with this center.*

(*d*) *If the surface have a diametral plane*, *i. e.* a plane bisecting all chords that are parallel to a certain direction, *the centroid lies in this plane.*

245. Homogeneous spherical solids can be treated by a method analogous to that used for circular areas (see Art. 232). Thus a **homogeneous spherical sector** can be resolved into infinitesmal elements, each of which is a pyramid whose vertex lies at the center of the sphere and whose base is an infinitesimal element of the spherical surface area of the sector. Such an element, regarded as a pyramid (Art. 243), has its centroid at the distance $\frac{3}{4}a$ from the center, if a be the radius of the sphere. We may regard its mass as concentrated at its centroid and have thus the solid sector replaced by a homogeneous segment of a spherical area, of radius $\frac{3}{4}a$. It has been shown in Art. 237 that the centroid of such a segment bisects its height.

Let 2α be the angle at the vertex of the given sector (Fig. 71); then the height of the segment of radius $\frac{3}{4}a$ is $\frac{3}{4}a(1 - \cos\alpha)$; hence the distance $\bar{x}$ of the centroid of the solid spherical sector from the center is

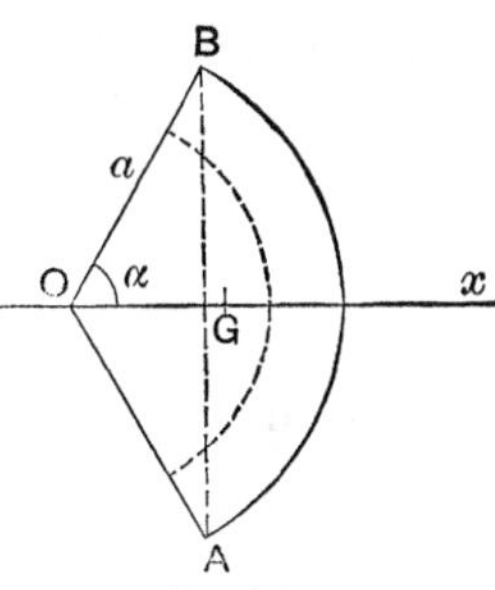

Fig. 71.

$$\bar{x} = \tfrac{3}{4}a\cos\alpha + \tfrac{3}{8}a(1 - \cos\alpha)$$
$$= \tfrac{3}{8}a(1 + \cos\alpha) = \tfrac{3}{4}a\cos^2\tfrac{1}{2}\alpha.$$

246. In a **homogeneous solid of revolution** the centroid lies on the axis of revolution, since this line is an axis of symmetry (Art. 244 (*b*)). Taking this line as the axis of x, the equation of the surface of the solid is determined by that of the meridian curve, that is, the curve bounding the generating area, say $y = f(x)$.

We select as element the circular plate of thickness dx contained between two sections of the solid at right angles to the axis of revolution (Fig. 66, p. 149). The centroid of each such element lies on the axis, and the volume of the element is $\pi y^2 dx$.

We have, therefore,

$$S = \pi\int_{x_1}^{x_2}[f(x)]^2 dx, \quad S\cdot\bar{x} = \pi\int_{x_1}^{x_2} x[f(x)]^2 dx.$$

Instead of x it may sometimes be more convenient to take y as independent variable.

It is easy to see how the formula has to be modified when more than one value of y corresponds to a given value of x.

247. In the most general case of **any solid** whatever the formulæ of Art. 214 assume different forms according to the system of co-ordinates used. Thus for rectangular cartesian co-ordinates the element of volume is $dv = dxdydz$, and we have:

$$M = \int\int\int\rho dxdydz, \quad M\cdot\bar{x} = \int\int\int\rho x dxdydz,$$
$$M\cdot\bar{y} = \int\int\int\rho y dxdydz, \quad M\cdot\bar{z} = \int\int\int\rho z dxdydz.$$

248. In polar co-ordinates, *i. e.* for the radius vector r, the co-latitude θ and the longitude ϕ (Fig. 72), the element of volume is an infinitesimal rectangular parallelepiped having the concurrent edges dr, $rd\theta$, $r\sin\theta d\phi$; hence

$$dv = r^2 \sin\theta dr d\theta d\phi.$$

As $x = r\cos\theta$, $y = r\sin\theta\cos\phi$, $z = r\sin\theta\sin\phi$, the centroid is determined by the equations:

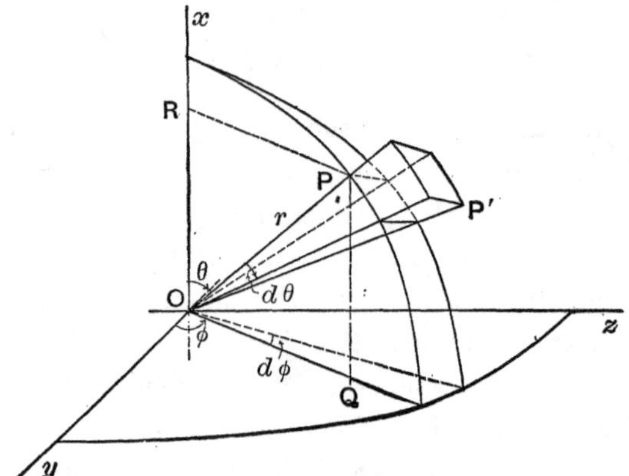

Fig. 72.

$$M = \int\int\int \rho r^2 \sin\theta dr d\theta d\phi,$$

$$M \cdot \bar{x} = \int\int\int \rho r^3 \sin\theta \cos\theta dr d\theta d\phi,$$

$$M \cdot \bar{y} = \int\int\int \rho r^3 \sin^2\theta \cos\phi dr d\theta d\phi,$$

$$M \cdot \bar{z} = \int\int\int \rho r^3 \sin^2\theta \sin\phi dr d\theta d\phi.$$

In the simple cases that occur most frequently in the applications the triple integrals can often be reduced at once to double or even simple integrals by selecting a convenient element.

249. As an illustration let us determine the centroid of the volume $OABCD$ (Fig. 73) bounded by the three co-ordinate planes and the warped quadrilateral (hyperbolic paraboloid) $ABCD$. The latter is generated by the line LM gliding along AB and CD so as to remain parallel to the plane yz. The data are $OA = CD = a$, $OB = b$, $OC = AD = c$.

We take as element an infinitesimal prism QP of base $dxdz$ and

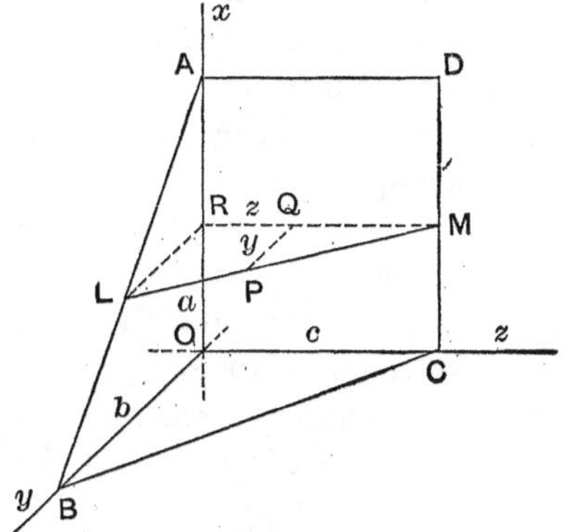

Fig. 73.

height y. From similar triangles we have $y/RL = (c - z)/c$, and $RL/b = (a - x)/a$; hence

$$y = b\frac{a-x}{a}\cdot\frac{c-z}{c}.$$

Thus we find, rejecting the constants which cancel in numerator and denominator,

$$\bar{x} = \frac{\int_0^a\int_0^c x(a-x)(c-z)dxdz}{\int_0^a\int_0^c (a-x)(c-z)\,dxdz} = \frac{\int_0^a x(a-x)\,dx\cdot(c^2-\frac{1}{2}c^2)}{\int_0^a (a-x)\,dx\cdot(c^2-\frac{1}{2}c^2)}$$

$$= \frac{\int_0^a x(a-x)dx}{\int_0^a (a-x)dx} = \frac{\frac{1}{2}a^3-\frac{1}{3}a^3}{a^2-\frac{1}{2}a^2} = \tfrac{1}{3}a;$$

$$\bar{y} = \frac{\frac{1}{2}\int_0^a\int_0^c b^2\frac{(a-x)^2}{a^2}\cdot\frac{(c-z)^2}{c^2}\,dxdz}{\int_0^a\int_0^c b\frac{a-x}{a}\cdot\frac{c-z}{c}\,dxdz} = \frac{b}{2ac}\,\frac{\int_0^a (a-x)^2dx\cdot\frac{1}{3}c^3}{\int_0^a (a-x)\,dx\cdot\frac{1}{2}c^2}$$

$$= \frac{1}{3}\cdot\frac{b}{a}\cdot\frac{\frac{1}{3}a^3}{\frac{1}{2}a^2} = \tfrac{2}{9}b.$$

Finally, $\bar{z} = \frac{1}{3}c$, by analogy with $\bar{x}$.

250. Exercises.

(1) Determine the centroid of a homogeneous solid hemisphere.

(2) Find the centroid of a frustum of a cone, the radii of the bases being r_1, r_2 $(r_1 < r_2)$, the height of the frustum h.

(3) Show that the formula for the frustum of the cone applies likewise to the frustum of any pyramid of the same height h if r_1, r_2 are any two homologous linear dimensions of the two bases.

(4) Find the centroid of a solid segment of a sphere of radius a, the height of the segment being h.

(5) Find the centroid of the paraboloid of revolution of height h, generated by the revolution of the parabola $y^2 = 4ax$ about its axis.

(6) The area bounded by the parabola $y^2 = 4ax$, the axis of x, and the ordinate $y = y_1$ revolves about the tangent at the vertex. Find the centroid of the solid of revolution so generated.

(7) The same area as in Ex. (6) revolves about the ordinate y_1. Find the centroid.

(8) Find the centroid of an octant of an ellipsoid

$$x^2/a^2 + y^2/b^2 + z^2/c^2 = 1.$$

(9) The equations of the common cycloid referred to a cusp as origin and the base as axis of x are $x = a(\theta - \sin\theta)$, $y = a(1 - \cos\theta)$. Find the centroid: (*a*) of the arc of the semi-cycloid (*i. e.* from cusp to vertex); (*b*) of the plane area included between the semi-cycloid and the base; (*c*) of the surface generated by the revolution of the semi-cycloid about the base; (*d*) of the volume generated in the same case; (*e*) of the surface generated by the revolution of the whole cycloid (from cusp to cusp) about its axis, *i. e.* the line through the vertex at right angles to the base; (*f*) of the volume so generated.

(10) Find the centroid of a solid hemisphere whose density varies as the nth power of the distance from the center.

(11) Show that, both for a triangular area and for a tetrahedral volume, the distance of the centroid from any plane is the arithmetic mean of the distances of the vertices from the same plane.

(12) Regarding the earth as a homogeneous sphere of density $\rho = 5.5$, how much would its centroid be displaced by superimposing over the area bounded by the arctic circle an ice-cap of a uniform thickness of 10 miles?

(13) From out of the right cone ABC a cone ABD is cut of the

same base and axis, but of smaller height. Find the centroid of the remaining solid.

(14) A triangle BC, whose sides are a, b, c, revolves about an axis situated in its plane. Find the surface area and volume of the solid so generated, if p, q, r are the distances of , A, B, C from the axis.

(15) "Water is poured gently into a cylindrical cup of uniform thickness and density. Prove that the locus of the center of gravity of the water, the cup, and its handle is a hyperbola." (Routh.)

(16) Prove that the volume of a truncated right cylinder (*i. e.* a right cylinder cut by a plane inclined at any angle to its base) is equal to the product of the area of its base into the height of the truncated cylinder at the centroid of its base.

(17) Prove that the volume of a doubly truncated cylinder, *i. e.* of any closed cylinder bounded by two planes intersecting in a line outside the cylinder, is equal to the product of the area of the cross-section into the distance of the centroids of the bases.

II. *Momentum; Force; Energy.*

251. Let us consider a point moving with constant acceleration from rest in a straight line. We know from Kinematics (Art. 71) that its motion is determined by the equations

$$v = jt, \quad s = \tfrac{1}{2} jt^2, \quad \tfrac{1}{2}v^2 = js, \qquad (1)$$

where s is the distance passed over in the time t, v the velocity, and j the acceleration, at the time t.

If, now, for the single point we substitute an m-tuple point, *i. e.* if we endow our point with the mass m, and thus make it a *particle* (see Art. 205), the equations (1) must be multiplied by m, and we obtain

$$mv = mjt, \quad ms = \tfrac{1}{2}mjt^2, \quad \tfrac{1}{2}mv^2 = mjs. \qquad (2)$$

The quantities mv, mj, $\frac{1}{2}mv^2$ occurring in these equations have received special names because they correspond to certain physical conceptions of great importance.

252. *The product mv of the mass m of a particle into its velocity v* is called the **momentum**, or the *quantity of motion*, of the particle.

In observing the behavior of a physical body in motion, we notice that the effect it produces — for instance, when impinging on another body, or more generally, whenever its velocity is changed — depends not only on its velocity, but also on its mass. Familiar examples are the following: a loaded railroad car is not so easily stopped as an empty one; the destructive effect of a cannon-ball depends both on its velocity and on its mass; the larger a fly-wheel, the more difficult is it to give it a certain velocity; etc.

It is from experiences of this kind that the physical idea of mass is derived.

The fact that any change of motion in a physical body is affected by its mass is sometimes ascribed to the so-called "*inertia*," or "force of inertia," of matter, which means, however, nothing else but the property of possessing mass.

253. Momentum, being by definition (Art. 252) the product of mass and velocity, has for its *dimensions* (Art. 57),

$$MV = MLT^{-1}.$$

The *unit of momentum* is the momentum of the unit of mass having the unit of velocity.

Thus in the C.G.S. system the unit of momentum is the momentum of a particle of 1 gram moving with a velocity of 1 cm. per second. There is no generally accepted name for this unit, although the name **bole** was proposed by the Committee of the British Association.

In the F.P.S. system, the unit is the momentum of a particle of one pound mass moving with a velocity of 1 ft. per second.

To find the relations between these two units, let there be x C.G.S. units in the F.P.S. unit; then

$$x \cdot \frac{\text{gm. cm.}}{\text{sec.}} = 1 \cdot \frac{\text{lb. ft.}}{\text{sec.}};$$

hence

$$x = \frac{\text{lb.}}{\text{gm.}} \cdot \frac{\text{ft.}}{\text{cm.}},$$

or, by Art. 202 and Art. 9,

$$x = 453.59 \times 30.48 = 13\,825.3;$$

i. e. 1 F.P.S. unit of momentum = 13 825.3 C.G.S. units, and 1 C.G.S. unit = 0.000 072 331 F.P.S. units.

254. Exercises.

(1) What is the momentum of a cannon-ball weighing 200 lbs. when moving with a velocity of 1500 ft. per second?

(2) With what velocity must a railroad-truck weighing 3 tons move to have the same momentum as the cannon-ball in Ex. (1)?

(3) Determine the momentum of a one-ton ram after falling through 4 feet.

255. *The product mj of the mass m of a particle into its acceleration j* is called **force.** Denoting it by F, we may write our equations (2) in the form

$$mv = Ft, \quad s = \frac{1}{2}\frac{F}{m}t^2, \quad \frac{1}{2}mv^2 = Fs. \tag{3}$$

As long as the velocity of a particle of constant mass remains constant, its momentum remains unchanged. If the velocity changes uniformly from the value v at the time t to v' at the time t', the corresponding change of momentum is

$$mv' - mv = mjt' - mjt = F(t' - t); \tag{4}$$

hence

$$F = \frac{mv' - mv}{t' - t}. \tag{5}$$

Here the acceleration, and hence the force, was assumed constant. If F be variable, we have in the limit as $t' - t$ approaches zero:

$$F = \frac{d(mv)}{dt} = m\frac{dv}{dt}. \tag{6}$$

Instead of defining force as the product of mass and acceleration, we may therefore define it as the *rate of change of momentum with the time.*

256. Integrating equation (6), we find

$$\int_t^{t'} Fdt = mv' - mv. \tag{7}$$

The product $F(t' - t)$ *of a constant force into the time* $t' - t$ *during which it acts*, and in the case of a variable force, *the time-integral* $\int_t^{t'} F dt$, is called the **impulse** of the force during this time.

It appears from the equations (4) and (7) that *the impulse of a force during a given time is equal to the change of momentum during that time.*

257. The idea of force is no doubt primarily derived from the sensation produced in a person by the exertion of his "muscular force." Like the sensations of light, sound, heat, etc., the sensation of exerting force is capable, in a rough way, of measurement. But the physiological and psychological phenomena attending the exertion of muscular force when analyzed more carefully are very complicated.

In popular language the term "force" is applied in a great variety of meanings. For scientific purposes it is of course necessary to attach a single definite meaning to it.

It is customary in physics to speak of force as *producing* or *generating velocity*, and to define force as the *cause of acceleration*. Thus observation shows that the velocity of a falling body increases during the fall; the cause of the observed change in the velocity, *i. e.* of the acceleration, is called the force of attraction, and is supposed to be exerted by the earth. Again, a body falling in the air, or in some other medium, is observed to increase its velocity less rapidly than a body falling *in vacuo;* a force of resistance is therefore ascribed to the medium as the cause of this change. In a similar way we speak of the expansive force of steam, of electric and magnetic forces, etc., because it is convenient to think of such agencies as producing changes of velocity.

Now, any change in the velocity v of a body of given mass m implies a change in its momentum mv; and it is this change of momentum, or rather the rate at which the momentum changes with the time, which is of prime importance in all the applications of mechanics. It is therefore convenient to have a special name for this rate of change of momentum, and that is what is called *force* in mechanics.

Thus, in using this term "force," it is not intended to assert anything as to the objective reality or actual nature of force and matter in the popular acceptation of these terms. With the ultimate causes science has nothing to do; it can observe only the phenomena themselves.

258. The definition of force (Art. 255) as the product of mass and acceleration gives the *dimensions* of force as

$$F = MJ = MLT^{-2}.$$

The *unit of force* is therefore the force of a particle of unit mass moving with unit acceleration.

Hence, in the C.G.S. system, it is the force of a particle of 1 gram moving with an acceleration of 1 cm./sec^2. This unit force is called a **dyne.**

The definition is sometimes expressed in a slightly different form.* We may say the dyne is the force which, acting on a gram uniformly for one second, would generate in it a velocity of 1 cm./sec.; or would give it the C.G.S. unit of acceleration; or it is the force which, acting on *any* mass uniformly for one second, would produce in it the C.G.S. unit of momentum.

That these various statements mean the same thing follows from the fundamental formulæ $F = mj$, $v = jt$, if F, m, t, v, j be expressed in C.G.S. units.

In the F.P.S. system, the unit of force is the force of a mass of 1 lb. moving with an acceleration of 1 ft./sec^2. It is called the **poundal.**

259. The dyne and the poundal are called the **absolute,** or scientific, units of force.

To find the relation between these two units, let x be the number of dynes in the poundal; then we have

$$x \cdot \frac{\text{gm. cm.}}{\text{sec.}^2} = 1 \cdot \frac{\text{lb. ft.}}{\text{sec.}^2};$$

hence, just as in Art. 253,

$$x = 13\,825.3;$$

* J. D. EVERETT, *C.G.S. system of units*, 1902, pp. 23, 24.

i. e. 1 poundal = 13 825.3 dynes, and 1 dyne = 0.000 072 331 poundals.

260. Another system of measuring force, the so-called **gravitation** (or engineering) system, is in very common use, and must be explained here.

Among the forces of nature the most common is the *force of gravity*, or the *weight*, *i. e.* the force with which any physical body is attracted by the earth. As we have convenient and accurate appliances for comparing the weights of different bodies at the same place, the idea suggests itself of selecting as unit force the weight of a certain standard mass.

In the metric gravitation system the *weight of a kilogram* has been selected as unit force; in the British gravitation system, the *weight of a pound* is the unit force.

261. There are two serious objections to the gravitation system of measuring force, one of a practical nature, the other theoretical. The former is that the words "kilogram" and "pound" are thus used in two different meanings: sometimes, and more correctly, as denoting a mass, sometimes as denoting a force. Wherever an ambiguity might arise from this double use, the word "mass" or "weight" must be added. The word "weight" itself which in scientific language denotes a force is often used by the engineer for "mass."

The other objection is more serious. The weight of a body, and hence the gravitation unit of force, is not a constant quantity; it changes from place to place as it depends on the value of g, the acceleration of gravity.

For, the weight W of any mass m being the force with which this mass is attracted by the earth, we have

$$W = mg,$$

where g is the acceleration produced by the earth's attraction. Now it is known from experiment that this acceleration varies from place to place; according to the law of gravitation, it is inversely proportional to the square of the distance from the center of the earth.

The weight of a body is therefore a meaningless term unless the place be specified where the body is situated, and the value of g at that place be given. It is true, however, that the value of g for different points on the earth's surface varies but little, so that for most practical purposes the gravitation system is accurate enough.

262. In the *absolute* system, then, the three fundamental units of mechanics are those of time, length, and mass, while force, as the product of mass and acceleration, is measured by a derived unit (see Arts. 203 and 258). On the other hand in the (older) *gravitation* system the fundamental units are those of time, length, and force, while mass is defined as the quotient obtained by dividing the force of gravity, or weight, of a body by the acceleration of gravity.

The general equations of mechanics are of course independent of the system of measurement adopted; they hold as well in the gravitation as in the absolute system. In applying them to any particular case, that is, in substituting numerical values for the general symbols, the following obvious rule must be observed: *express all the quantities involved in units of one and the same system.* See Art. 447.

In statics where we are mainly concerned with the *ratios* of forces and not with their absolute values it rarely makes any difference which system is used provided all forces are expressed in the same unit. And as elementary statics deals largely with the effects of gravity, the gravitation system will often be used in the present work in view of the practical applications.

263. The numerical relation between the absolute and gravitation measures of force is expressed by the equations

$$1 \text{ kilogram (force)} = 1000\, g \text{ dynes},$$

$$1 \text{ pound (force)} = g \text{ poundals},$$

where g is about 981 in metric units, and about 32.2 in British units. In most cases the more convenient values 980 and 32 may be used.

264. Exercises.

(1) What is the exact meaning of "a force of 10 tons"? Express this force in poundals and in dynes.

(2) Reduce 2 000 000 dynes to British gravitation measure.

(3) Express a pressure of 2 lbs. per square inch in kilograms per square centimeter.

(4) Prove that a poundal is very nearly half an ounce, and a dyne a little over a milligram, in gravitation measure.

(5) The numerical value of a force being 100 in (absolute) F.P.S. units, find its value for the yard as unit of length, the ton as unit of mass, and the minute as unit of time (see Art. 259).

265. *The quantity* $\frac{1}{2}mv^2$, *i. e., half the product of the mass of a particle into the square of its velocity*, is called the **kinetic energy** of the particle.

Let us consider again a particle of constant mass m moving with a constant acceleration, and hence with a constant force; let v be the velocity, s the space described, at the time t; v', s' the corresponding values at the time t'. Then the last of the three fundamental equations (see Arts. 251 and 255) gives

$$\tfrac{1}{2}mv'^2 - \tfrac{1}{2}mv^2 = F(s' - s); \tag{8}$$

hence

$$F = \frac{\frac{1}{2}mv'^2 - \frac{1}{2}mv^2}{s' - s}. \tag{9}$$

If F be variable, we have in the limit

$$F = \frac{d(\frac{1}{2}mv^2)}{ds} = mv\frac{dv}{ds}. \tag{10}$$

Force can therefore be defined as *the rate at which the kinetic energy changes with the space.* (Compare the end of Art. 255.)

266. Integrating the last equation (10), we find

$$\int_s^{s'} Fds = \tfrac{1}{2}mv'^2 - \tfrac{1}{2}mv.^2 \tag{11}$$

The product $F(s' - s)$ *of a constant force* F *into the space* $s' - s$ *described in the direction of the force*, and in the case of a variable force, *the space-integral* $\int_s^{s'} Fds$, is called the **work** of the force for this space.

The equations (8) and (11) show that *the work of a force is equal to the corresponding change of the kinetic energy.*

We have here assumed that the force acts in the direction of motion of the particle. A more general definition of work including the above as a special case will be given later (Art. 403).

The ideas of energy and work have attained the highest importance in mechanics and mathematical physics within comparatively recent times. Their full discussion belongs to Kinetics (see Part III).

267. According to their definitions, both *momentum* (Art. 252) and *force* (Art. 255) may be regarded mathematically as mere numerical multiples of velocity and acceleration, respectively. They are therefore so-called vector-quantities; *i. e.* a momentum as well as a force can be represented geometrically by a segment of a straight line of definite length, direction, and sense. Moreover, as they are referred to a particular point, viz., to the point whose mass is m, the line representing a momentum or a force must be drawn through this point; the line has therefore not only direction, but also position; *i. e., a momentum as well as a force is represented geometrically by a* **rotor** (compare Art. 181).

It follows that concurrent forces, for instance, can be compounded by geometrical addition, as will be explained more fully in Chapter IV.

On the other hand, kinetic energy and work are not vector-quantities.

268. The ideas of momentum, force, energy, work, with the fundamental equations connecting them, as given in the preceding articles, form the groundwork of the whole science of theoretical dynamics. The application of this science to the interpretation of natural phenomena gives results in exact agreement with observation and experiment. It is therefore important to inquire what are the physical assumptions and experimental data on which this application of dynamics is based.

These assumptions were formulated with remarkable clearness by Sir Isaac Newton in his *Philosophiæ naturalis principia mathematica*, first published in 1687, and have since been known as **Newton's laws**

of motion. As these three *axiomata sive leges motus*, as Newton terms them are very often referred to and, at least by English writers on dynamics, are usually laid down as the foundation of the science, they are given here in a literal translation :

I. Every body persists in its state of rest or of uniform motion along a straight line, except in so far as it is compelled by impressed (*i. e.*, external) forces to change that state.

II. Change of motion is proportional to the impressed moving force and takes place along the straight line in which that force acts.

III. To every action there is an equal and contrary reaction ; or, the mutual actions of two bodies on one another are always equal and directed in contrary senses.

269. Some explanation is necessary to understand correctly the meaning of these laws. Indeed, Newton's laws should not be studied by themselves ; they become intelligible only if taken in connection with the definitions preceding them in the *Principia*, and with the explanations and corollaries that Newton himself has appended to them.

The word "body" must be taken to mean particle ; the word "motion" in the second law means what is now called momentum.

All three laws imply the idea of *force as the cause of any change of momentum* in a particle.

270. With this definition of force the first law, at least in the ordinary form of statement, for a single particle, merely states that where there is no cause there is no effect. While this law may appear superfluous to us, it was not so in the time of Newton. Kepler and Galileo, less than a century before Newton, were the first to insist more or less clearly on this so-called **law of inertia**, viz. that there is no intrinsic power or tendency in moving matter to come to rest or to change its motion in any way.

271. The second law gives as the measure of a constant force the amount of momentum generated in a given time (see Art. 255) ; it can be called the **law of force.** If force be defined as the cause of any change of momentum, the second law follows naturally by assuming, as is usually done, that the effect is proportional to the cause.

The first two laws may thus be regarded from the mathematical point of view as nothing but a definition of force ; but they are certainly meant to emphasize the physical fact that the assumed definition of force is not arbitrary, but based on the characteristics of motion as observed in nature.

In the corollaries to his laws Newton tries to show how the composition and resolution of forces by the parallelogram rule follows from his definition. In deriving this result he tacitly assumes that the action of any force on a particle takes place independently of the action of any other forces that may be acting on the particle at the same time, a principle that would seem to deserve explicit statement. Some writers on mechanics, in particular French authors, prefer to replace Newton's second law by this *principle of the independence of the action of forces*.

272. The third law expresses the physical fact that in nature all forces occur in pairs of equal and opposite forces. In modern phraseology, two such equal and opposite forces in the same line are said to constitute a *stress*. Newton's third law is therefore called the **law of stress.**

This law, which was first clearly conceived in Newton's time, involves what may be regarded as the second fundamental property of matter or mass (the first being its indestructibility) ; viz. that *any two particles of matter determine in each other oppositely directed accelerations along the line joining them.*

The historical development of the fundamental ideas of mechanics is discussed in a very instructive manner by E. MACH, in his *Science of mechanics, a critical and historical account of its development*, translated by T. J. McCormack, 2d edition, Chicago, Open Court Publishing Co., 1900.

CHAPTER IV.

STATICS.

I. *Forces acting on the same particle.*

273. When a particle has two equal and opposite accelerations j, $-j$, its motion will not be changed. The same result must follow when a particle is acted on by two equal and opposite forces $F = mj$, $F' = -mj$. Their combined effect on the particle is *nil*, so that the particle, if originally at rest, will remain at rest; if originally moving with constant velocity in a straight line, it will continue to do so; and if originally moving under the action of any other forces in any way whatever, the introduction of the two equal and opposite forces will have no effect on its motion.

We say that two equal and opposite forces acting on a particle *balance*, or *are equivalent to zero*, or *are in equilibrium*. If no other forces act on the particle, the particle itself is said to be in equilibrium. It must be kept in mind that equilibrium is not synonymous with rest.

274. Let us next consider *any* two forces F_1, F_2 acting simultaneously on the same particle of mass m, and let j_1, j_2 be the accelerations produced by these forces so that

$$F_1 = mj_1, \quad F_2 = mj_2.$$

The resultant acceleration of the particle is found by geometrically adding the vectors j_1, j_2; let j be their geometric sum. Then the force

$$R = mj$$

producing the resultant acceleration is called the **resultant** of the forces F_1, F_2; these, or any other two or more forces having the same resultant R, are called the **components** of R.

Instead of compounding the accelerations and multiplying their resultant by m, we can evidently compound directly the vectors F_1, F_2 representing the forces (see Art. 267), the geometric sum of these vectors being the resultant force.

275. Let P, Q (Fig. 74) be two forces acting on the particle m at O, and let θ be the angle between their directions (taken with the proper sense as indicated by the arrow-heads).

By Art. 274, the resultant R of P and Q is found in magnitude, direction, and sense as the diagonal of the parallelogram con-

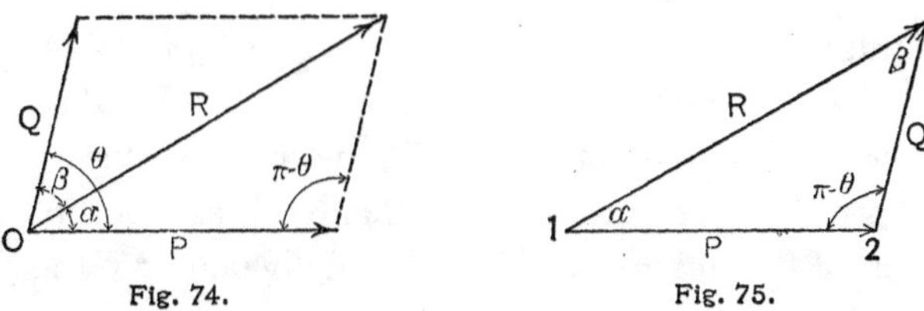

Fig. 74. Fig. 75.

structed on P and Q as adjacent sides. The figure gives for the *magnitude* of the resultant

$$(1) \qquad R = \sqrt{P^2 + Q^2 + 2P\cdot Q\cos\theta},$$

while its *direction* is determined by the angles α, β it makes with P, Q:

$$\frac{P}{\sin\beta} = \frac{Q}{\sin\alpha} = \frac{R}{\sin(\alpha+\beta)}, \qquad \alpha + \beta = \theta.$$

This proposition is known as the **parallelogram of forces.**

If the forces P, Q act on a particle *along the same line*, we have $\theta = 0$ or $\theta = 180°$ according as the forces are of equal or opposite sense; hence $R = P + Q$ in the former case, $R = P - Q$ in the latter case.

Graphically the resultant can also be found from the *triangle of forces* 123 (Fig. 75); that is, by drawing 12 geometrically equal to P (*i. e.* equal and parallel to, and of the same sense with, P), 23 geometrically equal to Q; then 13 represents R in magnitude, direction, and sense.

276. Conversely, any force R can be resolved into two components along any two lines passing through the particle on which R acts; the point O at which this particle is situated is called the *point of application* of R, as well as of the components.

To resolve R along two lines making angles α, β with it (Fig. 74) we have only to draw parallels to these lines through the extremities of R; these parallels cut off the components P, Q on the lines. But it is often convenient to construct the components in a separate diagram (Fig. 75), by making 13 geometrically equal to R and drawing through 1 and 3 parallels to the given lines; 12, 23 are the components P, Q.

277. Let any number n of forces $F_1, F_2, \dots F_n$ be applied at the same point O, *i. e.* act on the same particle at O. By Art. 275, we can find the resultant R_1 of F_1 and F_2, next the resultant R_2 of R_1 and F_3, then the resultant R_3 of R_2 and F_4, and so on. The resultant R of R_{n-2} and F_n is evidently equivalent to the whole system $F_1, F_2, F_3, \dots F_n$, and is called its **resultant.** It thus appears that *a system consisting of any number of forces acting on the same particle is equivalent to a single resultant.*

It may of course happen that this resultant is zero. In this case, the system is said to be **in equilibrium.** *The condition of equilibrium of a system of forces acting on the same particle* is therefore:

$$R = 0.$$

278. In practice, the process of finding the resultant indicated in Art. 277 is inconvenient when the number of forces is large. If the forces are given graphically, by their vectors, we have only to add these vectors geometrically; and this can best be done in a separate diagram, called the **force polygon,** or **stress diagram.** Thus, in Fig. 76, 1 2 is drawn equal and parallel to F_1, 2 3 equal and parallel to F_2, 3 4 to F_3, 4 5 to F_4, 5 6 to F_5. The closing line of the force polygon, viz. 1 6 in the figure, is equal and parallel to the resultant R, which is therefore obtained by drawing through the point of application of the forces a line equal and parallel to 1 6.

The graphical condition of equilibrium consists in the closing of the force polygon, that is, in the coincidence of its terminal point 6 with its initial point 1.

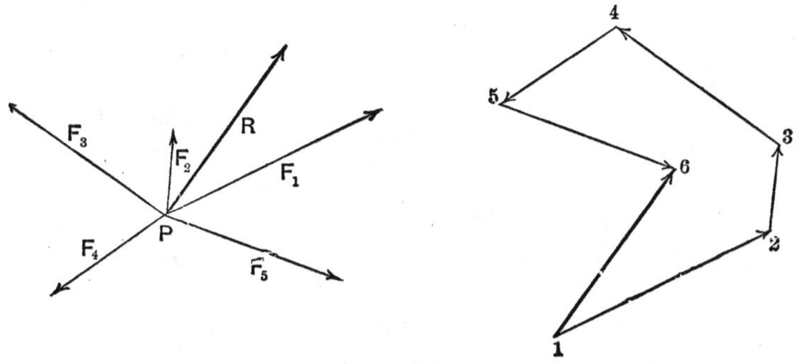

Fig. 76.

279. Analytically, a systematic solution is obtained by resolving each force F into three components X, Y, Z, along three rectangular axes passing through the particle, or point of application of the given forces. All components lying in the direction of the same axis can then be added algebraically, and the whole system of forces is found to be equivalent to three rectangular forces ΣX, ΣY, ΣZ, which, by the parallelogram law, can be combined into a single resultant

$$R = \sqrt{(\Sigma X)^2 + (\Sigma Y)^2 + (\Sigma Z)^2}.$$

The angles α, β, γ made by this resultant with the axes are given by the relations

$$\frac{\cos\alpha}{\Sigma X} = \frac{\cos\beta}{\Sigma Y} = \frac{\cos\gamma}{\Sigma Z} = \frac{1}{R}.$$

280. If the forces all lie in the same plane, only two axes are required and we have

$$R = \sqrt{(\Sigma X)^2 + (\Sigma Y)^2}, \quad \tan\theta = \frac{\Sigma Y}{\Sigma X},$$

where θ is the angle between the axis of X and R.

281. The condition of equilibrium (Art. 277) $R = 0$ becomes, by Art. 279, $(\Sigma X)^2 + (\Sigma Y)^2 + (\Sigma Z)^2 = 0$. As all terms in the left-hand member are positive, their sum can vanish only when each term is zero. *The analytical conditions of the equilibrium of any number of concurrent forces* are therefore:

$$\Sigma X = 0, \quad \Sigma Y = 0, \quad \Sigma Z = 0.$$

282. Exercises.

(1) Find the resultant of two equal forces acting at right angles to each other.

(2) Show that the resultant R of two equal forces P including an angle θ is $R = 2P \cos \frac{1}{2}\theta$.

(3) If the resultant of two equal forces be equal to P, what is the angle between the components?

(4) Find the magnitude and direction of the resultant of two forces of 100 and 150 lbs., including an angle of 60°.

(5) Resolve a force of 20 lbs. into two components making angles of 45° and 30° with the given force: (*a*) analytically; (*b*) graphically.

(6) Find the rectangular components of a force P if one of the components is to make an angle of 30° with P.

(7) The resultant R, one of the components P, and the angle between the two components, $\theta = 60°$, being given, find the other component Q.

(8) A particle is acted on by two forces P, Q lying in the same vertical plane and inclined to the horizon at angles p, q. Find their resultant in magnitude and direction, if $P = 527$ lbs., $Q = 272$ lbs., $p = 127° \, 52'$, $q = 32° \, 13'$.

(9) Two forces acting on a point are represented in magnitude and direction by the tangent and normal of a parabola passing through the point. Find their resultant, and show that it passes through the focus of the parabola.

(10) The magnitudes of two forces acting on a point are as 2 to 3. If their resultant be equal to their arithmetic mean, what is the angle between the forces?

(11) What is the angle between a force of 1 ton and a force of $\sqrt{3}$ tons if their resultant is 2 tons?

(12) Six forces, $F_1 \cdots F_6$ in the same plane, each of 10 lbs., are applied to a particle so that $\measuredangle F_1F_2 = F_2F_3 = F_3F_4 = F_4F_5 = F_5F_6 =$ 45°. Find the resultant in magnitude and direction.

(13) Six forces of 1, 2, 3, 4, 5, 6 lbs., respectively, act in the same plane on the same point, making angles of 60° with each other. Find their resultant in magnitude and direction: (*a*) graphically; (*b*) analytically.

(14) One of the vertices A of a regular hexagon is acted upon by 5 forces represented in magnitude and direction by the lines drawn from A to the other vertices of the hexagon. Find their resultant.

(15) Find the resultant of three equal forces P acting on a point, the angle between the first and second as well as that between the second and third being 45°.

(16) A mass m rests on a plane inclined to the horizon at an angle θ; it is kept in equilibrium: (*a*) by a force P_1 parallel to the plane; (*b*) by a horizontal force P_2; (*c*) by a force P_3 inclined to the horizon at an angle $\theta + \alpha$. Determine in each case the force P and the pressure R on the plane.

(17) Show that the three forces represented by the vectors OA, OB, OC are in equilibrium if O is the centroid of the triangular area ABC.

(18) Show that the three vectors OA, OB, OC have the same resultant as the three vectors OA', OB', OC', if A', B', C' are the middle points of the sides of the triangle ABC.

(19) Show that the resultant of the vectors OA, OB, OC is OO', if O is the center of the circle circumscribed to the triangle ABC and O' the intersection of the altitudes of the same triangle.

(20) Show that a force whose components P_1, P_2 include an angle θ can be resolved into two rectangular components $(P_1 + P_2) \cos\frac{1}{2}\theta$, $(P_1 - P_2) \sin\frac{1}{2}\theta$.

II. *Concurrent Forces; Moments.*

283. The range of applicability of the principles of the last section is considerably increased if we now introduce the assumption always made in the statics of rigid bodies that *the effect of a force is not changed if the force be transferred to any other position on its line of action.* This assumption is also expressed by saying that the force is supposed to act on a *rigid body*, or at least that its line of action is rigid, the distance of any two points of this line remaining unchanged. According to this principle the effect of the force F at P (Fig. 77) is the same as that of a force of equal magnitude, direction, and sense at P'; and *any two equal and opposite forces in the same line*, such as F at P and $-F$ at P' in Fig. 77, *are in equilibrium;* provided always that P and P' can be regarded as belonging to the same rigid body.

It follows that the only *essential* characteristics of a force acting on a rigid body are (a) its *magnitude* or intensity, (b) its *line* of action, (c) its *sense*, while the point of application may be taken anywhere on the line of action.

Strictly speaking, two forces should be called equal only when they agree in these essential characteristics. It is, however, cus-

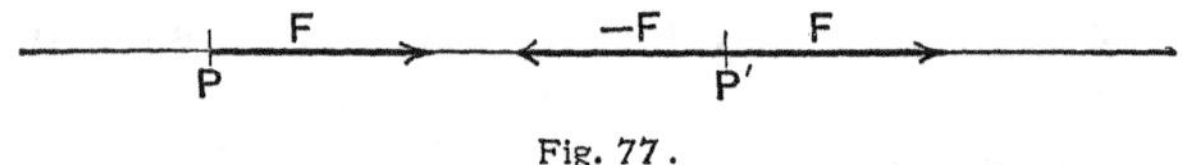

Fig. 77.

tomary to call two forces *equal* when they have merely equal magnitude; we shall call them *geometrically equal* when they agree in all three characteristics.

284. Forces whose lines of action all pass through the same point are called **concurrent.** As such forces can, by the principle of the last article, be transferred to the common point as point of application, all the results of the last section (Arts. 275–281) apply to concurrent forces. Thus, *any number of concurrent forces can be reduced to a single resultant* R, *and the condition of equilibrium of concurrent forces is* $R = 0$.

285. The projection of a closed polygon on any line being evidently zero, and the resultant being by definition the geometric sum of its components, it follows that *the projection of the resultant on any line equals the algebraic sum of the projections of its components.* This proposition is sometimes expressed in the following form: the resolved part of the resultant in any direction is equal to the algebraic sum of the resolved parts of the components.

Let l be the line on which we project (Fig. 78), and let (l, R) (l, P), (l, Q) denote the angles it makes with the resultant R and the components P, Q, respectively; then

$$R \cos(l, R) = P \cos(l, P) + Q \cos(l, Q).$$

286. Varignon's Theorem. Multiplying the last equation by any length $OS = s$ taken through the initial point O of R and at right angles to l, we obtain

$$R \cdot s \cos(l, R) = P \cdot s \cos(l, P) + Q \cdot s \cos(l, Q),$$

or since $s \cos(l, R) = r$, $s \cos(l, P) = p$, $s \cos(l, Q) = q$, where r, p, q are the perpendiculars let fall from S on R, P, Q, respectively,

$$Rr = Pp + Qq.$$

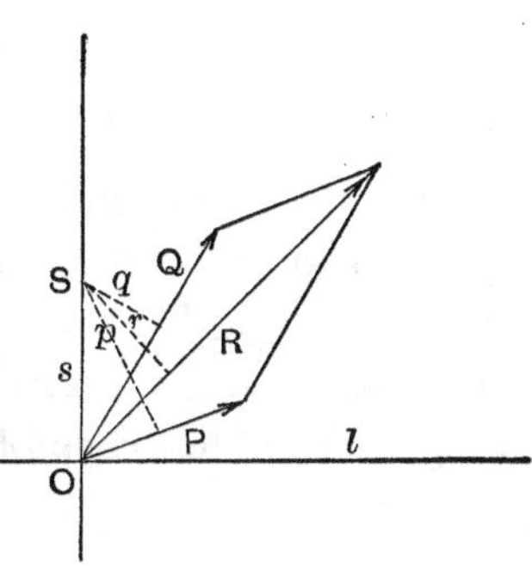

Fig. 78.

In this form the proposition is independent of the direction of the line l and holds for any point S in the plane of the parallelogram.

287. Moment of a Force. The product of a force into its perpendicular distance from a point is called the **moment** of the force about the point. The product is taken with the positive or negative sign according as the force tends to turn counter-clockwise or clockwise about the point.

The proposition of Art. 286, $Pp + Qq = Rr$, can now be stated in the following form: *the algebraic sum of the moments of any two intersecting forces about any point in their plane is equal to the moment of their resultant about the same point.*

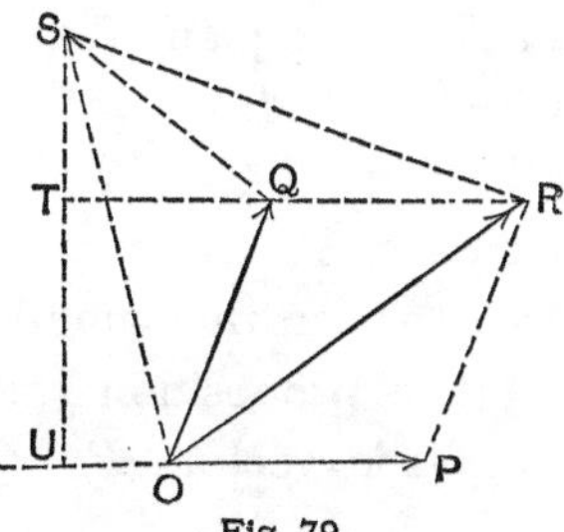

Fig. 79.

288. The product Rr represents twice the area of the triangle having R for its base and S for its vertex; Pp, Qq can be interpreted similarly. This remark leads to another simple proof of Varignon's theorem, which may serve to make its meaning better understood. With the notation of Fig. 79 we have

$$SOR = SOQ + SQR + QOR,$$

or

$$R \cdot r = Q \cdot q + P \cdot ST + P \cdot TU;$$

or since $ST + TU = SU = p$,

$$Rr = Qq + Pp.$$

If the point S be taken on the resultant R, we have $r = 0$, hence $Pp = -Qq$; *i. e. the sum of the moments of two forces about any point on their resultant is zero.*

289. The moment Fp of a force F about any point A (p being the perpendicular distance of A from F) can be represented by a vector of length $AB = Fp$, drawn through A at right angles to the plane (A, F), determined by the point A and the rotor F (Fig. 80). The sense of this vector, which is, strictly speaking, a

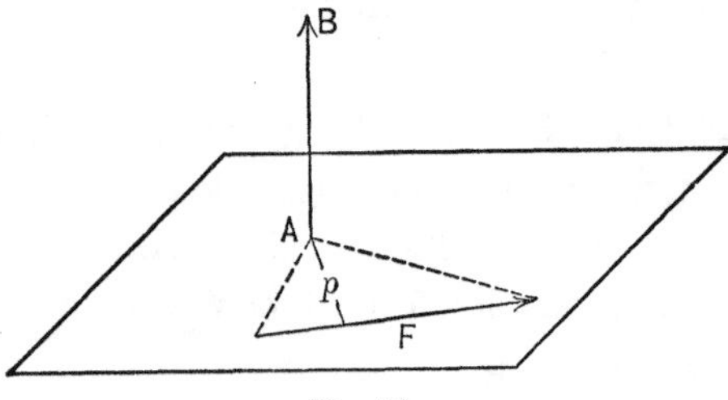

Fig. 80.

rotor since it must be drawn through the point A, is taken so that for a person looking from the arrowhead B on the plane (A, F), the force F tends to turn counter-clockwise about A.

In the proposition of Arts. 286–287, the forces P, Q, R all lie in the same plane with the point of reference S; the moments Pp, Qq, Rr are therefore represented by vectors on the same line through S, at right angles to the plane. The proposition states that the vector Rr is the algebraic sum of the vectors Pp, Qq.

290. As the projection on any line of any closed polygon, even when its sides do not lie in the same plane, is equal to zero, it follows that the proposition of Art. 285 holds for any number of concurrent forces.

Varignon's theorem (Arts. 286, 287), that the moment of the resultant about any point is equal to the sum of the moments of the components, can be shown to hold for any number of concurrent component forces, whether in the same plane or not; but in the latter case, the sum must be understood as a *geometric* sum; that is, each moment being represented by its vector, the geometric sum of the moments of the components is geometrically equal to the vector that represents the moment of the resultant. The proof, while elementary, is somewhat long;* it is omitted here as the general proposition will not be used in this work.

291. The forces of nature receive various special names according to the circumstances under which they occur. Thus the **weight** of a mass has already been defined (Art. 260), as the force with which the mass is attracted by the mass of the earth.

When a string carrying a mass at one end is suspended from a fixed point, it will be *stretched*, *i. e.* subjected to a certain tension. This means that if the string were cut it would require the application of a force along the line of the string to keep the weight in equilibrium. This force, which may thus serve to replace the action of the string, is called its **tension.**

When the surfaces of two physical bodies A, B are in contact, a **pressure** may exist between them; that is, if one of the bodies, say B, be removed, it may require the introduction of a force to keep A in the same state of rest or motion that it had before the removal of B. This force, which will obviously act along the common normal of the surfaces at the point of contact, is called the **resistance**, or **reaction,** of B, and a force equal and opposite to it is called the **pressure** exerted by A on B.

292. Exercises.

(1) A weight W is suspended from two fixed points A, B by means of a string ACB, C being the point of the string where the weight W is attached. If AC, BC be inclined to the vertical at angles α, β, find the tensions in AC, BC: (a) analytically; (b) graphically.

(2) Let $AB = c$ (Fig. 81) be the vertical post, $AC = b$ the jib, of a crane, the ends BC being connected by a chain of length a. If a

* See P. Appell, *Cours de mécanique à l'usage des candidats à l'Ecole centrale des arts et manufactures*, Paris, Gauthier-Villars, 1902, pp. 157–160.

weight W be suspended from C, find the tension T produced by it in the chain and the thrust P in AC.

(3) Let AC be hinged at A (Fig. 81) so as to turn freely in a vertical plane, and let the chain pass over a pulley at C and carry the weight W. In what position of AC will there be equilibrium?

(4) Find the resultant R of three concurrent forces A, B, C lying in the same plane and making angles α, β, γ with each other.

(5) Prove that the moments of the two components of a force about any point on the line of the force are equal and opposite.

(6) Prove that the moment of the resultant of any number of concurrent forces lying in the same plane about any point in this plane is equal to the sum of the moments of the forces about the same point.

(7) By means of Ex. (6), express the conditions of equilibrium of any number of concurrent forces in the same plane.

(8) When three concurrent forces are in equilibrium, show that they are proportional and parallel to the sides of a triangle.

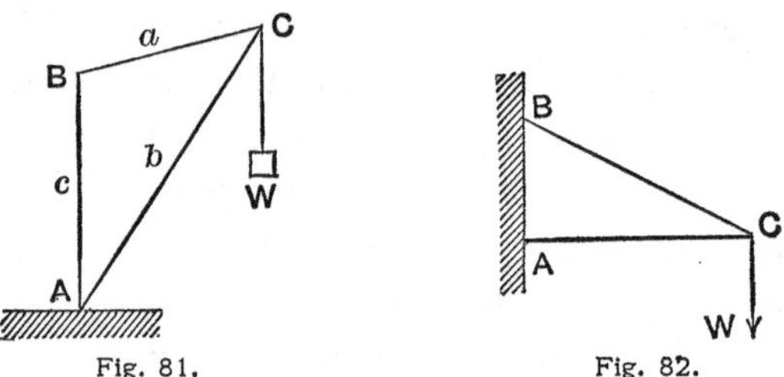

Fig. 81. Fig. 82.

(9) When any number of concurrent forces are in equilibrium, show that any one of them reversed is the resultant of all the others.

(10) A weightless rod AC (Fig. 82), hinged at one end A so as to be free to turn in a vertical plane, is held in a horizontal position by means of the chain BC. If a weight W be suspended at C, find the thrust P in AC and the tension T of the chain. Assume $AC = 8$ ft., $AB = 6$ ft.

(11) In Ex. (10), suppose the rod AC, instead of being hinged at A, to be set firmly into the wall in a horizontal position; and let the chain fastened at B run at C over a smooth pulley and carry the weight W. Find the tension of the chain and the magnitude and direction of the pressure on the pulley at C.

(12) In "tacking against the wind," let W be the force of the wind; α, β the angles made by the axis of the boat with the direction

in which the wind blows, and with the sail, respectively. Determine the force that drives the boat forward and find for what position of the sail it is greatest.

(13) A cylinder of weight W rests on two inclined planes whose intersection is horizontal and parallel to the axis of the cylinder. Find the pressures on these planes.

(14) Find the tensions in the string $ABCD$, fixed at A and D, and carrying equal weights W at B and C, if $AD = c$ is horizontal, $AB = BC = CD$, and the length of the string is $3l$.

(15) Let R be the effective piston pressure of a steam engine and φ the angle between the direction of motion of the piston and the connecting rod at any moment; show that the thrust in the connecting rod is $R \sec\varphi$ and the pressure on the guide-bars $R \tan\varphi$. For what position of the crank is the pressure on the guides greatest? What is the greatest pressure when the connecting rod is four times as long as the crank?

(16) The force $R \sec\varphi$ transmitted by the connecting rod is resolved at the crank pin tangentially and radially. Determine the tangential component which is the effective turning force and show that, if R be represented by the length of the crank, this tangential component is represented by the intercept made by the connecting rod on that radius of the crank circle which is perpendicular to the motion of the piston. Show also that when the connecting rod is long the tangential force is approximately $= R\left(\sin\theta + \frac{1}{2m}\sin 2\theta\right)$, where θ is the crank angle and m the ratio l/a of the length of the connecting rod to that of the crank. Compare Art. 103.

(17) Show that the force $R \sec\varphi$ will not differ more than about 3 per cent. from its minimum value R if $m = 4$. Regarding $R \sec\varphi$ as constant and representing it by the length of the crank, show that its tangential component is represented by the perpendicular let fall from the center of the crank circle on the connecting rod.

(18) In the toggle-joint press two equal rods CA, CB are hinged at C; a force F bisecting the angle 2α between the rods forces the ends A, B apart. If A be fixed, find the pressure exerted at B at right angles to F if $F = 100$ lbs. and $\alpha = 15°, 30°, 45°, 60°, 70°, 75°, 80°, 85°, 90°$.

(19) A stone weighing 800 lbs. hangs from a derrick by a chain 15

ft. long. If pulled 5 ft. away from the vertical by means of a horizontal rope attached to it, what are the tensions of the chain and the rope? What if pulled 9 ft. away?

(20) A rope 16 ft. long has its ends fastened to two points at the same height above the ground; a weight W is suspended from the rope by means of a ring free to slide along the rope. Find the tension of the rope.

(21) A string with equal weights W attached to its ends is hung over two smooth pegs A, B fixed in a vertical wall. Find the pressure on the pegs: (*a*) when the line AB is horizontal; (*b*) when it is inclined to the horizon at an angle θ. The weight of the string, its extensibility and stiffness, and the friction on the pegs are neglected in this problem as well as in those immediately following. For what position of the line AB are the pressures equal?

(22) The string being hung over three pegs A, B, C, determine graphically the pressures on the pegs. Let the vertical line through B lie between the vertical lines drawn through A and C; there will be a pressure on B only if B lies above the line CA. If B lies below AC, the pressure may be distributed over the three pegs by passing the string around the peg B from below.

(23) In Ex. (22), let AC be horizontal, and let α, β, γ denote the angles of the triangle ABC. What are the pressures on the pegs? What must be the position of B to make the pressures on the three pegs equal: (*a*) when B lies above AC? (*b*) when B lies below AC?

(24) If the string with the equal weights W attached to its ends be strung over any number of pegs, the pressures on the pegs are readily determined, either graphically or analytically, in magnitude and direction; these pressures depend only on the value of W and on the angles between the successive sides of the polygon formed by the string, but not on the distances between the pegs.

(25) Suppose the string be closed, its ends being fastened together. Let this string be hung over three pegs A, B, C forming an isosceles triangle in a vertical plane with its base AC horizontal, and let a weight W be suspended from the lowest point D of the string. If $AC = 4$ ft., $AB = BC = 2.5$ ft., and the length of the string $2l = 14$ ft., find the tension of the string and the pressures on the pegs.

(26) If, in Ex. (25), the triangles ABC and ADC be equilateral, what would be the tension and the pressures on the pegs?

(27) In Ex. (25), the triangles ABC and ADC being isosceles and their common base AC horizontal, what must be the relation between the angles 2β at B and 2δ at D to make the pressures on the three pegs A, B, C equal? The pressures being made equal, what angle gives the least pressure?

III. *Parallel Forces.*

293. The methods of the previous section make use of the point of concurrence of the forces and can therefore not be applied directly to parallel forces. But, *for parallel forces acting on a rigid body*, it is in general possible, with the aid of the principle of Art. 283, to find a single force which, acting on the same rigid body, will produce the same effect as the given parallel forces; and this single force is called the **resultant** of the parallel forces. We begin by showing how the resultant of *two* parallel forces can be found.

294. Resultant of two parallel forces. In the plane of the given parallel forces P, Q, resolve P, at any point p of its line of action,

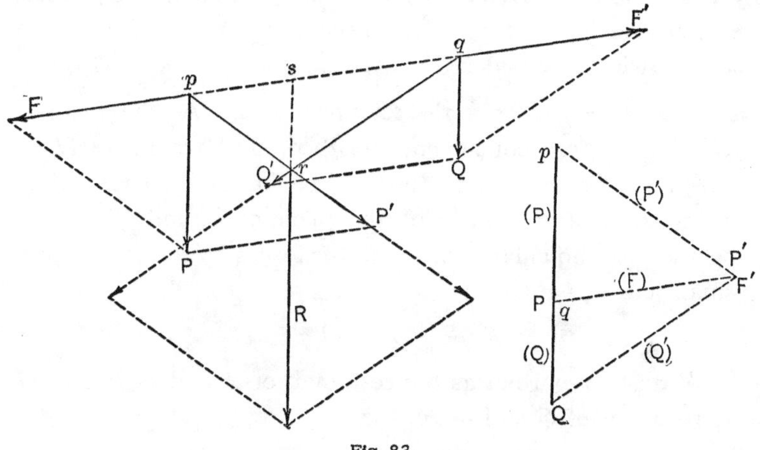

Fig. 83.

into any two components, say P' and F (Fig. 83); and at the point q where F meets the line of Q, resolve Q into two components F', Q', selecting for F' a force equal and opposite to, and in the same line with, F. Now, by Art. 283, the two equal and opposite forces F, F' in the same line pq have no effect on the rigid body so that the given forces P, Q are together equivalent to the two components P', Q' alone. The lines of P' and Q' will in general intersect at a point r and these forces can therefore be compounded into a resultant R passing through r.

By putting the triangles $pP'P$ and $qF'Q$ together so that their equal sides PP' and qF' coincide (as is done in Fig. 83, on the right) it appears at once that the resultant of P' and Q', and hence *the resultant R of P and Q, is parallel to P and Q and in magnitnde equal to the algebraic sum of P and Q*:

$$R = P + Q.$$

295. In Fig. 83 the two given parallel forces P, Q were assumed of the same sense. The construction applies, however, equally well to the case when they are of opposite sense. The resultant R will then be found to lie not between P and Q, but outside, on the side of the larger force. The construction fails only when the two given forces are equal and of opposite sense, since then the lines pP' and qQ' become parallel. This exceptional case will be considered in Art. 303.

296. *Varignon's theorem for parallel forces.* As the forces R, P', Q' (Fig. 83) are concurrent the theorem of moments (Arts. 286–287) can be applied to these three forces. Hence, taking moments about any point S of the plane of P' and Q' and denoting the perpendiculars from S to the forces by the corresponding small letters, we have:

$$Rr = P'p' + Q'q'.$$

Now P' can be regarded as the resultant of P and $-F$, and Q' as the resultant of Q and $-F'$; hence

$$P'p' = Pp - Ff, \quad Q'q' = Qq - F'f';$$

substituting these values and remembering that F and F' are equal and opposite and in the same line, we find

$$Rr = Pp + Qq.$$

This equation proves *Varignon's theorem of moments for two parallel forces.*

297. If in particular, the origin of moments be taken at the point s (Fig. 83) where pq is met by R, we have

$$0 = P \cdot ps - Q \cdot sq, \text{ or } \frac{P}{Q} = \frac{sq}{ps}.$$

This means that *the resultant of two parallel forces divides their distance in the inverse ratio of the forces.* As this proposition finds application in the theory of the lever, it is commonly referred to as the **principle of the lever.**

Dropping perpendiculars p, q from any point of the resultant R on the components P, Q, the relation can be expressed in the form

$$Pp = -Qq.$$

This important relation can also be obtained by observing that the triangles prs and $P'pP$, as well as the triangles qrs and $Q'qQ$, are similar, so that

$$\frac{ps}{sr} = \frac{F}{P}, \quad \frac{sq}{sr} = \frac{F}{Q},$$

whence, dividing,

$$\frac{ps}{sq} = \frac{Q}{P}.$$

298. It has been shown that two parallel forces P, Q acting on a rigid body, provided they are not equal and of opposite sense, have a resultant $R = P + Q$, parallel to P and Q, and that its position in the rigid body can be found either analytically from the fact that R divides the distance between P and Q in the inverse ratio of these forces, or graphically by the construction of Art. 294.

This *graphical construction* is best carried out in the following order (Fig. 84). The parallel forces P, Q being given in position, begin by constructing the **force polygon**, or **stress diagram** which here consists merely of a straight line on which the forces $P = 12$,

$Q = 2\ 3$ are laid off to scale; the closing line, 1 3, gives the resultant in magnitude, direction and sense; it only remains to find its position, and for this it suffices to find one point of its line of action.

Now, to resolve P and Q each into two components (as was done in Art. 294) so that one component of P and one of Q are equal and opposite and in the same line, it is only necessary to draw from an arbitrary point O, called the *pole*, the lines $O\ 1$, $O\ 2$, $O\ 3$; then $1\ O$, $O\ 2$ can be regarded as components of $P = 1\ 2$, and $2\ O$,

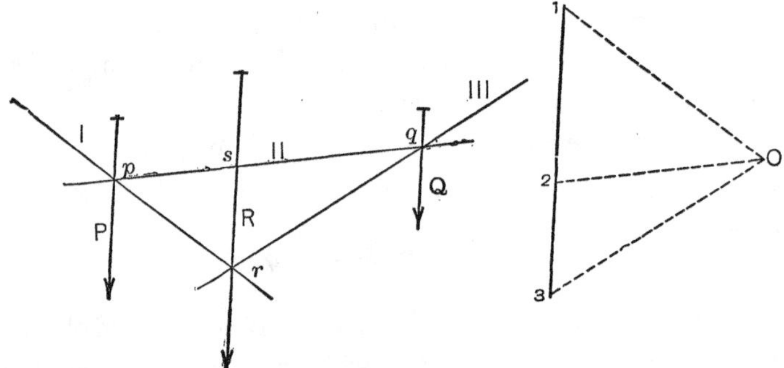

Fig. 84.

$O\ 3$ as components of $Q = 2\ 3$. Next construct the so-called **funicular polygon** (or **moment polygon**) by drawing a line I parallel to $O\ 1$, intersecting P say at p; through p a line II parallel to $O\ 2$ meeting Q say at q; through q a line III parallel to $O\ 3$.

The intersection r of I and III is a point of the resultant R as appears by comparing Figs. 84 and 83; Fig. 84 being the same as Fig. 83, with the superfluous lines left out.

299. Resultant of Any Number of Parallel Forces. The graphical method is readily extended to the general case of any number of parallel forces lying in the same plane. Whatever the number of the forces, the force polygon gives magnitude, direction, and sense of the resultant, which is simply the algebraic sum of the given forces; while the funicular polygon (formed by the lines I, II, III, etc.) gives the position of the resultant by furnishing one of its points, viz. the intersection of the first and last sides of the funicular polygon.

The process will best be understood from the following example.

The horizontal beam AB (Fig. 85) resting freely on the fixed supports A, B carries four weights W_1, W_2, W_3, W_4.

To determine the position of the resultant and the reactions A, B of the supports, construct the *force polygon* by laying off in succession on a vertical line $1\ 2 = W_1$, $2\ 3 = W_2$, $3\ 4 = W_3$, $4\ 5 = W_4$; select any point O as pole and join it to the points 1, 2, 3, 4, 5.

Now we may regard $1\ O$ and $O\ 2$ as components into which W_1 has been resolved; similarly $2\ O$ and $O\ 3$ as components of W_2, $3\ O$ and

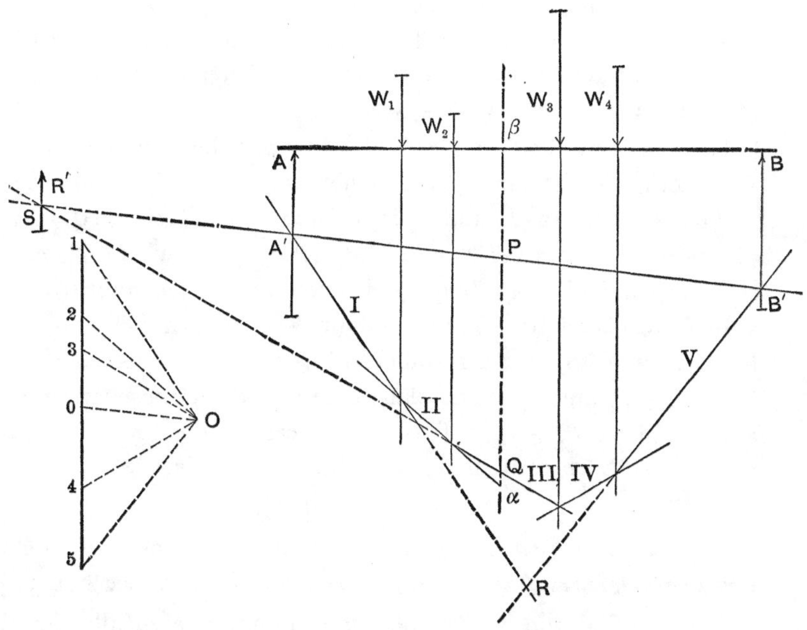

Fig. 85.

$O\ 4$ as components of W_3, and $4\ O$ and $O\ 5$ as components of W_4. This resolution of the weights into components is transferred into the main figure by constructing the *funicular polygon* as follows: through any point A' on the line of the reaction A draw a parallel I to $O\ 1$; through the intersection of I with W_1 draw a line II parallel to $O\ 2$; through the intersection of II with W_2 draw III parallel to $O\ 3$; through the intersection of III with W_3 draw IV parallel to $O\ 4$; and through the intersection of IV with W_4 draw V parallel to O 5; let B' be the intersection of V with the line of the reaction B.

If now each weight be regarded as resolved along the sides adjacent to it in the funicular polygon, since the two components falling on II are equal and opposite, and also those falling on III and IV, the system of weights is reduced to the two components along I and V. The intersection of these lines, *i. e.* of the first and last sides of the funicular polygon, gives a point, R, of the resultant of W_1, W_2, W_3, W_4.

Moreover, if the component on I be resolved along $A'B'$ and the vertical through A', and similarly the component on V along $B'A'$ and the vertical through B', the two components along $A'B'$ will be equal and opposite, each being equal to O 0, the parallel drawn to $A'B'$ in the force polygon. This parallel furnishes, therefore, the magnitudes of the reactions $A = 0\,1$, $B = 5\,0$.

300. Analytically, the resultant of n parallel forces F_1, F_2, $\cdots$ F_n, whether in the same plane or not, can be found as follows:

The resultant of F_1 and F_2 is a force $F_1 + F_2$ situated in the plane (F_1, F_2), so that $F_1 p_1 = F_2 p_2$ (Art. 297), where p_1, p_2 are the (perpendicular or oblique) distances of the resultant from F_1 and F_2, respectively. This resultant $F_1 + F_2$ can now be combined with F_3 to form a resultant $F_1 + F_2 + F_3$, whose distances from $F_1 + F_2$ and F_3 in the plane determined by these two forces are as F_3 is to $F_1 + F_2$. This process can be continued until all forces have been combined; the final resultant is

$$F_1 + F_2 + \cdots + F_n.$$

Any number of parallel forces are, therefore, *in general equivalent to a single resultant equal to their algebraic sum* (see Art. 303).

301. To find the *position* of this resultant analytically, let the points of application of the forces F_1, F_2, $\cdots$ F_n be (x_1, y_1, z_1), (x_2, y_2, z_2), $\cdots$ (x_n, y_n, z_n). The point of application of the resultant $F_1 + F_2$ of F_1 and F_2 may be taken so as to divide the distance of the points of application of F_1 and F_2 in the ratio $F_2 : F_1$; hence, denoting its co-ordinates by x', y', z', we have $F_1(x' - x_1) = F_2(x_2 - x')$, or

$$(F_1 + F_2)x' = F_1 x_1 + F_2 x_2,$$

and similarly for y' and z'.

The force $F_1 + F_2$ combines with F_3 to form a resultant $F_1 + F_2 + F_3$, whose point of application (x'', y'', z'') is given by

$$(F_1 + F_2 + F_3)x'' = F_1x_1 + F_2x_2 + F_3x_3,$$

and similar expressions for y'', z''.

Proceeding in this way, we find for the point of application $(\bar{x}, \bar{y}, \bar{z})$ of the resultant of all the given forces

$$(F_1 + F_2 + \cdots + F_n)\bar{x} = F_1x_1 + F_2x_2 + \cdots + F_nx_n,$$

with corresponding equations for $\bar{y}$ and $\bar{z}$. We may write these equations in the form:

$$\bar{x} = \frac{\Sigma Fx}{\Sigma F}, \quad \bar{y} = \frac{\Sigma Fy}{\Sigma F}, \quad \bar{z} = \frac{\Sigma Fz}{\Sigma F}, \tag{1}$$

unless $\Sigma F = 0$ (see Art. 303).

As these expressions for $\bar{x}, \bar{y}, \bar{z}$ are independent of the direction of the parallel forces it follows that the same point $(\bar{x}, \bar{y}, \bar{z})$ would be found if the forces were all turned in any way about their points of application, provided they remain parallel. The point $(\bar{x}, \bar{y}, \bar{z})$ is for this reason called the **center** of the system of parallel forces. It is nothing but the centroid of the points of application if these points are regarded as possessing masses equal to the magnitudes of the forces.

302. As the origin of co-ordinates in the last article is arbitrary, the equations (1) evidently express the proposition that *in any system of parallel forces whose algebraic sum is different from zero the sum of their moments about any point is equal to the moment of their resultant about the same point.* In particular, *the sum of the moments about any point on the resultant is zero.*

This proposition may be regarded as a generalization of the *principle of the lever* referred to in Art. 297. It furnishes the convenient method of "taking moments" for the purpose of determining the position of the resultant.

303. Couple of Forces. The construction given in Arts. 294 and 298 for the resultant of two parallel forces fails only when the two given forces are equal and of opposite sense. In this case, the lines pP' and qQ' in Fig. 83, and the lines I and III of the funicular polygon (Fig. 84), become parallel, so that their intersection r lies at infinity. The magnitude of the resultant is of course zero.

The combination of two equal and opposite parallel forces $(F, -F)$ acting on a rigid body is called a **couple.** *A couple is, therefore, properly speaking, not equivalent to a single force,* although it may be said to be equivalent to a force of magnitude zero at an infinite distance. The theory of couples will be considered in detail in Arts. 319–329.

A system of n parallel forces $F_1, F_2, \cdots F_n$ has been shown in Arts. 300 and 301 to reduce to a single resultant unless $\Sigma F = 0$. In this exceptional case the system may either reduce to a couple or it may be in equilibrium. To distinguish which of these cases arises it suffices to combine all the forces of the same sense; we obtain thus, since $\Sigma F = 0$, two equal and opposite forces. If these do not lie in the same line, the system reduces to a couple; otherwise it is in equilibrium.

304. Conditions of Equilibrium. It follows from the preceding article that *for the equilibrium of a system of parallel forces the condition $\Sigma F = 0$, or $R = 0$, though always necessary, is not sufficient.*

Now, if the resultant R of the n parallel forces $F_1, F_2, \cdots F_n$ is zero, the resultant R' of the $n - 1$ forces $F_1, F_2, \cdots F_{n-1}$ cannot be zero, and its point of application is found (by Art. 301) from $\bar{x} = (F_1x_1 + F_2x_2 + \cdots + F_{n-1}x_{n-1})/(F_1 + F_2 + \cdots + F_{n-1})$ and similar expressions for $\bar{y}$ and $\bar{z}$. The whole system of parallel forces is therefore equivalent to the two parallel forces R' and F_n. Two such forces can be in equilibrium only when they lie in the same straight line; *i. e.* F_n must lie in the same line with R' and must therefore pass through the point $(\bar{x}, \bar{y}, \bar{z})$, which is a point of R'.

The additional condition of equilibrium is, therefore,

$$\frac{\bar{x} - x_n}{\cos\alpha} = \frac{\bar{y} - y_n}{\cos\beta} = \frac{\bar{z} - z_n}{\cos\gamma},$$

where α, β, γ are the angles made by the direction of the forces with the axes.

305. For practical application it is usually best to replace the last condition by taking moments about a convenient point. Thus, the analytical conditions of equilibrium can be written in the form

$$\Sigma F = 0, \quad \Sigma Fp = 0.$$

Graphically, to the former corresponds the closing of the force-polygon, to the latter the closing of the funicular polygon.

306. Weight; Center of Gravity. The most important special case of parallel forces is that of the force of gravity which acts at any given place near the earth's surface in approximately parallel lines on every particle of matter.

If g be the acceleration of gravity, the force of gravity on a particle of mass m is

$$w = mg,$$

and is called the **weight** of the particle or of the mass m.

For a system of particles of masses m_1, m_2, $\cdots m_n$ we have

$$w_1 = m_1 g, \quad w_2 = m_2 g, \quad \cdots w_n = m_n g.$$

If the particles are rigidly connected, the resultant W of these parallel forces,

$$W = w_1 + w_2 + \cdots + w_n = (m_1 + m_2 + \cdots + m_n) g = Mg,$$

where M is the mass of the system, is called the weight of the system.

The center of the parallel forces of gravity of a system of rigidly connected particles has, by Art. 301, the co-ordinates

$$\bar{x} = \frac{\Sigma mgx}{\Sigma mg}, \quad \bar{y} = \frac{\Sigma mgy}{\Sigma mg}, \quad \bar{z} = \frac{\Sigma mgz}{\Sigma mg},$$

or since the constant g cancels,

$$\bar{x} = \frac{\Sigma mx}{\Sigma m}, \quad \bar{y} = \frac{\Sigma my}{\Sigma m}, \quad \bar{z} = \frac{\Sigma mz}{\Sigma m}.$$

This point is called the **center of gravity** of the system, and is evidently identical with the *center of mass*, or **centroid** (see Art. 212).

For continuous masses the same formulæ hold, except that the summations become integrations.

The *weight* W of a physical body of mass M is therefore a vertical force passing through the centroid of its mass.

307. Exercises.

(1) A straight rod (*lever*) of 6 feet length has suspended from its ends masses of 7 and 18 pounds, respectively. Find the point (*fulcrum*) on which it balances in a horizontal position: (*a*) if its own weight be neglected; (*b*) if it is homogeneous and weighs $1\frac{2}{3}$ pounds per running foot.

(2) A straight beam rests in a horizontal position on two supports A, B. The distance between the supports (the *span*) is 24 ft. The beam carries a weight of 11 tons at a distance of 8 ft. from A, and a weight of 6 tons at 16 ft. from A. Find the pressures on the supports (or the *reactions* of the supports): (*a*) when the proper weight of the beam is neglected; (*b*) when the beam weighs $\frac{1}{3}$ ton per running foot; (*c*) when the first third of the beam (from A) weighs $\frac{1}{8}$ ton, the second 1 ton, the third $\frac{1}{2}$ ton per running foot.

(3) A homogeneous circular plate weighing W pounds rests in a horizontal position on three equidistant supports near its edge. (*a*) What is the least weight P that will upset it when placed on the plate? (*b*) If there be four equidistant supports near the edge, what is the least weight that will upset the plate?

(4) Construct the resultant of two parallel forces of opposite sense by the graphical method of Art. 298.

(5) Solve exercises (1) and (2) by the graphical method.

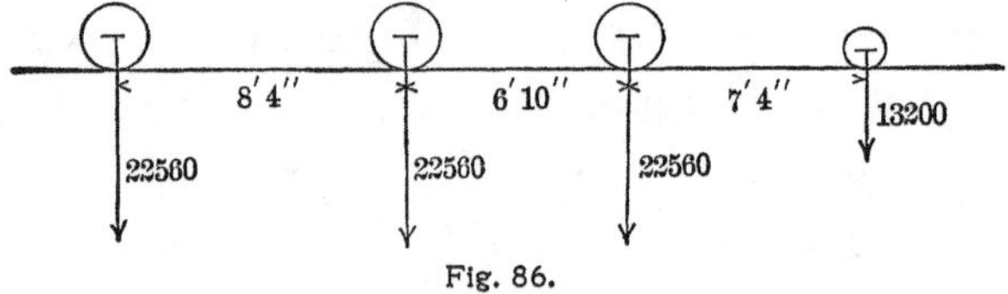

Fig. 86.

(6) Find the reactions of the supports of a bridge truss of 50 ft. span, produced by a freight locomotive whose weight is distributed

over the three pairs of driving wheels and the front truck, as indicated in Fig. 86: (*a*) when it stands in the middle of the span; (*b*) when its front truck stands over one support.

(7) Explain how the centroid of a plane area can be found graphically by dividing the area into narrow parallel strips.

(8) A homogeneous rectangular plate is pivoted on a horizontal axis through its center so as to turn freely in a vertical plane. If weights W_1, W_2, W_3, W_4 be suspended from its vertices, what is its position of equilibrium?

(9) The ends of a straight lever of length l are acted upon by two forces F_1, F_2 in the same plane with it, but inclined to the lever at angles α_1, α_2. Determine the position of the fulcrum.

(10) To resolve a force W graphically along two lines a, b parallel to W (*e. g.* to find the pressures on the two supports produced by a weightless beam carrying a single load W) let $W = 1\ 3$ represent the force polygon. Take any point A on a as pole, draw $A\ 1$ and take this line as the first side I of the funicular polygon. The third side III must then be the parallel $1\ B$ to $A\ 3$; and the second side II is found by joining A to the intersection B of III with b. As A is the pole, this line AB meets $1\ 3$ at a point 2 such that the required components are $1\ 2$ along a and $2\ 3$ along b. Draw the figure.

(11) The safety-valve of a steam-boiler consists of a horizontal arm, 32 in. long, hinged at one end A, and weighing 3 lbs.; the vertical stem of the valve is attached at 4 in. from A and weighs, with the valve disk, 1½ lb.; a ball weighing 20 lbs., suspended from the arm, 28 in. from A, presses the disk on the mouth of the valve. If the diameter of the mouth be 2 in., what is the blowing-off pressure? (*The Locomotive*, Vol. 16, pp. 113–120.)

(12) If in a safety-valve the arm is 66 in. long and weighs 18 lbs., while the valve-stem and disk, 3 in. from the hinge, weigh 7 lbs. and the diameter of the disk is 4 in., at what distance from the hinge must a 50-lb. weight be suspended to make the valve blow at a pressure of 100 lbs. / sq. in.? (*Ib.*)

308. Funicular Polygons and Catenaries. The *funicular polygon* in its original meaning represents the form of equilibrium assumed by a string or cord suspended from two fixed points and acted upon by any forces in the vertical plane. The "cord" is supposed to be perfectly flexible, inextensible, inelastic and without weight. When the number of forces is made infinite, the polygon becomes a continuous curve called a *catenary*.

The present discussion is confined to the case when the forces are all vertical so that they can be regarded as weights.

309. Let A, B (Fig. 87) be the fixed points, and let there be five weights, W_1, W_2, W_3, W_4, W_5, suspended from the cord which will form a polygon whose sides are denoted by I, II, III, IV, V, VI.

If the cord be cut on both sides of the first vertex, I II, and the corresponding tensions T_1, T_2 be introduced, this vertex must be

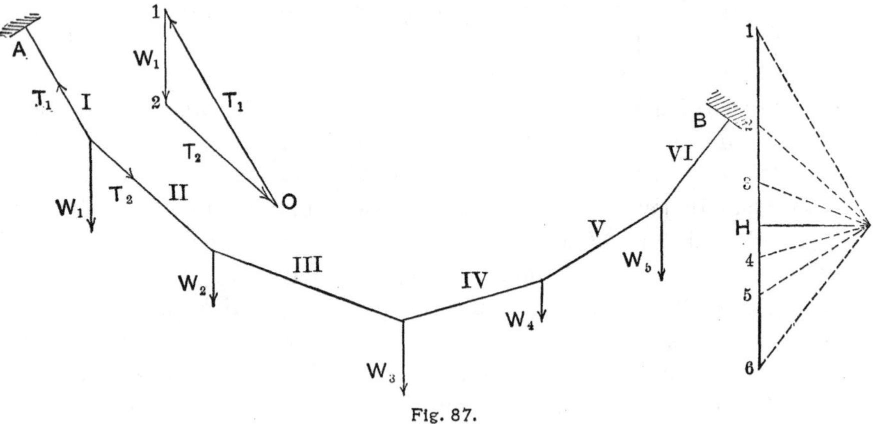

Fig. 87.

in equilibrium under the action of the three forces W_1, T_1, T_2. Hence drawing a line 1 2 to represent the weight W_1 and drawing through its ends 1, 2 parallels to I and II, respectively, we have the force polygon of the first vertex. Its sides O 1 and 2 O represent in magnitude, direction, and sense the tensions T_1, T_2; in other words, the weight W_1 has thus been resolved into its components along the adjacent sides.

The same can be done at every vertex of the polygon I II ··· VI, and all the tensions can thus be found. But as the tension T_2 in II occurs again (with sense reversed) in the force polygon for the next vertex, and so on, the successive force polygons can be fitted together, every triangle having one side in common with the next one. Thus the complete force polygon of the whole cord is formed, as shown on the right, in Fig. 87. Its vertical line represents the successive weights $W_1 = 12$, $W_2 = 23$, $W_3 = 34$, $W_4 = 45$, $W_5 = 56$, while the lines radiating from the pole O represent on the same scale the *tensions* or *stresses* in I, II, III, IV, V, VI.

310. The polygon I II ⋯ VI is called the *funicular polygon.* It will be noticed that if we have given the fixed points A, B, the magnitudes of the weights, their horizontal distances, say from A, and the directions of the first and last sides (whatever may be the number of the forces), the remaining sides of the funicular polygon can be found by laying off on a vertical line the weights $W_1 = 12$, $W_2 = 23$, etc., in succession, drawing through 1 a parallel to the first side, through the end of the last weight (6 in Fig. 87) a parallel to the last side, and joining the intersection O of these parallels to the points 2, 3, etc. The sides of the funicular polygon must be parallel to the lines radiating from O; at the same time these lines represent the tensions in these sides.

311. For the analytical investigation, let P_i be that vertex of a funicular polygon of any number of sides at which the ith and $(i + 1)$th sides intersect; let α_i, α_{i+1} be the angles at which these sides are inclined to the horizon, and W_i the weight suspended from the vertex P_i (Fig. 88).

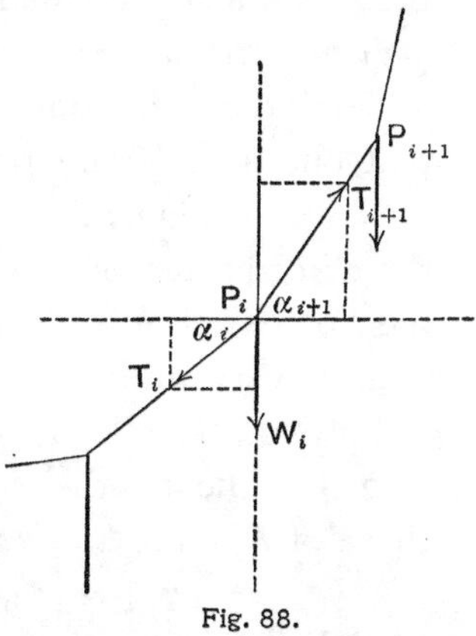

Fig. 88.

Cutting the cord on both sides of P_i, and introducing the tensions T_i and T_{i+1}, the conditions of equilibrium of the point P_i are found by resolving the three forces W_i, T_i, T_{i+1} horizontally and vertically (Art. 280):

$$T_{i+1} \cos\alpha_{i+1} = T_i \cos\alpha_i, \tag{1}$$

$$T_{i+1} \sin\alpha_{i+1} = T_i \sin\alpha_i + W_i. \tag{2}$$

The former of these equations shows that, whatever the weights W and the lengths and inclinations of the sides, *the horizontal components of the tensions T are all equal.* Denoting this constant value by H, we have

$$T_1 \cos\alpha_1 = T_2 \cos\alpha_2 = \cdots = T_i \cos\alpha_i = \cdots = H. \qquad (3)$$

Substituting the values of T_i and T_{i+1} as obtained from these relations, in (2), this equation becomes

$$\tan\alpha_{i+1} = \tan\alpha_i + \frac{W_i}{H}, \qquad (4)$$

which shows that as soon as all the weights and the inclination and tension of any one side are given, the inclinations and tensions of all the other sides can be found.

312. Let us now assume that the weights W are all equal. Then the values of $\tan\alpha_{i+1}$ given by (4) form an arithmetical progression. If, in addition, we assume that *the sides of the polygon are such as to have equal horizontal projections, i. e.*, if we assume the weights to be equally spaced horizontally, *the vertices of the polygon will lie on a parabola* whose axis is vertical.

To find its equation, let us suppose, for the sake of simplicity, that one side of the polygon, say the kth, is horizontal so that $\alpha_k = 0$. Taking this side as axis of x, its middle point O as origin, the co-ordinates of the vertex P_k are $\frac{1}{2}a$, 0, if a be the length of the horizontal side and hence also that of the horizontal projection of every side.

Putting $W/H = \tau$, we have $\tan\alpha_k = 0$, $\tan\alpha_{k+1} = \tau$, $\tan\alpha_{k+2} = 2\tau, \cdots$; hence the co-ordinates of P_{k+1} are $x = \frac{3}{2}a$, $y = a\tau$; those of P_{k+2} are $x = \frac{5}{2}a$, $y = a\tau + 2a\tau = 3a\tau$; those of P_{k+3} are $x = \frac{7}{2}a$, $y = a\tau + 2a\tau + 3a\tau = 6a\tau$, etc.; those of the nth vertex after P_k are

$$x = \frac{2n+1}{2}a, \quad y = \frac{n(n+1)}{2}a\tau.$$

Eliminating n, we find the equation

$$x^2 = \frac{2a}{\tau}\left(y + \frac{a\tau}{8}\right),$$

which represents a parabola whose axis is the axis of y, and whose vertex lies at the distance $\frac{1}{8}a\tau = \frac{1}{8}aW/H$ below the origin O.

313. Let the number of sides be increased indefinitely, the length a and the weight W approaching the limit 0, but so that the quotient a/W remains finite, say $\lim (a/W) = 1/w$. Then $\lim (a/\tau) = H/w$, $\lim (a\tau) = 0$; so that the equation of the parabola becomes

$$x^2 = \frac{2H}{w}y,$$

where w is evidently the weight of the cord, or chain, per unit of horizontal length.

The parabola is, therefore, *the form of equilibrium of a cord suspended from two points when the weight of the cord is uniformly distributed over its horizontal projection.* This is, for instance, the case approximately in a suspension bridge with uniformly loaded roadbed, the proper weight of the chains being neglected.

314. This result can easily be derived independently of Art. 312, by considering the equilibrium of any portion OP of the chain beginning at the lowest point O (Fig. 89). The forces acting on this portion are the horizontal tension H at O, the tension

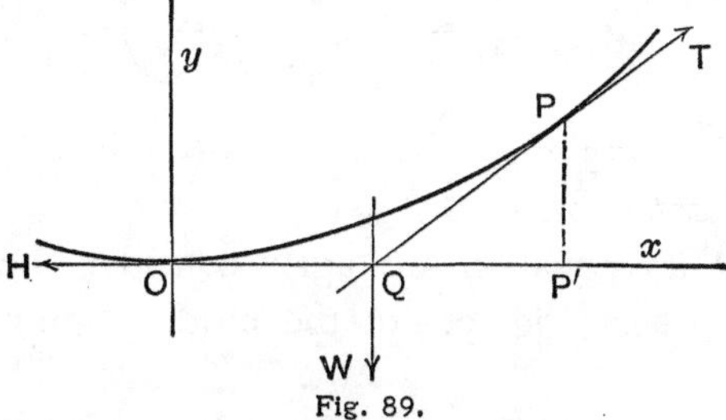

Fig. 89.

T along the tangent at P, and the proper weight W of the chain. As this weight is assumed to be uniformly distributed over the horizontal projection $OP' = x$ of OP, the weight is $W = wx$, and bisects OP'. For equilibrium the three forces W, H, T must pass through a point; hence T must pass through the middle point Q of OP'.

Resolving the forces in the horizontal and vertical directions, we find, as conditions of equilibrium,

$$-H+T\frac{dx}{ds}=0, \quad -wx+T\frac{dy}{ds}=0;$$

whence, eliminating ds,

$$\frac{dy}{dx}=\frac{w}{H}x.$$

Integrating and considering that $x=0$ when $y=0$, we find the equation of the parabola as above,

$$y=\frac{w}{2H}x^2.$$

This relation can be directly read off from the figure; for, taking moments about P, we have $0=-Hy+wx\cdot\frac{1}{2}x$, whence $y=(w/2H)x^2$.

315. The force polygon of H, T, W is evidently a triangle similar to the triangle QPP'.

Hence, if the height of a suspension bridge be h, its span $2l$, its total weight $2W$, we have for the horizontal tension H and the tension T at the point of support:

$$\frac{H}{\frac{1}{2}l}=\frac{T}{\sqrt{h^2+\frac{1}{4}l^2}}=\frac{W}{h}.$$

316. *The form of equilibrium assumed by a homogeneous cord is a common catenary.*

To find its equation, we again consider the equilibrium of a portion $OP=s$ (Fig. 90) of the cord, beginning at the lowest point O.

The weight of this portion is now $W=ws$, and if α be the angle made by the tangent at P with a horizontal line, we have the conditions of equilibrium

$$T\cos\alpha=T\frac{dx}{ds}=H, \quad T\sin\alpha=T\frac{dy}{ds}=ws.$$

Dividing and putting $H/w=c$, we have the differential equation of the curve in the form

$$\frac{dx}{dy} = \frac{c}{s}.$$

Substituting this value of dx/dy in the relation $ds^2 = dx^2 + dy^2$, we obtain

$$\left(\frac{ds}{dy}\right)^2 = 1 + \frac{c^2}{s^2}, \text{ or } dy = \pm \frac{sds}{\sqrt{s^2 + c^2}},$$

which gives by integration $y + C = \sqrt{s^2 + c^2}$, the minus sign being rejected since y increases with s.

The constant C can be made to disappear by taking the origin

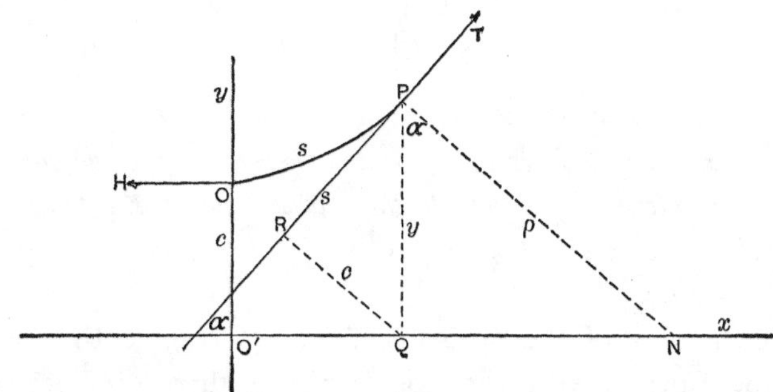

Fig. 90.

O' on the vertical through O at the distance $O'O = c$ below the lowest point O. We have, therefore,

$$y^2 = s^2 + c^2.$$

By means of this relation, s can be eliminated from the original differential equation, and the result,

$$\frac{cdy}{\sqrt{y^2 - c^2}} = dx,$$

can be integrated:

$$c \log (y + \sqrt{y^2 - c^2}) = x + C.$$

As $y = c$ when $x = 0$, we find $C = c \log c$; hence

$$y + \sqrt{y^2 - c^2} = ce^{\frac{x}{c}}.$$

Taking reciprocals and rationalizing the denominator we find

$$y - \sqrt{y^2 - c^2} = ce^{-\frac{x}{c}};$$

hence, adding and subtracting,

$$y = \tfrac{1}{2}c\left(e^{\frac{x}{c}} + e^{-\frac{x}{c}}\right) = c \cosh (x/c),$$

$$s = \tfrac{1}{2}c\left(e^{\frac{x}{c}} - e^{-\frac{x}{c}}\right) = c \sinh (x/c).$$

317. The first equations of Art. 316, $T \cos\alpha = H = wc$, $T \sin\alpha = ws$, give for the total tension T at any point P

$$T^2 = w^2(c^2 + s^2) = (wy)^2.$$

Thus, *while the horizontal component is constant, the vertical component at any point P is equal to the weight of the portion of the cord from the lowest point O to the point P, and the total tension is equal to the weight of a portion of the cord equal to the ordinate of the point P.*

Let Q be the foot of the ordinate of P (Fig. 90), N the intersection of the normal with the axis $O'x$, and draw QR perpendicular to the tangent. Then $PR = y \sin\alpha = s$, since $T \sin\alpha = ws$ and $T = wy$; also $QR = y \cos\alpha = c$. Dividing, we have $\tan\alpha = s/c$; hence, differentiating,

$$\frac{1}{\cos^2\alpha}\frac{d\alpha}{ds} = \frac{1}{c}; \quad \text{and } \rho = \frac{ds}{d\alpha} = \frac{c}{\cos^2\alpha}.$$

The figure shows that the radius of curvature ρ is equal to the length of the normal PN.

The relation $\rho \cos^2\alpha = c$ shows further that at the vertex $(\alpha = 0)$ the radius of curvature is $\rho_0 = c$. It follows that for a cord or chain suspended from two points B, C in the same horizontal line, c (and consequently H) is large when ρ_0 is large, *i. e.* when the curve is flat at the vertex; in other words, when B and C are far apart.

318. Exercises.

(1) A weightless cord $ABCDEF$ is suspended from the fixed points A, F, and carries weights at the intermediate points B, C, D, E. Taking A as origin, the axis of x horizontal, the axis of y vertically upwards, the co-ordinates of the points B, C, D, E, F are $(2, -1)$, $(4, -1.5)$, $(7, -1.5)$, $(8.5, -1)$, $(10, 2)$. If the weight at B be one pound, what are the weights at C, D, E? What are the tensions of the segments of the cord? What are the reactions of the fixed points A, F?

(2) The total weight of a suspension bridge is $2W = 60$ tons; the span is $2l = 250$ ft.; the height is $h = 25$ ft. Find the tension of the chain at the ends and in the middle, both graphically and analytically.

(3) A uniform wire of length $2s$ is stretched between two points in the same horizontal line whose distance $2x$ is very nearly equal to $2s$. Find an approximate expression for the parameter c of the catenary and hence for the tension of the wire.

(4) Find the tension of a telegraph wire No. 8 (diameter = 0.165 in., weight 378.1 lbs. per mile) stretched between poles 100 ft. apart, if the length of the wire between two poles is 100 ft. 4 in. Determine also the sagging of the wire $(y - c)$.

(5) A chain, 64 ft. long, and weighing 1 lb./ft. is suspended between two points in the same horizontal line, 60 ft. apart; find the tensions at the lowest point and at the ends, and the depth of the vertex below the points of suspension.

(6) If the load of a suspension bridge is proportional to the distance x from the middle point, show that the cable forms a cubical parabola.

(7) Show that when x/c is so small that its third and higher powers can be neglected the arc of the catenary in Art. 316 can be regarded as an arc of a parabola.

IV. *Theory of Couples.*

319. *The combination of two equal forces of opposite sense F, $-F$, acting on a rigid body along parallel lines, is called a couple of forces,* or simply a **couple** (Art. 303).

The perpendicular distance $AB = p$ (Fig. 91) of the forces of the couple is called the **arm**, and the product Fp of the force F

into the arm p is called the **moment** of the couple. The moment, or the couple itself, is also called a **torque.**

If we imagine the couple (F, p) to act upon an invariable plane figure in its plane, and if the middle point of its arm be a fixed point of this figure, the couple will evidently tend to turn the figure about this middle point. (It is to be observed that it is *not* true, in general, that a couple acting on a rigid body produces rotation about an axis at right angles to its plane.) A couple of the type (F, p) or (F', p') (see Fig. 91) will tend to rotate counter-clockwise, while a couple of the type (F'', p'') tends to turn clockwise. Couples in the same plane, or in parallel planes, are therefore distinguished as to their **sense;** and this sense is expressed by the algebraic sign attributed to the moment. Thus, the moment of the couple (F, p) in Fig. 91, is $+Fp$, that of the couple (F'', p'') is $-F''p''$.

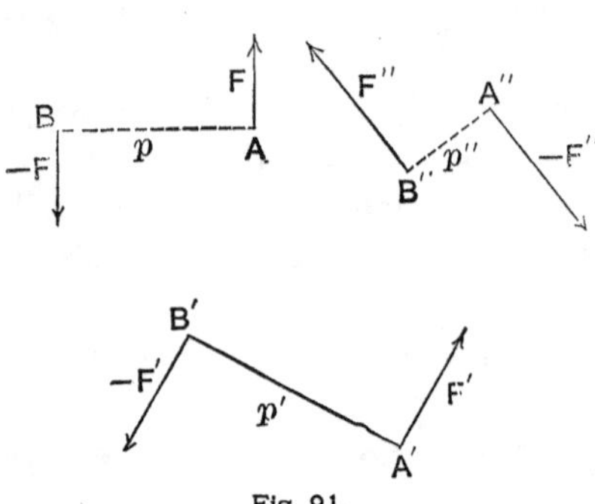

Fig. 91.

320. *The effect of a couple is not changed by translation, i. e.* by moving its plane parallel to itself without rotating it.

Let $AB = p$ (Fig. 92) be the arm of the couple (F, p) in its original position, and $A'B'$ the same arm in a new position parallel to the original one in the same plane, or in any parallel plane. By introducing at each end of the new arm $A'B'$ two opposite forces $F, -F$, each equal and parallel to the original forces F, the given system is not changed (Art. 273). But the two equal and parallel forces F at A and B' form a resultant $2F$ at the middle point O of the diagonal AB' of the parallelogram

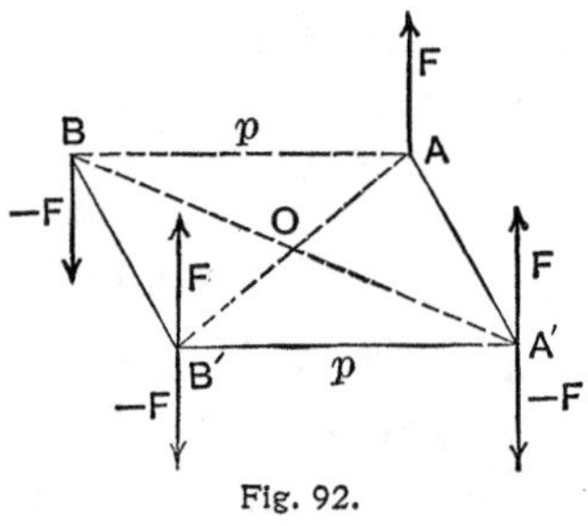

Fig. 92.

$ABB'A'$. Similarly, the two forces $-F$ at B and A' are together equivalent to a resultant $-2F$ at the same point O. These two resultants, being equal and opposite and acting in the same line, are together equivalent to zero. Hence the whole system reduces to the force F at A' and the force $-F$ at B', which form, therefore, a couple equivalent to the original couple at AB.

321. *The effect of a couple is not changed by rotation in its plane.*

Let AB (Fig. 93) be the arm of the couple in the original position, C its middle point, and let the arm be turned about C into the position $A'B'$. Applying again at A', B' equal and opposite forces each equal to F, the forces $-F$ at A' and F at A will form a resultant acting along CD, while F at B' and $-F$ at B give an equal and opposite resultant along CE. These two resultants destroy each other and leave nothing but the couple formed by F at A' and $-F$ at B', which is therefore equivalent to the original couple.

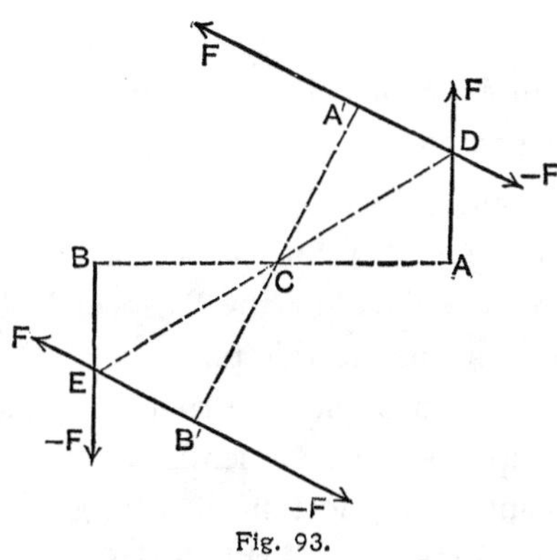

Fig. 93.

Any other displacement of the couple in its plane, or to a parallel plane, can be effected by a translation combined with a rotation in its plane about the middle point of its arm. *The effect of a couple is therefore not changed by any displacement in its plane or to a parallel plane.*

322. *The effect of a couple is not changed if its force F and its arm p be changed simultaneously in any way, provided their product Fp remain the same.*

Let $AB = p$ be the original arm (Fig. 94), F the original force of the couple; and let $A'B' = p'$ be the new arm. The introduction of two equal and opposite forces F' at A', and also at

B', will not change the given system F, $-F$. Now, selecting for F' a magnitude such that $F'p' = Fp$, the force F at A and the force $-F'$ at A' combine (Art. 297) to form a parallel resultant through C, the middle point of the arm, since for this point $F \cdot \frac{1}{2}p + (-F') \cdot \frac{1}{2}p' = 0$. Similarly, $-F$ at B and F' at B' give a resultant of the same magnitude, in the same line through C, but of opposite sense. These two resultants thus destroying each other, there remains only the couple formed by F' at A' and $-F'$ at B', for which $Fp = F'p'$.

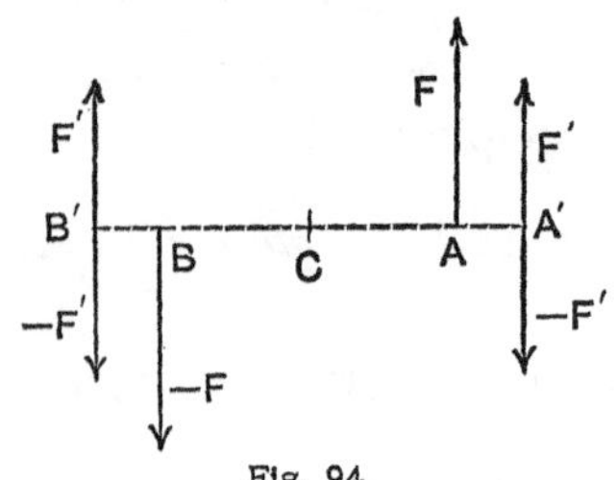

Fig. 94.

323. It results from the last three articles that the only essential characteristics of a couple are: (*a*) the numerical value of the moment; (*b*) the sense, or direction of rotation; and (*c*) what has been called the "aspect" of its plane, *i. e.* the direction of any normal to this plane.

It is to be noticed that the plane of the two forces forming the couple is not an essential characteristic of the couple; just as the point of application of a force is not an essential characteristic of the force (see Art. 283); provided, of course, that the couple (or force) is acting on a *rigid* body.

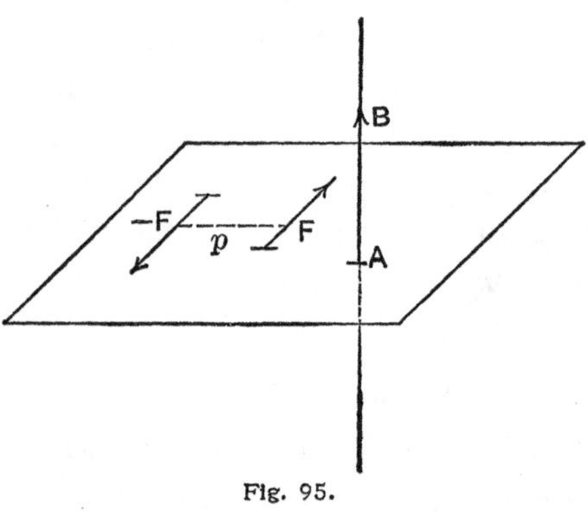

Fig. 95.

Now the three characteristics enumerated above can all be indicated by a *vector* which can therefore serve as the geometrical representative of the couple. Thus, the couple formed by the forces F, $-F$ (Fig. 95), whose perpendicular distance is p, is represented by the vector $AB = Fp$ laid off on any normal to the plane of the couple. The sense is indicated by drawing the vector toward that side of the plane from which the couple is seen to rotate counter-clockwise.

We shall call this geometrical representative AB of the couple simply the **vector** of the couple. It is sometimes called its *moment*, or its *axis*, or its *axial moment*.

324. As was pointed out in Art. 303, a couple can be regarded as the limit of a force whose magnitude approaches zero while its line of action is removed to infinity. Similarly, in kinematics a rotation whose angle approaches zero while its axis is removed to infinity, reduces to a *translation*, and an angular velocity whose magnitude tends to zero while its axis is removed indefinitely becomes in the limit a *velocity of translation*.

Just as, in kinematics (see Art. 184), two equal and opposite angular velocities about parallel axes produce a velocity of translation, so in statics two equal and opposite forces along parallel lines form a new kind of quantity called a *couple*.

It should, however, be noticed that while rotations, angular velocities, and forces are represented by *rotors*, *i. e.* by vectors confined to definite lines, translations, velocities of translation, and couples have for their geometrical representatives vectors not confined to particular lines.

Just as in the case of couples of angular velocities, the vector representing a couple of forces has for its magnitude and sense those of the moment of the couple, and for its direction that perpendicular to the plane of the couple.

It is due to this analogy between the two fundamental conceptions that a certain dualism exists between the theories of statics and kinematics, so that a large portion of the theory of kinematics of a rigid body might be made directly available for statics by simply substituting for angular velocity and velocity of translation the corresponding ideas of force and couple.

325. It is easily seen how, by means of Arts. 320–322, any number of couples acting on a rigid body can be reduced to a single resultant couple. It can also be proved without much difficulty that the vector of the resultant couple is the geometric sum of the vectors of the given couples; in other words, *vectors representing couples acting on the same rigid body are combined by the parallelogram law* (Art. 40).

In the particular case when the couples all lie in parallel planes, or in the same plane, their vectors may be taken in the same line, and can, therefore, be added algebraically.

Hence, *the resultant of any number of couples is a single couple whose vector is the geometric sum of the vectors of the given couples.*

Conversely, a couple can be resolved into components by resolving its vector into components.

326. To combine a single force P with a couple (F, p) lying in the same plane it is only necessary to place the couple in its plane in such a position (Fig. 96) that one of its forces, say $-F$, shall lie in the same line and in opposite sense with the single force P, and to transform the couple (F, p) into a couple (P, p'), by Art. 322, so that $Fp = Pp'$. The original force P and the force $-P$ of the transformed couple destroying each other at A, there remains only the other force P, at A', of the transformed couple, that is, a force parallel and equal to the original single force P, at the distance

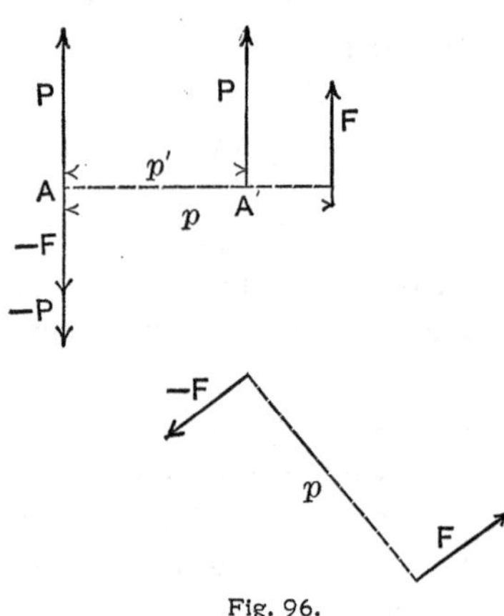

Fig. 96.

$$p' = \frac{F}{P} p$$

from it.

Hence, *a couple and a single force in the same plane are together equivalent to a single force equal and parallel to, and of the same sense with, the given force, but at a distance from it which is found by dividing the moment of the couple by the single force.*

327. Conversely, *a single force P applied at a point A of a rigid body can always be replaced by an equal and parallel force P of the same sense, applied at any other point A' of the same body, in combination with the couple formed by P at A and $-P$ at A'.*

This follows at once by applying at A' two equal and opposite forces each equal and parallel to P.

328. The proposition of Art. 326 applies even when the force lies in a plane parallel to that of the couple, since the couple can be transferred to any parallel plane without changing its effect.

If the single force intersects the plane of the couple, it can be resolved into two components, one lying in the plane of the couple, while the other is at right angles to this plane. On the former component the couple has, according to Art. 326, the effect of transferring it to a parallel line. We thus obtain *two non-intersecting*, or *skew*, *forces at right angles to each other.*

Let P be the given force, and let it make the angle α with the plane of the given couple, whose force is F and whose arm is p. Then $P \sin\alpha$ is the component at right angles to the plane of the couple, while $P \cos\alpha$ combines with the couple whose moment is Fp into a force $P \cos\alpha$ in the plane of the couple; this force $P \cos\alpha$ is parallel to the projection of P on the plane, and has the distance $Fp/P \cos\alpha$ from this projection.

Hence, in the most general case, *the combination of a single force and a couple can be replaced by the combination of two single forces crossing each other* (*without meeting*) *at right angles;* it can be reduced to a single force only when the force is parallel to the plane of the couple.

329. Exercises.

(1) Show that the moment of a couple can be represented by the area of the parallelogram formed by the two forces of the couple, or by twice the area of the triangle formed by joining any point on the line of one of the forces to the ends of the other force.

(2) Show that the sum of the moments of two forces forming a couple is the same for any point in the plane of the couple (comp. Art. 289).

(3) Show, by means of Arts. 320–322, how to combine any number of couples situated in the same plane, or in parallel planes.

(4) Find the resultant of two couples situated in non-parallel planes, without using the vectors of the couples.

(5) Prove that the vector of the resultant of two couples in different planes is the geometric sum of the vectors of the given couples.

V. *Plane Statics.*

I. THE CONDITIONS OF EQUILIBRIUM.

330. Suppose a rigid body to be acted upon by any number of forces, all of which are situated in the same plane. To reduce such a *plane system of forces* to its simplest form the proposition of Art. 327 may be used. This proposition enables us to transfer all the forces to a common origin, by introducing, in addition to each force, a certain couple in the same plane. The concurrent forces can then be combined into their resultant by geometric addition, *i. e.*, by forming their force polygon (Art. 277); and the couples lying all in the same plane combine by algebraic addition of their moments into a resultant couple (Art. 325).

Thus, let F (Fig. 97) be one of the forces of the given plane system, P its point of application. Selecting any point O in the plane as origin, apply at O two equal and opposite forces F, $-F$, each equal and parallel to the given force F; and let p be the perpendicular distance of the origin O from the line of action of the given force F. The force F at P is equivalent to the force F at O together with the couple formed by F at P and $-F$ at O; the moment of this couple is Fp, and its vector is perpendicular to the plane of the system.

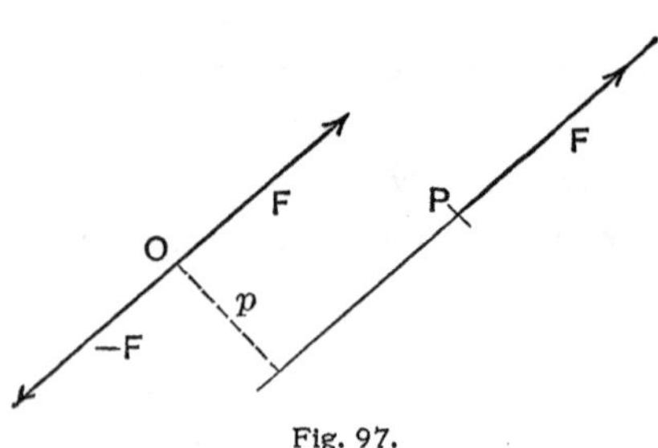

Fig. 97.

Proceeding in the same way with every force of the given system, all forces are transferred to the common origin O. The whole system is therefore equivalent to the resultant R passing through O, together with the resulting couple $H = \Sigma Fp$.

331. The given system of forces is said to be in **equilibrium** if the two following **conditions of equilibrium** are fulfilled:

$$R = 0, \quad H = 0.$$

It will be noticed that the moment Fp of the couple introduced by transferring the force F to the point O is the moment of the force F with respect to this point O.

Hence, *a plane system of forces is in equilibrium if* (a) *its resultant is zero, and* (b) *the algebraic sum of the moments of all its forces is zero with respect to any point in its plane.*

332. It is evident that the magnitude and direction of the resultant R do not depend on the selection of the origin O. But the position of this resultant and the magnitude of the resulting couple H will in general differ with the point selected as origin. Indeed, the origin can be so taken as to make the couple H vanish (unless the resultant R be zero); that is, the whole system can be reduced to a single resultant.

To do this (see Art. 326), it is only necessary, after determining R and H for some point O, to transfer R to a parallel line at such a distance r from its original position as to make the moment Rr of the couple introduced by the transfer equal and opposite to the moment ΣFp; *i. e.*, we must take (Art. 326)

$$r = -\frac{H}{R}.$$

The line along which this single resultant acts is called the **central axis** of the given system of forces.

333. For a purely analytical reduction of a plane system of forces the system is referred to rectangular axes Ox, Oy, arbitrarily assumed in the plane (Fig. 98). Every force F is resolved at its point of application P (x, y) into two components X, Y, parallel to the axes, so that

$$X = F\cos\alpha, \quad Y = F\sin\alpha,$$

α being the angle made by F with the axis Ox. At the origin O two equal and opposite forces X, $-X$ are applied along Ox, and two equal and opposite forces Y, $-Y$ along Oy. Thus, X at P is equivalent to X at O together with the couple formed by X at P and $-X$ at O; the moment of this couple is evidently

$-yX$. Similarly, Y at P is replaced by Y at O together with a couple whose moment is xY. The force F at P is therefore equivalent to the two forces X, Y at O together with a couple whose moment is $xY - yX$.

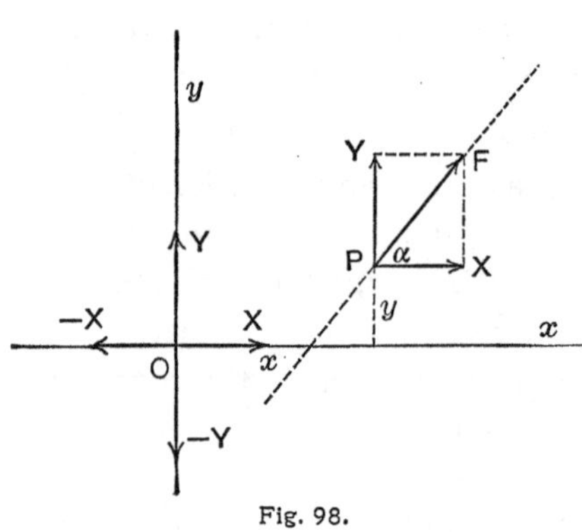

Fig. 98.

Proceeding in the same way with every given force, we obtain a number of forces X along Ox whose algebraic sum we call ΣX, and a number of forces Y along Oy which give ΣY. These two rectangular forces form the resultant

$$R = \sqrt{(\Sigma X)^2 + (\Sigma Y)^2}$$

whose direction is given by

$$\tan\alpha = \frac{\Sigma Y}{\Sigma X},$$

where α is the angle between Ox and R.

In addition to this, we obtain a number of couples $xY - yX$ whose algebraic sum forms the resulting couple

$$H = \Sigma(xY - yX).$$

The whole system is thus found equivalent to a resultant force R together with a resultant couple H in the same plane with R. The *conditions of equilibrium* $R = 0$, $H = 0$ (Art. 331) can therefore be expressed analytically by the three equations

$$\Sigma X = 0, \quad \Sigma Y = 0, \quad \Sigma(xY - yX) = 0.$$

334. If R be not zero, R and H can be combined into a single resultant R' equal and parallel to R at the distance $-H/R$ from it (see Art. 332). The equation of the line of this single resultant R', *i. e.*, the *central axis* of the system of forces, is found by considering that it makes the angle α with the axis of x and that its distance from the origin is

$$H/R = \Sigma(xY - yX)/\sqrt{(\Sigma X)^2 + (\Sigma Y)^2}.$$

Hence its equation is

$$\xi \cdot \Sigma Y - \eta \cdot \Sigma X - \Sigma(xY - yX) = 0.$$

If $R = 0$, the system is equivalent to the couple

$$H = \Sigma(xY - yX).$$

If H itself be also zero, the system is in equilibrium.

335. The following examples will illustrate the application of the conditions of equilibrium. To establish these conditions in any particular problem it will generally be found best to resolve the forces along two rectangular directions and equate the sums of the components to zero; and then to "take moments," *i. e.*, equate to zero the sum of the moments of all the forces with respect to some point conveniently selected as origin.

The principle that three forces acting on a rigid body can be in equilibrium only if they are concurrent or parallel and lie in the same plane is often used to advantage although its proof can only be derived from the general conditions of equilibrium of a rigid body.

336. *A homogeneous straight rod* $AB = 2l$ (Fig. 99) *of weight* W *rests with one end* A *on a smooth horizontal plane* AH, *and with the point* E ($AE = e$) *on a cylindrical support, the axis of the cylinder being at right angles to the vertical plane containing the rod. Determine what horizontal force* F *must be applied at a given point* F *of the rod* ($AF = f > e$) *to keep the rod in equilibrium when inclined to the horizon at an angle* θ.

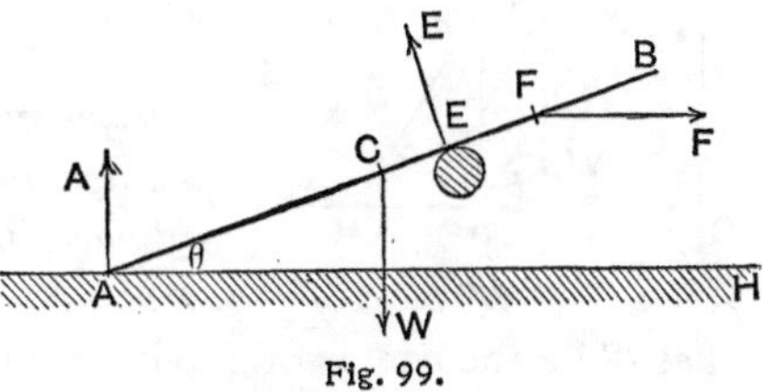

Fig. 99.

The rod exerts a certain unknown pressure on each of the supports at A and E, in the direction of the normals to the surfaces of contact, provided there be no friction, as is here assumed. The supports may therefore be imagined removed if forces A, E, equal and opposite to these pressures, be introduced; these forces A, E are called the *reactions* of the supports. The rod itself is here regarded as a straight line; its weight W is applied at its middle point C.

Taking A as origin and AH as axis of x, the resolution of the forces gives

$$\Sigma X \equiv F - E\sin\theta = 0, \qquad (1)$$

$$\Sigma Y \equiv A - W + E\cos\theta = 0. \qquad (2)$$

Taking moments about A, we find

$$E \cdot e - W \cdot l\cos\theta - F \cdot f\sin\theta = 0. \qquad (3)$$

Eliminating F from (1) and (3), we have

$$E = \frac{l\cos\theta}{e - f\sin^2\theta} W;$$

hence from (2),

$$A = \left(1 - \frac{l\cos^2\theta}{e - f\sin^2\theta}\right) W$$

and finally from (1),

$$F = \frac{l\sin\theta\cos\theta}{e - f\sin^2\theta} W.$$

337. *A cylinder of length $2l$ and radius r rests with the point A* (Fig. 100) *of the circumference of its lower base on a horizontal plane and with the point B of the circumference of its upper base against a vertical wall. The vertical plane through the axis of the cylinder contains the points A, B, and is perpendicular to the intersection of the vertical wall and the horizontal plane. If there be no friction at A and B, what horizontal force F applied at A will keep the cylinder in equilibrium? When is this force $F = 0$?*

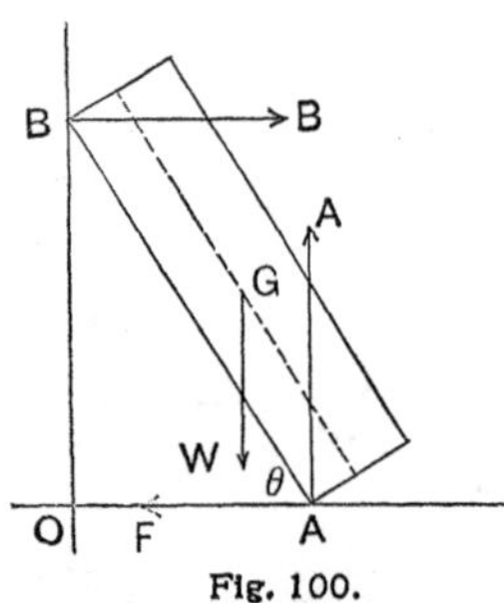

Fig. 100.

Let G be the center of gravity of the cylinder; W its weight; A, B the reactions at A, B; and θ the given angle between AB and the horizontal plane. Then $B - F = 0$, $A - W = 0$, and taking moments about A,

$$W(l\cos\theta - r\sin\theta) = B \cdot 2l\sin\theta.$$

Hence

$$A = W,$$
$$B = F = W \cdot \frac{l\cos\theta - r\sin\theta}{2l\sin\theta}$$
$$= \tfrac{1}{2}\left(\cot\theta - \frac{r}{l}\right)\cdot W.$$

If either the dimensions of the cylinder, or the angle θ, be such as to make $\tan\theta = l/r$, no force F will be required to maintain equilibrium; G and A will then lie in the same vertical line.

338. *The homogeneous rod $AB = 2l$ of weight W is jointed at A, so as to turn about A in a vertical plane. A cord BC attached to the end B of the rod runs at C over a smooth pulley, and carries a weight P. The axis of the pulley C is parallel to, and in the same vertical plane with, the axis of the joint A; $AC = h$. Find the position of equilibrium and the pressure on the axis of the joint A.* (Fig. 101.)

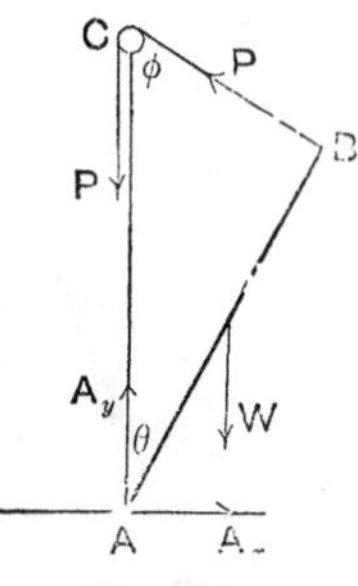

Fig. 101.

To *make the rod AB free*, cut the cord between B and C and introduce the tension, which is $= P$; also, replace the pressure A (since not only its magnitude but also its direction is unknown) by its horizontal and vertical components A_x, A_y. Then, putting $\angle ACB = \varphi$, $\angle BAC = \theta$, the conditions of equilibrium are

$$A_x = P\sin\varphi, \quad A_y = W - P\cos\varphi,$$
$$P \cdot h\sin\varphi = W \cdot l\sin\theta.$$

From the last equation

$$\frac{\sin\varphi}{\sin\theta} = \frac{l}{h}\cdot\frac{W}{P},$$

while from the triangle ABC

$$\frac{\sin\varphi}{\sin\theta} = \frac{2l}{BC};$$

hence $BC = 2hP/W$, *i.e.* if we take h to represent W, P will be represented by $\tfrac{1}{2}BC$.

For the total pressure A we have

$$A = A_x^2 + A_y^2 = W^2 + P^2 - 2W\cdot P\cos\varphi,$$

i. e. A is the third side of the triangle having W and P for the other two sides and φ for the included angle. The magnitude of A is therefore represented by the median from A in the triangle ABC on the same scale on which W is represented by h. But this median gives also the direction of A; for we have

$$\frac{A_y}{A_x} = \frac{W - P\cos\varphi}{P\sin\varphi} = \frac{h - \frac{1}{2}BC\cos\varphi}{\frac{1}{2}BC\sin\varphi}.$$

339. *A weightless rod AB rests without friction on two planes inclined to the horizon at angles α, β, and carries a weight W at the point D. The intersection C of these planes is horizontal and at right angles to the vertical plane through AB. Find the inclination θ of AB to the horizon, and the pressures at A and B.* (Fig. 102.)

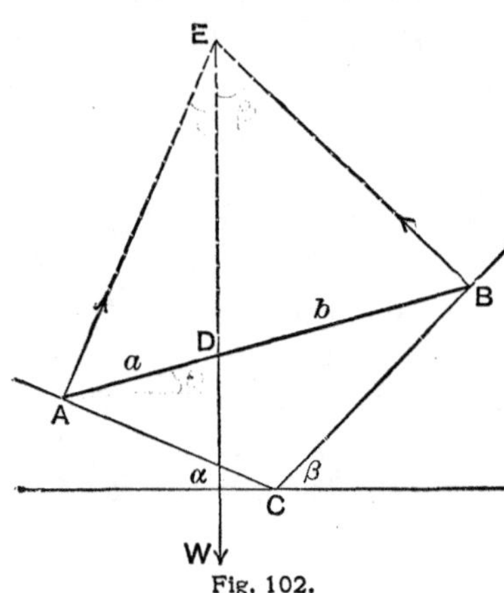

Fig. 102.

As there are only three forces, viz. the weight W and the reactions A and B, their lines must intersect at a point E. Resolving horizontally and vertically, we have

$$A\sin\alpha = B\sin\beta,$$

$$A\cos\alpha + B\cos\beta = W,$$

whence

$$A = \frac{\sin\beta}{\sin(\alpha+\beta)}W,$$

$$B = \frac{\sin\alpha}{\sin(\alpha+\beta)}W.$$

Taking moments about D, we find, with $AD = a$, $DB = b$,

$$A \cdot a\sin DAE = B \cdot b\sin DBE,$$

or

$$Aa\cos(\alpha+\theta) = Bb\cos(\beta-\theta).$$

To eliminate A and B divide by the first equation above:

$$a\frac{\cos(\alpha+\theta)}{\sin\alpha} = b\frac{\cos(\beta-\theta)}{\sin\beta};$$

solving for θ, we finally obtain

$$\tan\theta = \frac{a\cot\alpha - b\cot\beta}{a+b}.$$

340. Exercises.

(1) A homogeneous rod $AB = 2l = 8$ ft., weighing $W = 20$ lbs., rests with one end A on a horizontal plane AH, and with the point E on a support whose height above AH is $DE = h = 3$ ft. A horizontal cord $AD = d = 4$ ft. holds the rod in equilibrium. Find the tension T of this cord, and the reactions at A and E.

(2) A weightless rod AB of length l can turn freely about one end A in a vertical plane. A weight W is suspended from a point C of the rod; $AC = c$. A cord BD attached to the end B of the rod holds it in equilibrium in a horizontal position, the angle ABD being $\alpha = 150°$. Find the tension T of the cord and the resulting pressure A on the hinge at A.

(3) A homogeneous rod $AB = 2l$ of weight W rests with its upper end A against a smooth vertical wall, while its lower end B is fastened by a cord of given length, $BC = 2b$, to a point C in the wall. The rod and the cord are in the vertical plane at right angles to the wall. Find the position of equilibrium, *i. e.* the angle $\varphi = ACB$, the tension T of the cord, and the pressure A against the wall.

(4) A homogeneous rod $AB = 2l$ of weight W rests with one end A on a smooth horizontal plane AC, with the other end B against a smooth vertical wall BC, the vertical plane through AB being at right angles to the intersection C of the wall with the horizontal plane. The rod is kept in equilibrium by a cord EC. Find the tension T of this cord if the angles $CAB = \theta$ and $ECA = \varphi$ are given.

(5) A weightless rod $AB = l$ can revolve in a vertical plane about a hinge at A; its other end B leans against a smooth vertical wall whose distance from A is $AD = a$. At the distance $AC = c$ from A, a weight W is suspended. Find the horizontal thrust A_x at A and the normal pressures A_y and B at A and B.

(6) The same as (5) except that at B the rod rests on a smooth horizontal cylinder whose axis is at right angles to the vertical plane through AB. In which of the two problems is the horizontal thrust A_x at A least?

(7) In the problem of Art. 336, if the force F be at right angles to AB, find F and the reactions A, E for equilibrium: (*a*) when $f > e$; (*b*) when $f < e$.

(8) A smooth weightless rod $AB = l$ rests at C on a smooth horizontal cylinder whose axis is at right angles to the vertical plane through the rod; its lower end A leans against a smooth vertical wall whose distance from C is $CD = a$; from its upper end B a weight W is suspended. Determine the distance $AC = x$ for equilibrium, and the reactions at A and C.

(9) A homogeneous rod of weight W is hinged at its lower end A, while its upper end B leans against a smooth vertical wall. The rod is inclined at an angle θ to the vertical, and carries three weights, each equal to w, at three points dividing the rod into four equal parts. Determine the pressure on the wall and the reaction of the hinge.

(10) A homogeneous rod $AB = 2l$ of weight W rests with one end A on the inside of a fixed hemispherical bowl of diameter $2a$ and leans at C on the horizontal rim of the bowl, so that the other end B is outside. Determine the inclination to the horizon θ in the position of equilibrium.

(11) A homogeneous rod $AB = l$, of weight W, is hinged at its upper end A so as to turn in a vertical plane; a cord attached to its lower end B runs over a fixed pulley C (regarded as a point, on the same level with A) and carries a weight P. If $AC = c$ and $\measuredangle CAB = \theta$ be given find P and the reaction at A for equilibrium.

(12) A homogeneous rod $AB = 2l$ is free to turn in a vertical plane about a horizontal axis C dividing its length into segments $AC = a$, $CB = b$. It is in equilibrium in a horizontal position under the action of its own weight W and two forces F at A and F' at B, inclined downward and outward at angles α, α' to the horizon. Find the ratio $m = a/b$ and the pressure on the axis at C.

(13) A homogeneous rod $AB = 2l$, of weight W, rests at A on a smooth floor and leans at B against a smooth wall inclined to the horizon at an angle α. A cord attached at B runs over a pulley C at the top of the wall and carries a weight P. Find P and the reactions at A and B for equilibrium.

(14) What horizontal force P must be applied to the axle of a wheel of weight W to just pull it over an obstacle B rising one mth of the radius a above the horizontal road? What is the pressure P' on the obstacle?

(15) Determine the force P which, applied at the center of a wheel and inclined at a given angle α to the horizon, will just start the

wheel over a fixed cylindrical log, whose diameter is one nth of that of the wheel. For what value of a is P least?

2. STABILITY.

341. The equilibrium of the forces acting on a rigid body may subsist while the body is in motion. Thus, if the motion consist in a mere translation with constant velocity, the equilibrium will not be disturbed during the motion if the forces remain equal and parallel to themselves.

If, however, the body be subjected to a rotation, this will in general not be the case. The present considerations are restricted to the case of plane motion; the forces are supposed to lie in the plane of the motion and to remain equal and parallel to themselves and applied at the same points of the body.

342. Let A_1A_2 (Figs. 103 and 104) be a rigid rod having two equal and opposite forces F_1, F_2 applied at its extremities in the direction of the line A_1A_2. Let this rod be turned through an

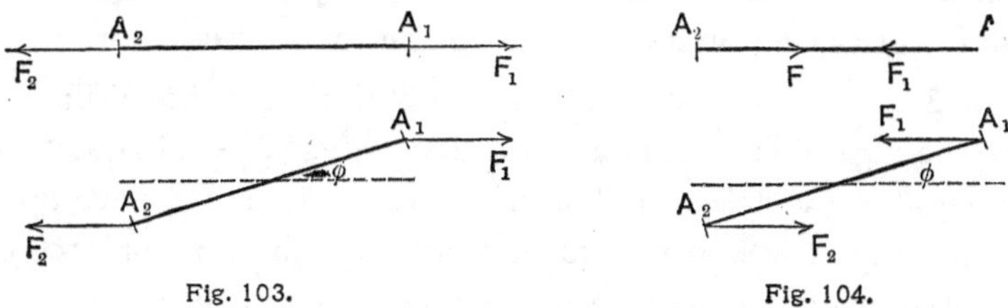

Fig. 103. Fig. 104.

angle ϕ about an axis at right angles to A_1A_2. In the new position the forces F_1, F_2, instead of being in equilibrium, form a couple whose moment is $\pm F_1 \cdot A_1A_2 \sin\phi$.

If in the original position of the rod the forces tend to increase the distance A_1A_2 (Fig. 103), the couple in the new position will tend to bring the rod back to its position of equilibrium. In this case the original position of the rod is said to be a position of **stable equilibrium.** The effect of the earth's magnetism on the needle of a compass offers a familiar example.

If, however, in the original position the forces tend to diminish the distance A_1A_2 (Fig. 104), the couple arising after displacement tends to increase the displacement and thus to remove the rod still farther from equilibrium. The original position in this case is said to be one of **unstable equilibrium.** The weight of a rod balanced in a vertical position and the reaction of the support may be taken as an illustration.

Finally, a third case would arise if the forces F_1, F_2, being still equal and opposite, were applied at one and the same point of the rod. The forces would then remain in equilibrium after a displacement of the rod; such equilibrium is called **neutral** or **astatic.**

Generally, the equilibrium of a rigid body is said to be **stable** if, after a sufficiently small displacement of the body, the forces tend to bring the body back to its position of equilibrium; **unstable** if, after a sufficiently small displacement, the forces tend to remove the body still farther from its position of equilibrium; **neutral** if, after a small displacement, the body remains in equilibrium.

These three types of equilibrium are also well illustrated by a homogeneous sphere placed within a hollow sphere (stable equilibrium), or balanced on top of another sphere (unstable equilibrium), or placed on a horizontal plane (neutral equilibrium).

343. The different cases of equilibrium can be distinguished by the algebraic sign of the product $A_1A_2 \cdot F_1 = A_2A_1 \cdot F_2$, which is negative for stable equilibrium, since A_1A_2 and F_1 have opposite sense (Fig. 103), positive for unstable equilibrium (Fig. 104), and indeterminate (since $A_1A_2 = 0$) for neutral equilibrium.

It is to be noticed that these considerations will hold whether the rotation of angle ϕ take place in the positive or negative sense. But they hold only within certain limits for the angle of rotation. Thus, in the example illustrated by Figs. 103 and 104, if ϕ were taken as large as π, a new position of equilibrium would be reached; but it would be unstable for Fig. 103, stable for Fig. 104.

344. Exercises.

(1) Explain the nature of the equilibrium of a body of weight W supported at a single point according to the position of that point above the centroid G, below G, and at G (*common balance*).

(2) A body of weight W is placed on a horizontal plane. Show that the equilibrium is stable if W meets the horizontal plane at a point A within the area of contact and that it is unstable if A lies on the contour of this area. If the actual area of contact have re-entrant angles, or consist of several detached portions, the area bounded by a thread drawn tightly around the actual area, or areas, of contact must be substituted.

(3) An oblique cylinder rests with its circular base on a horizontal plane in unstable equilibrium. If the length of its axis be twice the diameter of its base, what is the inclination of the axis to the horizon?

(4) Show how to determine graphically the stability of a retaining wall against toppling over the front edge of the base, the pressure of the earth behind the wall being given in magnitude, direction and position.

(5) A cube, a hemisphere, a right circular cone of height h and base radius a, and a right pyramid of height h and square base of side a are placed on a plane inclined to the horizon at an angle θ, the cube and the pyramid so that one side of the base is horizontal. If the lowest line or point be fixed, for what inclination θ will they topple over?

(6) A solid is formed by gluing the base of a homogeneous hemisphere of radius a and density ρ_1 on to the base of a homogeneous right cone whose base has the radius a and whose density is ρ_2. The solid is placed on a horizontal table, with the axis of the cone vertical. What must be the height h of the cone to make the equilibrium neutral?

(7) In Ex. (6), substitute a cylinder for the cone.

3. JOINTED FRAMES.

345. The equations of equilibrium are derived on the suppositions that all the forces of the given system act on one and the same rigid body and that this body is perfectly free to move. Hence, in applying these equations to determine the equilibrium of an engineering structure, a machine, etc., each rigid body must be considered separately, and the reactions required to make the body free must be introduced. It will be shown in a subsequent section how the principle of virtual work makes it possible to dispense with some of these precautions.

When two rigid rods are connected by a frictionless pin-joint whose axis is perpendicular to the plane of the rods, the action of either rod on the other at the joint is represented by a single force whose direction is in general unknown. Sometimes considerations of symmetry will enable us to determine this direction.

If a rigid rod, in equilibrium, be hinged at both ends and not acted upon by any other forces, the reactions of the hinges must of course be along the rod, and must be equal and opposite.

346. *Two rods AC, BC (Fig. 105) in a vertical plane, hinged together at C, rest with the ends A, B on a horizontal plane, and carry a weight W suspended from the joint C. If the proper weight of the rods be neglected, determine the normal pressures A_y, B_y and the horizontal thrusts A_x, B_x at A, B.*

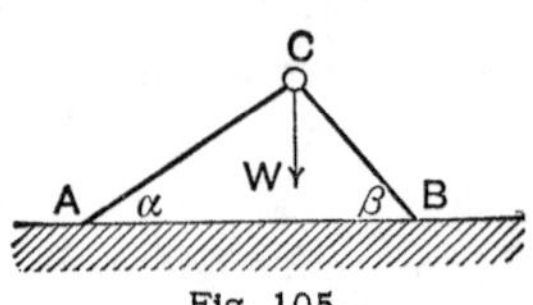

Fig. 105.

Resolving the weight W along CA, CB into W_A, W_B and considering the rod AC alone, it appears that the total reaction at A is along AC and $= W_A$; hence resolving W_A in the horizontal and vertical directions, A_x and A_y are found; similarly for BC. If α, β be the angles at A and B in the triangle ABC, we find

$$W_A = \frac{\cos\beta}{\sin(\alpha+\beta)} W, \quad W_B = \frac{\cos\alpha}{\sin(\alpha+\beta)} W;$$

$$A_x = B_x = \frac{\cos\alpha \cos\beta}{\sin(\alpha+\beta)} W, \quad A_y = \frac{\sin\alpha \cos\beta}{\sin(\alpha+\beta)} W, \quad B_y = \frac{\cos\alpha \sin\beta}{\sin(\alpha+\beta)} W.$$

As the horizontal thrusts at A and B are equal it makes no difference whether the rods be hinged to the supports at A and B, or whether the thrust is taken up by lateral supports, or by a string connecting the ends A, B of the rods.

347. *Two equal homogeneous rods AC, BC (Fig. 106) are hinged at A, B, C so as to form a triangle whose height h is vertical and whose base $AB = 2b$ is horizontal. The weight of each rod being W, find the reactions at the joints.*

Owing to the symmetry of the figure, the reactions at C must be equal and opposite and horizontal. The rod AC is subject to three forces only, viz. the horizontal reaction C, the weight W, and the reaction A; the latter must therefore pass through the intersection D of C and W.

If the direction of W intersect AB at E and the scale of forces be taken so as to have W represented by $DE = h$, DEA will be the force polygon; hence EA represents C and AD represents A on the same scale on which W is represented by h.

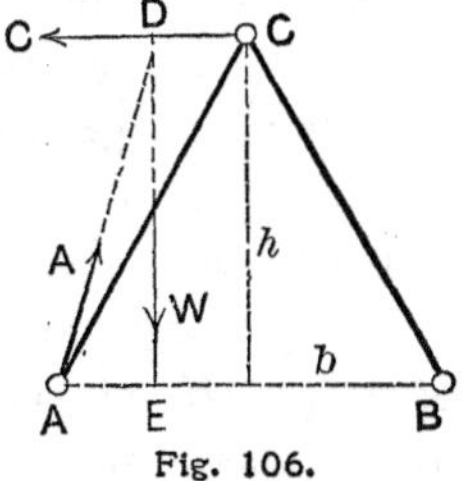

Fig. 106.

Analytically, the reactions are found by resolving the forces horizontally and vertically and taking moments about A:

$$A_x = C, \quad A_y = W, \quad C \cdot h = W \cdot \tfrac{1}{2} b;$$

whence

$$C = mW, \; A = \sqrt{A_x^2 + A_y^2} = \sqrt{m^2 + 1} \cdot W,$$

where $m = b/2h$.

348. *Two equal homogeneous rods AC, BC, each of weight W, are hinged at C; their ends A, B rest on a smooth horizontal plane; a third rod DE is hinged to them, connecting their middle points* (Fig. 107).

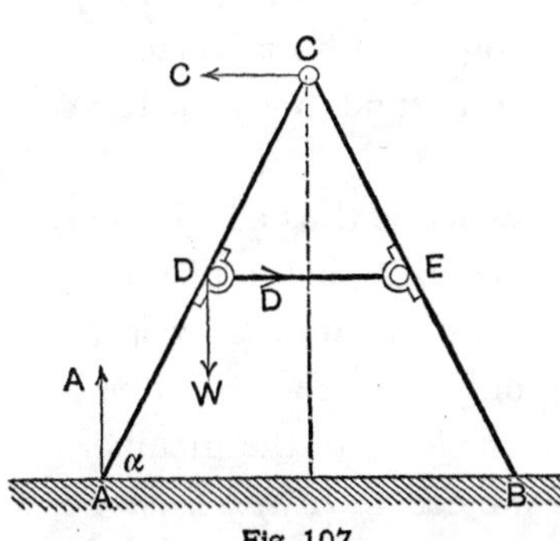

Fig. 107.

The plane AB being smooth, the reaction at A is vertical; the reaction at C is horizontal owing to the symmetry; that at D is likewise horizontal if the weight of the rod DE be neglected, for then this rod is subject only to the reactions at its ends.

Resolving horizontally and vertically and taking moments about D, we find in this case

$$A = W, \quad C = D = W \cot\alpha,$$

where

$$\alpha = \measuredangle BAC.$$

If, however, the weight w of the rod DE cannot be neglected, we have at D a horizontal reaction D_x and a vertical reaction D_y. The equilibrium of DE requires that $2D_y = w$. Hence resolving and taking moments as before, we find

$$A = W + \tfrac{1}{2} w, \quad C = D_x = (W + \tfrac{1}{2} w) \cot\alpha, \quad D_y = \tfrac{1}{2} w.$$

349. Exercises.

(1) Solve the problem of Art. 347 by replacing each weight W by $\frac{1}{2}W$ at each end of each rod, then resolving the load W at C along the rods, etc.

(2) Apply to the problem of Art. 348 the principle that three forces in equilibrium must pass through a point.

(3) Two homogeneous rods AC, BC of equal weight, but unequal length are hinged together at C while their other ends are attached to fixed hinges A, B in the same vertical line. Show that the line of action of the reaction at C bisects AB.

(4) Two homogeneous rods $AC = BC = 2l$, in a vertical plane, each of weight W, are connected by a pin-joint at C while at A and B there are fixed pin-joints, the point A lying a ft. above B. Find the reactions of A and B if the inclinations α, β of AC, BC to the horizon are given.

350. A triangular frame formed of rigid rods is rigid as a whole, even when the connections are pin-joints. A quadrangular frame with pin-joints becomes rigid only by the insertion of a diagonal.

The iron and steel trusses used for roofs and bridges generally consist of a system of triangles, or quadrangles with diagonals, so that the whole truss can be regarded as one rigid body, at least in first approximation.

Any one rod, or *member*, of the frame-work is thus acted upon by two equal and opposite forces, *i. e.*, by a *stress*, in the direction of its length, the external forces, including the proper weight, being regarded as applied at the joints only. If the stress be a *tension*, *i. e.*, if the forces tend to stretch or elongate the member, the latter is called a **tie**; a member subject to *compression* or crushing is called a **strut**.

Strictly speaking, owing to elasticity, the members of the frame slightly change their length and the supports yield. The points of application of the forces change therefore their position under any loading. These changes are neglected in the first approximation, the supports being assumed fixed and the frame rigid.

351. For the purpose of dimensioning the members, it is necessary to know the stress in every member. The following example illustrates a simple method for finding these stresses when the external forces are given.

Let the frame-work represented in Fig. 108 be cut in two along any line $\alpha\beta$; the portion on either side of this line must be in equilibrium under the action of its external forces and the

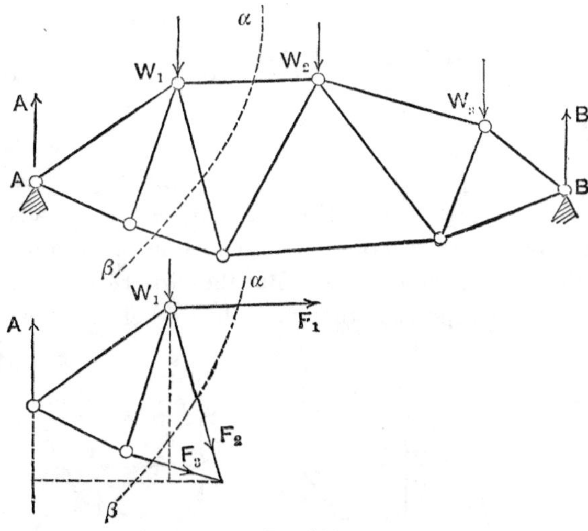

Fig. 108.

stresses in the members intersected by $\alpha\beta$. Thus, in the figure, the forces A, W_1, F_1, F_2, F_3 form a system in equilibrium; hence, the sum of the moments of these forces with respect to any point must vanish.

To determine F_1, take moments about the intersection of F_2 and F_3; thus F_2 and F_3 are eliminated from the equation of moments, and F_1 is found. Similarly F_2 is obtained by taking moments about the intersection of F_3 and F_1. The arms of the moments are best taken from an accurately drawn diagram of the frame-work.

If only two members be intersected by $\alpha\beta$, the origin for the moments is taken first on one, then on the other, of the two members intersected.

By beginning at one of the supports and taking sections through the successive panels, it will in the more simple cases be possible to draw the line $\alpha\beta$ so as to intersect not more than three members whose stresses are unknown. Thus the stresses in all the members can be determined.

352. Exercises.

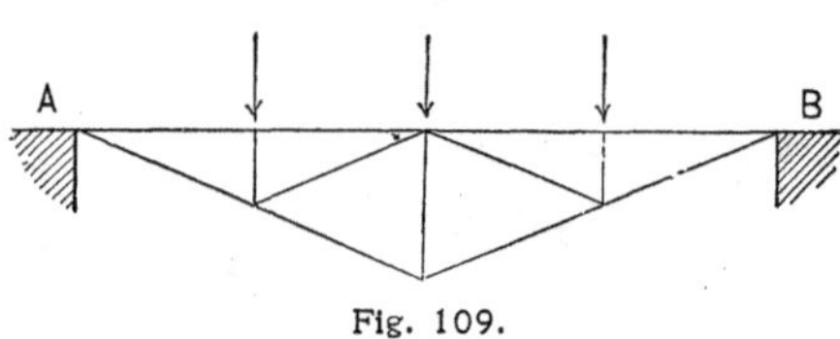

Fig. 109.

(1) Find the stresses in the braced beam AB (Fig. 109), carrying a weight of 6 tons at each joint of the upper chord. The horizontal width of the panels is 12 ft., the middle vertical is 9 ft.

(2) In Fig. 110, the dimensions are in feet, the loads in tons. After the first panel the sections cannot be so taken as to intersect not more than three unknown stresses. But the girder can be regarded as obtained by the superposition of two girders (each carrying half the

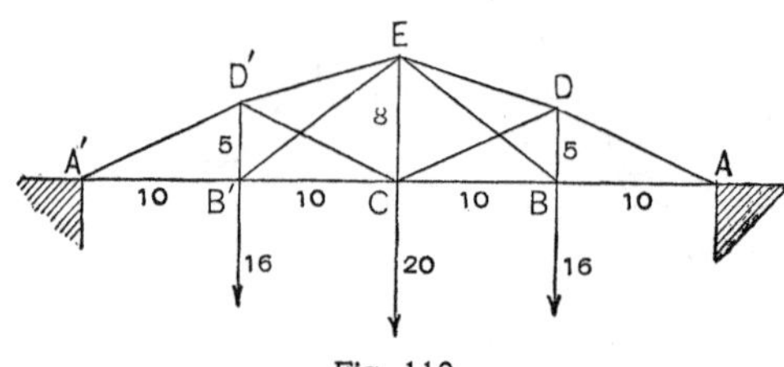

Fig. 110.

load), in one of which the diagonals $B'E$, CD are wanting, while in the other $D'C$, EB are wanting. Each of these can readily be computed.

4. GRAPHICAL METHODS.

353. The graphical method explained in Art. 299 for determining the resultant of a system of parallel forces can be extended without difficulty to the general case of a plane system of forces. The only difference will appear in the form of the force polygon, which for parallel forces collapses into a straight line, while in the general case it is an ordinary (unclosed) polygon whose closing line represents the resultant in magnitude and direction. In other words, when the forces are not parallel, they must be added geometrically, and not algebraically.

The construction of the funicular polygon and its properties are the same as for parallel forces.

If the force polygon does not close, the given system is equivalent to a single resultant represented in magnitude, direction, and sense by the closing line; its position is obtained from the funicular polygon whose initial and final lines must intersect on the resultant.

If, however, the force polygon closes, the system may be equivalent to a couple, or it may be in equilibrium. The distinction between these two cases is indicated by the funicular polygon. If the initial and final lines of this polygon coincide, the system is in equilibrium; if they are merely parallel, these lines are the directions of the forces of the couple to which the whole system reduces. The magnitude and sense of the forces of the resulting couple are obtained from the force polygon.

354. Thus it follows from the graphical as well as from the analytical method that *a plane system may be equivalent to a single force, or to a couple, or to zero.* In the first case, the force polygon does not close, and the initial and final sides of the funicular polygon intersect at a finite distance. In the second case, the force polygon closes, and the initial and final sides of the funicular polygon are parallel. In the third case, the force polygon closes, and the initial and final sides of the funicular polygon coincide.

The *graphical conditions of equilibrium* of a plane system are therefore, two: (1) the force polygon must close; (2) the funicular polygon must have its initial and final sides coincident.

355. To every *vertex* of the force polygon corresponds a *side* of the funicular polygon, and *vice versa.* The force polygon is said to close if the last vertex coincides with the first; similarly, the funicular polygon might be said to close when its last side coincides with the first. With this convention, we may say that *the conditions of equilibrium of a plane system require the closing of both the force polygon and the funicular polygon.*

356. One of the most important applications of the graphical methods is found in the *determination of the stresses in the frame-works* used for bridges, roofs, cranes, etc. The following example will illustrate the method.

Fig. 111 represents the skeleton frame of a roof truss subjected to the "loads" W_1, W_2, W_3 and the reactions of the supports A, B. The members of the frame in connection with the lines of action of these forces (imagined as drawn from infinity up to the points of application) divide the whole plane into a number of compartments marked in the figure by the letters $a, b, c, d, \ldots$. The external forces as well as the members of the frame (or the stresses acting along them) can thus be designated by the two letters of the two portions of the plane separated by the force or stress. For instance, the reaction A is denoted by ab, and the stresses in the two members concurring at A are bc and ca. The figure just described may be called the *frame diagram*; and we proceed now to construct its *stress diagram*.*

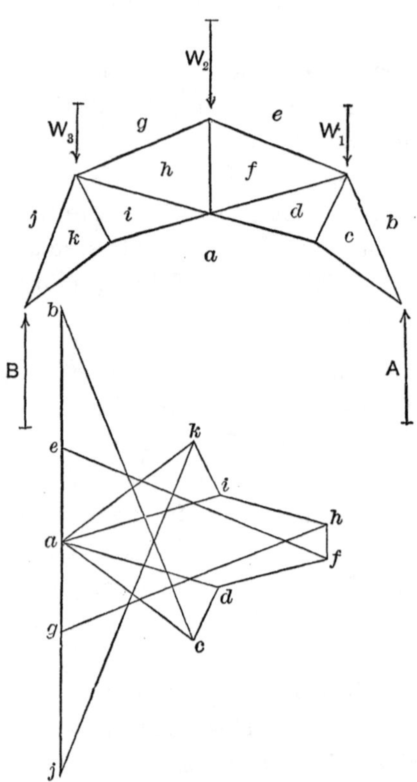

Fig. 111.

Laying off on a vertical line $be = W_1$, $eg = W_2$, $gi = W_3$, and bisecting bj at a, we have the polygon of the external forces which gives the reactions $A = ab$, $B = ja$.

* The student is advised to draw the stress diagram himself step by step as indicated in the text.

Next, beginning at the vertex A the stresses in the two members intersecting at A are found by resolving the reaction A along the directions of these members; and this is done in the stress diagram by drawing parallels to these directions through the points a and b. The intersection is denoted by c.

357. It will be noticed that the three lines meeting at A have corresponding to them, in the stress diagram, the three sides ab, bc, ca of a triangle. The force $A = ab$ is represented by ab; the stress in the member bc (*i. e.* in the member separating the compartments b, c in the frame diagram) is represented in magnitude, direction, and sense by the side bc in the stress diagram; and the stress in the member ca is given by the side ca of the triangle abc. To obtain the sense of each stress correctly, the triangle abc in the stress diagram must be traversed in the sense of the known force $A = ab$; this shows that the member bc is compressed, the stress at A acting towards A, while ca is subject to tension.

It will be found in general that *the lines of the stress diagram corresponding to all the lines meeting at any one vertex of the frame diagram form a closed polygon.* The reason is obvious: the forces at the vertex must be in equilibrium.

358. To continue the construction of the stress diagram, we pass to another vertex of the frame diagram, selecting one at which not more than two stresses are unknown. Thus at the vertex acd the stress in ac is known, being represented by ac in the stress diagram. Hence drawing through a a parallel to da, through c a parallel to cd, we find the point d of the stress diagram.

The vertex $dcbef$ can now be attacked; dc, cb, be are already drawn, and it only remains to draw ef parallel to ef and df parallel to df.

The rest explains itself. Considerations of symmetry are frequently helpful in affording checks.

359. Exercises.

(1) Check the computed stresses of Exercises (1) and (2), Art. 352, by constructing the stress diagrams.

(2) Find the stresses in the frame (Fig. 112) if $AA' = 40$ ft., $BB' = 12$ ft., the distance between these lines is 3 ft., and $BA = BC$.

(3) Determine the stresses in the frame, Fig. 113, if the load consists of seven weights, each of 2 tons, applied at the joints of the upper chord. Owing to the symmetry of the figure, it is sufficient to construct the stress diagram for half the frame. Beginning at A a difficulty arises at B and E since three members with unknown stresses pass through each of these points. The difficulty can be overcome in various ways; for instance, by observing that, owing to symmetry, the stresses in EG and EH must be equal; or by taking a vertical section near C and moments about C, whereby the stress in BB' can be found.

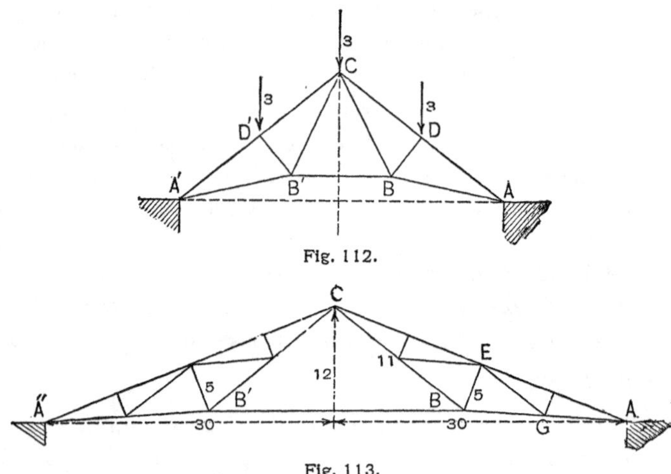

Fig. 112.

Fig. 113.

360. Shearing Force and Bending Moment. Consider a horizontal beam fixed at one end A (Fig. 114), and acted upon at the other end B by a vertical force F. If the beam be cut at any point C and the equilibrium of the portion AC be considered, the action on AC of the portion removed must be replaced by its equivalent. Now the force F at B is equivalent, by Art. 327, to an equal and parallel force F at C in connection with a couple whose moment is $F \cdot BC$.

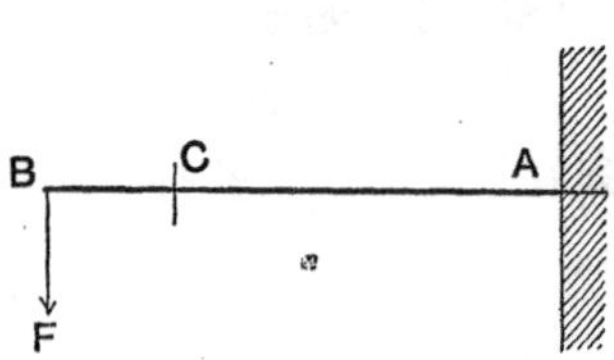

The force F at C is called the **shearing force** of the cross-section C, and the moment $F \cdot BC$ the **bending moment** at C. Both are of great importance in engineering, as their combined effect

Fig. 114.

represents what must be overcome by the resistance of the material of the beam, *i. e.* by the internal forces holding together its fibers.

These definitions are readily generalized. Let any beam or girder, supported in any manner, and acted upon by any number of vertical forces, be divided by a vertical cross-section into two portions A and B. For the portion A the shearing force at the cross-section is the sum of all the external forces acting on B; and the bending moment is the sum of the moments of all these forces with respect to some point in the cross-section.

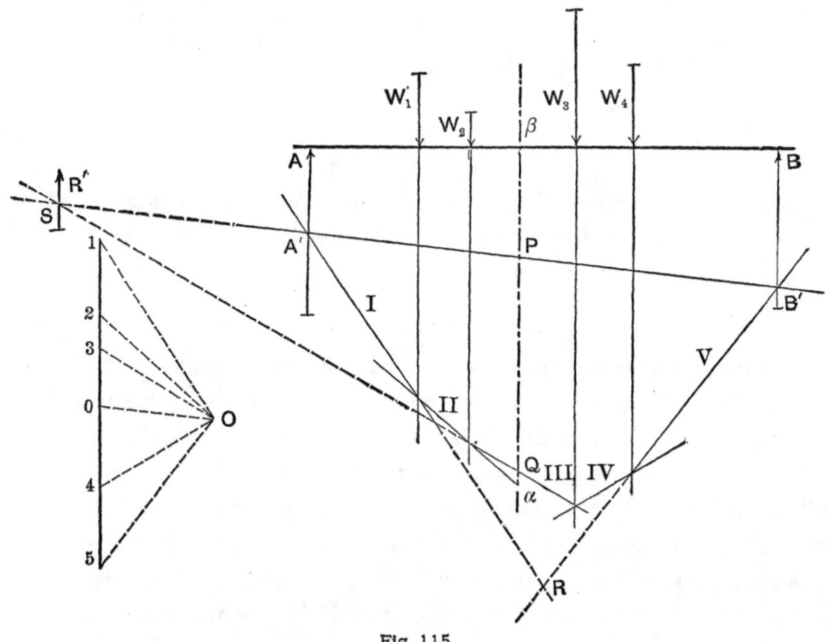

Fig. 115.

361. According to its definition the bending moment of a beam at any cross-section is found by adding the moments, with respect to the cross-section, of all the external forces on one side of the section.

Graphically, the bending moment is readily derived from the funicular polygon. Thus in Fig. 115, for the cross-section $\alpha\beta$, the resultant of the forces on the left is $R' = A - W_1 - W_2 = 03$

in the force polygon. Its position is found by determining the intersection of the two sides $A'B'$ and III of the funicular polygon met by the section $\alpha\beta$. For the funicular polygon resolves A along $A'B'$ and I, W_1 along I and II, W_2 along II and III. The components falling into the same line being equal and opposite (as appears from the force polygon), the forces A, W_1, W_2 are together equivalent to the components along $A'B'$ and III; their resultant R' must therefore pass through the intersection S of these lines.

Now if p be the horizontal distance of the point S from $\alpha\beta$, the bending moment at $\alpha\beta$ is $R' \cdot p = 0\,3 \cdot p$. If $\alpha\beta$ intersect $A'B'$ at P, III at Q, the triangles SPQ and $O\,0\,3$ are similar, so that their altitudes p and H are as the corresponding sides PQ and $0\,3$; hence

$$p = \frac{PQ}{0\,3} \cdot H,$$

and the value of the bending moment is $H \cdot PQ$. As H is constant, we find that *the bending moment is proportional to the vertical height, or ordinate, of the funicular polygon.*

5. FRICTION.

362. The reaction between two surfaces in contact has so far been regarded as directed along the common normal of the surfaces. This is true when the surfaces are *perfectly smooth.*

The surfaces of physical bodies are *rough, i. e.* they present small elevations and depressions; when two such surfaces are "in contact" the projections of one will more or less enter into depressions of the other; the greater the normal pressure between the surfaces, the more will this be the case. Hence when a tangential force acting on one of the bodies tends to *slide* its surface over that of the other body, a resistance will be developed whose magnitude must depend on the roughness of the surfaces and on the normal pressure between them. This resistance is called the **force of sliding friction,** or simply the **friction.**

The study of friction belongs properly to applied mechanics, and will here only be touched upon very briefly.

363. Imagine a body resting with a plane surface on a horizontal plane. Let a small horizontal force P be applied at its centroid (which is supposed to be situated so low that the body is not overturned), and let the force P be gradually increased until motion ensues. At any instant before motion sets in, the friction is equal to the value of P at that instant. The value of P at the moment when motion just begins is equal and opposite to the *frictional resistance* F between the surfaces at this moment, and this resistance is called the **limiting static friction.**

Careful experiments with dry solids in contact have shown this force to be subject to the following laws:

(1) *The magnitude of the limiting friction F bears a constant ratio to the normal pressure N between the surfaces in contact;* that is

$$F = \mu N,$$

where μ is a constant depending on the condition and nature of the surfaces in contact. This constant which must be determined experimentally for different substances and surface conditions is called the **coefficient of static friction.** It is in general a proper fraction; for perfectly smooth surfaces $\mu = 0$.

(2) *For a given normal pressure the limiting static friction, and hence the coefficient of static friction, is independent of the area of contact, provided the pressure be not so great as to produce cutting or crushing.*

364. The frictional resistance between two surfaces in relative motion is called **kinetic friction.** It is subject, in addition to the two laws just mentioned, to the third law:

(3) *For moderate velocities, kinetic friction is nearly independent of the velocities of the bodies in contact.*

The coefficient of static friction is somewhat greater than that of kinetic friction. A slight jarring will often reduce the coefficient from its static to its kinetic value.

It must not be forgotten that these so-called **laws of friction** are experimental laws, and therefore true only approximately and within the limits of the experiments from which they were deduced. When the relative velocity of the surfaces in contact is high, or when, as is usually the case in machinery, a lubricating material is introduced between the two surfaces, the frictional resistance is found to depend on a number of other circumstances, such as the temperature, the form of the surfaces, the velocity, the nature of the lubricator, etc. Indeed, when the supply of the lubricant is sufficient, the two solid surfaces are kept by it out of actual contact; the coefficient of friction in this case varies with the pressure, area of contact, velocity, and temperature.

365. Consider again a body resting on a horizontal plane (Fig. 116) and acted upon by a horizontal force P just large enough to equal the limiting friction F. The normal reaction N of the plane is equal and opposite to the weight W. The body is thus in equilibrium under the action of the two pairs of equal and opposite forces; but motion will ensue as soon as P is increased. If P be decreased, F will decrease at the same rate, so that the equilibrium remains undisturbed.

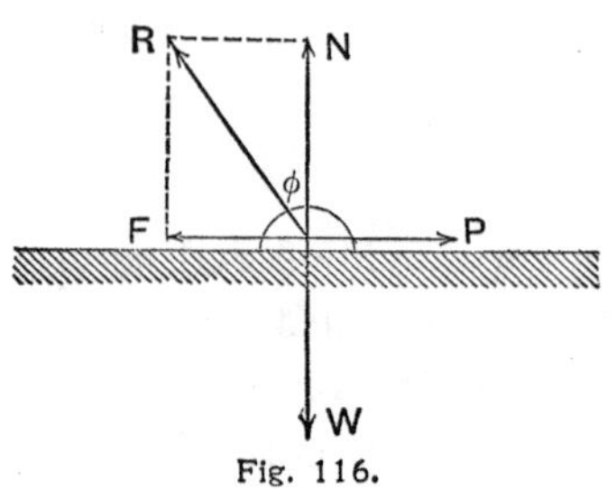

Fig. 116.

The force of friction F can be combined with the normal reaction N to form a resultant,

$$R = \sqrt{F^2 + N^2} = \sqrt{P^2 + W^2},$$

which represents the *total reaction* of the horizontal plane.

If ϕ be the angle between N and R when F has its limiting value $F = \mu N$ (Art. 363), we have, since $\tan\phi = F/N$,

$$\tan\phi = \mu.$$

The angle ϕ thus furnishes a graphical representation for the coefficient of friction μ; it is called the **angle of friction.**

366. If the plane be not horizontal, but inclined to the horizon at an angle θ, the weight W of the body (regarded as a particle) resting on the plane can be resolved into a component $W \sin\theta$ along the plane, and a component $W \cos\theta$ perpendicular to it (Fig. 117). Hence, if no other forces act on the body it will be in equilibrium, provided the component $W \sin\theta$ be not greater than the limiting friction $F = \mu W \cos\theta$. The limiting condition of equilibrium is, therefore,

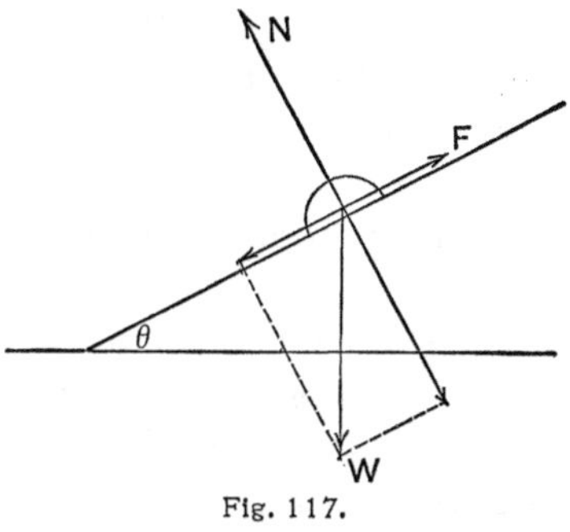

Fig. 117.

$$\mu W \cos\theta = W \sin\theta, \quad \text{or} \quad \mu = \tan\theta;$$

in other words, if the angle θ be gradually increased, the body will not slide down the plane until $\theta > \phi$. This furnishes an experimental method of determining the angle of friction ϕ, which on this account is sometimes called the **angle of repose.**

367. A particle P (Fig. 118) will be in equilibrium on any rough surface, if the total reaction of the surface, *i. e.* the resultant R of the normal reaction N and the friction F, is equal and opposite to the resultant R' of all the other forces acting on the particle.

The limiting value of the angle between N and R is ϕ, so that the particle can be in equilibrium only if the resultant R' makes with the normal an angle $\overline{\gtrless}\, \phi$. Hence, if about the normal PN as axis, and with P as vertex, a cone be described whose vertical angle is 2ϕ, the condition of equilibrium is that R' must lie within this cone. The cone is called the **cone of friction.**

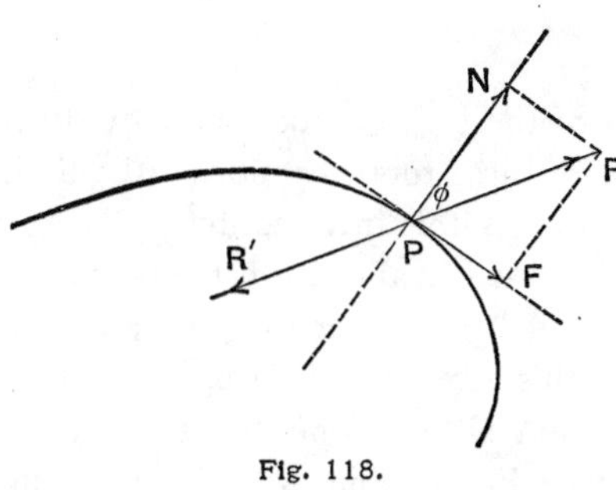

Fig. 118.

368. The idea of the angle of friction suggests a *graphical method* for problems on equilibrium with friction.

The case of a rod resting on two inclined planes, Art. 339, Fig. 102, may serve as an example. If the intersection E, of the normal reactions A and B lies on the vertical through D, the rod will be in equilibrium whether there be friction at A and B or not. When this condition is not fulfilled, the rod may still be in equilibrium if there be sufficient friction between the ends of the rod and the supporting planes.

Let $\mu = \tan\varphi$ be the coefficient of friction on the plane CA, $\mu' = \tan\varphi'$ that on CB; then the total reactions at A and B will, by Art. 365, make angles not greater than φ and φ', respectively, with the normals to the planes. Hence the two limiting positions of equilibrium for

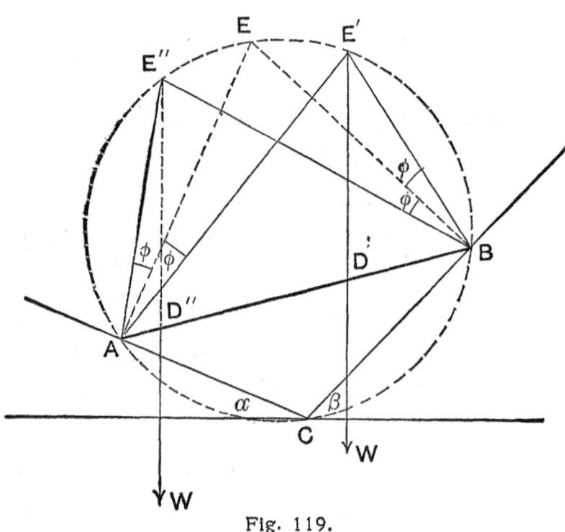

Fig. 119.

the weight W, in a given position of the rod, can be found by determining the intersection of the lines of these total reactions; the limiting position of W is the vertical through this intersection. Thus, to prevent the rod from sliding up the plane CA and down the plane CB, the friction angles φ, φ' must be applied in the negative sense (clockwise) to the normals at A and B; this gives one limiting position D' for the point D. The other position D'' is found by applying the friction angles in the positive sense. Equilibrium will therefore subsist if the weight be placed anywhere between D' and D''.

The construction is somewhat simplified when $\varphi = \varphi'$, since then the intersections of the total reactions lie on the circle described about ABC (Fig. 119).

369. As another example consider the ordinary jack intended to raise an eccentric load W acting vertically downwards through A (Fig. 120) by a force P passing vertically upwards through the pitch line B of the rack. Near C and D the rack is pressed against the casing. The directions of the total reactions C, D at these points are found by applying the friction angle to the normals.

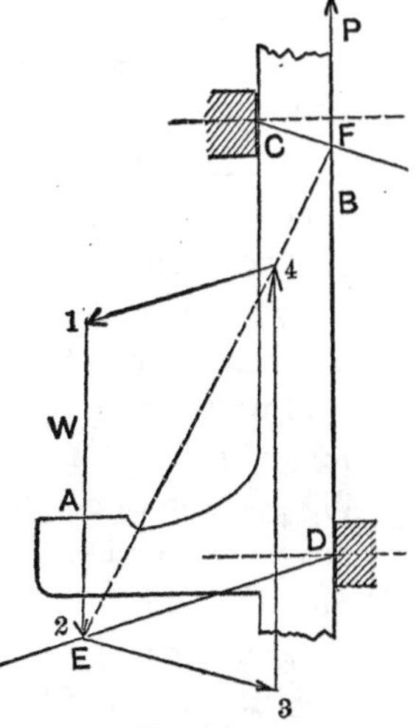

Fig. 120.

The four forces W, P, C, D can be in equilibrium only if the resultant of W and D is equal and opposite to the resultant of P and C; hence, if E be the intersection of W and D, F that of P and C, each of these resultants must act along EF.

If the load W be known, the other forces can now be found by constructing the force polygon. Draw $1\,2 = W$ in position (*i. e.* through A); draw $2\,3$ parallel to C; $4\,1$ parallel to D; and through the intersection 4 of $4\,1$ with EF draw the vertical $3\,4$ to the intersection 3 with $2\,3$.

370. Exercises.

(1) Determine the tractive force required to haul a train of 160 tons with constant velocity up a grade of 2 per cent. if the coefficient of friction is 1/200.

(2) A weight W is to be hauled along a horizontal plane, the coefficient of friction being $\mu = \tan\varphi$. Determine the required tractive force P if it is to act at an inclination α to the horizon, and show that this force is least when $\alpha = \varphi$.

(3) A particle of weight W is in equilibrium on a rough plane inclined to the horizon at an angle θ, under the action of a force P parallel to the plane along its greatest slope. Determine P: (a)

when $\theta > \varphi$, (*b*) when $\theta = \varphi$, (*c*) when $\theta < \varphi$, $\varphi = \tan^{-1}\mu$ being the angle of friction.

(4) Solve Ex. (3) (*a*) graphically by means of the friction angle and determine what part of P is required to overcome friction.

(5) A body weighing 240 pounds is pulled up a plane inclined at 45°, by means of a rope. If $\mu = \frac{1}{3}$, find the tension of the rope. What portion of it is due to friction?

(6) A particle of weight W is kept in equilibrium on a plane inclined at an angle θ to the horizon by a force P making an angle α with the line of greatest slope (in the vertical plane at right angles to the intersection of the inclined plane with the horizon). Find the conditions of equilibrium when the particle is on the point of moving (*a*) down the plane, (*b*) up the plane.

(7) A homogeneous straight rod $AB = 2l$ of weight W rests with one end A on a horizontal floor, with the other end B against a vertical wall whose plane is at right angles to the vertical plane of the rod. If there be friction of angle φ at both ends, determine the limiting position of equilibrium.

(8) Two particles whose weights are W, W' are in equilibrium on an inclined plane, being connected by a string directed along the line of greatest slope. If the coefficients of friction are μ, μ', determine the inclination of the plane.

(9) A straight homogeneous rod $AB = 2l$, of weight W, rests with the lower end A on a rough horizontal plane and with the point C ($AC = c$) on a smooth cylindrical support. The rod is in equilibrium when inclined at a given angle θ to the horizon; determine the coefficient of friction at A and the reactions at A and C.

(10) If in Ex. (9) there be friction both at A and C, the friction angle φ being the same, find the position of equilibrium and the reactions at A and C.

(11) A solid homogeneous hemisphere is placed with its curved surface on a rough inclined plane; investigate the conditions of equilibrium.

371. Journal Friction. A journal, or trunnion, is the cylindrical end of a horizontal shaft, by means of which the shaft is supported in its bearing. In Fig. 121, which shows the cross-section of shaft and bearing, the difference of the diameters is exaggerated for the sake of clearness; the fit between journal and bearing, even if it were perfect originally, would not remain so owing to unequal wearing.

When the shaft is at rest the pressure W of the journal on the bearing acts vertically downward through the center O and the lowest point A of the circle. The rotation produces at A a linear velocity to the right if ω be counter-clockwise; hence the frictional resistance μW is to the left. The moment of this force about O is $= \mu Wr$; and an equal and opposite moment is required to overcome this friction moment. But this is not sufficient for equilibrium; the sum of the forces in the horizontal direction should be zero, and this would not be the case if the frictional force acted at the lowest point A. Indeed, the journal will move up the side of the bearing to a point D such that the horizontal components of the normal reaction N and of the friction μN are equal and opposite. This will be the case (Fig. 122), when the radius OD makes with the vertical an angle α such that

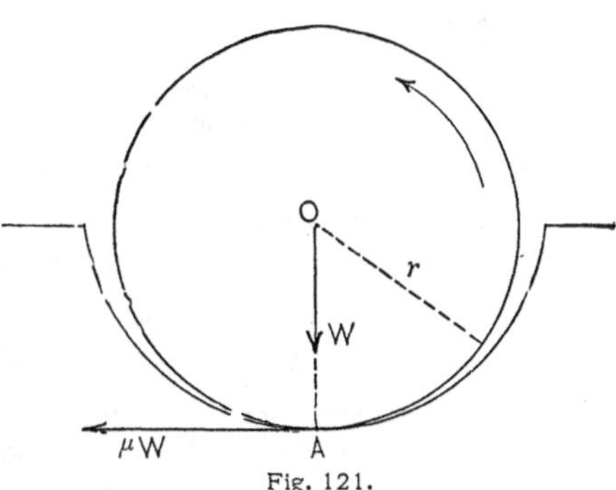

Fig. 121.

$$N \sin\alpha = \mu N \cos\alpha,$$

whence $\tan\alpha = \mu = \tan\varphi$, *i. e.* $\alpha = \varphi$. As the total reaction R of the bearing, *i. e.* the resultant of N and μN, is also inclined at the friction angle φ to the normal DO, it appears that, in the position of equilibrium, the reaction R is parallel to OA and hence vertical.

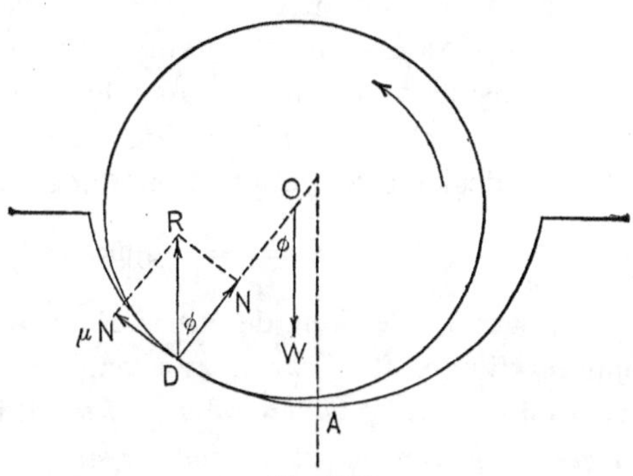

Fig. 122.

The two equal and opposite forces R and W form the friction couple whose moment is $= W \cdot r \sin\varphi$. Owing to the smallness of the lateral displacement of the center O, the weight W can still be regarded as passing through the lowest point A of the bearing.

For lubricated journals we have $\mu = \tan\varphi < 0.1$, so that we can replace $\tan\varphi$ by $\sin\varphi$; the moment of the friction couple is then $= \mu Wr$, *i. e.* the frictional force μW can be regarded as acting at the lever-arm r (Fig. 121).

372. The effect of journal friction can also be illustrated as follows (Fig. 123). The shaft and journal may be regarded as rotating uniformly about their common horizontal axis under the action of a driving force whose moment about O would have to be equal and opposite to that of the resistance, or load, if there were no journal friction. For, in this case, the reaction of the bearing to the weight W of the shaft would act vertically upwards through the axis of the shaft, so that its moment would be zero. The existence of friction between journal and bearing requires an increase of the driving force, which may be regarded as a small tangential force P applied at any point B such that its moment $P \cdot OB$ equals the moment about O of the frictional resistance.

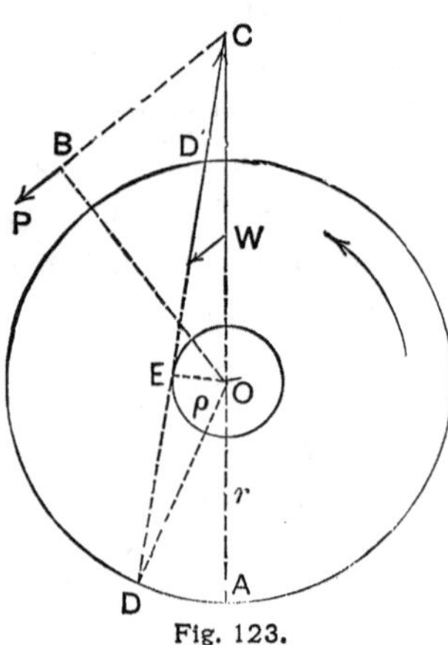

Fig. 123.

Let C be the intersection of the direction of this force P with the vertical through O and A, which is the line of action of the weight W of the shaft. The resultant of P and W passes through C, and intersects the circumference of the journal at a point D near A; the total reaction of the bearing is equal and opposite to this resultant. As this total reaction must make an angle φ with the normal at D, we have for the perpendicular OE dropped from O on CD:

$$OE = \rho = r \sin\varphi,$$

where r is the radius of the journal. A circle described about O, with ρ as radius, has the total reaction of the bearing as a tangent. This circle is called the **friction circle.** As φ is generally very small in the case of journal friction, $\mu = \tan\varphi$ can be substituted for $\sin\varphi$, and we have for the radius ρ of the friction circle

$$\rho = \mu r.$$

As soon as any one point is known through which the total reaction must pass (as the point C in Fig. 123), its direction is found by drawing through this point a tangent to the friction circle.

373. If the shaft revolved in the opposite sense, *i. e.* clockwise (instead of counter-clockwise, as assumed in Fig. 123), the tangent to the friction circle would have to be drawn through C on the *other* side of the friction circle.

In the case of **axle-friction,** *i. e.* when the journal, or axle, is fixed, while the bearing, or hub, revolves about it, the same considerations would apply, except that the point of application of the total reaction would now be at the top, at D', instead of D.

374. Pin-friction, as it occurs in link-work and jointed frames that are not absolutely stiff, is not different from journal friction or axle-friction, and can be treated in the same way. Thus, a link connected to other parts of a machine by means of a pin at each end would transmit the force along the line joining the centers of the pins if there were no friction. To take account of pin-friction, we have only to draw the friction circles about the center of each pin; the direction in which the force is transmitted by the link is tangent to both these circles.

Which one of the four common tangents represents this direction must be decided in each particular case by considering that the reaction exerted by one link on another connected with it by a pin is in the

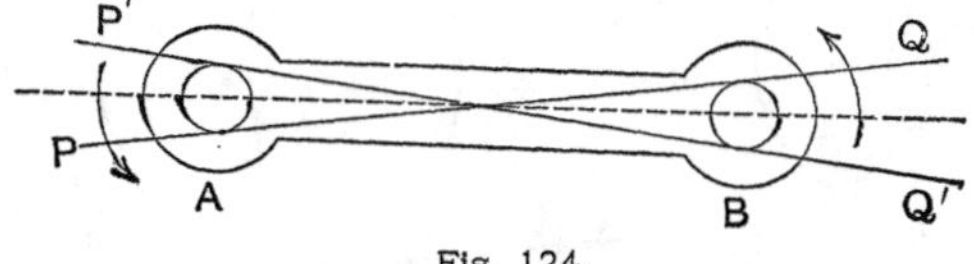

Fig. 124.

direction of the motion of the former relative to the other. Thus if the link AB (Fig. 124) be subject to tension, and its motion relative to the adjoining links at A and B be as indicated by the arrows in the figure, the contact between the link and pin will be on the outside both at A and B; the friction is, therefore, directed downwards at A and upwards at B, and the line PQ along which the force is transmitted touches the friction circle at A below, at B above.

If the link were under compression, with the same relative motions, the line of the force would have the direction $P'Q'$.

375. The simplest case of **pivot-friction** is that of a vertical shaft of weight W resting with its circular end on a plane horizontal support. If a be the radius of the end of the shaft, the pressure per unit of area is $W/\pi a^2$, and the pressure on a polar element of area is $(W/\pi a^2) \cdot rdrd\theta$.

The friction at this element, $\mu(W/\pi a^2) \cdot rdrd\theta$, is directed along the tangent to the circle of radius r; its moment with respect to the center O of the circle is therefore $\mu(W/\pi a^2)r^2drd\theta$. Hence the whole moment of friction about O is

$$\frac{\mu W}{\pi a^2}\int_0^{2\pi} d\theta \int_0^a r^2dr = \tfrac{2}{3}\mu Wa = \mu W \cdot \tfrac{2}{3}a.$$

This may be regarded as the moment of a force μW applied at a distance $\frac{2}{3}a$ from the center. The result is of little practical value since the pressure of a vertical shaft cannot be regarded as distributed uniformly over the area of contact, as here assumed.

376. Belt-friction. A belt running over two pulleys and stretched so tight as to prevent slipping is a common means of transferring the rotary motion about the axis of one pulley, say A, to the axis of the other pulley B; A is called the driver, B is the driven pulley. We assume the axes parallel and the rotation counter-clockwise.

When the pulleys are at rest the tension in DE (Fig. 125) is of

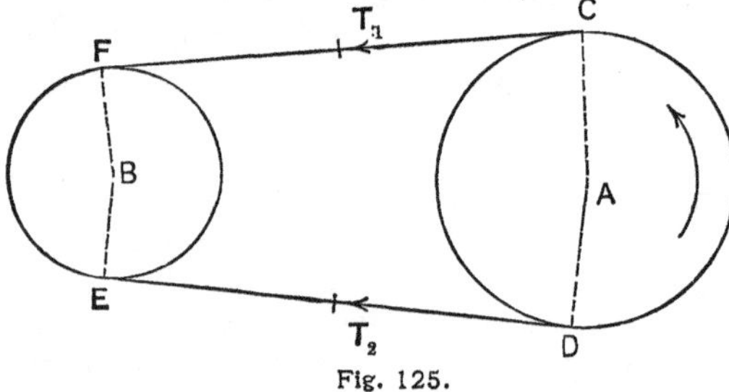

Fig. 125.

course equal to the tension in CF. But if the pulley A be set in motion, say by a tangential driving force P acting at a lever arm p, while the pulley B experiences a resistance Q whose arm is q, the tension in DE will increase to a certain value T_2, and the tension in CF will decrease to a value T_1 until the difference $T_2 - T_1$ is sufficient to overcome the resistance Q. This difference is due to the friction along the surface of contact CD. If the resistance Q be too great this friction might not be sufficient, and slipping of the belt on the driver would occur.

377. Let us try to determine the condition which T_1 and T_2 must satisfy to prevent slipping. To do this we determine the equilibrium of the belt at the moment when slipping is just on the point of taking place.

The tension of the belt increases gradually along the arc of contact CD from the value T_1 at C to the value T_2 at D. Let it be T at the point P and $T + \Delta T$ at the near point P' (Fig. 126). The portion PP' of the belt is in equilibrium under the action of the forces T, $T + \Delta T$ and the reaction ΔR of the pulley; hence ΔR must pass through the

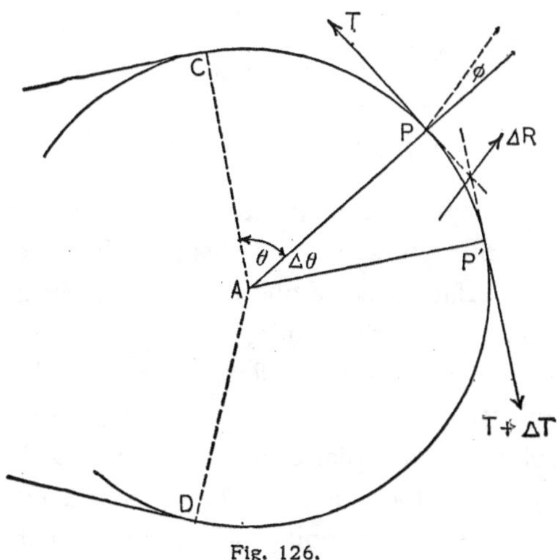

Fig. 126.

intersection of T and $T + \Delta T$ and be equal and opposite to their resultant. As the angle $PAP' = \Delta\theta$ approaches zero, the reaction ΔR approaches as limiting direction the line through P inclined to AP at the friction angle φ. Resolving therefore $T + \Delta T$ along T and at right angles to it we have:

$$\tan\varphi = \lim_{\Delta\theta = 0} \frac{(T + \Delta T)\cos\Delta\theta - T}{(T + \Delta T)\sin\Delta\theta}$$

$$= \lim \frac{1}{T + \Delta T} \cdot \frac{\Delta\theta}{\sin\Delta\theta} \cdot \frac{(T + \Delta T)\cos\Delta\theta - T}{\Delta\theta}$$

$$= \lim \frac{1}{T + \Delta T} \cdot \frac{\Delta\theta}{\sin\Delta\theta} \cdot \left(\frac{\Delta T}{\Delta\theta}\cos\Delta\theta + T \cdot \frac{\cos\Delta\theta - 1}{\Delta\theta}\right).$$

As $\lim \Delta\theta / \sin\Delta\theta = 1$, $\lim \Delta T / \Delta\theta = dT/d\theta$, $\lim(\cos\Delta\theta - 1) / \Delta\theta = 0$, we have finally

$$\tan\varphi = \frac{1}{T}\frac{dT}{d\theta}.$$

Writing μ for $\tan\varphi$ and integrating over the whole arc of contact CD we find, if θ_1 be the angle of this arc,

$$\log T_2 - \log T_1 = \mu\theta_1, \quad \text{or} \quad \frac{T_2}{T_1} = e^{\mu\theta_1}.$$

Taking common logarithms we have:

$$\text{Log}\frac{T_2}{T_1} = 0.4343\mu\theta_1,$$

where θ must be expressed in radians.

378. Rolling Friction. Kinematically the difference between sliding and rolling motion consists in this: in the case of sliding the same point or surface area of one body comes in contact with different points or areas of the other; in rolling, the points that come successively in contact are different for both bodies. We shall consider only the simplest case, that of a homogeneous circular cylinder rolling over a plane surface. The motion is called *pure rolling* only when the line of contact is the instantaneous axis of rotation; the lowest point of the cross-section of the cylinder is therefore instantaneously at rest in the case of rolling. To bring about this result there must exist sufficient sliding friction between cylinder and plane to prevent sliding; rolling therefore always presupposes the existence of sliding friction.

If both cylinder and plane were perfectly rigid the slightest horizontal force (say, at right angles to the axis of the cylinder) would suffice to produce rolling; in this case there would be no resistance to rolling.

If the cylinder alone be regarded as rigid while the material of the supporting plane yields (unelastically) to the pressure W of the cylinder (Fig. 127), the case is somewhat similar to that of pulling a wheel over an obstacle in the road (see Ex. 14, Art.

340). The reaction of the ground will act a short distance c in front of the vertical radius $CA = r$; for equilibrium this reaction must pass through the center C. The horizontal force F necessary to produce rolling is equal and opposite to the horizontal component of this reaction; and as we assume the yielding of the ground to be very slight, we have $F/W = c/r$, or

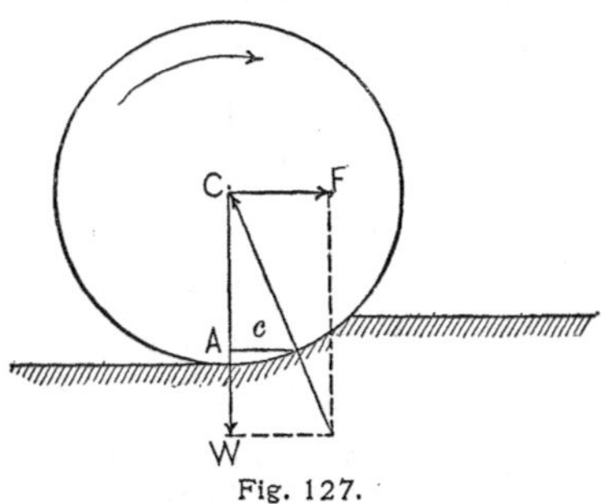

Fig. 127.

$$F = c\frac{W}{r}.$$

The formula shows the advantage of a large radius.

The force F is called the *force of rolling friction;* a couple of moment Fr is often spoken of as the *couple of rolling friction.*

As it would be difficult to determine the point of application of the reaction, the constant c must be found experimentally; it is called the *coefficient of rolling friction.* The numerical value of this coefficient for hard substances is very much smaller than that of the coefficient of sliding friction.

379. If the material of the supporting plane is partly elastic it will rise behind the wheel after having been compressed, thus diminishing the amount of rolling friction. It would seem to follow that there is no resistance to rolling for perfectly elastic materials. This is, however, not the case as careful experiments have shown. The vertical compression which makes the supporting surface bulge out in front of the wheel as well as behind it is accompanied by a horizontal extension which makes the wheel slip slightly; this slipping calls forth sliding friction both behind and in front of the lowest point.

VI. *Solid Statics.*

I. THE CONDITIONS OF EQUILIBRIUM.

380. The equilibrium of a rigid body in the most general case, that is, when acted upon by any number of forces F in a space of three dimensions, can be investigated in a manner similar to that adopted for the plane system in Art. 330.

Selecting as origin any point O rigidly connected with the body, let two equal and opposite forces F, $-F$ be applied at O, for every one of the given forces F. The effect of the given system of forces on the body is not changed by the introduction of these forces at O. But we may now regard the given force F acting at its point of application P as replaced by the equal and parallel force F at O, in combination with the couple formed by the original force F at P and the force $-F$ at O. All the forces of the given system are thus transferred to a common point of application O, and can therefore be compounded into a single resultant R, passing through O and represented in magnitude and direction by the geometric sum of the forces. In addition to this resultant R, we obtain as many couples $(F, -F)$ as there were forces given; and their resultant is found by geometrically adding the vectors of the couples (Art. 325).

Thus the given system of forces is seen to be equivalent to a resultant R in combination with a couple whose vector we shall call H; in other words, it has been proved that *any system of forces acting on a rigid body can be reduced to a single resultant force in combination with a single resultant couple.*

381. A further reduction is in general not possible. The general *conditions of equilibrium* are, therefore,

$$R = 0, \quad H = 0.$$

Under special conditions it may of course happen that R is perpendicular to the vector H. In this case R and H can be combined into a single force R (Art. 326), so that the whole system reduces to a single resultant.

382. It is to be noticed that in the general reduction of forces (Art. 380), the magnitude, direction, and sense of the resultant force R are entirely independent of the position of the origin O, the resultant being simply the geometric sum of all the given forces. The resultant couple H, on the other hand, will in general differ according to the origin selected.

To investigate this dependence, let R, H (Fig. 128) be the *elements of reduction* for the origin O; *i. e.* let R be the resultant, H the vector of the resulting couple of a given system of forces when O is selected as origin. To find the elements of reduction of the same system of forces when some other point O' is taken as origin, it is only necessary to apply at O' two equal and opposite forces R, $-R$, each equal and parallel to the original resultant R. The given system of forces being equivalent to R and H at O will also be equivalent to the resultant R at O', the couple whose vector is H (which may be drawn through O' without changing its effect), and the couple formed by R at O and $-R$ at O'. If l be the line of R through O, l' the line of R through O', and r the distance of these parallels, the moment of the latter couple is Rr and its vector is at right angles to the plane (l, O'). Combining the vectors H and Rr into a resultant vector H' by geometric addition, we have found the elements of reduction R, H' for the origin O'.

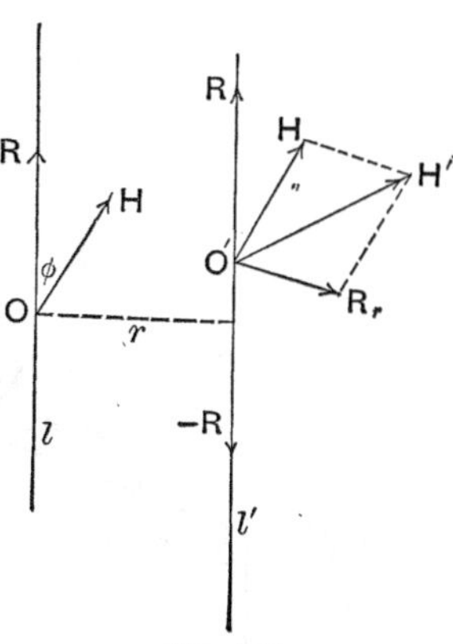

Fig. 128.

383. If the new origin O' had been selected on the line l of the original resultant, no new couple (R, r) would have been introduced, and H would not have been changed. But whenever the line of action l of the resultant is changed, the vector of the resultant couple H is changed.

By increasing the distance r between l and l' the moment Rr of the additional couple is increased. The effect of combining this additional couple Rr with H is, in general, to vary both the magnitude of the resulting vector H' and the angle ϕ it makes with the direction of the resultant R. It can be shown that the line l' of the new resultant can always be selected so as to reduce the angle ϕ to zero. The line l_0 for which $\phi = 0$, *i. e.* for which the vector H of the resultant couple is parallel to the resultant force R, is called the **central axis** of the given system of forces. We proceed to show how it can be found.

384. Let the vector H be resolved at O into a component $H_0 = H \cos\phi$ along l, and a component $H_1 = H \sin\phi$, at right angles to l (Fig. 129). In the plane passing through l at right angles to H_1, it is always possible to find a line l_0 parallel to l at a distance r_0 from l, such as to make $Rr_0 = -H_1$.

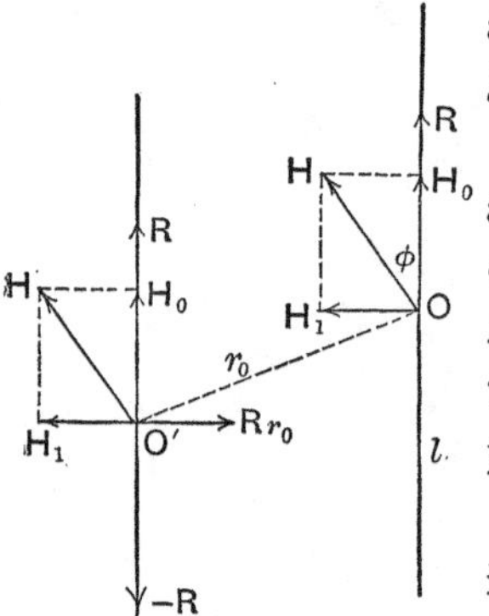

Fig. 129.

The line l_0 so determined is the central axis. For, if this line be taken as the line of the resultant R, the additional couple Rr_0 destroys the component H_1, so that the resulting couple H_0 has its vector parallel to R.

385. As the direction of the vector H is always changed in passing from line to line, there can be but one central axis for a given system of forces.

It appears from the construction of the central axis given in Art. 384, that the vector of the resulting couple for this axis l_0 is $H_0 = H \cos\phi$; it is, therefore, less than for any other line.

It is instructive to observe how the vector H increases and changes its direction as we pass from the central axis l_0 to any parallel line l.

The transformation from l_0 to l requires the introduction of a couple (R, r_0) whose vector Rr_0 (Fig. 130) is at right angles to the plane (l_0, l) and combines with H_0 to form the resulting couple H for l. As the distance r_0 of l from l_0 is increased, both the magnitude of H and the angle ϕ it makes with l increase until, for an infinite r_0, the angle ϕ becomes a right angle.

386. It is evident that since $H_0 = H \cos\phi$, the product $RH \cos\phi$ is a constant quantity for a given system of forces. It has been called the **invariant** of the system.

If the elements of reduction for the central axis (R, H_0) be given, those for any parallel line l at the distance r_0 from the central axis are determined by the equations

$$H^2 = H_0^2 + R^2 r_0^2, \quad \tan\phi = \frac{Rr_0}{H_0}.$$

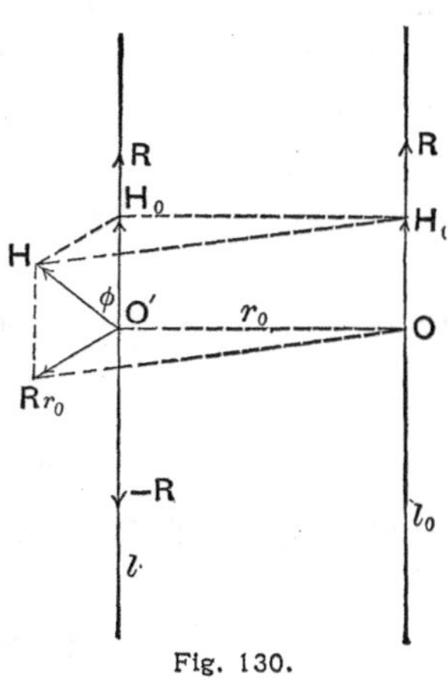

Fig. 130.

To sum up the results of the preceding articles, it has been shown that *any system of forces acting on a rigid body can be reduced, in an infinite number of ways, to a resultant R in combination with a couple H.* For all these reductions the magnitude, direction, and sense of the resultant R are the same, but the vector H of the couple changes according to the *position* assumed for the line of R. There is one, and only one, position of R, called the central axis of the system, for which the vector H is parallel to R and has at the same time its least value, H_0; this value H_0 is equal to the projection of any other vector H on the direction of the resultant R.

387. While, in general, a system of forces cannot be reduced to a single resultant, it can always be reduced to *two non-intersecting forces.* This easily follows by considering the system reduced to its resultant R and resulting couple H for any origin O (Fig. 131). Let F, $-F$ be the forces, p the arm of the couple H, and place this couple so that one of the forces, say $-F$, intersects R at O. Then, combining R and $-F$ into their resultant F', the given system of forces is evidently equivalent to the two non-intersecting forces F, F' (compare Art. 328).

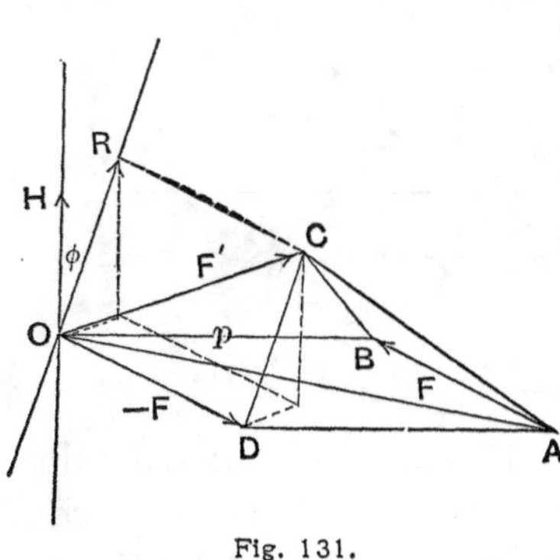

Fig. 131.

388. The two forces F, F' determine a tetrahedron $OABC$; and it can be shown that *the volume of this tetrahedron is constant and equal to one sixth of the invariant of the system* (Art. 386). The proof readily appears from Fig. 131. The volume of the tetrahedron $OABC$ is evidently one half of the volume of the quadrangular pyramid whose vertex is C and whose base is the parallelogram $OBAD$. The area of this parallelogram is $Fp = H$; and the altitude of the pyramid is $= R\cos\phi$, being equal to the perpendicular let fall from the extremity of R on the plane of the couple; hence the volume of the tetrahedron

$$= \tfrac{1}{6}RH\cos\phi = \tfrac{1}{6}RH_0.$$

389. To effect the reduction of a given system of forces analytically, it is usually best to refer the forces F and their points of application P to a rectangular system of co-ordinates Ox, Oy, Oz (Fig. 132). Let x, y, z be the co-ordinates of P and X, Y, Z the components of F parallel to the axes.

To transfer these components to O as common origin, we proceed as in Art. 333. Thus to transfer, say X, we introduce at P', the foot of the perpendicular let fall from P on the plane zx,

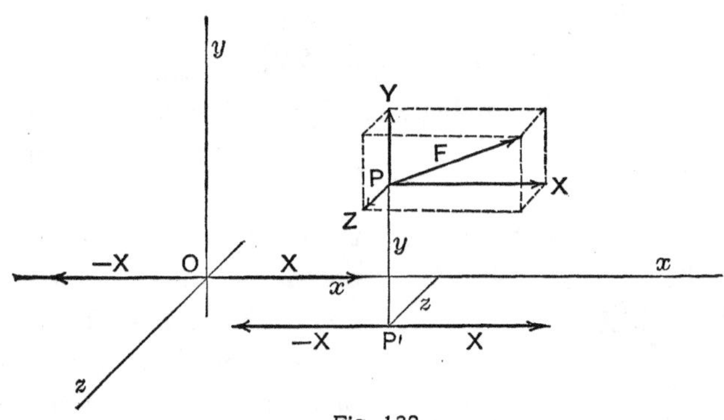

Fig. 132.

two equal and opposite forces X, $-X$; and we do the same thing at O. Then the single force X at P is replaced by the force X at O in combination with the two couples formed by X at P, $-X$ at P', and X at P', $-X$ at O. The vector of the former couple is parallel to Oz, its moment is $-yX$; the negative sign being used because for a person looking on the plane of the couple from the positive side of the axis Oz the couple rotates clockwise. The vector of the latter couple is parallel to Oy, and its moment is zX.

The transfer of Y to the origin O requires the introduction of two couples, $-zY$ having its vector parallel to Ox and xY having its vector parallel to Oz.

Finally, transferring Z to O, we have to introduce the couples $-xZ$ with a vector parallel to Oy, and yZ with a vector parallel to Ox.

Thus each force F is replaced by three forces X, Y, Z along the axes of co-ordinates and applied at O, in combination with three couples whose vectors are $yZ-zY$ parallel to Ox, $zX-xZ$ parallel to Oy, $xY-yX$ parallel to Oz.

390. Doing the same thing for every force of the given system and adding the components having the same direction, the system will be found equivalent to the three rectangular forces

$$\Sigma X, \ \Sigma Y, \ \Sigma Z,$$

applied at O, together with the three couples

$$\Sigma(yZ-zY), \quad \Sigma(zX-xZ), \quad \Sigma(xY-yX),$$

whose vectors are at right angles.

The three forces can now be compounded into a single resultant

$$R=\sqrt{(\Sigma X)^2+(\Sigma Y)^2+(\Sigma Z)^2},$$

whose direction is determined by the angles α, β, γ which it makes with the axes Ox, Oy, Oz:

$$\cos\alpha=\frac{\Sigma X}{R}, \quad \cos\beta=\frac{\Sigma Y}{R}, \quad \cos\gamma=\frac{\Sigma Z}{R}.$$

In the same way the three couples can be compounded into a single resulting couple whose moment is

$$H=\sqrt{[\Sigma(yZ-zY)]^2+[\Sigma(zX-xZ)]^2+[\Sigma(xY-yX)]^2}.$$

391. Since R^2, as well as H^2, is thus found as the sum of three squares, each of these quantities can vanish only if the three squares composing it vanish separately. The **conditions of equilibrium of a rigid body** (Art. 381) are therefore expressed analytically by the following six equations:

$$\Sigma X = 0, \quad \Sigma Y = 0, \quad \Sigma Z = 0,$$
$$\Sigma(yZ - zY) = 0, \quad \Sigma(zX - xZ) = 0, \quad \Sigma(xY - yX) = 0.$$

As the system of co-ordinates can be selected arbitrarily, the meaning of the first three equations is that the sum of the components of all the forces along any three lines not in the same plane must vanish. The last three equations express that the sum of the moments of all the forces about any three axes not in the same plane must also vanish. The *moment of a force about an axis* must be understood as meaning the moment of its projection on a plane at right angles to the axis with respect to the point of intersection of the axis with the plane. This definition is in accordance with the somewhat vague notion of the moment of a force as representing its "turning effect." For, regarding the force as acting on a rigid body with a fixed axis, the force can be resolved into two components, one parallel, the other perpendicular, to the axis; the former component evidently does not contribute to the turning effect, which is therefore measured by the moment of the latter alone.

392. The *equations of the central axis* (Art. 383) can be found by a transformation of co-ordinates.

Let the system be reduced for any origin O to its resultant R, whose rectangular components we denote by

$$A = \Sigma X, \quad B = \Sigma Y, \quad C = \Sigma Z,$$

and to the vector H of its resulting couple with the components

$$L = \Sigma(yZ - zY), \quad M = \Sigma(zX - xZ), \quad N = \Sigma(xY - yX).$$

If a point O' whose co-ordinates are ξ, η, ζ be taken as new origin and the co-ordinates of any point with respect to parallel axes through O' be denoted by x', y', z', we have $x = \xi + x'$, $y = \eta + y'$, $z = \zeta + z'$. Substituting these values, we find

$$L = \Sigma[(\eta + y')Z - (\zeta + z')Y] = \eta\Sigma Z - \zeta\Sigma Y + \Sigma(y'Z - z'Y)$$
$$= \eta C - \zeta B + L',$$

where L' is the x-component of the couple H' resulting for O' as origin. Similar expressions hold for M and N. The components of H' are therefore

$$L' = L - \eta C + \zeta B,\ M' = M - \zeta A + \xi C,\ N' = N - \xi B + \eta A;$$

and its direction cosines are

$$\lambda = \frac{L'}{H'},\quad \mu = \frac{M'}{H'},\quad \nu = \frac{N'}{H'}.$$

The central axis being defined (Art. 383) as that line for which the vector of the resulting couple is parallel to the direction of the resultant, the point $O'(\xi, \eta, \zeta)$ will lie on the central axis if the direction cosines of H' are equal to those of R, viz., to

$$\alpha = \frac{A}{R},\quad \beta = \frac{B}{R},\quad \gamma = \frac{C}{R}.$$

Hence the equations of the central axis are

$$\frac{L'}{A} = \frac{M'}{B} = \frac{N'}{C},$$

or

$$\frac{L - \eta C + \zeta B}{A} = \frac{M - \zeta A + \xi C}{B} = \frac{N - \xi B + \eta A}{C}.$$

393. To show the application of the conditions of equilibrium, let us consider the apparatus called *wheel and axle*. It consists of a horizontal shaft (Fig. 133) resting with its ends on the supports or bearings A, B, and is intended to raise a weight W, suspended vertically by means of a rope wound around the shaft. The driving force F is applied at the circumference of the "wheel" *i. e.* of a circular disk at right angles to the axis of the shaft. It is required to find the relation between F and W for equilibrium, and the pressures on the bearings A, B.

Let r be the radius of the shaft, R that of the wheel, *i. e.* the lever-arm of the force F; and let F be inclined to the vertical at an angle θ. Then with the co-ordinates and notations of the figure, the conditions $\Sigma X = 0$, $\Sigma Y = 0$, $\Sigma Z = 0$, give

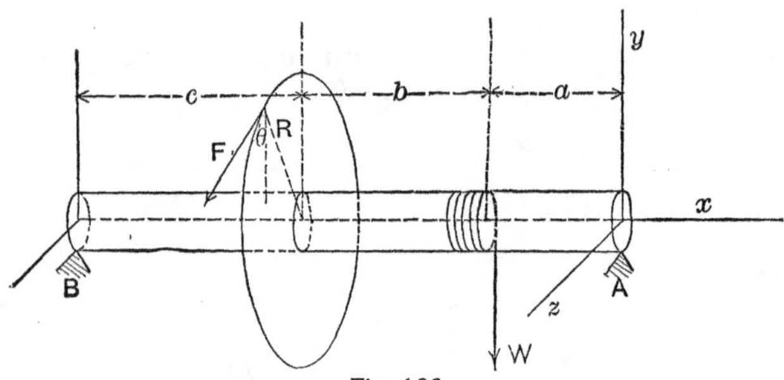

Fig. 133.

$A_x + B_x = 0$. $A_y + B_y - W - F\cos\theta = 0$, $A_z + B_z + F\sin\theta = 0$, where A_x, A_y, A_z are the components of the unknown reaction at A; B_x, B_y, B_z those at B.

Taking moments about each of the co-ordinate axes, we find

$$FR = Wr,\ (a + b)F\sin\theta + lB_z = 0,\ aW + (a + b)F\cos\theta - lB_y = 0,$$

where $l = a + b + c$ is the length of the shaft.

Solving the equations we find

$$F = \frac{R}{r} \cdot W,$$

$$A_y = \frac{1}{l}\left(b + c + \frac{cr}{R}\cos\theta\right) \cdot W, \quad B_y = \frac{1}{l}\left(a + \frac{(a + b)r}{R}\cos\theta\right) \cdot W,$$

$$A_z = -\frac{1}{l}\frac{cr}{R}\sin\theta \cdot W, \qquad B_z = -\frac{1}{l}\frac{(a + b)r}{R}\sin\theta \cdot W$$

394. As another example, consider *a rigid body of weight W, supported at three points A_1, A_2, A_3*; and let it be required *to determine the distribution of the pressure between the three supports.*

Let the vertical through the centroid of the body meet the plane of the triangle $A_1A_2A_3$ at a point G, whose distances from the sides A_2A_3, A_3A_1, A_1A_2 we may denote by p_1, p_2, p_3. Then, if A_1, A_2, A_3 be the unknown reactions, and h_1, h_2, h_3 the altitudes of the triangle, we have

$$A_1 + A_2 + A_3 = W,$$

and, taking moments about A_2A_3, A_3A_1, A_1A_2,

$$A_1 \cdot h_1 = W \cdot p_1, \quad A_2 \cdot h_2 = W \cdot p_2, \quad A_3 \cdot h_3 = W \cdot p_3.$$

Hence,

$$A_1 = \frac{p_1}{h_1} W, \quad A_2 = \frac{p_2}{h_2} W, \quad A_3 = \frac{p_3}{h_3} W.$$

Substituting these values in the first equation, we find the condition,

$$\frac{p_1}{h_1} + \frac{p_2}{h_2} + \frac{p_3}{h_3} = 1.$$

If G falls outside the triangle, one or two of the points A_1, A_2, A_3 will be subject to pressures vertically upwards. If G be the centroid of the triangular area $A_1A_2A_3$, we have $p_1/h_1 = p_2/h_2 = p_3/h_3 = \frac{1}{3}$; hence in this case the three reactions are equal. (Comp. Art. 401.)

395. *The axis of the hinges of a door is inclined at an angle θ to the horizon. The door is turned out of its position of equilibrium through an angle φ, and held in this position by a force F perpendicular to the plane of the door. Determine F and the reaction of the hinges A, B* (Fig. 134).

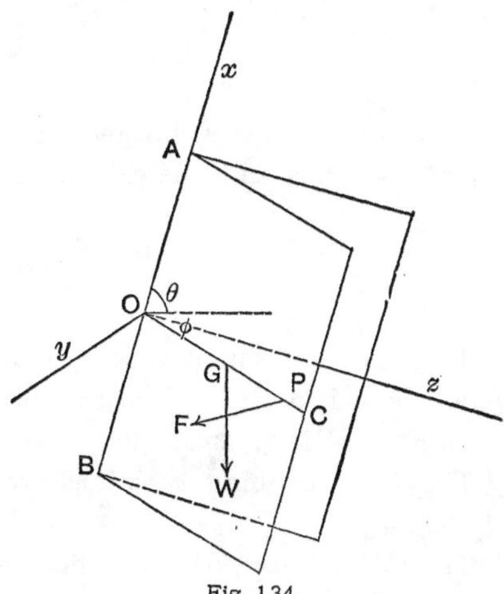

Fig. 134.

Let the axis of the hinges be taken as the axis of x, the vertical plane through it as the plane zx, and the point midway between the hinges A, B as the origin O. Regarding the door as a homogeneous rectangular plate whose dimensions are $AB = 2a$, $OC = 2b$, the co-ordinates of its centroid G are 0, $b\sin\varphi$, $b\cos\varphi$. If the force F be applied at a point P on the middle line OC at the distance $OP = p$ from O, the co-ordinates of its point of application P are 0, $p\sin\varphi$, $p\cos\varphi$.

To proceed systematically, we may tabulate the components of the forces, and the co-ordinates of their points of application, and then form the component couples, as shown below. The components of the unknown reactions A, B of the hinges are called A_x, A_y, A_z, B_x, B_y, B_z.

Forces.	Components			Co-ordinates.			Couples.		
	X	Y	Z	x	y	z	$yZ-zY$	$zX-xZ$	$xY-yX$
W	$-W\sin\theta$	o	$W\cos\theta$	o	$b\sin\phi$	$b\cos\phi$	$Wb\cos\theta\sin\phi$	$-Wb\sin\theta\cos\phi$	$Wb\sin\theta\sin\phi$
F	o	$F\cos\phi$	$-F\sin\phi$	o	$p\sin\phi$	$p\cos\phi$	$-Fp(\sin^2\phi+\cos^2\phi)$	o	o
A	A_x	A_y	A_z	a	o	o	o	$-A_z a$	$A_y a$
B	B_x	B_y	B_z	$-a$	o	o	o	$B_z a$	$-B_y a$

From this table the six conditions of equilibrium are at once obtained:

$$-W\sin\theta \qquad\qquad + A_x + B_x = 0, \tag{1}$$

$$F\cos\varphi + A_y + B_y = 0, \tag{2}$$

$$W\cos\theta - F\sin\varphi + A_z + B_z = 0, \tag{3}$$

$$Wb\cos\theta\sin\varphi - Fp \qquad\qquad = 0, \tag{4}$$

$$-Wb\sin\theta\cos\varphi + (-A_z + B_z)a = 0, \tag{5}$$

$$Wb\sin\theta\sin\varphi + (A_y - B_y)a \qquad = 0. \tag{6}$$

If the reactions were not required, equation (4) alone would be sufficient, as it furnishes the value of F, viz.,

$$F = \frac{b}{p}\cos\theta\sin\varphi \cdot W.$$

This relation can of course be found directly by taking moments about the axis of the hinges. It shows that, for a given inclination of the hinges, F is greatest when $\varphi = \pi/2$.

The five remaining equations are sufficient to determine $A_x + B_x$, A_y, A_z, B_y, B_z.

To find the reactions for a door with vertical axis, we have to put $\theta = \pi/2$, which gives of course $F = 0$, and

$$A_x + B_x = W, \quad A_y + B_y = 0, \quad A_z + B_z = 0,$$

$$A_y - B_y = -\frac{b}{a}W\sin\varphi, \quad A_z - B_z = -\frac{b}{a}W\cos\varphi;$$

as φ may be assumed = o in this case, we find

$$A_y = -B_y = 0, \quad A_z = -B_z = -\frac{b}{2a}W.$$

The signs indicate that the upper hinge A is pulled out while the lower one B is pressed in.

2. CONSTRAINTS.

396. It has been shown in Art. 391 that the number of the conditions of equilibrium is six, for a rigid body that is perfectly free. This number will be diminished whenever the body is subject to conditions restricting its possible motions. Such conditions, or **constraints**, may be of various kinds; the body may have a fixed point, or a fixed axis, or one of its points may be constrained to move along a given curve or to remain on a given surface, etc.

As explained in Art. 31, a free rigid body is said to have six degrees of freedom. The most general form of motion that it can have is a screw-motion, or twist, consisting of a rotation about a certain axis, and a translation along this axis; each of these resolves itself analytically into three rectangular components, and these six components may be regarded as constituting the six possible motions of the body, on account of which it is said to have six degrees of freedom.

Equilibrium will exist only when these six possible motions are prevented; hence there must be six conditions of equilibrium.

397. We proceed to consider some forms of constraint and the corresponding changes in the equations of equilibrium.

It is generally convenient in dynamics to replace such restraining conditions by forces, usually called **reactions.** Whenever it is possible to introduce such forces having the same effect as the given conditions, the body may be regarded as free, and the general equations of equilibrium can be applied.

398. Rigid Body with a Fixed Point. A body that is free to turn about a fixed point A can be regarded as free if the reaction

A of this point be introduced and combined with the other forces acting on the body.

Let A_x, A_y, A_z be the components of A; then, taking the fixed point A as origin, the six equations of equilibrium (Art. 391) are

$$\Sigma X + A_x = 0, \qquad \Sigma Y + A_y = 0, \qquad \Sigma Z + A_z = 0,$$
$$\Sigma(yZ - zY) = 0, \quad \Sigma(zX - xZ) = 0, \quad \Sigma(xY - yX) = 0.$$

The first three of these equations serve to determine the reaction of the fixed point; the last three are the actual conditions of equilibrium corresponding to the three degrees of freedom of a body with a fixed point.

Hence, *a rigid body having a fixed point is in equilibrium if the sum of the moments of all the forces vanishes for any three axes passing through the fixed point and not situated in the same plane.*

399. Rigid Body with a Fixed Axis. A body with a fixed axis has but one degree of freedom; indeed, the only possible motion consists in rotation about this axis.

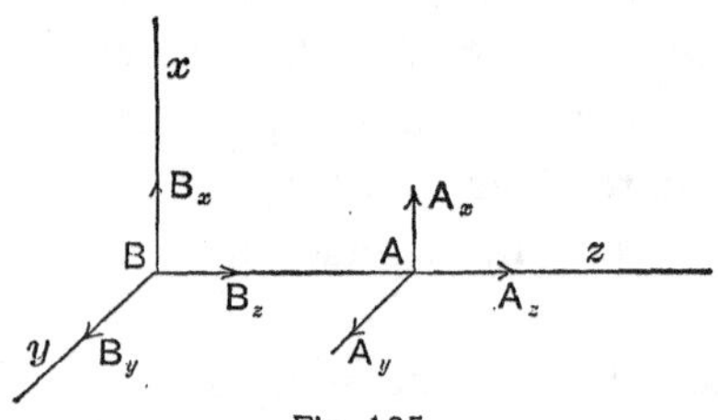

Fig. 135.

An axis is fixed as soon as two of its points, say A, B, are fixed. Hence, after introducing the reactions A_x, A_y, A_z, B_x, B_y, B_z of these points, the body can be regarded as free. If the point B be taken as origin, the line BA as axis of z (Fig. 135), the equations of equilibrium become

$$\Sigma X + A_x + B_x = 0, \quad \Sigma Y + A_y + B_y = 0, \quad \Sigma Z + A_z + B_z = 0,$$
$$\Sigma(yZ - zY) - A_y a = 0, \quad \Sigma(zX - xZ) + A_x a = 0,$$
$$\Sigma(xY - yX) = 0,$$

where $a = BA$.

The last of the six equations is the only independent condition of equilibrium of the constrained body; the first five determine A_x, B_x, A_y, B_y, $A_z + B_z$. The two z-components cannot be found separately, since they act in the same straight line.

Hence, *a rigid body having a fixed axis is in equilibrium if the sum of the moments of all the forces vanishes for the fixed axis.*

400. If, in the preceding article, the axis be not absolutely fixed, but only fixed in direction so that *the body can rotate about the axis and also slide along it,* we have evidently

$$A_z = 0, \quad B_z = 0;$$

hence, by the third equation of equilibrium,

$$\Sigma Z = 0,$$

as an additional condition of equilibrium.

The body has in this case two degrees of freedom.

401. Rigid Body with a Fixed Plane. A body constrained to slide on a fixed plane (that is, to move so that the paths of all its points lie in parallel planes) has three degrees of freedom. At every point of contact between the body and the plane, the latter exerts a reaction. As all these reactions are parallel, they can be combined into a single resultant N. Taking the fixed plane as the plane xy, N will be parallel to the axis of z; hence, if a, b, 0 be the co-ordinates of its point of application, the six equations of equilibrium are

$$\Sigma X = 0, \quad \Sigma Y = 0, \quad \Sigma Z + N = 0,$$

$$\Sigma(yZ - zY) + bN = 0, \quad \Sigma(zX - xZ) - aN = 0,$$

$$\Sigma(xY - yX) = 0.$$

The third, fourth, and fifth equations determine the reaction N and the co-ordinates a, b of its point of application. The three other equations are the actual conditions of equilibrium; they agree, of course, with the three conditions of equilibrium of a plane system as found in Art. 333.

If there be not more than three points of contact (or supports) between the body and the fixed plane, the reactions of these points can be found separately. Let A_1, A_2, A_3 be the three points of contact; N_1, N_2, N_3 the required reactions; a_1, b_1, a_2, b_2, a_3, b_3

the co-ordinates of A_1, A_2, A_3; then N must be resolved into three parallel forces passing through these points, and the conditions are

$$N_1 + N_2 + N_3 = N,$$
$$a_1N_1 + a_2N_2 + a_3N_3 = aN,$$
$$b_1N_1 + b_2N_2 + b_3N_3 = bN.$$

These three equations always determine N_1, N_2, N_3. For if the determinant of the coefficients of N_1, N_2, N_3 vanished,

$$\begin{vmatrix} 1 & 1 & 1 \\ a_1 & a_2 & a_3 \\ b_1 & b_2 & b_3 \end{vmatrix} = 0,$$

the three points A_1, A_2, A_3 would lie in a straight line, and hence the body would not be properly constrained.

The reactions become indeterminate whenever there are more than three points of contact.

VII. *The Principle of Virtual Work.*

402. Work has been defined in Art. 266 as the product of a constant force into the displacement of its point of application in the direction of the force.

Thus the expansive force F of the steam in the cylinder of a steam-engine, in pushing the piston through a distance s, is said to *do work*, and this work is measured by the product Fs, if F is constant. Similarly the force of gravity, *i. e.* the attractive force of the earth's mass, does work on a falling body.

The resistance to be overcome by the engine, in the former case, and the resistance of the air in the latter, are also forces acting on the body during its displacement. But as the sense of the displacement is opposite to that of these forces, their work is negative; *work is done against these forces.* Thus the muscular force of a man who raises a weight does work against gravity; if the weight he holds is so heavy as to pull him down, gravity does work against his force; if he merely tugs at a weight without being able to lift it, the work is zero, because the displacement is zero.

403. In general, the point of application of a force F may be acted upon by a number of different forces, so that the displacement s of this point will not necessarily take place in the direction of F. In this general case *the work of a constant force is defined as the product of the force into the projection of the displacement of its point of application on the direction of the force.*

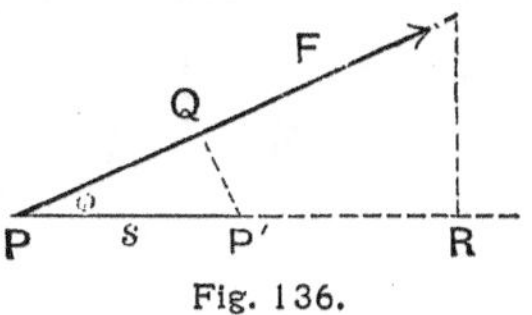

Fig. 136.

In Fig. 136, for instance, the particle P while acted upon by the force F (and any number of other forces) is displaced from P to P'; hence if $PP' = s$, and $\measuredangle P'PQ = \phi$, the work of the force F is

$$W = F \cdot s \cos\phi. \tag{1}$$

It is obvious that this work might also be defined as the product of the displacement into the projection of the force on the displacement; for we have

$$Fs \cos\phi = F \cdot PQ = s \cdot PR.$$

The work of a force is evidently positive or negative according as the angle ϕ is less or greater than $\frac{1}{2}\pi$, provided we select for ϕ always that angle between F and s which is not greater than π.

404. The above definition of work assumes that the force F remains constant, both in magnitude and direction, while the displacement s takes place, and that this displacement is rectilinear. If either, or both, of these conditions be not fulfilled, the definition can be applied for small displacements, approximately. In the language of infinitesimals, the infinitesimal work done by a (finite) force F during an infinitesimal displacement ds is denoted by dW so that

$$dW = Fds \cos\phi, \tag{2}$$

and the total work done by any variable force F, while its point of application is displaced along any straight or curvilinear path PQ, is obtained by integrating from P to Q:

$$W = \int_P^Q F \cos\phi ds. \quad (3)$$

405. Since work can always be regarded as the product of a force into a length, its dimensions are found by multiplying those of force, MLT^{-2} (Art. 258), by L; hence, the *dimensions of work* are

$$W = ML^2T^{-2}.$$

The unit of work is the work of a unit force (poundal, dyne) through a unit distance (foot, centimeter). The unit of work in the F.P.S. system is called the **foot-poundal**; in the C.G.S. system, the **erg**. Thus, the erg is the amount of work done by a force of one dyne acting through a distance of one centimeter. These are the *absolute* units.

In the gravitation system where the pound, or the kilogram, is taken as unit of force, the British unit of work is the **foot-pound,** while in the metric system it is customary to use the **kilogram-meter** as unit.

406. The numerical relations between these units are obtained as follows. Let x be the number of ergs in the foot-poundal, then (comp. Art. 259),

$$x \cdot \frac{\text{gm. cm.}^2}{\text{sec.}^2} = 1 \cdot \frac{\text{lb. ft.}^2}{\text{sec.}^2},$$

hence

$$x = \frac{\text{lb.}}{\text{gm.}} \cdot \left(\frac{\text{ft.}}{\text{cm.}}\right)^2 = 453.59 \times 30.4797^2 = 4.2139 \times 10^5;$$

i. e. 1 foot-poundal $= 4.2139 \times 10^5$ ergs, and 1 erg $= 2.3721 \times 10^{-6} = 0.000\,002\,372\,1$ foot-poundal.

Again, let x be the number of kilogram-meters in 1 foot-pound, then

$$x \text{ kg. m.} = 1 \text{ ft. lb.},$$

hence

$$x = \frac{\text{lb.}}{\text{kg.}} \cdot \frac{\text{ft.}}{\text{m.}} = 0.45359 \times 0.3048 = 0.13825,$$

i. e. 1 foot-pound = 0.13825 kilogram-meters.

Finally, 1 foot-pound = g foot-poundals (Art. 262); hence 1 foot-pound = 1.356×10^7 ergs, and 1 erg = 7.3737×10^{-8} foot-pounds, if $g = 981$.

407. Exercises.

(1) A *joule* being defined as 10^7 ergs, show that 1 foot-pound = 1.356 joules, and that 1 joule is about 3/4 foot-pound.

(2) Show that a kilogram-meter is nearly 10^8 ergs.

(3) What is the work done against gravity in raising 300 lbs. through a height of 25 ft.: (*a*) in foot-pounds, (*b*) in ergs?

(4) Find the work done against friction in moving a car weighing 3 tons through a distance of 50 yards on a level road, the coefficient of friction being 0.02.

(5) A mass of 12 lbs. slides down a smooth plane inclined at an angle of 30° to the horizon, through a distance of 25 ft.; what is the work done by gravity?

408. It follows from the definition of work that, if any number of forces $F_1, F_2, \ldots F_n$ act on a particle P, *the sum of their works for any displacement* $PP' = ds$ *is equal to the work of their resultant* R *for the same displacement.* For, the resultant R being the closing line of the polygon constructed by adding the forces F_1, $F_2, \ldots F_n$ geometrically, the projection of R on any direction, such as PP', is equal to the sum of the projections of the forces F on the same line (Art. 285); that is, if $\alpha_1, \alpha_2, \ldots \alpha_n$ be the angles made by $F_1, F_2, \ldots F_n$ with PP', and α the angle between R and PP', we have

$$F_1 \cos\alpha_1 + F_2 \cos\alpha_2 + \cdots + F_n \cos\alpha_n = R \cos\alpha;$$

multiplying this equation by ds, we obtain the above proposition

$$F_1 \cos\alpha_1 ds + F_2 \cos\alpha_2 ds + \cdots + F_n \cos\alpha_n ds = R \cos\alpha ds,$$

which expresses the so-called **principle of work** for a single particle.

409. When the particle is in equilibrium, so that the forces do not actually change the motion, we may derive from this proposition a convenient expression for the conditions of equilibrium by considering displacements that *might be* given to the particle. Such displacements are called *virtual*, and the corresponding work of any of the forces is called **virtual work.**

It is customary to denote a virtual displacement by δs, the letter δ being used to distinguish from the actual displacement ds; this distinction becomes of importance in kinetics.

410. The resultant being zero in the case of equilibrium, *the sum of the virtual works of all forces acting on the particle must be zero for any virtual displacement, i. e.*

$$F_1 \cos\alpha_1 \delta s + F_2 \cos\alpha_2 \delta s + \cdots + F_n \cos\alpha_n \delta s = 0. \qquad (4)$$

As the resultant must vanish if its three projections vanish for any three axes not lying in the same plane, the necessary and sufficient conditions of equilibrium of a single particle are that *the sum of the virtual works of all forces must be zero for any three virtual displacements not all in the same plane.*

This is the **principle of virtual work** for a single particle.

If the particle be referred to a rectangular system of co-ordinates, its displacement δs can be resolved into three component displacements δx, δy, δz, parallel to the axes. The forces acting on the particle being replaced by their components X, Y, Z, the sum of their virtual works, for the displacement δs is $\Sigma X \cdot \delta x + \Sigma Y \cdot \delta y + \Sigma Z \cdot \delta z$. Hence the analytical expression for the principle of virtual work:

$$\Sigma X \cdot \delta x + \Sigma Y \cdot \delta y + \Sigma Z \cdot \delta z = 0. \qquad (5)$$

As the displacements δx, δy, δz are independent of each other, and perfectly arbitrary, this single equation is equivalent to the three equations

$$\Sigma X = 0, \quad \Sigma Y = 0, \quad \Sigma Z = 0,$$

which are the ordinary conditions of equilibrium of a single particle.

411. The principle of virtual work is particularly useful in eliminating the unknown reactions arising from constraints.

Suppose the particle be constrained to a smooth surface or curve. After introducing the normal reaction of the surface or curve the particle can be regarded as free; and the equation of virtual work can be used to express the conditions of equilibrium. This equation will, in general, contain the unknown reaction. But as this reaction has the direction of the normal, it will be eliminated if the virtual displacement be selected along a tangent. Hence, *in the case of constraint to a surface, the two conditions of equilibrium independent of the reaction are found by forming the equation of virtual work for virtual displacements along any two tangents to the surface; and in the case of constraint to a curve the one such condition is found from a virtual displacement along the tangent.*

If it be required to find the normal pressure on the surface or curve, which is of course equal and opposite to the reaction, it can be found from a virtual displacement along the normal.

412. If the equation (4) which expresses the principle of virtual work be divided by the element of time δt, during which the displacement δs would take place, the factor $\delta s / \delta t = v$ represents a *virtual velocity*, and the equation becomes

$$F_1 \cos\alpha_1 \cdot v + F_2 \cos\alpha_2 \cdot v + \cdots + F_n \cos\alpha_n \cdot v = 0.$$

On account of this form, the proposition is often called the *principle of virtual velocities.*

The product of a force into the virtual velocity of its point of application in the direction of the force, $F \cos\alpha \cdot v$, is sometimes called the *virtual moment* of the force.

413. The principle of virtual work can readily be extended to the case of a rigid body acted upon by any number of forces.

The forces acting on a rigid body can always be reduced to a resultant R and a resulting couple H (Art. 380). This reduction is based on the supposition (Art. 283) that the point of application of a force can be displaced arbitrarily along the line of the force. It can be shown that such a displacement of the point of application P of a force F (Fig. 137), from P to Q along the line of the force, does not affect the work done by the force in any infinitesimal displacement of the body. Let $PP' = \delta s$ be the displacement of P, $QQ' = \delta s'$ that of Q; let p and q be the projections of P' and Q' on the line of the force F; then, since the body is rigid, $P'Q' = PQ$; and consequently Qq will differ from Pp only by an infinitesimal of an order higher than the order of the displacement $PP' = \delta s$. Hence,

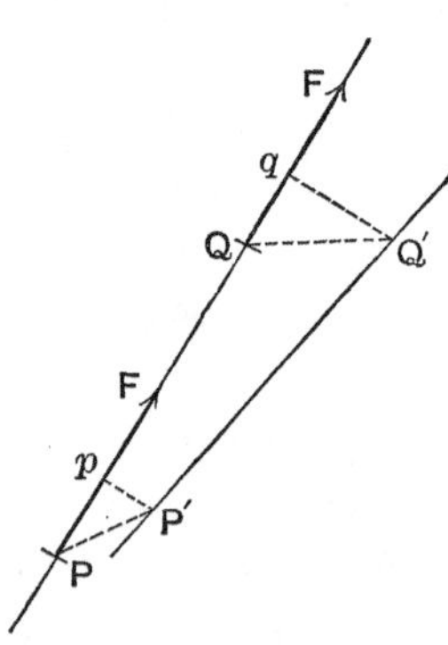

Fig. 137.

$$F \cdot Pp = F \cdot Qq.$$

It may here be noted that, in general, the principle of virtual work must be understood to mean that the sum of the works of the forces differs from the work of their resultant by an infinitesimal of an order higher than that of the virtual displacement. It does not mean that the difference is absolutely zero.

414. Owing to the proposition proved in the preceding article, the sum of the works of all the forces acting on a rigid body is equal to the sum of the works of the resultant R and the resulting couple H for any infinitesimal displacement of the body, and the work of the forces is not changed by such a displacement.

It follows that the necessary and sufficient conditions of equilibrium of a rigid body (Art. 381), viz.

$$R = 0, \quad H = 0,$$

can be expressed by saying that *the sum of the virtual works of all the forces must be zero for any infinitesimal displacement of the body.*

For when the forces are in equilibrium, this condition is evidently fulfilled. To prove that there must be equilibrium whenever this condition is fulfilled, it is only necessary to show that both R and H must vanish if the sum of their works is zero for any infinitesimal displacement.

To see this, consider first a displacement of translation, δs, parallel to R. The work of R will be $R\delta s$ while the works of the two forces constituting H are equal and opposite, so that the work of H is zero. As the sum of the works of R and H must vanish by hypothesis, it follows that $R = 0$.

Next consider a displacement of rotation $\delta\theta$ about an axis parallel to the vector H. Taking this axis so as to intersect R and bisect the arm p of the couple H, the work of R will be zero while that of each of the forces F of the couple H will be $\frac{1}{2}p\delta\theta \cdot F$; hence the whole work of H is $Fp\delta\theta = H\delta\theta$. As the sum of the works of R and H must vanish by hypothesis, it follows that $H = 0$.

The two conditions $R = 0$, $H = 0$ are, therefore, both fulfilled.

415. The following examples may serve to illustrate the application of the principle of virtual work.

To find the force just necessary to move a cylinder of radius r and weight W up a plane inclined at an angle α to the horizon by means of a crow-bar of length l set at an angle β to the horizon (Fig. 138).

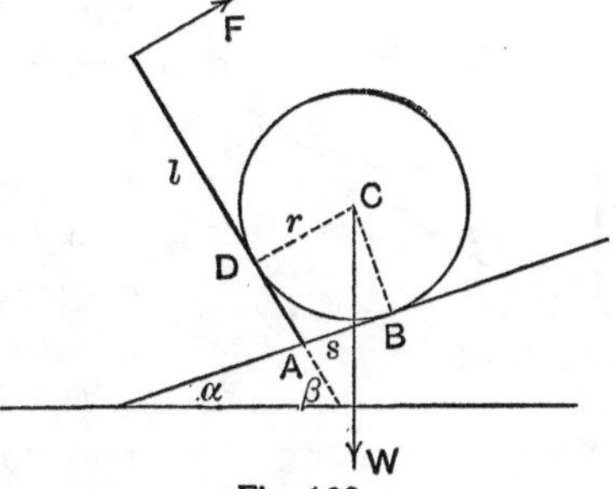

Fig. 138.

Let s be the distance from the fulcrum A of the crow-bar to the point of contact B of the cylinder with the plane.

If the crow-bar be turned about A through an angle $\delta\beta$, the work of the force F acting at the end of the bar is $F \cdot l\delta\beta$. The corresponding displacement of the center C of the circle, which is the point of application of the force W, is parallel to the inclined plane, and may be regarded as the differential δs of the distance $AB = s$. The work of W is, therefore, $W\delta s \sin\alpha$. This gives the equation of work

$$F . l\, \delta\beta = W . \delta s \sin\alpha;$$

hence

$$F = \frac{W}{l} \sin\alpha . \frac{\delta s}{\delta\beta}.$$

The relation between s and β can be found by projecting $ABCDA$ on the vertical line; this gives

$$r \cos\alpha + s \sin\alpha = r \cos\beta + s \sin\beta,$$

whence

$$s = r \frac{\cos\beta - \cos\alpha}{\sin\alpha - \sin\beta}.$$

Differentiating the former equation, we find

$$\sin\alpha\, \delta s = - r \sin\beta\, \delta\beta + s \cos\beta\, \delta\beta + \sin\beta\, \delta s,$$

$$\therefore \frac{\delta s}{\delta\beta} = \frac{s \cos\beta - r \sin\beta}{\sin\alpha - \sin\beta} = r \frac{\cos^2\beta - \cos\alpha \cos\beta - \sin\alpha \sin\beta + \sin^2\beta}{(\sin\alpha - \sin\beta)^2},$$

or

$$\frac{1}{r} \frac{\delta s}{\delta\beta} = \frac{1 - \cos(\alpha - \beta)}{(\sin\alpha - \sin\beta)^2} = \frac{1}{1 + \cos(\alpha + \beta)}$$

Hence, finally,

$$F = \frac{r}{l} \frac{\sin\alpha}{1 + \cos(\alpha + \beta)} . W.$$

416. *A weightless rod of length $AB = l$ rests at C on a horizontal cylinder whose axis is at right angles to the vertical plane through the rod; its lower end A leans against a vertical wall, and from its upper end B a weight W is suspended. Determine the reactions at A and C, and the distance $AC = x$ for equilibrium, if the distance $CD = a$ of the point of support from the vertical wall is given* (Fig. 139).

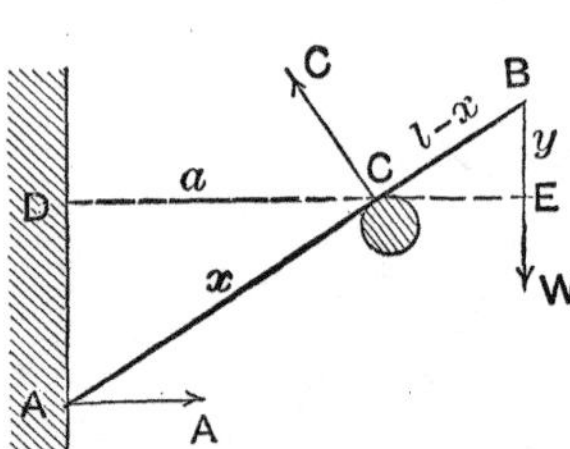

Fig. 139.

(*a*) Let A glide vertically upwards, C remaining in contact. At A as well as at C the forces are perpendicular to the displacements; hence, putting $EB = y$, we have $W\, \delta y = 0$.

$$\therefore \delta y = 0.$$

Also,

$$\frac{y}{l - x} = \frac{\sqrt{x^2 - a^2}}{x}.$$

$$\therefore \delta y = \delta\left\{\left(\frac{l}{x} - 1\right)\sqrt{x^2 - a^2}\right\} = 0,$$

whence $(l - x)x^2 - l(x^2 - a^2) = 0$,

or $\quad x^3 = a^2 l.$

(*b*) Give the rod a vertical displacement to a parallel position:

$$-W\delta y + C\frac{a}{x}\delta y = 0; \therefore C = \frac{x}{a}W = \sqrt[3]{\frac{l}{a}}\cdot W.$$

(*c*) Give the rod a displacement in its own direction:

$$A\frac{a}{x}\delta s + C\cos\tfrac{1}{2}\pi\cdot\delta s - W\frac{\sqrt{x^2 - a^2}}{x}\cdot\delta s = 0.$$

$$\therefore A = \frac{\sqrt{x^2 - a^2}}{a}\cdot W = \sqrt{\left(\frac{l}{a}\right)^{\frac{2}{3}} - 1}\cdot W.$$

417. *In a parallelogram formed by four rods with hinges at the vertices, elastic strings are stretched along the diagonals. Determine the ratio of the tensions in these strings.*

Let m, m' be the lengths of the diagonals, T, T' the tensions, and δm, $\delta m'$ the changes of length of the diagonals when the parallelogram is slightly deformed; then by the principle of virtual work

$$T\,\delta m + T'\delta m' = 0. \qquad (1)$$

From geometry we have, if a, b are the sides of the parallelogram,

$$m^2 + m'^2 = 2a^2 + 2b^2,$$

hence, differentiating,

$$m\,\delta m + m'\,\delta m' = 0. \qquad (2)$$

From (1) and (2) we find

$$T/T' = m/m'. \qquad (3)$$

418. As any body or system of bodies can be regarded as made up of particles and as the work of such a system is the algebraic sum of the works of the forces acting on each particle it follows from Art. 410 that *it is a necessary condition of equilibrium of any* *system that the sum of the virtual works of all the forces must be zero for any virtual displacement.*

Without entering into an elaborate discussion of the limitations of this very general principle it will be well to show its meaning and usefulness by means of some simple illustrations of its application.

419 The systems that we shall consider are mostly mechanisms and simple machines.

The object of a *mechanism* in machinery is to change the direction and magnitude of an available force P so as to make it more useful in overcoming a resistance Q. The ratio Q/P of the resistance Q to the given force P is called the *mechanical advantage* of the mechanism.

It is the object of every *machine* to do work in a certain prescribed way, *i. e.* to exert force, or overcome a resistance, through a certain distance. The various forces of nature, such as the muscular force of man and other animals, the force of gravity, the pressure of the wind, electricity, the expansive force of steam or gas, etc., are called upon for this purpose. In most cases it would not do to apply these forces directly; they must be controlled, guided, and transformed in various ways to become useful, and this is done by interposing the machine between the given *driving force*, sometimes called the *power*, and the force against which the final work is to be done, usually called the *resistance*, *load*, or *weight*. We shall in general denote the driving force by P, the resistance by Q.

The term *power* is objectionable in this connection, being here used to denote a force, while in kinetics it is used for the *rate of doing work* (see Art. 496).

Under the action of the driving force P its own point of application as well as that of the resistance Q is displaced. The corresponding work of the force P may be called the *available* or *total work*; that done against the force Q is called the *useful work*.

The ratio of the useful work to the total work is called the **efficiency** of the machine.

In all machines this efficiency is a proper fraction, owing to the fact that the work done by P must balance not only the useful work, but also the so-called wasteful work due to friction, stiffness of ropes, slipping of belts, lack of rigidity, etc.

420. In the straight **lever** AB, with fulcrum at C (Fig. 140), let $CA = a$, $CB = b$. If the lever be turned through a small angle

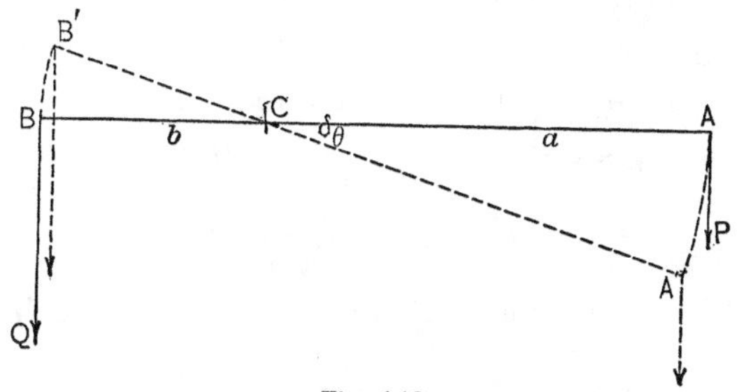

Fig. 140.

$ACA' = \delta\theta$, the work of the force P at A is $= P \cdot a\,\delta\theta$, that of the resistance Q at B is $= - Q \cdot b\,\delta\theta$. According to the principle of virtual work (Art. 418) the sum of the works must be zero for equilibrium; hence:

$$Pa\,\delta\theta - Q\,b\,\delta\theta = 0, \text{ or } \frac{Q}{P} = \frac{a}{b},$$

the well-known law of the lever (Art. 297). The mechanical advantage of the lever is therefore the ratio of its arms.

421. In writing down the sum of the works in this case it is not necessary to consider among the forces acting on the lever the unknown reaction of the fulcrum at C because, for the displacement selected, the work of this force is evidently zero (the point of application of the force remaining fixed). The main advantage of the principle of virtual work lies in this fact that if the displacement be selected so as to be *compatible with the conditions and constraints* of the system, the unknown reactions and internal forces of the system will, for the most part, do no work and need not be considered. The most important exception to this rule is formed by the force of friction; if not so small as to be negligible, its work must be included in the equation of virtual work.

Thus, if the displacement be selected so as to be compatible with the conditions, the equation of work will in general not contain the unknown reactions; in this form it can therefore not serve to determine these reactions. But by selecting the displacement differently, the principle of virtual work can be used to find these unknown forces. Thus, in the case of the lever (Fig. 140), if instead of turning the rod AB about C, it be given a vertical translation δs, say downward, the equation of virtual work will be

$$P\delta s + Q\,\delta s - C\delta s = 0,$$

where C is the reaction of the support at C which is thus found to be

$$C = P + Q.$$

422. Let one face BC of the **wedge** (Fig. 141) slide along a fixed plane; the vertical force P applied at right angles to AB can

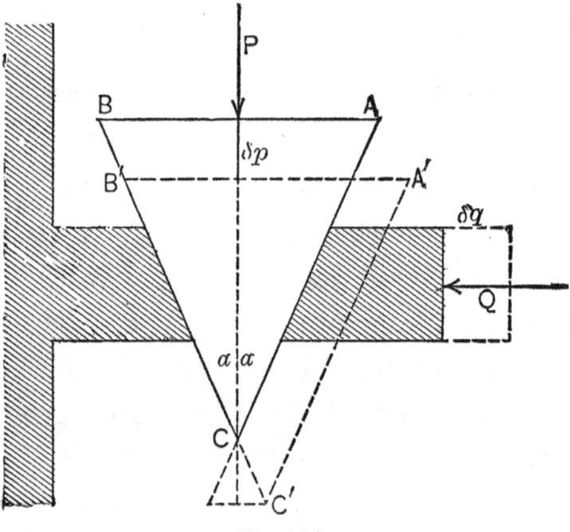

Fig. 141.

then serve to overcome a horizontal resistance Q acting on the face AC. If the wedge be given a displacement compatible with the condition that BC remains in its plane, the equation of virtual work (neglecting friction) is

$$P\,\delta p - Q\,\delta q = 0,$$

where δp, δq are the displacements of the points of application of P and Q in the directions of these forces. It is easily seen that $\delta q = 2\,\delta p\,\tan\alpha$, where α is half the angle of the wedge. Substituting and dropping the factor δp, we find for the mechanical advantage of the wedge

$$\frac{Q}{P} = \tfrac{1}{2}\cot\alpha.$$

The normal pressure on each of the faces AC, BC does no work since the displacement is at right angles to the force. By resolving P at right angles to the faces, this pressure is found $= \frac{1}{2}P\sin\alpha$.

If the friction on the faces is taken into account the equation of virtual work is

$$P\,\delta p - 2Q\,\delta p\,\tan\alpha - 2 \times \tfrac{1}{2}\mu\,P\sin\alpha\,\frac{\delta p}{\cos\alpha} = 0,$$

whence

$$\frac{Q}{P} = \tfrac{1}{2}(\cot\alpha - \mu).$$

423. In the **clamp** or **binding screw** (Fig. 142), the driving force P is applied tangentially to the rim of the screw head A; the resistance Q is along the axis of the screw. If the piece B be fixed, the screw in turning exerts a pressure on the piece C.

If the screw be turned through an angle $\delta\theta$ and the radius of the head be a, the work of P is $P \cdot a\,\delta\theta$. If s be the *pitch* of the screw, that is, the displacement along the axis corresponding to one revolution of the head, we have for the displacement δq of the piece C corresponding to the rotation $\delta\theta$:

$$\frac{s}{2\pi a} = \frac{\delta q}{a\,\delta\theta}, \quad \therefore \delta q = \frac{s\,\delta\theta}{2\pi}.$$

Hence, by the principle of virtual work,

$$P \cdot a\,\delta\theta = Q \cdot \frac{s\,\delta\theta}{2\pi},$$

whence

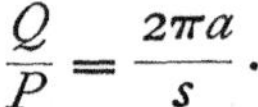

$$\frac{Q}{P} = \frac{2\pi a}{s}.$$

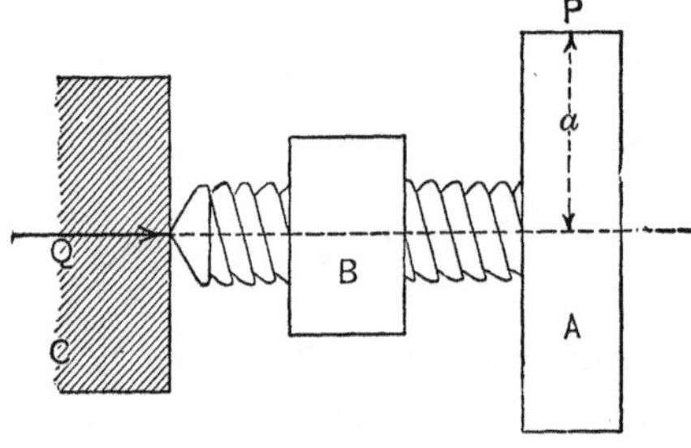

Fig. 142.

This shows the advantage of a large head and a small pitch.

424 In the mechanism formed by the crank and connecting rod of a steam-engine (Fig. 143), known as slider-crank, the driving force P acts at P in the direction through the center O of

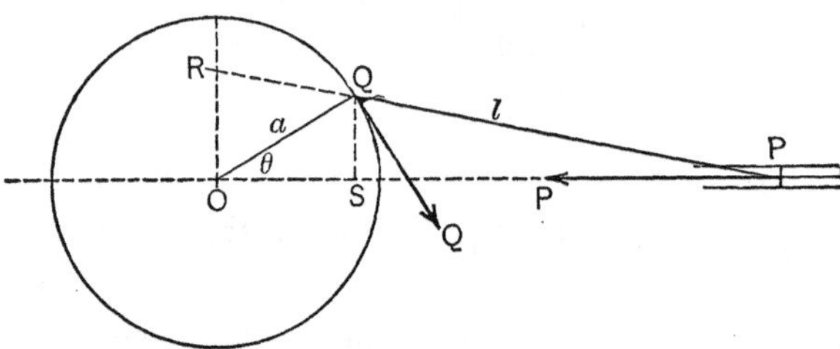

Fig. 143.

the crank circle while the resistance Q is applied at Q tangentially to the crank circle. For the only displacement compatible with the conditions of the mechanism we must again have

$$P \cdot \delta p = Q \cdot \delta q,$$

where δp is the distance through which P advances along PO and δq is the infinitesimal circular arc $a\,\delta\theta$ described by Q. To express δp in terms of $\delta\theta$ and the given lengths $OQ = a$, $PQ = l$, we regard it as the differential of the distance $OP = p$ for which we have the relation

$$l^2 = a^2 + p^2 - 2a\,p\cos\theta\,;$$

differentiating this equation we find

$$0 = 2p\,dp - 2a\cos\theta\,dp + 2a\,p\sin\theta\,d\theta,$$

whence

$$\delta p = -\,dp = \frac{a\,p\sin\theta}{p - a\cos\theta}\,\delta\theta.$$

Substituting this value in the equation of virtual work we find:

$$\frac{Q}{P} = \frac{\delta p}{\delta q} = \frac{p\sin\theta}{p - a\cos\theta}.$$

If through O we draw the radius perpendicular to OP and intersecting PQ produced at R, and also drop from Q the perpendicular SQ on OP, we have $p - a\cos\theta = SP$, $a\sin\theta = SQ$; hence

$$\frac{Q}{P} = \frac{p \cdot a\sin\theta}{a\,(p - a\cos\theta)} = \frac{p \cdot SQ}{a \cdot SP},$$

or since $SQ/SP = OR/p$:

$$\frac{Q}{P} = \frac{p}{a}\,\frac{OR}{p} = \frac{OR}{a}.$$

This result follows at once by the principle of virtual velocities (Art. 412) from Art. 101, where it was proved that the velocities of P and Q are as OR is to a.

The effect of friction at P can be taken into account by observing that the normal pressure on the guides at P is $P\tan\phi$ (see Art. 292, Ex. (15)), where ϕ is the angle OPQ. Hence, if the coefficient of friction be μ, the frictional force is $\mu P\tan\phi$, and the equation of virtual work becomes:

$$P\delta p = Q\delta q + \mu P \tan\phi \, \delta p,$$

whence

$$\frac{Q}{P} = (1 - \mu \tan\phi)\frac{\delta p}{\delta q},$$

or with the above value of $\delta p / \delta q$:

$$\frac{Q}{P} = (1 - \mu \tan\phi)\frac{p \sin\theta}{p - a \cos\theta}$$

$$= (1 - \mu \tan\phi)\frac{OR}{a}.$$

PART III:

KINETICS.

CHAPTER V.

KINETICS OF THE PARTICLE.

I. *Impulses; Impact of Homogeneous Spheres.*

425. Momentum and Impulse. A particle of mass m, moving with the velocity v, is said to have the *momentum* mv (see Art. 252). As long as this momentum remains constant, the particle will move in a straight line with constant velocity v (Newton's first law of motion, Art. 268). Any change occurring in the momentum is ascribed to the action of a force F on the particle.

If the motion is rectilinear, *the rate at which the momentum changes with the time t* is taken as the measure of the *force* (Newton's second law of motion, Arts. 255, 268, 271):

$$F = \frac{d(mv)}{dt}. \tag{1}$$

Integrating from the time t_0 when the velocity is v_0 to the time t when the velocity is v, we have

$$\int_{t_0}^{t} F dt = mv - mv_0. \tag{2}$$

If, in particular, the momentum varies *uniformly* with the time, the force F is constant, and equation (2) reduces to

$$F(t - t_0) = mv - mv_0. \tag{2'}$$

The left-hand member of (2′) for a constant force, of (2) for a variable force, is called the **impulse** of the force F during the time $t - t_0$ (Art. 256). Hence, in rectilinear motion, *the impulse of the force during any time is equal to the change of momentum during that time.*

426. It appears, from equations (2) and (2′), that a very large force may produce a finite change of momentum in a very short interval of time, but that it would require an infinite force to produce an *instantaneous* change of momentum of finite amount. The impact of one billiard ball on another, the blow of a hammer, the stroke of the ram of a pile driver, the shock imparted by a falling body, by a projectile, by a railway train in motion, by the explosion of the powder in a gun, are familiar instances of large forces acting for only a very short time and yet producing a very appreciable change of velocity. The time of action, $t - t_0$, of such a force is the very brief period during which the colliding bodies are in contact. The force, F, is a pressure or an elastic stress exerted by either body on the other during this time.

Forces of this kind are called *impulsive*, or *instantaneous, forces*.

427. In the case of such impulsive forces, it is generally difficult or impossible by direct observation or experiment to determine separately the very brief time of action, $t - t_0$, as well as the magnitude F of the impulsive force. Moreover, what is of most practical importance and interest in such cases of impact is, generally, not the force itself, but the change of momentum produced, *i. e.*, the impulse of the impulsive force.

In the present section, which is devoted to the study of the simplest cases of impact, we shall therefore deal rather with impulses and momenta than with forces.

428. It should be observed that some authors use the name *impulsive*, or *instantaneous, force* for what has here been called the impulse of the impulsive force. They define an impulsive force as the limiting value of the integral $\int_{t_0}^{t} F dt$ when F increases indefinitely, while at the same time the difference of the limits, $t - t_0$, is indefinitely diminished; in other words, as the impulse of an infinite force producing a finite change of momentum in an infinitesimal time.

According to this definition, an impulsive or instantaneous force is a magnitude of a character different from that of an ordinary force, and is measured by a different unit. Its dimensions are MLT^{-1}, and not MLT^{-2}. Its unit is the same as the unit of momentum. Indeed, it is not a force, but an impulse.

429. The momentum mv of a particle, being a mere multiple of its velocity, can be represented by a vector (more correctly a vector confined to a line, or rotor), *i. e.*, by a rectilinear segment drawn through the particle and representing by its length the magnitude of the momentum, by its direction and sense the direction and sense of the velocity (Art. 267). Hence *the momenta of rigidly connected particles are compounded and resolved according to the rules that hold for forces in statics*, the transferability of a momentum along its line being assumed (comp. Art. 283).

Thus, in particular, by Arts. 294, 297, for two rigidly connected particles m, m' whose velocities are equal and parallel (Fig. 144), the momenta mv, $m'v$ have a resultant momentum $(m + m')v$ which is parallel to the common velocity v of the particles and divides their distance in the inverse ratio of the masses m, m'. It follows that this resultant momentum passes through the centroid G of the masses m, m' (see Art. 219).

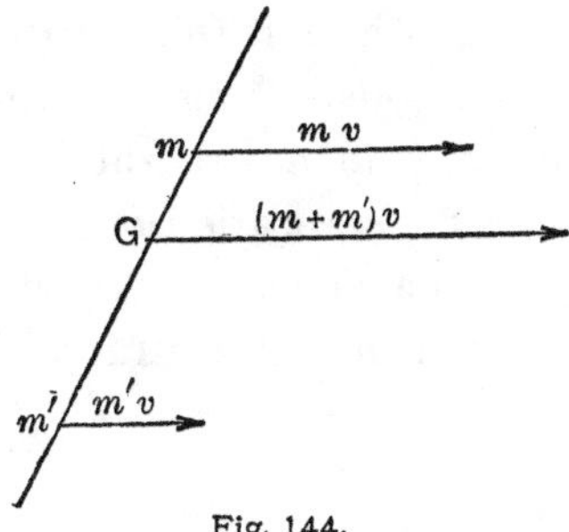

Fig. 144.

430. This result is readily generalized. By reasoning as in Art. 306, it can be proved that for a rigid body or mass M, having a velocity v of translation (so that all its particles have equal and parallel velocities v), the resultant of the momenta of all its particles is equal to Mv, is parallel to v, and passes through the centroid G of the mass M of the body. This momentum Mv, applied at G, is called the *momentum of the rigid body*, or the *momentum of its centroid.*

431. The term *momentum of the centroid* is also used in a more general sense. It has been shown in Art. 212 that, for any system of particles, whether rigidly connected or not, whether at rest or in motion, there exists at every instant a certain point G, the *centroid*, whose co-ordinates with respect to a fixed set of rectangular axes are

$$\bar{x} = \frac{\Sigma mx}{\Sigma m}, \quad \bar{y} = \frac{\Sigma my}{\Sigma m}, \quad \bar{z} = \frac{\Sigma mz}{\Sigma m}.$$

If the particles are in motion, this point G will, in general, change its position in the course of time. Differentiating the equations with respect to the time, and putting $\Sigma m = M$, we find:

$$M\frac{d\bar{x}}{dt} = \Sigma m\frac{dx}{dt}, \quad M\frac{d\bar{y}}{dt} = \Sigma m\frac{dy}{dt}, \quad M\frac{d\bar{z}}{dt} = \Sigma m\frac{dz}{dt}. \tag{3}$$

These equations determine the rectangular components $d\bar{x}/dt$, $d\bar{y}/dt$, $d\bar{z}/dt$ of the velocity $\bar{v}$ of the centroid G. The product $M\bar{v}$ is called the *momentum of the centroid* of the system of particles.

If each of the particles moves with constant velocity in a straight line, it follows from the equations (3) that the centroid has a constant momentum; *i. e.*, it will also move with constant velocity in a straight line. This proposition can be regarded as a generalization of Newton's first law of motion.

In particular, if two particles m, m' move with the velocities v, v' in the same straight line, the velocity of their centroid is

$$\bar{v} = \frac{mv + m'v'}{m + m'}. \tag{4}$$

432. Direct Impact. We proceed to consider the particular case of two homogeneous spheres of masses m, m', whose centers C, C' move with velocities u, u' in the same straight line. The spheres are supposed not to rotate, but to have a motion of pure translation; then their momenta are mu, $m'u'$, and can be represented by two vectors drawn from the centers C,

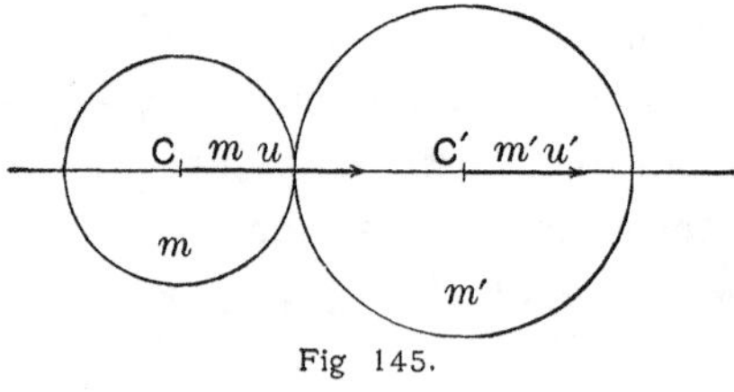

Fig 145.

C' along the line CC' (Fig. 145). To fix the ideas, we assume the velocities u, u' to have the same sense and $u > u'$, so that m will finally impinge upon m'. The case when the velocities are of opposite sense will not require special investigation, as only the sign of u' would have to be changed.

It is our object to determine the velocities v, v' of m, m' immediately after impact, when the velocities u, u' immediately before impact are given.

The results here derived for homogeneous spheres hold generally, whatever the shape of the impinging bodies, provided that they do not rotate, and that the common normal at the point of contact passes through both centroids.

433. If the spheres were perfectly rigid, the problem would be indeterminate, for there is no way of deciding how the velocities would be affected by the collision.

Natural bodies are not perfectly rigid. The effect of the impact will, in general, consist in a compression of the portions of the bodies brought into contact. Moreover, all natural bodies possess a certain degree of elasticity; the compression will therefore be followed by an extension, each sphere tending to regain its shape at least partially.

The compression acts as a retarding force on the impinging sphere m, and as an accelerating force on m'. It will last until the velocities u, u' have become equal, say $= w$. During the subsequent period of extension, or *restitution*, the elastic stress still further diminishes the velocity of m, and increases that of m', until they become, say, v, v'.

434. The stress varies, of course, during the whole time τ of compression and restitution. But, according to Newton's third law, the pressure F exerted at any instant by m on m' must be equal and opposite to the pressure F' exerted by m' on m at the same instant. Since $F = m du/dt$, $F' = m' du'/dt$, and $F = -F'$ at any instant during the time τ, we have

$$\int_0^T F dt = -\int_0^T F' dt, \text{ or } m\int_0^T du = -m'\int_0^T du',$$

whence $$mv - mu = -(m'v' - m'u'),$$

or $$mv + m'v' = mu + m'u'; \tag{5}$$

that is, *the total momentum after impact is equal to that before impact.*

435. This proposition will evidently hold for any number of spheres whose centers move in the same line, and can then be expressed in the form

$$\Sigma mv = \Sigma mu.$$

It can be regarded as a special case of the so-called principle of the conservation of the motion of the centroid to be proved hereafter for any system not acted upon by external forces. On the other hand, the proposition can be looked upon as a further generalization of Newton's first law of motion. While the latter asserts that the momentum of a *particle* remains unchanged as long as no external forces act upon it, our law of impact asserts the same thing for the momentum of a *system*.

436. If the spheres were *perfectly non-elastic*, there would be only compression and no subsequent extension. As at the end of the period of compression, the velocities u, u' have both become equal, viz., $= w$ (Art. 433), the spheres after impact would have the common velocity

$$w = \frac{mu + m'u'}{m + m'}.$$

437. If the spheres were *perfectly elastic*, *i. e.*, if the elastic stress following the compression, or the so-called *force of restitution*, were just equal to the preceding stress of compression, the spheres would completely regain their original shape. In this case, the elastic stress causes the impinging sphere m to lose during the period of restitution an amount of momentum

$m(w - u)$ equal to that lost during the period of compression. Hence, the final velocity of m after impact would be

$$v = w - (u - w) = 2w - u.$$

Similarly, we have for the other sphere m'

$$v' = w + (w - u') = 2w - u'.$$

As w is known from Art. 436, the velocities after impact can be determined for perfectly elastic spheres by means of these formulæ.

438. In general, physical bodies are *imperfectly elastic*, the force of restitution being less than that of the original compression; that is, we have

$$w - v = e(u - w),$$
$$v' - w = e(w - u'),$$

where e is a proper fraction whose limiting values are 0 for perfectly inelastic bodies and 1 for perfectly elastic bodies. This fraction e, whose value for different materials must be determined experimentally, is called the *coefficient of restitution* (or, less properly, the coefficient of elasticity).

439. To eliminate w we have only to add the last two equations; this gives

$$v' - v = e(u - u'); \qquad (6)$$

that is, *the ratio of the relative velocity after impact to the relative velocity before impact is constant and equal to the coefficient of restitution.*

This proposition, in connection with the proposition of Art. 434, expressed by formula (5), is sufficient to solve all problems of so-called *direct* impact, *i. e.*, when the centers of the spheres move in the same line.

440. As the coefficient e is frequently difficult to determine, the limiting cases $e = 0$, $e = 1$ are important as giving approximate solutions for certain classes of substances.

Thus, for nearly inelastic bodies (such as clay, lead, etc.) we may put $e = 0$, whence, by (6), $v' = v$, *i. e.*, the velocities of the spheres become equal after impact; and the value of the common velocity is found from (5) as

$$v = \frac{mu + m'u'}{m + m'},$$

which agrees with the result found in Art. 436. For perfectly elastic bodies $e = 1$, and formula (6) shows that in this case the relative velocity after impact is numerically equal to that before impact, but of opposite sense.

441. Exercises.

(1) From what height must a 5-lb. mass fall to acquire the momentum of a 1-oz. bullet moving 1200 ft./sec.?

(2) A shell bursts into two pieces of 9 and 6 lbs. which move on in the same line at 1100 and 1800 ft./sec., respectively; what was the velocity of the shell just before bursting?

(3) A body weighing 50 lbs. moving at 10 ft./sec. impinges on another of 20 lbs. moving in the same line at 6 ft./sec. If the bodies are inelastic, what is the velocity after impact?

(4) Two balls of clay ($e = 0$) weighing 7 and 3 oz. move in the same direction. The heavier ball impinges from behind upon the lighter ball at the moment when the latter moves at the rate of 15 ft./sec. If the velocity of the lighter ball is doubled by the impact, what was the original velocity of the heavier ball?

(5) Two glass balls ($e = 1$) weighing 1 lb. and 10 oz., respectively, move in the same line with velocities of 5 and 4 ft./sec. What are their velocities after impact: (*a*) if their original velocities were of the same sense? (*b*) if they were of opposite sense?

(6) A ball weighing 5 lbs., while moving with a speed of 51 ft./sec., overtakes a ball of 7 lbs. moving in the same line at the rate of 40 ft./sec. If the coefficient of restitution be $\frac{1}{3}$, what are the velocities of the two balls after impact?

(7) With the data of Ex. (6), show that the velocities after impact would be equal if the balls were perfectly inelastic, and that these velocities would differ more than in Ex. (6) if the balls were perfectly elastic.

(8) Find the velocity with which an elastic ball rebounds from a fixed surface after impinging upon it perpendicularly with a velocity u.

(9) To determine the coefficient of restitution, a ball is dropped from a height H on a fixed horizontal plate of the same material, and the height of rebound h is measured. Show that $e = \sqrt{h/H}$.

(10) A ball is dropped from a height $H = 12$ ft. on a fixed horizontal plate. Find the height h to which it will rebound if $e = \frac{5}{6}$.

(11) If not disturbed, the ball in Ex. (10) will continue to fall and rebound alternately. (*a*) What height does it reach at the tenth rebound? (*b*) In what time does it come to rest? (*c*) What is the whole space described?

(12) A number of equal, perfectly elastic balls are placed in contact so that their centers are in a straight line. An equal ball impinges with a velocity u along this line on the first ball of the row. Show that the last ball of the row will move off with the velocity u, while all the other balls will remain at rest.

(13) Find the velocity of the last (nth) ball in Ex. (12), when the coefficient of restitution is e.

(14) An inelastic ball of 8 lbs. is moving with a velocity of 12 ft./sec. (*a*) With what velocity must a ball of 20 lbs. meet it to arrest its motion? (*b*) With what velocity would the ball of 20 lbs. have to impinge from behind on the ball of 8 lbs. to double its velocity?

(15) A ball m impinges upon a ball m' from behind with a velocity u. Determine the velocities after impact, both for inelastic and for perfectly elastic balls: (*a*) when m' is originally at rest; (*b*) when m' is at rest, and very large in comparison with m; (*c*) when m' has the initial velocity u', and is very large in comparison with m.

442. Kinetic Energy. A particle of mass m, moving with the velocity v, has the kinetic energy $\frac{1}{2}mv^2$ (Art. 265). As this is not a vector quantity, the kinetic energy of a system consisting of any number of free particles is simply the algebraic sum, $\Sigma \frac{1}{2}mv^2$, of the kinetic energies of these particles. It is an essentially positive quantity, provided the masses are all positive.

The kinetic energy of a rigid body having a motion of pure translation is evidently $= \frac{1}{2}mv^2$, if m be the mass of the body and v the common velocity of all its points.

In rectilinear motion, *the rate at which the kinetic energy increases with the distance* s is, by Art. 265, equal to the force F:

$$F = \frac{d(\frac{1}{2}mv^2)}{ds}. \tag{7}$$

It is easy to verify that this definition of force agrees with the definition of force as the rate at which the momentum increases with the time (Art. 425). For we have

$$\frac{d(\frac{1}{2}mv^2)}{ds} = mv\frac{dv}{ds} = mv\frac{dv}{dt} \Big/ \frac{ds}{dt} = m\frac{dv}{dt} = \frac{d(mv)}{dt}.$$

Integrating the equation (7) from $s = s_0$ to $s = s$, we have

$$\int_{s_0}^{s} Fds = \tfrac{1}{2}mv^2 - \tfrac{1}{2}mv_0^2. \tag{8}$$

If, in particular, the force F is constant, we have

$$F(s - s_0) = \tfrac{1}{2}mv^2 - \tfrac{1}{2}mv_0^2. \tag{8'}$$

In (8) and (8′) the quantity in the left-hand member is called the **work** done by the force F in the distance $s - s_0$ (Art. 266). Thus a particle of mass m, falling from rest through a distance s, acquires its kinetic energy owing to the work done upon it by the constant attractive force, $F = mg$, of the earth, and we have, with $s_0 = 0$, $v_0 = 0$,

$$\tfrac{1}{2}mv^2 = Fs = mgs.$$

The kinetic energy $\frac{1}{2}mv^2$, possessed by a particle of mass m, moving with the velocity v, can therefore always be regarded as equivalent to a certain amount of work. If the motion of this particle be *opposed* by a constant force or resistance $-F$, the distance s through which it will go on moving until it comes to rest is of course determined from the same equation,

$$\tfrac{1}{2}mv^2 = Fs. \tag{9}$$

This results from the equation (8′) by putting $s_0 = 0$, $v = 0$, and writing $-F$ for F, v for v_0. It is then said that the kinetic energy of the particle is spent in overcoming the resistance $-F$, or in doing work against the force $-F$ (see Art. 402).

443. In the case of the direct impact of spheres, as considered in Art. 432, the velocity, and hence also the kinetic energy, of each sphere is in general changed by the impact; a transfer of kinetic energy can be said to take place. Thus, when a sphere at rest is struck by a moving sphere, kinetic energy is imparted to the former by the impulsive force, and this energy can then be spent in doing work against a resistance. Impact is therefore frequently used for the purpose of performing useful work.

444. For instance, to drive a nail into a wooden plank, the resistance $-F$ of the wood must be overcome through a certain distance s. This might be done by applying a pressure F equal and opposite to the resistance $-F$; as, however, this pressure would have to be very large, it is more convenient to impart to the nail, by striking it with a hammer, an amount of kinetic energy, $\frac{1}{2}mv^2$, equivalent to the work Fs that is to be done. Neglecting elasticity, and denoting the mass of the hammer by m, that of the nail by m', the velocity of the hammer at the moment when it strikes the head of the nail by u, we have, by (5),

$$mv + m'v' = mu,$$

or since, by (6), for inelastic impact $v' = v$,

$$v = \frac{m}{m + m'} u.$$

This is the common velocity of hammer and nail after the stroke. We find, therefore, by (9),

$$\tfrac{1}{2}(m + m') \cdot \left(\frac{mu}{m + m'}\right)^2 = Fs,$$

or

$$\frac{m}{m + m'} \cdot \tfrac{1}{2} mu^2 = Fs. \qquad (10)$$

445. It will be noticed that while the total kinetic energy of hammer and nail just before striking was $\frac{1}{2}mu^2 + 0$, the kinetic energy utilized for driving the nail is only the fraction

$m/(m+m')$ of this total kinetic energy. The remaining portion of the original kinetic energy, viz.,

$$\frac{m'}{m+m'} \cdot \tfrac{1}{2}\, mu^2, \qquad (11)$$

must be regarded as spent in producing the slight deformations of hammer and nail and such accompanying phenomena as vibrations of the plank, sound, heat, etc. For it is an experimental result of modern physical research that, wherever kinetic energy disappears as such, there is done an exactly equivalent amount of work. The apparently disappearing kinetic energy may either be transferred to some other body, as in the case of the vibrations of the plank, or it may reappear in the form of vibrations, causing sound or heat; or it may be transformed into an equivalent amount of so-called potential energy. This physical fact is known as *the principle of the conservation of energy.*

446. In our example the total original kinetic energy, $\frac{1}{2} mu^2$, resolves itself into two portions, the portion (10) used for driving the nail, and the "wasted" or, as it is often called, "lost" portion (11). It may, however, happen that the portion (11) does the useful work, while (10) is wasteful. This would be the case, for instance, in molding a rivet with a hammer, or in forging a piece of iron under the blows of a steam-hammer. The useful work here consists in the deformation of the bolt or piece of iron.

It appears from the expressions (10) and (11) that, for the purpose of driving the nail, m should be large in comparison with m', while for molding a rivet it is of advantage to have m' large in comparison with m.

447. It may be well to recall here (comp. Art. 262) that in applying the formulæ (8) to (11), and indeed any of the formulæ of theoretical mechanics, *the same system of units must be used consistently throughout the same formula.* In theoretical mechanics, as in physics and electrical engineering, the absolute system is generally used. With British units, the mass is then expressed in pounds, the force in poundals. The engineer, however, commonly uses the gravitation system; that is, he is accustomed to express a force in pounds. Hence, if for instance in the formula (9),

$$\tfrac{1}{2}mv^2 = Fs,$$

m be expressed in pounds, the numerical value obtained for F must be divided by g to reduce it to pounds. The same result is evidently obtained by substituting for m not the number that expresses the mass in pounds, but this number divided by g, whereupon the force F will result directly in pounds. For this reason engineers often write the above formula in the form

$$\frac{1}{2}\frac{w}{g}v^2 = Fs,$$

w being the "weight" in pounds, and F the force in pounds.

448. Let us now consider the *change of the total kinetic energy* produced by direct impact in two partially elastic spheres. With the notations of Art. 432, we have for the excess of the kinetic energy after impact over that before impact:

$$\tfrac{1}{2}(mv^2 + m'v'^2) - \tfrac{1}{2}(mu^2 + m'u'^2).$$

To eliminate v and v' from this expression, square the equations (5) and (6),

$$(mv + m'v')^2 = (mu + m'u')^2,$$

$$(v - v')^2 = e^2(u - u')^2;$$

multiply the latter by mm', and write it in the form

$$mm'(v - v')^2 + (1 - e^2)mm'(u - u')^2 = mm'(u - u')^2;$$

finally add the former equation,

$$(m + m')(mv^2 + m'v'^2) + (1 - e^2)\,mm'(u - u')^2$$
$$= (m + m')(mu^2 + m'u'^2),$$

whence

$$\tfrac{1}{2}(mv^2 + m'v'^2) - \tfrac{1}{2}(mu^2 + m'u'^2) = -\tfrac{1}{2}(1 - e^2)\frac{mm'}{m + m'}(u - u')^2. \quad (12)$$

As the right-hand member of this equation is essentially negative, it appears that *while in impact the total momentum remains unchanged, the total kinetic energy is in general dimin-*

ished; only in the limiting case of perfectly elastic bodies ($e = 1$) does the kinetic energy remain the same after as before impact. The "lost" kinetic energy (12) mainly represents the amount of energy spent in producing the permanent deformation of the impinging bodies, but sometimes also vibrations, heat, etc.

449. Exercises.

(1) Compare the momenta and kinetic energies of a train of 200 tons, running 60 miles an hour, and of the shot of a large gun, the shot weighing 1000 lbs., and the muzzle velocity being 1760 ft./sec.

(2) The head of a hammer, weighing $2\frac{1}{2}$ lbs. and moving with a speed of 40 ft./sec., is stopped in $\frac{1}{1000}$ sec. What is the average "force of the blow"?

(3) A ball of $5\frac{1}{4}$ oz. strikes a bat with a velocity of $12\frac{1}{2}$ ft./sec., and returns in the same line with a velocity of 32 ft./sec. If the blow lasts $\frac{1}{20}$ sec., what is the force exerted by the striker?

(4) A 2-lb. hammer strikes a 2-lb. chisel, used in cutting metal, with a velocity of 10 ft./sec. If the impact lasts $\frac{1}{10000}$ sec., what is the average force exerted on the metal?

(5) A hammer weighing 1.5 lbs. strikes a nail weighing $\frac{1}{2}$ oz. with a velocity of 20 ft./sec., and drives it $\frac{1}{4}$ in. Find the mean resistance of the wood, and determine the useful and wasteful work.

(6) In Art. 441, Ex. (6), find the loss of kinetic energy due to the impact.

(7) A train of 120 tons runs, with a speed of 15 miles an hour, into a train of 80 tons at rest. Neglecting elasticity, determine the destructive work of the collision, and the velocity along the track after impact.

(8) A pile weighing m' lbs. is driven into the ground by a ram of m lbs., falling from a height of h ft. If the pile sinks s in. into the ground after n falls of the ram, show that the resistance of the ground (assumed as uniform) is $= \frac{12n}{s}\left(\frac{1+e}{1+m'/m}\right)^2 m'h$ pounds.

(9) If, in Ex. (8), the elasticity of ram and pile be neglected, ram and pile will have equal velocities after impact, and move together. Hence, the factor m' should be replaced by $m + m'$, and the resistance is $= \frac{12n}{s} \cdot \frac{1}{1 + m'/m} \cdot mh$ pounds.

(10) Ten blows of a ram of 450 lbs., falling from a height of 6 ft., sink a pile of 350 lbs. 6 in. If the permanent load of a pile be taken as one fifth of the resistance, what permanent load can the pile bear?

(11) A steam hammer of 3 tons is used in forging. It has a fall of $4\frac{1}{2}$ ft. If the weight of the anvil be 18 tons, what is the useful and what the wasteful work?

(12) The block of a ram weighs 900 lbs., the pile 450 lbs. The block falls from a height of 5 ft. If, at the last blow, it drives the pile $\frac{1}{4}$ in., what is the resistance of the ground?

(13) A pile driver of 300 lbs. has a fall of 16 ft. and is stopped in $\frac{1}{10}$ sec.; determine the average force exerted on the pile.

(14) * A column of water in a 6-in. pipe, 30 ft. long, is moving behind a piston at 15 ft./sec. The motion of the piston being stopped in $\frac{1}{10}$ sec., what is the average pressure per square inch exerted on the piston?

(15) * Water is flowing through a service pipe at the rate of 60 ft./sec. If the water be brought to rest uniformly in $\frac{1}{10}$ sec. by closing the stop valve, what will be the increase of the pressure of the water near the valve, the pipe being taken as 50 ft. long, the resistance of the pipe and the compressibility of the water being neglected?

(16) * Thirty gallons of water per second enter a wheel in a direction AB from a horizontal 4-in. pipe, the shortest distance from the axis of the pipe to the vertical axis of the wheel being 1.3 ft. The water leaves the wheel in a horizontal direction CD, with a velocity of 3 ft./sec., the shortest distance of CD from the axis of the wheel being 0.8 ft. Find the useful turning moment.

(17) A ball of $5\frac{1}{4}$ oz. moving at 50 ft./sec. is caught, being brought to rest in a distance of 6 in. What is the average pressure on the hand?

450. Recoil. The explosion of the powder in a gun produces an impulsive pressure both on the shot and on the body of the gun. Assuming the line of motion of the centroid of the shot to pass through the centroid of the gun, we may apply equation (5), with $u=0$, $u'=0$. Hence, denoting by m the mass of the gun, by m' that of the shot, we find for the velocity of recoil

$$v = -\frac{m'}{m}v'. \qquad (13)$$

---

* From J. Perry, *Applied Mechanics*, New York, Van Nostrand, 1898.

The kinetic energies $\frac{1}{2}mv^2$ and $\frac{1}{2}m'v'^2$ of gun and shot are in the ratio m'/m; hence, the energy of recoil is the fraction $m'/(m+m')$ of the total energy $\frac{1}{2}mv^2+\frac{1}{2}m'v'^2$ of the explosion of the powder, while the energy of the shot is $=m/(m+m')$ of the total energy. In large guns the recoil is diminished by a special elastic cushion or "compressor." Moreover, the mass of the powder gases cannot be entirely neglected in all cases.

451. Oblique Impact. In the case of oblique impact, *i. e.*, when the centers of the colliding spheres do not move in the same straight line, the velocities after impact can be found without difficulty, provided that the velocities of the centers before impact lie in the same plane and that the spheres are perfectly smooth.

With these assumptions, let m, m' be the masses of the two spheres; C, C' their centers (Fig. 146); u, u' the velocities before impact; α, α' the angles made by u, u' with the line CC'; v, v' the velocities after impact; and β, β' the angles they make with CC'.

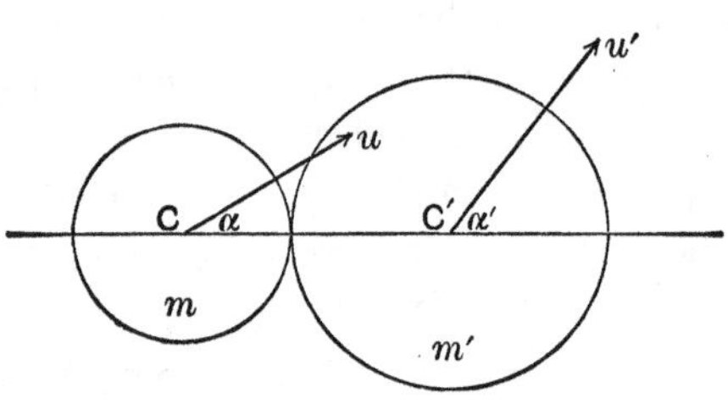

Fig. 146.

As there is no friction, the forces of impact act along the line CC' that joins the centers. Hence, resolving each velocity along and at right angles to CC', the components perpendicular to CC' must remain unchanged by the collision; that is, we must have

$$v\sin\beta=u\sin\alpha,\quad v'\sin\beta'=u'\sin\alpha'. \tag{14}$$

The components of the velocities along CC' must satisfy the equations (5) and (6). Hence, substituting $u\cos\alpha$, $u'\cos\alpha'$, $v\cos\beta$, $v'\cos\beta'$ for u, u', v, v', respectively, we must have

$$mv\cos\beta+m'v'\cos\beta'=mu\cos\alpha+m'u'\cos\alpha', \tag{15}$$

$$v'\cos\beta'-v\cos\beta=e(u\cos\alpha-u'\cos\alpha'). \tag{16}$$

452. The particular case of the oblique impact of a homogeneous sphere against a smooth fixed plane deserves special mention. In this case, u' and v' are zero; and the angles α, β made by the velocities u, v with the normal to the plane, are called the *angle of incidence* and *of reflection*, respectively.

The equations (14) and (16) reduce to the following:

$$v \sin \beta = u \sin \alpha, \quad v \cos \beta = -eu \cos \alpha, \tag{17}$$

where the minus sign indicates that the projections of u and v on the normal have opposite sense. Dividing the former of these equations by the latter, we find

$$\tan \alpha = -e \tan \beta,$$

where the minus sign merely indicates that the angles α, β lie on opposite sides of the normal.

For perfectly elastic bodies the last equation shows that the angles of incidence and reflection are equal.

453. In the **kinetic theory of gases** the pressure of a gas on the walls of the containing vessel is regarded as due to the impacts of the molecules of the gas on these walls. The gas is supposed to consist of a large number of molecules, all in very rapid rectilinear motion in all possible directions. The impact must be assumed as perfectly elastic, since otherwise each impact would diminish the velocity of the molecule; this would mean that the pressure of a gas in a closed vessel diminishes in the course of time, which is contrary to experience.

Let the vessel be a cylinder with a movable piston. As the mass of the piston is very large in comparison with the mass m of the molecule, it follows easily from equations (5) and (6) that *the momentum imparted to the piston by a molecule m impinging on it at right angles with velocity u is $= 2\,mu$;* that is, twice the momentum of the molecule.

454. Next, let the vessel be a cube whose edge is 1 cm. A molecule m, moving at right angles to a face, has to travel 2 cm. from one impact to the next impact on the same face; hence, if it strikes the face n times per second, we have $2 \cdot n = u \cdot 1$, whence $n = \frac{1}{2} u$.

The pressure on the face produced by the molecule being the rate at which it imparts momentum, *i. e.*, in our case the momentum imparted in one second, is therefore

$$n \cdot 2\,mu = mu^2.$$

Now, let u be the velocity of a molecule m moving in *any* direction, and let u_x, u_y, u_z be its components along the edges of the cube; then the kinetic energy of the molecule is

$$\tfrac{1}{2}\,mu^2 = \tfrac{1}{2}\,mu_x^{\,2} + \tfrac{1}{2}\,mu_y^{\,2} + \tfrac{1}{2}\,mu_z^{\,2};$$

i. e., it is the same as that of three molecules, each of mass m, whose velocities u_x, u_y, u_z are perpendicular to the faces.

The total kinetic energy of the gas contained in the cubic centimeter divides itself similarly into three parts:

$$\Sigma\,\tfrac{1}{2}\,mu^2 = \Sigma\,\tfrac{1}{2}\,mu_x^{\,2} + \Sigma\,\tfrac{1}{2}\,mu_y^{\,2} + \Sigma\,\tfrac{1}{2}\,mu_z^{\,2}.$$

If these parts are assumed equal, each is $= \tfrac{1}{3}\,\Sigma\,\tfrac{1}{2}\,mu^2$.

Let the cube contain N molecules, each of mass m; then the total kinetic energy is

$$\Sigma\,\tfrac{1}{2}\,mu^2 = \tfrac{1}{2}\,m\Sigma u^2 = \tfrac{1}{2}\,Nm\bar{u}^2,$$

where $\bar{u} = \sqrt{\Sigma u^2/N}$ is the so-called *mean square velocity*. The pressure p on each face of the cube is, therefore,

$$p = \tfrac{1}{3}\,\Sigma mu^2 = \tfrac{1}{3}\,Nm\bar{u}^2.$$

As Nm is the total mass in one cubic centimeter, *i. e.*, the density ρ of the gas, we have also

$$p = \tfrac{1}{3}\,\rho\bar{u}^2.$$

This formula gives the pressure per square centimeter on the walls of a vessel of any shape or size; to make this generalization it is only necessary to imagine the vessel divided up into cubic centimeters. The formula is known as *Boyle's law*, which expresses that the pressure of a gas is proportional to its density, if the temperature is constant, *i. e.*, if the mean square velocity $\bar{u}$ is constant.

For hydrogen, at atmospheric pressure, *i. e.*, $p = 1033$ grams per square centimeter, and at 0° C., the density $\rho = 0.000\,089\,6$; hence the mean square velocity of a hydrogen molecule (with $g = 981$) is found = 1842 meters per second.

455. Exercises.

(1) A baseball weighing $5\tfrac{1}{4}$ oz., while moving with a velocity of 100 ft./sec., is struck by the bat in a direction at right angles to its line of motion. Find the momentum imparted by the blow if it deflects the ball through an angle of 60°.

(2) Determine the velocity of recoil of a gun weighing 1500 lbs. when a 12-lb. shot is fired from it with an initial velocity of 2000 ft./sec.

(3) The heavier one of two ivory balls ($e = 0.88$), whose centroids are C, C' and whose masses are 1 lb. and $\frac{3}{4}$ lb., impinges upon the lighter. The velocity of the heavier ball is 15 ft./sec. and makes an angle of 30° with the line CC', while the velocity of the lighter ball is 5 ft./sec. and makes an angle of 60° with the line CC' (produced). Find the velocities after impact in magnitude and direction.

(4) A 40-lb. shot is fired from a gun weighing 5 tons, with an initial velocity of 1600 ft./sec.; find the velocity of recoil and the average force of the explosion if the gun is 10 ft. long.

(5) A stone is thrown with a velocity of 33 ft./sec. at right angles to a train running at 45 miles per hour; compare the kinetic energy with which the train is struck with that expended by the thrower.

(6) If a stone is thrown with a velocity of 33 ft./sec. from one train into another train while they pass, both trains running at 45 miles per hour, how does the kinetic energy with which the train is struck compare with that expended by the thrower?

(7) Two equal perfectly elastic spheres meet at right angles; the velocities u, u' before impact being given, find the velocities v, v' after impact.

II. *Rectilinear Motion of a Particle.*

456. It has been shown in Statics (Arts. 273–277) that any number of forces acting on the same particle, of mass m, can be replaced by a single force, their resultant R, which is found by adding the given forces geometrically. This force R gives to the particle m an acceleration j which is the geometric sum, or resultant, of the accelerations due to the given forces separately. The vectors j and R have the same direction and sense and are such that

$$mj = R.$$

This relation is called the *equation of motion of the particle.*

It should be observed that it is not essential that the mass m of the "particle" be concentrated at a point; what is essential is that the moving mass m has a mere motion of translation so that the motion of any one point determines the motion of the whole mass, and that all the forces can be regarded as applied at one and the same point which, if the force of gravity acts, must be the centroid of the mass m. If these conditions are satisfied, the "particle" may be a body of any size.

457. We begin by considering the simplest case, that of *rectilinear motion;* both j and R are then directed along the line of motion.

As shown in Kinematics (Art. 65), the acceleration in rectilinear motion is $j = dv/dt = d^2s/dt^2$, where v is the velocity of the particle and s its distance from some fixed point, or origin, taken on the line of motion. The *equation of rectilinear motion* can therefore be written

$$m\frac{dv}{dt} = R, \text{ or } m\frac{d^2s}{dt^2} = R. \tag{1}$$

It is here assumed that the mass m (as is usually the case) remains constant during the motion. If this were not the case, it would be necessary to use the more general definition of force (Art. 255) as the time-rate of change of momentum and to write the equation of motion in the form

$$\frac{d(mv)}{dt} = R.$$

458. *If the force R is constant,* the acceleration $j = R/m$ is constant, and the motion is uniformly accelerated. Integrating the former of the equations (1), we find in this case

$$mv - mv_0 = R(t - t_0); \tag{2}$$

i. e., the increase of momentum in any time $t - t_0$ is equal to the impulse of the force R in this time (comp. Arts. 256, 425).

If the given "initial" velocity v_0 be zero and the time t be counted from the corresponding instant, the relation (2) reduces to the more simple form

$$mv = Rt. \tag{2'}$$

459. In the case of a constant force R, the latter of the two equations (1) can also be integrated; for, after multiplying both members by ds/dt, the left-hand member becomes the exact time-derivative for the quantity $\frac{1}{2}m(ds/dt)^2 = \frac{1}{2}mv^2$. We thus find:

$$\tfrac{1}{2}mv^2 - \tfrac{1}{2}mv_0^2 = R(s - s_0); \tag{3}$$

i.e., the increase of the kinetic energy in any distance $s - s_0$ *is equal to the work done by the force* R *in this distance* (see Arts. 266, 442).

If the initial velocity v_0 be zero and the distance s be counted from the corresponding initial position, the relation (3) reduces to

$$\tfrac{1}{2}mv^2 = Rs. \tag{3'}$$

The proposition expressed by equation (3) or (3′) is known as the **principle of kinetic energy and work** for the case of the rectilinear motion of a particle under a constant force.

It should be observed that the important relations (2) and (3) are nothing but the two fundamental definitions of a constant force (comp. Art. 442). The following examples of rectilinear motion under a constant force will further illustrate the meaning and use of equations (1) to (3′). With regard to the units of mass and force, see the remark in Art. 447.

460. Falling Body. With the assumptions of Art. 72, the dynamical equation of motion of a particle of mass m falling *in vacuo* is

$$m\frac{dv}{dt} = W, \text{ or } m\frac{d^2s}{dt^2} = W,$$

where $W = mg$ is the *weight* (Art. 306) of the particle, *i.e.*, the force of attraction exerted by the earth on the mass m. Integrating, we find as above

$$mv - mv_0 = W(t - t_0),$$

$$\tfrac{1}{2}mv^2 - \tfrac{1}{2}mv_0^2 = W(s - s_0).$$

Upon division by m, all these equations reduce to results found previously in Kinematics (Arts. 68–74). The advantage of the present equations lies in their dynamic interpretation. Thus, the last equation represents the increase of the kinetic energy of the body during its fall as due to the work of its weight W, *i. e.*, of the attraction of the earth. The work $W(s - s_0)$ done by gravity is said to be transformed into the gained kinetic energy of the body.

For a body thrown vertically upward with an initial velocity v_0, we have

$$\tfrac{1}{2} mv^2 - \tfrac{1}{2} mv_0^2 = - Ws,$$

if s be counted from the starting point and positive upward. The kinetic energy here decreases; the initial kinetic energy, $\tfrac{1}{2} mv_0^2$, is gradually spent in doing work against the force of gravity.

This dynamic language is often useful in eliciting a mental picture of the phenomena described and in giving to them a physical meaning. But the analogies suggested by it should not be pressed too far.

461. Inclined Plane. When a particle of mass m is moved uniformly up a smooth inclined plane from P_0 to P_1 (Fig. 147), the work done against gravity is equal to the work that would have to be done in raising the particle m through the vertical height PP_1 of P_1 above the initial point P_0. For, putting $P_0P_1 = s$, $PP_1 = h$, and denoting the inclination of the plane to the horizon by θ, the normal component $mg \cos \theta$ of the weight $W = mg$ is balanced by the normal reaction of the plane, and the work of the component $mg \sin \theta$ parallel to the plane is

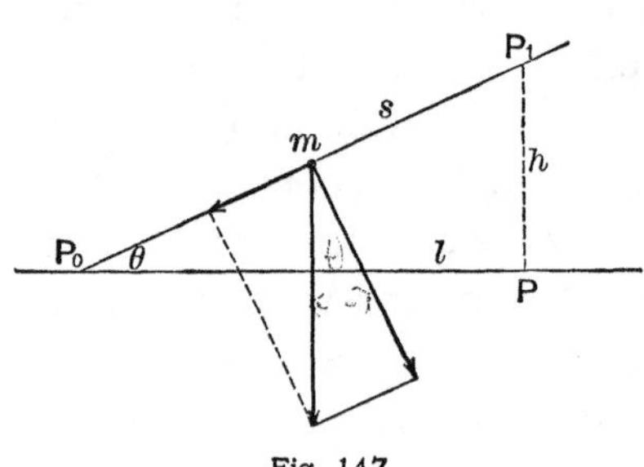

Fig. 147.

$$mg \sin \theta \cdot s = mg \cdot s \sin \theta = mg \cdot h.$$

If the plane be rough, the coefficient of friction being μ, the resultant force for motion upwards is $= mg \sin \theta + \mu mg \cos \theta$; hence, the work done in moving the mass m from P_0 to P_1 is

$$mgs \sin \theta + \mu mgs \cos \theta = mg \cdot h + \mu mg \cdot l = mg(h + \mu l),$$

where $l = P_0P$ is the horizontal distance of the final position P_1 from the starting point P_0. The total work is, therefore, the sum of the work of overcoming gravity through the *vertical* distance h and the work of overcoming friction through the *horizontal* distance l.

462. Work against Gravity for a System of Particles. Let there be given any number of particles of masses m_1, m_2, $\cdots$ m_n at the distances s_1, s_2, $\cdots$ s_n above a fixed horizontal plane; and let these masses be raised vertically against gravity so that their distances from the same plane become s_1', s_2', $\cdots$ s_n'. The centroid of the masses in their original position has a distance $\bar{s} = \Sigma ms/\Sigma m$ from the fixed plane, while in the final position it has the distance $\bar{s}' = \Sigma ms'/\Sigma m$ from the same plane. It has, therefore, been raised through a distance $\bar{s}' - \bar{s}$. It follows that *the total work done in raising the separate masses*, viz.,

$$m_1 g(s_1' - s_1) + m_2 g(s_2' - s_2) + \cdots + m_n g(s_n' - s_n) = g(\Sigma ms' - \Sigma ms),$$

is equal to the work that would be done in raising the total mass Σm *through the distance* $\bar{s}' - \bar{s}$ *traversed by the centroid, i. e.*, to

$$g\Sigma m \cdot (\bar{s}' - \bar{s}).$$

463. Let us consider a mass m that is being raised or lowered by means of a rope or chain (Fig. 148), such as a building stone suspended from a derrick. The rope acts as a *constraint*, conditioning the motion of the stone. To make the stone free we may imagine the rope cut just above the stone and the **tension** of the rope, T, introduced as a substitute. The stone then moves under the action of two forces, viz., its weight $W = mg$ and the tension T of the rope. Taking the downward sense as positive, we have the equation of motion,

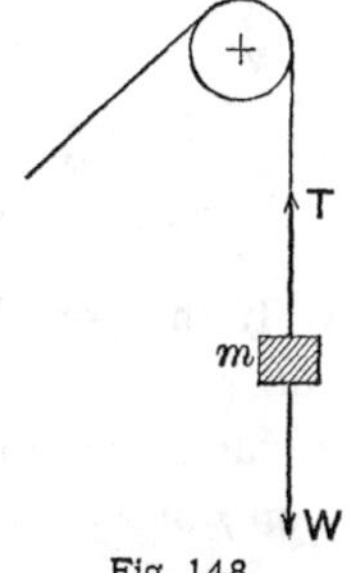

Fig. 148.

$$m\frac{d^2s}{dt^2} = mg - T.$$

Writing j for the acceleration d^2s/dt^2 with which the stone is being lowered or raised, we find for the tension T of the rope

$$T = m(g - j).$$

This equation shows that the tension is equal to the weight of the stone, not only when it is hanging at rest, but also whenever it is raised or lowered with constant velocity; and that the tension is zero if the stone is lowered with an acceleration equal to that of gravity, as is otherwise evident.

464. Let us next consider two particles, m_1, m_2, connected by a cord hung over a vertical fixed pulley, as in the apparatus known as *Atwood's machine* (Fig. 149). If $m_1 > m_2$, m_1 will descend while m_2 ascends. The force producing the acceleration of the system formed by the two particles is evidently the difference of the weights of the particles, viz., $W_1 - W_2 = (m_1 - m_2)g$, while the whole mass moved (neglecting the mass of the cord and of the pulley) is $m_1 + m_2$. Hence, we have for the acceleration j of the system,

$$j = \frac{m_1 - m_2}{m_1 + m_2} g. \qquad (4)$$

This acceleration is constant, and the relations between space, time, and velocity are found just as for a single particle falling freely, except that the acceleration of gravity g is replaced by the fraction $(m_1 - m_2)/(m_1 + m_2)$ of g. It follows that if the masses m_1, m_2 be selected nearly equal, the acceleration will be small, and the motion can be observed more conveniently than that of a freely falling body.

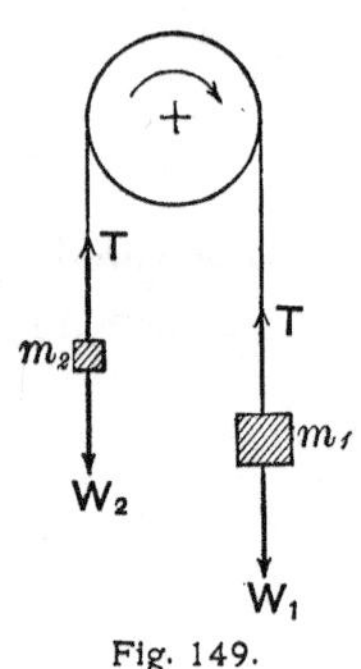

Fig. 149.

The tension T of the cord is, of course, the same at every point of the cord if, as is here assumed, the weight of the cord, the inertia of the pulley, and the axle-friction of the pulley be neglected. To determine this tension, we have only to consider either particle separately.

If the cord be cut just above m_1, and the tension T be introduced, the particle m_1 will move like a free particle under the action of the resultant force $W_1 - T = m_1 g - T$. Hence, as the sense of the acceleration j of m_1 agrees with that of g,

$$j = \frac{m_1 g - T}{m_1}. \qquad (5)$$

Similarly, we have for the acceleration of m_2, whose sense is opposite to that of g,

$$j = -\frac{m_2 g - T}{m_2}. \qquad (6)$$

Eliminating j between any two of the equations (4), (5), (6), we find the tension

$$T = \frac{2 m_1 m_2}{m_1 + m_2} g. \qquad (7)$$

465. If the two particles of Art. 464 move on inclined planes intersecting in the horizontal axis of the pulley (Fig. 150), it is only necessary to resolve the weights $m_1 g$ and $m_2 g$ into two components, one parallel, the other perpendicular, to the inclined plane. If the planes be smooth,

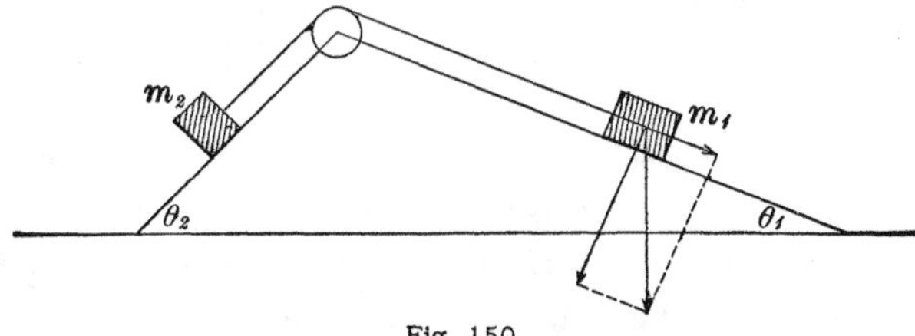

Fig. 150.

the system formed by the two particles is made free by introducing the normal reactions of the planes which counterbalance the perpendicular components of the weights. The resultant force is therefore the difference of the parallel components, and the acceleration is

$$j = \frac{m_1 \sin\theta_1 - m_2 \sin\theta_2}{m_1 + m_2} g, \tag{8}$$

where θ_1, θ_2 are the angles of inclination of the planes to the horizon.

The tension T of the connecting cord is again determined by equating this value of j to the one obtained by considering either of the two particles separately. Thus, m_1 taken by itself, becomes free if we introduce not only the normal reaction of the plane, but also the tension of the string. This gives

$$j = \frac{m_1 g \sin\theta_1 - T}{m_1}. \tag{9}$$

Similarly, we have for m_2

$$j = \frac{T - m_2 g \sin\theta_2}{m_2}. \tag{10}$$

Hence,
$$T = \frac{m_1 m_2}{m_1 + m_2} g (\sin\theta_1 + \sin\theta_2). \tag{11}$$

With $\theta_1 = \theta_2 = \pi/2$ the formulæ (8) to (11) reduce, of course, to the formulæ (4) to (7).

466. Exercises.

(1) A stone weighing 200 lbs. is raised vertically by means of a chain running over a fixed pulley. Determine the tension of the chain: (*a*) when the motion is uniform; (*b*) when the motion is uniformly accelerated upwards at the rate of 8 ft./sec.^{2}; (*c*) when the acceleration is 32 ft./sec.^{2} downwards. Neglect the weight of the chain and the axle-friction of the pulley.

(2) A railroad car weighing 4 tons is pushed by four men over a smooth horizontal track. If each man exerts a constant pressure of 100 lbs.: (*a*) what is the velocity acquired by the car at the end of 5 sec.? (*b*) what is the distance passed over in these 5 sec.?

(3) (*a*) Determine the constant force required to give a train of 100 tons a velocity of 30 miles an hour in 6 min. after starting from rest. (*b*) How far does the train go in this time? (*c*) If the same velocity is to be acquired at the end of the first mile, what must be the tractive force of the engine?

(4) If in Atwood's machine (Fig. 149) the two masses are each 2 lbs., and an additional mass of 1 oz. be placed on one of these masses, how long will it take this mass to descend 6 ft.? ($g = 32.2$.)

(5) A cord passing over a pulley, fixed 40 ft. above the ground, carries a mass of 50 lbs. at one end and a mass of 51 lbs. at the other. If initially both masses are 20 ft. above the ground, where will they be after 10 sec.? What is the tension of the cord? When does the heavier mass reach the ground? ($g = 32$.)

(6) Solve the problem of Art. 465 when the inclined planes are rough, the coefficients of friction being μ_1, μ_2.

(7) A mass of 9 lbs. rests on a smooth horizontal table, and has a cord attached which runs over a smooth pulley on the edge of the table. If a mass of 2 lbs. be suspended from the cord, find the acceleration and the tension of the cord.

(8) A sleigh weighing 450 lbs. is drawn over a horizontal road, the coefficient of friction being $\frac{1}{50}$. Find the pull exerted by the horses when the motion is uniform.

(9) When the U.S.S. *Raritan* was launched she was observed to pass in 11 sec. over an incline of 3° 40′, 54 ft. long. Find the coefficient of friction.

(10) A coaster, after coming down a hill, runs up another hill, of

slope 1 : 10, and comes to rest in 12 sec. How far up did it go if the coefficient of friction is $\frac{1}{40}$?

(11) A train of 120 tons is running 35 miles an hour. Find what constant force is required to bring it to rest: (*a*) in 4 min.; (*b*) in half a mile.

(12) If it takes 1 min. to coast down a hill on a uniformly sloping road of 1 mile length, and the coefficient of friction be 0.02, what is the height of the hill?

(13) If a man is able to lift 150 lbs. on the ground, how much can he lift in an elevator (*a*) ascending, (*b*) descending, with an acceleration of 8 ft./sec.^{2}?

(14) A train of 180 tons reaches its regular velocity of 45 miles per hour 3 min. after starting. What is the accelerating force if regarded as constant?

(15) The barrel of a rifle is 30 in. long; the ball, weighing 1 oz., leaves the barrel with a velocity of 1200 ft./sec. What is the average pressure of the powder gases?

(16) In a coal-pit shaft, a cage (*a*) ascends, (*b*) descends, with an acceleration of 10 ft./sec.^{2} Find the tension of the rope when the cage is empty and when it carries two men, of 165 lbs. each. What is the pressure exerted by each man on the bottom of the cage?

(17) A train of 60 tons runs one mile with constant speed; if the resistances be 8 lbs. per ton, find the work done by the engine: (*a*) on a level track; (*b*) on an average grade of 1%. (*c*) On a 1% grade, what is the ratio of the work done against gravity to that done against the resistances? (*d*) If, while the train is running 30 miles per hour, the steam is shut off, how far on the level will the train go before its speed is reduced to 10 miles per hour?

(18) A train of 100 tons (excluding the engine) starting from rest acquires a velocity of 30 miles an hour on a level road at the end of the first mile. Determine the average tractive force of the engine, that is, the pull on the drawbar between engine and train: (*a*) if the frictional resistances be neglected; (*b*) if these resistances be estimated at 8 lbs. per ton. (*c*) What tractive force is required to haul the same train at a constant speed over a level road? (*d*) up a grade of 1 in 160?

(19) Determine the work expended in raising from the ground the materials for a brick wall 30 ft. high, 40 ft. long, and 2 ft. thick, the weight of a cubic foot of brickwork being 112 lbs.

(20) A chain 350 ft. long and weighing 10 lbs. per foot is hanging down a shaft, with a weight of 520 lbs. attached at the end; what work is done by a capstan in winding it up?

(21) An elevator weighing 800 lbs. is raised by a hoisting engine; if the steam is shut off when the upward speed is 6 ft./sec., how much higher will the elevator rise?

(22) To what height could a train be lifted vertically by the work spent in giving it a speed of 30 miles per hour?

(23) The interior diameter of a chimney is 10 ft.; the exterior diameter is 28 ft. at the base, 14 ft. at the top; the height is 300 ft. Find the work done in raising the building material from the ground to its place.

467. Variable Force. The two fundamental propositions of Arts. 458, 459 are true for rectilinear motion, even when the resultant force R is variable. For in this more general case we find by integrating the equation of motion (1) Art. 457, in each of its two forms:

$$mv - mv_0 = \int_{t_0}^{t} R dt, \tag{12}$$

$$\tfrac{1}{2} mv^2 - \tfrac{1}{2} mv_0^2 = \int_{s_0}^{s} R ds; \tag{13}$$

and according to the definitions of the impulse (Art. 256) and of the work (Art. 266) of a variable force, these equations mean that *the increase of momentum in any time is equal to the impulse of the force during this time*, and that *the increase in kinetic energy in any distance is equal to the work done by the resultant force in this distance* (comp. Arts. 425, 442).

If R is constant, the equations (12), (13) evidently reduce to (2), (3) respectively.

It should also be observed that the equations (12) and (13) are equivalent, respectively, to

$$R = \frac{d(mv)}{dt}$$

and

$$R = \frac{d(\frac{1}{2} mv^2)}{ds};$$

that is, they merely express the two fundamental definitions of *force* as the *time-rate of change of momentum* and the *space-rate of change of kinetic energy* (comp. Art. 442).

468. The equation (13) expresses the **principle of kinetic energy and work** for the case of the rectilinear motion of a particle (comp. Art. 459). The integral in the right-hand member, $\int_{s_0}^{s} R ds$, which represents the work of the force R in the distance $s - s_0$, can be evaluated analytically only when R is a given function of the distance s in the interval $s - s_0$. Graphical or mechanical methods are therefore often used to find its value approximately.

We can always put $\int_{s_0}^{s} R ds = \bar{R} \cdot (s - s_0)$, where $\bar{R}$ is the space-average of the force R for the distance $s - s_0$; in other words, the **average,** or **mean, force** $\bar{R}$ is that constant force which acting through the distance $s - s_0$ does the same work as the variable force R.

Similar considerations apply to the relation (12) and lead to the time-average of the force R.

469. Pressure of a Piston. The work of a variable force is well illustrated by the expansion of gas or steam in a cylinder with a movable piston (Fig. 151). Let r be the radius of the cylinder, p the pressure (in pounds) at any instant of the gas per square inch of surface; then the total pressure of the gas on the inside of the piston is $P = \pi r^2 p$ pounds, and if P_0 be the pressure on the outside (say the atmospheric pressure), the resultant force acting on the piston is $R = P - P_0$, friction being neglected.

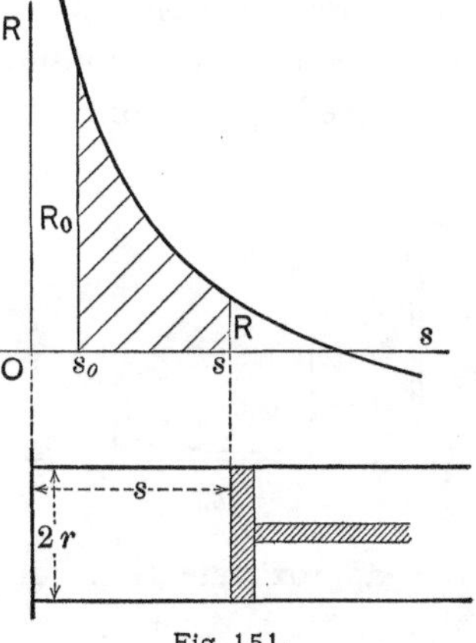

Fig. 151.

The force R is variable, since the pressure p varies with the volume v occupied by the gas. This volume being in the present case proportional to the distance s of the piston from the fixed base of the cylinder, the force R is a function of s. The variation of R can therefore be represented graphically by a curve having s for abscissa and R for ordinate (Fig. 151); and *the area of this curve, i.e.*, the area contained between the curve, the axis of s, and two ordinates whose abscissas are s_0 and s, being given by the integral $\int_{s_0}^{s} Rds$, *represents the work done on the piston when pushed through the distance* $s - s_0$.

470. In the case of a perfect gas, Boyle's law gives the relation $pv = k$, where k is constant if the temperature remains constant. Hence,

$$R = \frac{K}{s} - P_0,$$

where K and P_0 are constants. This equation represents an equilateral hyperbola, whose asymptotes are the axis of R and a line parallel to the axis of s. For steam, the law connecting pressure and volume is more complicated, but the curve may be taken as very near hyperbolic.

471. The Steam-engine Indicator is an apparatus for measuring the pressure of the steam in the cylinder and at the same time recording it automatically on a drum revolving as the piston moves. Thus, if the indicator be put in connection with the interior of the cylinder, the curve traced by the indicator has for its abscissas the distances s of the piston from one end, and for its ordinates the corresponding pressures P of the steam on the inside of the piston.

At the beginning of the stroke, steam is admitted and acts with nearly constant pressure on the piston; the line AB (Fig. 152) traced by the indicator will therefore be nearly parallel to the axis of s. As soon as the steam is shut off by the slide-valve, the steam, being now confined within the cylinder, begins to expand nearly according to the law $pv =$ const., or $Ps =$ const.; the curve traced by the indicator is therefore approximately an equilateral hyperbola BC, having the axes as asymptotes. When the slide-valve connects the cylinder with the condenser, a partial vacuum is established behind the piston, and the pressure curve is approximately a line CD, parallel to the axis of P.

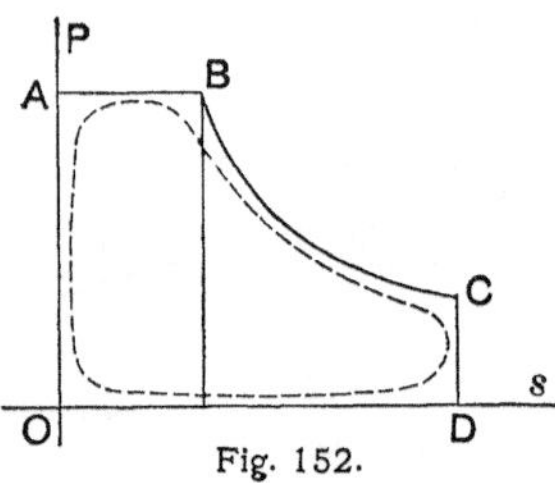

Fig. 152.

The area $ABCDO$ evidently represents approximately the work of the pressure on the inside of the piston in a double (forward and backward) stroke. In reality, various circumstances produce deviations from the regular shape $ABCDO$, and the actual trace, obtained by means of an indicator for a double (forward and backward) stroke, usually called the *indicator diagram*, forms a loop somewhat like that indicated by the dotted curve in Fig. 152. The area of this loop, which represents the work in question, can readily be found approximately by dividing it up into narrow rectangular strips, or with the aid of a planimeter.

The resultant or *effective* piston pressure is of course the difference between the pressures on the two sides of the piston. A diagram should therefore be obtained for each side of the piston; from these two diagrams the curve of effective piston pressure is then derived by constructing the curve whose ordinates are the differences of the corresponding pressures on the two sides. By dividing the area contained between this curve and the axes by the length of the stroke, the average, or *mean, piston pressure* is finally found (see Art. 468).

For details the student is referred to special works on the steam engine.

472. Attractive and Repulsive Forces. According to Newton's law of universal gravitation any particle of matter, of mass m, is attracted by any other such particle, of mass m', with a force R, directed along the line joining the particles and in magnitude directly proportional to the masses m, m' and *inversely proportional to the square of the distance* s of the particles. If we regard the attracting mass m' as fixed, say at the point O, and take this point as origin of the distance $s = OP$ of m from m' (Fig. 153), the force R with which m' attracts m is directed toward O, *i.e.*, in the negative sense of s; we have therefore

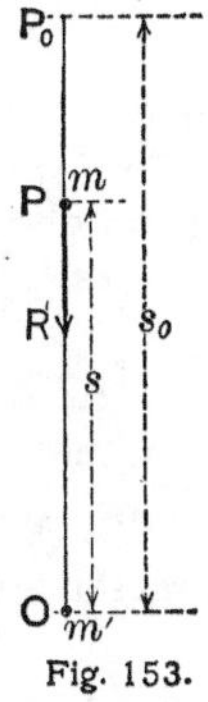

Fig. 153.

$$R = -\kappa \frac{mm'}{s^2}, \tag{14}$$

where κ is a positive constant, called the **constant of gravitation.**

473. The constant κ evidently represents the force with which two particles, each of mass 1, attract each other when at the distance 1. It is a physical constant to be determined by experiment, and its numerical value depends on the units of measurement adopted. What can be directly observed is of course not the force itself, but the acceleration it produces. Dividing the force R, as given by formula (14), by the mass m of the particle on which it acts, we find for the *acceleration* j produced by the attraction of the mass m' in the mass m at the distance r from m':

$$j = \kappa \frac{m'}{r^2}.$$

It is shown in the theory of attraction that the attraction of a homogeneous sphere at an external point is the same as if the mass of the sphere were concentrated at its center. Thus, if m' be the mass of the earth (here assumed as a homogeneous sphere), the acceleration j it produces in any mass m situated at a point P above its surface, at the distance $OP = r$ from the center O, is $= \kappa m'/r^2$.

Now for points P near the earth's surface this acceleration j is known from experiments; it is the acceleration of gravity, usually denoted by g. As the radius of the earth, $r = 6.37 \times 10^8$ centimeters, and its mean density $\rho = 5\frac{1}{2}$, are also known, the value of the constant κ can be found from the formula

$$g = \kappa \frac{m'}{r^2},$$

whence

$$\kappa = \frac{3}{4} \frac{g}{\pi \rho r}.$$

With $g = 980$ we find in C. G. S. units

$$\kappa = \frac{1}{1.50 \times 10^7} = 0.000\,000\,067.$$

This, then, is approximately the force in dynes with which two masses of 1 gram each would attract each other if concentrated at two points 1 centimeter apart.

474. The force between two electric charges as well as that between two magnetic poles follows Newton's law (Art. 472); that is, the force is directly proportional to the charges, or pole-strengths, and inversely proportional to the distance. But the constant κ has a very different value. It is customary in electricity and magnetism to select the units for electric charges and magnetic pole-strengths so that the constant $\kappa = 1$. In astronomy and in the general theory of attraction the unit of mass is also often selected so as to make $\kappa = 1$.

475. Let us now return to the problem (Art. 472) of the motion of a particle m when attracted according to Newton's law by a fixed mass m' (Fig. 153). This motion will be rectilinear if the initial velocity v_0 of m is directed along the line OP, and of course also when $v_0 = 0$.

The equation of motion is in this case

$$m\frac{d^2s}{dt^2} = R = -\kappa\frac{mm'}{s^2}.$$

Upon division by m it reduces to the equation discussed in Kinematics, Arts. 77–80. It is there shown how the integration can be carried through; the only thing to be added here is the interpretation of the equation of kinetic energy and work (13), which, with the above value of R, reduces to

$$\tfrac{1}{2}mv^2 - \tfrac{1}{2}mv_0^2 = -\kappa mm\int_{s_0}^{s}\frac{ds}{s^2} = \kappa m\left(\frac{m'}{s} - \frac{m'}{s_0}\right). \qquad (15)$$

476. The quantity m'/s, or, in absolute measure, $\kappa m'/s$, is called the **potential** at P, due to the attracting mass m'; so that $\kappa m'/s_0$ is the value of the same potential at P_0. The right-hand member of equation (15), which represents the work done upon the particle m by the attraction of the mass m' in the distance $PP_0 = s - s_0$, is therefore m times the difference of the potentials at P and P_0, due to m'. In other words, the *difference of potential between two positions, P and P_0, is the work that would have to be done by the attraction of m' to bring a unit mass from P_0 to P.*

It also follows from equations (14) and (15) that *the force of attraction exerted by a mass m' (at O) on a unit mass at any point, P, is the space-derivative of the potential at P:*

$$\frac{1}{m}R = \frac{d}{ds}\left(\kappa\frac{m'}{s}\right).$$

477. The negative of the potential multiplied by the mass m, *i. e.*, the quantity $-\kappa mm'/s$, is called the **potential energy** of the moving particle m. Denoting this by V, and the kinetic energy by T, the last equation becomes

$$T - T_0 = -V + V_0,$$

or
$$T + V = T_0 + V_0 = \text{const.}; \qquad (16)$$

i. e., the sum of the kinetic and potential energies remains constant during the motion. This is **the principle of the conservation of energy** for this particular problem.

478. The physical idea to which the term *potential energy* is due may perhaps require some explanation. The region surrounding an attracting mass m' is called the **field** of the force of attraction R of m'. Wherever in this field a particle m be placed (say, with zero velocity), it will become subject to the attraction R of m' and move toward m' with increasing velocity, thus acquiring kinetic energy; at the same time the force R does an amount of work on m which is exactly equivalent to the kinetic energy gained by m. It follows that, the farther away from m' the particle m is placed, initially, the greater will be the amount of work that m' can do upon it. It is this "potentiality" for doing work, due to the distance of m from m', which is denoted as *energy of position*, or *potential energy*. The equation (16), or the equation (15) which differs from (16) merely in notation, shows that *what the particle m in moving toward m' gains in kinetic energy it loses in potential energy* so that the sum of kinetic and potential energy always remains constant.

479. By properly generalizing the idea of potential and potential energy the principle of the conservation of energy (Art. 477) can be extended to the more general case of any force R that is a function of the distance s alone. For, if $R = F(s)$, the principle of kinetic energy and work (Arts. 467, 468) gives for rectilinear motion

$$\tfrac{1}{2} mv^2 - \tfrac{1}{2} mv_0^2 = \int_{s_0}^{s} F(s)\,ds; \qquad (17)$$

hence, putting $\int F(s)ds = mf(s)$ and defining $f(s)$ as the *potential*, and $-mf(s)$ as the *potential energy*, due to the force $R = F(s)$, we have

$$\tfrac{1}{2}mv^2 - \tfrac{1}{2}mv_0^2 = mf(s) - mf(s_0),$$

or, with the notation of Art. 477,

$$T + V = T_0 + V_0 = \text{const.} \tag{18}$$

It appears from these definitions that (just as in the particular case of Art. 476) the force exerted on unit mass at any point is the space-derivative of the potential at that point:

$$\frac{1}{m}R = \frac{d}{ds}f(s).$$

480. Free Oscillations. Among the forces of the form $R = F(s)$, next to the Newtonian forces (Art. 472) which are inversely proportional to the square of the distance s, the most important on account of their applications are *forces directly proportional to the distance* s from a fixed point O.

With the origin at O, the equation of rectilinear motion under such a force is

$$m\frac{d^2s}{dt^2} = -m\kappa^2 s, \tag{19}$$

if the force be attractive, *i. e.*, directed toward O. The less important case of repulsion for which the minus sign would have to be replaced by plus, will not be considered here.

It has been shown in Kinematics (Arts. 81–84, 120–128, see especially Art. 125) that the rectilinear motion defined by this equation (19) is a simple harmonic oscillation or vibration, about the point O as *center*. This point O, at which the force $R = -m\kappa^2 s$ is zero, is therefore a position of equilibrium for the particle.

The potential energy V due to the force $R = -m\kappa^2 s$ is, by Art. 479,

$$V = -\int R\,ds = m\kappa^2 \int s\,ds = \tfrac{1}{2}m\kappa^2 s^2 + C.$$

Hence the principle of the conservation of energy gives

$$v^2 + \kappa^2 s^2 = \text{const.}$$

If the initial velocity be zero for $s = s_0$, we have

$$v = \mp \kappa \sqrt{s_0^2 - s^2}.$$

481. As in the applications the moving particle m is generally subject to the constant force of gravity, it is important to notice that the introduction of a constant force F along the line of motion does not essentially change the character of the motion. For, the equation of motion

$$m\frac{d^2 s}{dt^2} = -m\kappa^2 s + F = -m\kappa^2\left(s - \frac{F}{m\kappa^2}\right)$$

reduces, with

$$s - F/m\kappa^2 = x,$$

to

$$m\frac{d^2 x}{dt^2} = -m\kappa^2 x,$$

which is of the same form as (17). The only change in the results is that the *center* of the oscillations, *i.e.*, the position of equilibrium of the particle m, is not the point O, but a point at the distance $e = F/m\kappa^2$ from O.

482. Forces proportional to a distance, or length, are directly observed in the stretching of so-called *elastic* materials. Thus, a homogeneous straight steel wire when suspended vertically from one end and weighted at the other end is found to stretch; and careful measurements have shown that the extension, or change of length, is directly proportional to the weight applied (the weight of the wire itself being assumed, for the sake of simplicity, as very small in comparison with the load applied). Conversely, the *tension*, or *elastic stress*, of the wire is proportional to the *extension* produced. Moreover, when the weight is removed the wire is found to contract to its original length.

This physical law, known as *Hooke's law of elastic stress*, holds only within certain limits. If the weight exceeds a certain limiting value, the extension is no longer proportional to the weight, and after removing the weight, the wire does not regain its original length, but is found to have acquired a *permanent set*, or lengthening; it is said in this case that the elastic limits have been exceeded.

Materials for which Hooke's law holds exactly within certain limits of tension and extension are called *perfectly elastic*. Strictly speaking, such materials probably do not exist; but many materials follow Hooke's law very closely within proper limits. Thus, elastic strings, such as rubber bands, and spiral steel springs show these phenomena very clearly on account of the large extensions allowable within the elastic limits.

483. The elastic constant $m\kappa^2$. Let an elastic string whose natural length is l assume the length $l+x$ when the tension is F, so that, according to Hooke's law,

$$F = -m\kappa^2 x.$$

To determine the factor of proportionality $m\kappa^2$ for a given string, we may observe the length l_1 assumed by the string under a known tension, *e.g.*, the tension $-m_1 g$ produced by suspending a given mass m_1 from the string (the weight of the string itself being neglected).

We then have

$$-m_1 g = -m\kappa^2 (l_1 - l),$$

whence

$$m\kappa^2 = \frac{m_1 g}{l_1 - l},$$

and

$$F = -\frac{m_1 g}{l_1 - l} x.$$

484. Let the same string be placed on a smooth horizontal table, one end being fixed at a point O (Fig. 154), while a particle of mass m is attached to the other end. Stretch the string to a length $OP_0 = l + x_0$ (within the limits of elasticity) and let go; the particle m will move under the action of the tension F alone, its weight being balanced by the reaction of the table. The equation of motion is

Fig. 154.

$$m\frac{d^2x}{dt^2} = -\frac{m_1 g}{l_1 - l} x,$$

the distance $OP=x$ being counted from the point Q at the distance $OQ=l$ from the fixed point O. Putting again (Art. 483)

$$\kappa=\sqrt{\frac{m_1 g}{m\,(l_1-l)}},$$

and integrating, we find

$$x=c_1\cos\kappa t+c_2\sin\kappa t,$$

whence

$$=\frac{dx}{dt}=-\kappa c_1\sin\kappa t+\kappa c_2\cos\kappa t.$$

As $x=x_0$ and $v=0$ for $t=0$, we have $c_1=x_0$, $c_2=0$; hence

$$x=x_0\cos\kappa t,\quad v=-\kappa x_0\sin\kappa t.$$

It should be noticed that these equations hold only as long as the string is actually stretched, *i. e.*, as long as $x>0$. The subsequent motion is, however, easily determined from the velocity for $x=0$.

485. It was assumed, in the preceding article, that the particle m is let go from its initial position P_0 with zero velocity. This can be brought about by pulling the particle from Q to P_0 with a gradually increasing force which at any point P is just equal and opposite to the corresponding elastic tension, or *stress*, $P=-m\kappa^2 x$. The work thus done against the tension, *i. e.*, in stretching or *straining* the string, is stored in the particle m as potential energy, or *strain energy*, V. To find its amount, observe that, as the particle m is pulled through the short distance Δx, the work of the force is $=m\kappa^2 x\,\Delta x$; this being the potential energy ΔV gained in the distance Δx, we have $\Delta V=m\kappa^2 x\Delta x$; hence

$$V=\int_0^{x_0} m\kappa^2 x\,dx=\tfrac{1}{2}\,m\kappa^2 x_0^2.$$

Thus, in the initial position P_0 the particle m possesses this potential energy, but no kinetic energy. During its motion from P_0 to Q, the particle gains kinetic energy and loses potential energy. At any intermediate point P, for which $QP=x$, the kinetic energy is $T=\frac{1}{2}\,mv^2$, while the potential energy is $V=\frac{1}{2}\,m\kappa^2 x^2$. By the principle of the conservation of energy (Art. 479), the sum of these two quantities, the so-called *total energy*, E, remains constant as long as no other forces besides the elastic stress act on the particle:

$$\tfrac{1}{2}\,mv^2+\tfrac{1}{2}\,m\kappa^2 x^2=\text{const.}$$

The value of the constant is $= \frac{1}{2} m\kappa^2 x_0^2$, since this is the total energy at P_0; hence,

$$v^2 + \kappa^2 x^2 = \kappa^2 x_0^2.$$

(Comp. Art. 480.) This relation also follows from the values of x and v given in Art. 484, upon eliminating t.

When the particle arrives at the position of equilibrium Q, the potential, or strain, energy has been consumed, having been converted completely into kinetic energy.

486. Exercises.

(1) In a steam engine, let $p = 35$ lbs. per square inch be the mean piston pressure during one stroke, $s = 15$ in. the length of the stroke, and $d = 1.5$ ft. the diameter of the cylinder. (*a*) What is the work in one stroke? (*b*) To what height could a mass of 500 lbs. be raised by this work?

(2) The work done by an ideal gas in expanding from a volume v_1 to a volume v_2 is $W = \int_{v_1}^{v_2} p dv$, where $pv =$ const. if the change goes on at constant temperature. Find the final pressure and the work done when 12 cu. ft. of air expand at constant temperature to 60 cu. ft., the initial pressure being 30 lbs. per square inch.

(3) Show that, in F. P. S. units, the constant of gravitation is about $1/9.3 \times 10^8$.

(4) Knowing that on the surface of the earth the attraction per unit of mass is $g = 32$, find what it would be on the sun if the density of the sun be $\frac{1}{4}$ of that of the earth, and its diameter 108 times that of the earth.

(5) Describe in words the motion of the particle in the problem of Art. 484; determine the time of one complete (back and forth) oscillation, and the work done by the tension in a quarter oscillation.

(6) In the problem of Art. 484, let the string be a rubber band whose natural length of 1 ft. is increased 3 in. when a weight of 4 oz. is suspended from it; determine the motion of a 1-oz. particle attached to one end, the band being initially stretched to a length of $1\frac{1}{2}$ ft.; find (*a*) the greatest tension of the band, (*b*) the greatest velocity of the particle, (*c*) the period, (*d*) the work done by the tension in a quarter oscillation.

(7) Discuss the effect of friction, of coefficient μ, in the problem of Art. 484.

(8) The length $OQ = l$ of an elastic string is increased to $OQ_1 = l_1 = l + e$ if a mass m is suspended from its lower end, the upper end O being fixed (Fig. 155). The mass m is pulled down to the distance $Q_1P_0 = x_0$ from the position of equilibrium Q_1 and then released. Prove the following results: With Q_1 as origin the equation of motion of m is

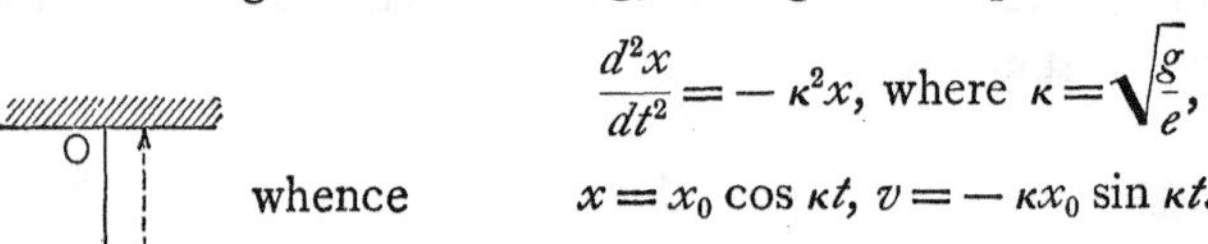

$$\frac{d^2x}{dt^2} = -\kappa^2 x, \text{ where } \kappa = \sqrt{\frac{g}{e}},$$

whence $$x = x_0 \cos \kappa t, \quad v = -\kappa x_0 \sin \kappa t.$$

If $x_0 < e$, the tension never vanishes, and m performs isochronous oscillations of period $2\pi\sqrt{e/g}$, the period being the same as for the small oscillations of a pendulum of length e. If $x_0 > e$, the tension vanishes for $x = -e$, *i. e.*, at Q; the velocity at this point is $v_1 = -\kappa\sqrt{x_0^2 - e^2}$, and the particle rises to the height $h = (x_0^2 - e^2)/2e$ above Q. The total time of one up and down motion is

Fig. 155.

$$2\sqrt{e/g}[\tfrac{1}{2}\pi + \cos^{-1}(e/x_0) + \sqrt{(x_0/e)^2 - 1}].$$

(9) How is the motion of Ex. (8) modified if the elastic string be replaced by a spiral spring suspended vertically from one end? Assume the resistance of the spring to compression equal to its resistance to extension.

(10) The particle in Ex. (8) is let fall from a height h above Q; determine the greatest extension of the string.

(11) An elastic string whose natural length is l is suspended from a fixed point. A mass m_1 attached to its lower end stretches it to a length l_1; another mass m_2 stretches it to a length l_2. If both these masses be attached and then the mass m_2 be cut off, what will be the motion of m_1?

(12) If a straight smooth hole be bored through the earth, connecting any two points A, B on the surface, in what time would a particle slide from A to B? The attraction in the interior is directly proportional to the distance from the center of the earth.

(13) A straight rod of length l ft. and cross-section A sq. ft. is loaded at one end so as to swim upright when h ft. of its length are immersed in water. The weight mg of the rod is then balanced by its *buoyancy*, *i. e.*, a force equal and opposite to the weight, 62.5 Ah lbs., of the water

displaced by the immersed part. If $h_0 (> h)$ ft. were immersed, the resultant force acting at the centroid vertically upward, would be $= 62.5\ Ah_0 - mg$. Show that the centroid will perform simple harmonic oscillations provided the rod remains always partly immersed.

(14) If the rod in Ex. (13) be dropped upright into the water, its lower end being initially h_1 ft. above the water, find, by the principle of the conservation of energy, the depth x_1 of immersion, provided the upper end does not pass below the surface. Determine the relation between l and h necessary for the latter condition. Take $l = 3$ ft., $h = 1$ ft., $h_1 = 1$ ft.

487. Resistance of a Medium. It is known from observation that the velocity v of a rigid body moving in a liquid or gas is continually diminished, the medium apparently exerting on the body a retarding force which is called the *resistance of the medium*. This force F is found to be roughly proportional to the density ρ of the medium, the greatest cross-section A of the body (at right angles to the velocity v), and generally, at least for large velocities, to the square of the velocity v:

$$F = k\rho A v^2,$$

where k is a coefficient depending on the shape and physical condition of the surface of the body.

This expression for the resistance F can be made plausible by the following consideration. As the body moves through the medium, say with constant velocity v, it imparts this velocity to the particles of the medium it meets. The portion of the medium so affected in the unit of time can be regarded as a cylinder of cross-section A and length v, and hence of mass $\rho A v$. To increase the velocity of this mass from 0 to v in the unit of time requires, by equation (2′) of Art. 458, a force

$$\frac{\rho A v \cdot v}{1} = \rho A v^2.$$

The retarding force of the medium must be equal and opposite to this force multiplied by a coefficient k to take into account various disturbing influences.

For small velocities, however, the resistance can be assumed proportional to the velocity, $F=kv$, the coefficient k to be determined by experiment.

488. The above consideration is only a very rough approximation. Thus the particles of the medium are not simply given the velocity v in the direction of motion; they are partly pushed aside and move in curves backwards, causing often whirls or eddies alongside and behind the body. If the medium is a gas, it is compressed in front, and rarefied behind the body; indeed, when the velocity is great (greater than that of sound in the gas), a vacuum will be formed behind the body. Moreover, a layer of the medium adheres to and moves with the body, thus increasing the cross-section. It is therefore often found necessary to assume a more general expression for the resistance; and this is, in ballistics, generally written in the form

$$F = \kappa\rho A v^2 f(v).$$

The careful experiments that have been made to determine the resistance offered by the air to the motion of projectiles have shown that for velocities up to about 250 meters per second, as well as for velocities above 420 m./sec., $f(v)$ can be regarded as constant, *i.e.*, the resistance is proportional to the square of the velocity. But for velocities between 250 and 420 m./sec., *i. e.*, in the vicinity of the velocity of sound in air (330–340 m./sec.), the law of resistance is more complicated.

489. Falling Body in Resisting Medium. Assuming the resistance proportional to the square of the velocity, the equation of motion for a body falling (without rotating) in a medium of constant density is

$$m\frac{d^2s}{dt^2} \equiv m\frac{dv}{dt} = mg - mkv^2, \tag{20}$$

where k is a positive constant. To simplify the resulting formulæ, put

$$k = \frac{\mu^2}{g};$$

then the separation of the variables v and t gives

$$dt = \frac{gdv}{g^2 - \mu^2 v^2},$$

whence
$$t = \frac{1}{2\mu} \log \frac{g + \mu v}{g - \mu v}, \tag{21}$$

the constant of integration being zero if the initial velocity is zero. Solving for v, we have

$$v = \frac{g}{\mu} \cdot \frac{e^{\mu t} - e^{-\mu t}}{e^{\mu t} + e^{-\mu t}} = \frac{g}{\mu} \tanh \mu t. \tag{22}$$

Writing ds/dt for v and integrating again, we find, since $s = 0$ for $t = 0$,

$$s = \frac{g}{\mu^2} \log \tfrac{1}{2}(e^{\mu t} + e^{-\mu t}) = \frac{g}{\mu^2} \log \cosh \mu t. \tag{23}$$

The relation between v and s can be obtained by eliminating t between the expressions for v and s, or more conveniently by eliminating t from the original differential equation by means of the relation

$$\frac{dv}{dt} = \frac{dv}{ds}\frac{ds}{dt} = v\frac{dv}{ds}.$$

This gives
$$ds = \frac{gvdv}{g^2 - \mu^2 v^2},$$

whence, with $v = 0$ for $s = 0$,

$$s = \frac{g}{2\mu^2} \log \frac{g^2}{g^2 - \mu^2 v^2}. \tag{24}$$

490. Exercises.

(1) Show that, as t increases, the motion considered in Art. 489 approaches more and more a state of uniform motion without ever reaching it.

(2) Determine the motion of a body projected vertically upward in the air with given initial velocity v_0, the resistance of the air being proportional to the square of the velocity.

(3) In Ex. (2) find the whole time of ascent and the height reached by the particle.

(4) Show that, owing to the resistance of the air, a body projected vertically upward returns to the starting point with a velocity less than the initial velocity of projection.

(5) A ball, 6 in. in diameter, falls from a height of 300 ft.; find how much its final velocity is diminished by the resistance of the air, if $k = 0.00090$.

(6) Determine the rectilinear motion of a body in a medium whose resistance is proportional to the velocity, when no other forces act on it.

(7) A body falls from rest in a medium whose resistance is proportional to the velocity; find v and s in terms of t, v in terms of s.

491. Damped Oscillations. Let a particle of mass m be attracted by a fixed center O, with a force proportional to the distance from O, and move in a medium whose resistance is proportional to the velocity. If the initial velocity be directed through O (or be zero), the motion will be rectilinear, and the equation of motion is

$$m\frac{d^2s}{dt^2} = -m\kappa^2 s - mkv,$$

or, putting

$$k = 2\lambda,$$

$$\frac{d^2s}{dt^2} + 2\lambda\frac{ds}{dt} + \kappa^2 s = 0. \tag{25}$$

This is a homogeneous linear differential equation of the second order with constant coefficients, which can be integrated by a well-known process. The roots of the auxiliary equation,

$$-\lambda \pm \sqrt{\lambda^2 - \kappa^2},$$

are real or imaginary according as $\lambda > \kappa$, or $\lambda < \kappa$. The limiting cases $\lambda = \kappa$, $\lambda = 0$, $\kappa = 0$, also deserve special mention.

(*a*) If $\lambda > \kappa$, the roots are real and different, and as λ is positive, both roots are negative; denoting them by $-a$ and $-b$, so that a and b are positive constants, and $b > a$, the general solution is

$$s = c_1 e^{-at} + c_2 e^{-bt}.$$

As the force has a finite value at the center O, we can take $s = 0$, $v = v_0$ for $t = 0$ as initial conditions. This gives

$$s = \frac{v_0}{b-a}(e^{-at} - e^{-bt}), \qquad v = \frac{v_0}{b-a}(be^{-bt} - ae^{-at}).$$

The velocity reduces to zero at the time

$$t_1 = \frac{1}{b-a}\log\frac{b}{a}.$$

As a and b are positive and $b > a$, s has always the sign of v_0, *i. e.*, the particle remains always on the same side of O; it reaches its elongation at the time t_1, for which v vanishes, and then approaches the point O asymptotically.

Hence, in this case, the damping effect of the medium is sufficiently great to prevent actual oscillations. Such motions are sometimes called **aperiodic.**

(*b*) If $\lambda = \kappa$, the roots are real and equal, viz., $= -\lambda$, and the general solution is

$$s = (c_1 + c_2 t)e^{-\lambda t}.$$

With $s = 0$, $v = v_0$ for $t = 0$, we find

$$s = v_0 t e^{-\lambda t}, \quad v = v_0(1 - \lambda t)e^{-\lambda t}.$$

The velocity vanishes for $t_1 = 1/\lambda$, and then only. The nature of the motion is essentially the same as in the previous case.

(*c*) If $\lambda < \kappa$, the roots are complex, say $= -\alpha \pm \beta i$, where α and β are positive constants. The general solution

$$s = e^{-\alpha t}(c_1 \cos\beta t + c_2 \sin\beta t)$$

gives with $s = 0$, $v = v_0$ for $t = 0$:

$$s = \frac{v_0}{\beta}e^{-\alpha t}\sin\beta t, \quad v = \frac{v_0}{\beta}e^{-\alpha t}(\beta\cos\beta t - \alpha\sin\beta t).$$

Here v vanishes whenever $\tan\beta t = \beta/\alpha = \sqrt{(\kappa/\lambda)^2 - 1}$; s vanishes (*i. e.*, the particle passes through O) whenever t is an integral multiple of π/β; s has an infinite number of maxima and minima whose absolute values rapidly diminish.

The resistance of the medium, while not sufficient to extinguish the oscillations, continually shortens their amplitude; this is the typical case of *damped oscillations*.

(*d*) If $\lambda = 0$, the roots are purely imaginary, viz., $= \pm \kappa i$. In this case, the second term in equation (25) is zero; there is no damping effect, and we have the case of *free oscillations* (see Arts. 480–486).

(*e*) If $\kappa = 0$, one of the roots is zero, the other is $= -2\lambda$. The attracting (or elastic) force being zero, we have the case of Ex. (6), Art. 490.

492. As shown in Arts. 479, 485, the *principle of the conservation of energy* holds for the *free* oscillations of a particle (under a force proportional to the distance). In the case of *damped* oscillations (Art. 491), this principle, in the restricted sense in which it has been stated so far, is not applicable, the resistance of the medium not being given as a function of the distance s. The total energy $E = T + V$ of the particle, or rather the energy stored in the system formed by the spring with the particle attached (in the example used above), diminishes in the course of time because the spring has to do work against the resistance of the medium, thus transferring part of its energy to the medium (setting it in motion, heating it, etc.). Thus, in a generalized meaning, the principle of the conservation of energy can be said to hold for the larger system, formed by the spring, together with the medium (see Art. 498).

493. The rate at which the total energy E diminishes with the *time* is here proportional to the *square* of the velocity:

$$\frac{dE}{dt} = -2m\lambda v^2;$$

for, substituting for E its value $E = T + V = \frac{1}{2}mv^2 + \frac{1}{2}mk^2s^2$ (Art. 485) and reducing, we find the equation of motion (25).

The *space*-rate of change of the total energy E is proportional to the velocity, and is nothing else but the resistance of the medium:

$$\frac{dE}{ds} = -2m\lambda v;$$

for we have

$$\frac{dE}{dt} = \frac{dE}{ds}\frac{ds}{dt} = v\frac{dE}{ds}.$$

494. Forced Oscillations. In the case of free simple harmonic oscillations, while the force regarded as a function of the distance s is directly proportional to s, the same force regarded as a function of the time is of the form

$$R = -m\kappa^2 s_0 \cos \kappa t,$$

since $s = s_0 \cos \kappa t$. Conversely, a particle acted upon by a single force $R = mk \cos \mu t$, or $R = mk \sin \mu t$, directed toward a fixed center O, will, if the initial velocity passes through O, have a simple harmonic motion.

Suppose that such a force in the line of motion be superimposed in the case of Art. 491 so that the equation of motion becomes

$$m\frac{d^2s}{dt^2} = -m\kappa^2 s - 2m\lambda v + mk \cos \mu t,$$

or

$$\frac{d^2s}{dt^2} + 2\lambda\frac{ds}{dt} + \kappa^2 s = k \sin \mu t. \qquad (26)$$

The particle is then said to be subject to *forced oscillations*. For a particle suspended from a spiral spring this could be realized by subjecting the point of suspension to a vertical simple harmonic motion of amplitude k and period $2\pi/\mu$.

The non-homogeneous linear differential equation (26) with constant coefficients can be integrated by well-known methods.

495. Exercises.

(1) With $\mu = 2$, $v_0 = 4$, sketch the curves representing s as a function of t in the five cases of Art. 491; take (*a*) $\lambda = 3$, (*b*) $\lambda = 2$, (*c*) $\lambda = \frac{1}{4}$, (*e*) $\lambda = 2$.

(2) Compare the cases (*c*) and (*d*) of Art. 491; show that the oscillations in a resisting medium are isochronous, but of greater period than *in vacuo*. The ratio of the amplitude at any time to the initial amplitude is called the *damping ratio;* show that the logarithm of this ratio, the so-called *logarithmic decrement*, is proportional to the time.

(3) Derive the equation of motion in the case of free oscillations from the principle of the conservation of energy.

(4) Integrate and discuss the equation $\frac{d^2s}{dt^2} + \kappa^2 s = a \sin \mu t$; show that the amplitude of the forced oscillation becomes very large if the periods of the free and forced oscillations are nearly equal. Discuss the limiting case when $\mu = \kappa$.

(5) Integrate (26), assuming a particular integral of the form $c \cos \mu t + c' \sin \mu t$ and determining the constants c, c' by substituting this expression in (26). Discuss the result.

496. Power. It has been shown that the time-effect of a force is measured by its *impulse* (Arts. 425, 467), while the space-effect is measured by its *work* (Arts. 442, 467). In applied mechanics it is of great importance to take time and space into account simultaneously. *The time-rate at which work is performed by a force* has therefore received a special name, **power**, or **activity**. The source from which the force for doing useful work is derived is commonly called the *agent*, or *motor*; and it is customary to speak of the power of an agent, this meaning the rate at which the agent is capable of supplying work.

497. The *dimensions* of power are evidently ML^2T^{-3}. The *unit of power* is the power of an agent that does unit work in unit time. Hence, in absolute measure, it is the power of an agent doing one erg per second in the C. G. S. system, and one foot-poundal per second in the F. P. S. system. As, however, the idea of power is of importance mainly in engineering practice, power is usually measured in gravitation units. In this case, the unit of power is the power of an agent doing one foot-pound per second in the F. P. S. system, and one kilogram-meter in the metric system.

A larger unit is frequently found more convenient. For this reason, the name **horse-power** (H. P.) is given to the power of doing 550 foot-pounds of work per second, or $550 \times 60 = 33,000$ foot-pounds per minute.

498. Work and Efficiency of Machines. The principle of the conservation of energy has been *proved* in Art. 479 for a particle in rectilinear motion under the action of a force which is a given function of the distance. In this particular case, the principle states that *the total energy* E, which is the algebraic sum of the kinetic energy T and the potential energy V of the particle, *remains constant throughout the motion* as long as no other forces act on the particle; and the truth of the principle in this restricted sense follows directly from the fundamental definitions of dynamics.

By a generalization as bold and far-reaching as was Newton's extension of the property of mutual attraction to all matter (Arts. 543–546), modern physics has been led to the assumption that *work and energy are quantities which can never be destroyed*, but can be transformed in a variety of ways. This assumption, the **general principle of the conservation of energy**, while fully borne out so far by the results deduced from it, is of course not capable of mathematical proof. Indeed, it may be said that in defining the various forms of energy, such as heat, chemical energy, radio-activity, etc., the *definitions* are so formulated as to conform to this principle; it has always been found possible to do this. The general principle of the conservation of energy cannot be fully discussed here, since this would require a study of all the forms of energy known to physics.

In its application to machines, the principle states that the *total work* W supplied to a machine in a given time by the agent, or motor, driving it (such as animal force, the expansive force of steam, the pressure of the wind, the impact of water, etc.) is equal to the sum of the *useful work* W_u, done by the machine in the same time and the so-called *lost*, or *wasteful, work* W_w spent in overcoming friction and other passive resistances of the machine:

$$W = W_u + W_w.$$

While W and W_u can be determined with considerable accuracy, it is difficult to determine W_w directly with equal precision; but it is found that the more accurately in any given machine W_w is determined, the more nearly will the above equation be found satisfied. This serves as a verification of the principle of the conservation of energy in its application to machines.

As explained in Art. 419, the ratio W_u/W of the useful work to the total work is called the **efficiency** of the machine. The term *modulus* is sometimes used for efficiency.

499. Exercises.

(1) In electrical engineering a *watt* is defined as the power of doing one joule, *i. e.*, 10^7 ergs, per second. Find the relation between the watt and the horse-power.

(2) In countries using the metric system of weights and measures the horse-power is defined as 75 kilogram-meters per second. Find its relation to the watt and to the British horse-power.

(3) Find the horse-power of the engine in Art. 486, Ex. (1), if it make 1 stroke per second.

(4) The cylinder of a steam engine has a diameter of 15 in.; the stroke is $2\frac{3}{4}$ ft.; the number of strokes per minute is 70; the mean pressure of the steam is 40 lbs. per square inch. What is the horse-power of the engine?

(5) Find the horse-power required of the locomotive to haul a train of 100 tons at the rate of 30 miles an hour, the resistances amounting to 8 lbs. per ton: (*a*) on a level road; (*b*) up a 1% grade; (*c*) up a 2% grade.

(6) How much water can an engine furnishing 40 H. P. raise per minute from the bottom of a mine 840 ft. deep?

(7) The diameter of the cylinder of a steam engine is 30 in.; the stroke 4 ft.; the mean pressure 15 lbs. per square inch; the number of revolutions 24 per minute. If the efficiency of the engine be $\frac{3}{5}$, what is the amount of water raised per hour from a depth of 250 ft.?

(8) In what time would an engine yielding 2 H. P. perform the work of raising the brickwork in Art. 466, Ex. (19)?

(9) A shaft of 8 ft. diameter is to be sunk to a depth of 320 ft. through a material whose specific gravity is 2.2. Determine: (*a*) the total work of raising the material to the surface; (*b*) the time in which it can be done by an engine yielding 3.5 H. P.; (*c*) the time in which it can be done by 4 men working in a capstan, if each laborer does 2500 ft.-lbs. per minute, working 8 hr. per day.

(10) A steam engine (diameter of piston = 9 in., stroke = 1 ft., number of revolutions = 150 per minute, mean effective piston pressure = 50 lbs. per square inch) drives a circular saw of 3 ft. diameter, making 4000 rev./min. Neglecting the frictional resistances, determine the force exerted by the teeth of the saw.

(11) A saw of 10 in. diameter makes 4000 rev./min.; a planer whose head has a diameter of 6 in. makes 5000 rev./min. If the resistance at the teeth of the saw be 10 lbs., at the planer 15 lbs., and if both are driven by an engine making 170 rev./min., with piston diameter = 8 in., stroke = 10 in., what is the mean effective piston pressure?

(12) Find the horse-power of the wheel in Ex. (16), Art. 449, if it makes 150 rev./min.

(13) A water-wheel weighing (with the water in it) 21,000 lbs. makes 10 rev./min.; its horizontal shaft rests in bearings 8 in. in diameter. The coefficient of friction being 0.1, determine the horse-power lost in friction.

(14) Determine the indicated horse-power of a gas engine working at 150 rev./min., if there is an explosion every two revolutions; diameter of piston = 12 in., length of crank = 8 in.; mean effective pressure in one cycle = 62 lbs.*

III. *Free Curvilinear Motion.*

1. GENERAL PRINCIPLES.

500. Let j be the acceleration of a particle of mass m at the time t; R the resultant of all the forces acting on the particle; then its equation of motion is (Art. 456)

$$mj = R.$$

In curvilinear motion (Fig. 156) the direction of j and R differs from the direction of the velocity v; and the angle ψ between j

* For further applied problems of a similar character, the student is referred to J. PERRY, *Applied Mechanics*, New York, Van Nostrand, 1898, pp. 46-49, and F. B. SANBORN, *Mechanics Problems*, New York, The Engineering News Publishing Company, 1902.

and v varies in general in the course of time. As shown in Kinematics (Art. 112), the acceleration can be resolved into a tangential component $j_t = dv/dt = d^2s/dt^2$ and a normal component $j_n = v^2/\rho$, where ρ is the radius of curvature of the path. Hence, if the resultant force R which has the direction of j be resolved into a *tangential force* $R_t = R \cos \psi$, and a *normal force* $R_n = R \sin \psi$, the above equation of motion will be replaced by the following two equations:

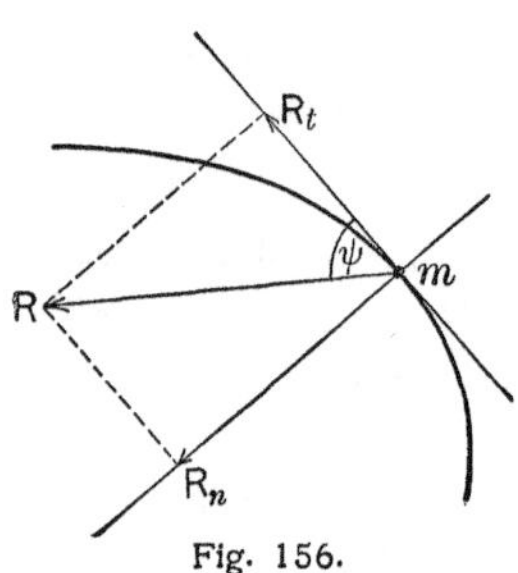

Fig. 156.

$$m\frac{dv}{dt} = R_t, \quad m\frac{v^2}{\rho} = R_n. \tag{1}$$

501. The formulæ (1) show how the force R affects the velocity of the particle and the curvature of the path. The change of the *magnitude* of the velocity is due to the tangential force R_t alone. If this component be zero, *i. e.*, if the resultant force R be constantly normal to the path, the velocity v will remain of constant magnitude. The curvature of the path, $1/\rho$, and hence the *direction* of v, depends on the normal component R_n. If this component be zero, the curvature is zero; *i. e.*, the path is rectilinear.

502. Instead of resolving the resultant force R along the tangent and normal, it is often more convenient to resolve it into three components, $R \cos \alpha = X$, $R \cos \beta = Y$, $R \cos \gamma = Z$, parallel to three fixed rectangular axes of co-ordinates Ox, Oy, Oz, to which the whole motion is then referred. If x, y, z be the co-ordinates of the particle m at the time t, the equations of motion assume the form (comp. Art. 113)

$$m\frac{d^2x}{dt^2} = X, \quad m\frac{d^2y}{dt^2} = Y, \quad m\frac{d^2z}{dt^2} = Z. \tag{2}$$

Thus, the curvilinear motion is replaced by three rectilinear motions.

503. If the components X, Y, Z were given as functions of the time t alone, each of the three equations (2) could be integrated separately. In general, however, these components will be functions of the co-ordinates, and perhaps also of the velocity and of the time. No general rules can be given for integrating the equations in this case. By combining the equations (2) in such a way as to produce exact derivatives in the resulting equation, it is sometimes possible to effect an integration. Two methods of this kind have been indicated for the case of two dimensions in a particular example in Kinematics, Arts. 163–165. We now proceed to study these *principles* of integration from a more general point of view, and to point out the physical meaning of the expressions involved.

504. The Principle of Kinetic Energy and Work. Let us combine the equations of motion (2) by multiplying them by dx/dt, dy/dt, dz/dt respectively, and then adding. The left-hand member of the resulting equation will be the derivative with respect to t of

$$\tfrac{1}{2}m\left[\left(\frac{dx}{dt}\right)^2+\left(\frac{dy}{dt}\right)^2+\left(\frac{dz}{dt}\right)^2\right]=\tfrac{1}{2}mv^2.$$

We find, therefore,

$$\frac{d(\tfrac{1}{2}mv^2)}{dt}=X\frac{dx}{dt}+Y\frac{dy}{dt}+Z\frac{dz}{dt}.$$

Let v_0 be the velocity at any point P_0 of the path, v the velocity at any other point P, and let arc $P_0P=s-s_0$, the distances s_0, s being counted along the path. Then, integrating from P_0 to P, we find:

$$\tfrac{1}{2}mv^2-\tfrac{1}{2}mv_0^2=\int_{s_0}^{s}(Xdx+Ydy+Zdz). \qquad (3)$$

The left-hand member represents the increase in the kinetic energy of the particle; the right-hand member represents the work done by the resultant force R, since its work is equal to the sum of the works of its components X, Y, Z (comp. Arts. 404, 408). Equation (3) states, therefore, that *the amount by which the kinetic energy increases, as the particle passes from* P_0 *to* P, *is equal to the work done by the resultant force* R *on the particle* (comp. Arts. 467, 468).

505. The principle of kinetic energy and work can also be deduced from the former of the two equations (1). Multiplying this equation by $v = ds/dt$, we have

$$\frac{d(\frac{1}{2}mv^2)}{dt} = R_t \frac{ds}{dt} = R \cos\psi \frac{ds}{dt};$$

hence, integrating as in Art. 504:

$$\tfrac{1}{2}mv^2 - \tfrac{1}{2}mv_0^2 = \int_{s_0}^{s} R \cos\psi ds. \tag{4}$$

506. The evaluation of the work integral in the right-hand member of the equations (3) or (4) requires in general a knowledge of the path, since the integration is to be extended along this path. As in many problems the path is not known beforehand, but is one of the things to be determined, it is very important to notice that *the work integral*

$$\int_{s_0}^{s} (Xdx + Ydy + Zdz)$$

has a value independent of the path connecting the initial and final positions P_0, P, *whenever the expression* $Xdx + Ydy + Zdz$ *is the exact differential of a one-valued function* U *of* x, y, z; for in this case

$$\int_{s_0}^{s} (Xdx + Ydy + Zdz) = \int_{s_0}^{s} dU = U - U_0,$$

where U_0 is the value of $U(x, y, z)$ at P_0, U that at P.

507. Now the expression $Xdx + Ydy + Zdz$ is certainly an exact differential whenever there exists a function U of the co-ordinates x, y, z alone (*i. e.*, not involving the velocities or the time explicitly), such that

$$\frac{\partial U}{\partial x} = X, \quad \frac{\partial U}{\partial y} = Y, \quad \frac{\partial U}{\partial z} = Z; \tag{5}$$

for then

$$Xdx + Ydy + Zdz = \frac{\partial U}{\partial x}dx + \frac{\partial U}{\partial y}dy + \frac{\partial U}{\partial z}dz = dU.$$

The function U is called the **force-function**, and forces for which a force-function exists are called **conservative forces.**

The conditions (5) for the existence of a force-function U can be put into a different analytical form which is frequently useful. Differentiating the second of the equations (5) with respect to z, the third with respect to y, we find

$$\frac{\partial^2 U}{\partial z \partial y} = \frac{\partial Y}{\partial z}, \quad \frac{\partial^2 U}{\partial y \partial z} = \frac{\partial Z}{\partial y},$$

whence $\partial Y/\partial z = \partial Z/\partial y$. If we proceed in a similar way with the other equations (5), it appears that they can be replaced by the following conditions:

$$\frac{\partial Y}{\partial z} = \frac{\partial Z}{\partial y}, \quad \frac{\partial Z}{\partial x} = \frac{\partial X}{\partial z}, \quad \frac{\partial X}{\partial y} = \frac{\partial Y}{\partial x}. \tag{6}$$

The equations (5) or the equivalent equations (6) are therefore *sufficient* conditions for the existence of a force-function U.

508. The dynamical meaning of the existence of a force-function U lies mainly in the fact that, if a one-valued force-function exists, the work done by the forces as the particle passes from its initial to its final position depends only on these positions, and not on the intervening path. The equation (3) then gives

$$\tfrac{1}{2} mv^2 - \tfrac{1}{2} mv_0^2 = U - U_0; \tag{7}$$

the work done on the particle as it passes from P_0 to P is $= U - U_0$.

It follows that the work of conservative forces is zero if the particle returns finally to its original position, that is, if it describes a closed path, provided that the force-function U is one-valued, an assumption which will here always be made.

In the case of central forces depending only on the distance, for which a force-function can always be shown to exist, the force-function, divided by the mass of the particle, is usually called the *potential* (see Arts. 476, 479). The negative of the force-function, say

$$V = - U,$$

is called the **potential energy**. If this quantity be introduced, and the kinetic energy be denoted by T (as in Art. 477), the equation (7) assumes the form

$$T + V = T_0 + V_0, \tag{8}$$

which expresses the **principle of the conservation of energy** for a particle: *the total energy*, i. e., *the sum of the kinetic and potential energies, remains constant throughout the motion whenever there exists a force-function.* In other words, whatever is gained in kinetic is lost in potential energy, and *vice versa.*

509. As the force-function U is a function of the co-ordinates x, y, z alone, an equation of the form

$$U = c,$$

where c is a constant, represents a surface which is the locus of all points of space at which the force-function has the same value c. By giving to c different values, a family of surfaces is obtained, and these surfaces are called **level**, or **equipotential, surfaces.**

The values of the derivatives of U at any point P (x, y, z) are proportional to the direction-cosines of the normal to the equipotential surface $U = c$ at P. But, by (5), they are also proportional to the direction-cosines of the resultant force R at this point. It follows that *the resultant force R at any point P is always normal to the equipotential surface passing through P.*

If the equation of the equipotential surfaces be given, the resultant force R at any point (x, y, z) is readily found, both in magnitude and direction, from its components (5):

$$R^2 = X^2 + Y^2 + Z^2 = \left(\frac{\partial U}{\partial x}\right)^2 + \left(\frac{\partial U}{\partial y}\right)^2 + \left(\frac{\partial U}{\partial z}\right)^2.$$

510. The force-function U determines, as has been shown, a family of equipotential surfaces $U =$ const. Starting from a point P on one of these surfaces, say $U = c$ (Fig. 157), let us draw through P the direction of the resultant force, which is normal to the surface $U = c$. Let this direction intersect at P' a near surface, $U = c'$. At P' draw the normal to $U = c'$, and let it intersect a near surface, $U = c''$, at P''. Proceeding in this way, we obtain a series of points $P, P', P'', P''', \cdots$, which in the limit will form a continuous curve whose direction at any point coincides with the direction of the resultant force at that point. Such a line is called a **line of force**.

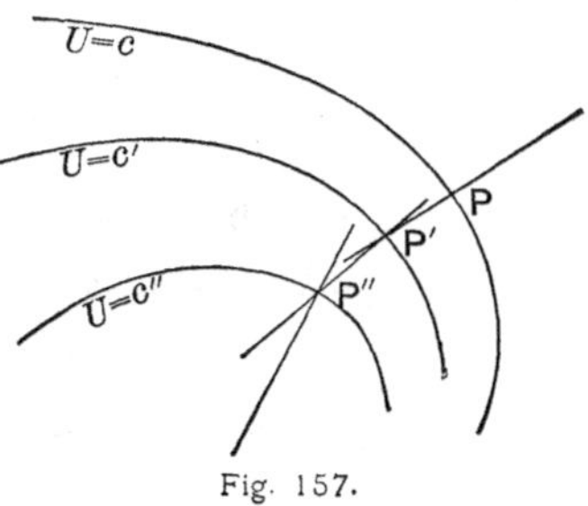

Fig. 157.

The lines of force evidently form the orthogonal system to the family of equipotential surfaces. The differential equations of the lines of force are therefore:

$$\frac{dx}{\frac{\partial U}{\partial x}} = \frac{dy}{\frac{\partial U}{\partial y}} = \frac{dz}{\frac{\partial U}{\partial z}}.$$

511. Exercises.

(1) Show that a force-function exists when the resultant force is constant in magnitude and direction.

(2) Find the force-function in the case of a free particle moving under the action of the constant force of gravity alone (projectile *in vacuo*); determine the equipotential surfaces and the potential energy.

(3) Show the existence of a force-function when the direction of the resultant force is constantly perpendicular to a fixed plane, say the xy-plane, and its magnitude is a given function $f(z)$ of the distance z from the plane.

(4) Find the force-function, the equipotential surfaces, and the kinetic energy when the force is a function $f(r)$ of the perpendicular distance r from a fixed line, and is directed towards this line at right angles to it.

(5) Show the existence of a force-function for a central force, *i. e.*, a force passing through a fixed point (x_0, y_0, z_0), if the force is a function of the distance r from this point. What are the level surfaces?

(6) Show that a force-function exists when a particle moves under the action of any number of such central forces as in Ex. (5).

512. The Principle of Angular Momentum or of Areas. We confine ourselves to the case of plane motion, so that we have only two equations of motion:

$$m\frac{d^2x}{dt^2}=X,\quad m\frac{d^2y}{dt^2}=Y.$$

Combining these by multiplying the former by y, the latter by x, and subtracting the former from the latter, we find

$$mx\frac{d^2y}{dt^2}-my\frac{d^2x}{dt^2}=xY-yX.$$

The right-hand member is evidently the moment (about the origin) of the resultant force R, whose components are X and Y. The left-hand member is an exact derivative, viz., the derivative with respect to the time of $mxdy/dt-mydx/dt$, as is easily verified by differentiating this quantity. The equation can therefore be written

$$\frac{d}{dt}\left(mx\frac{dy}{dt}-my\frac{dx}{dt}\right)=xY-yX. \qquad (9)$$

Integrating from t_0 to t, we find

$$mx\frac{dy}{dt}-my\frac{dx}{dt}=\int_{t_0}^{t}(xY-yX)dt. \qquad (10)$$

This equation expresses the *principle of angular momentum or of areas* for plane motion.

513. The name is due to the following interpretation of the left-hand member of equation (10). Divided by m, this left-hand member is, by Art. 97, twice the sectorial velocity. Equation (9), after division by m, can therefore be expressed in words as follows: The time-rate of change of twice the sectorial velocity about any point is equal to the moment of the acceleration about that point. This kinematical interpretation accounts for the name principle of areas.

514. The dynamical meaning of equation (10) appears by considering that mdx/dt, mdy/dt are the components of the momentum mv of the moving particle (Fig. 158). The product mvp of the momentum and its perpendicular distance from the origin is called the **moment of momentum**, or the **angular momentum**, of the particle about the origin.

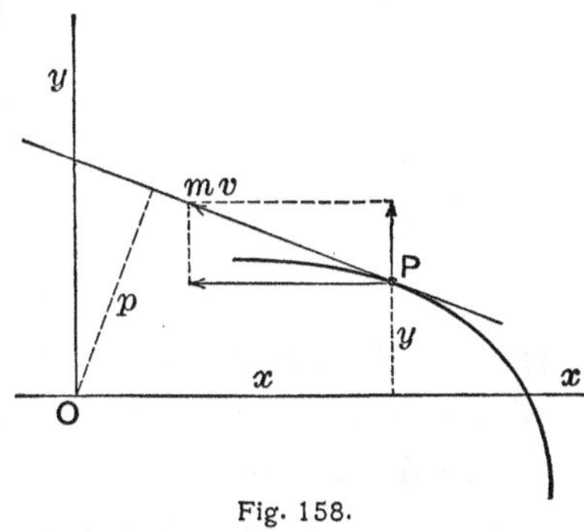

Fig. 158.

As the moment of mv is equal to the algebraic sum of the moments of its components, we have

$$mvp = mx\frac{dy}{dt} - my\frac{dx}{dt}.$$

The angular momentum is evidently nothing but twice the sectorial velocity multiplied by the mass, just as linear momentum is linear velocity times mass.

The dynamical meaning of equation (9) can therefore be expressed as follows: *the time-rate of change of angular momentum about any fixed point is equal to the moment of the resultant force about the same point.*

515. The most important case in which the integration in (10) can be performed is the case when

$$xY - yX = 0,$$

which evidently means that the direction of the resultant force R passes through the origin. If this condition be fulfilled, equation (10) reduces to the form

$$mx\frac{dy}{dt} - my\frac{dx}{dt} = c, \qquad (11)$$

where c is a constant of integration to be determined from the initial position and velocity.

Kinematically, this equation means that the sectorial velocity remains constant. It can be put into the form

$$\frac{dS}{dt} = \frac{c}{2m} = c',$$

whence, by integration, we find

$$S - S_0 = c'(t - t_0). \tag{12}$$

Hence, *if the acceleration passes constantly through a fixed point, the sector* $S - S_0$ *described about this point in any time* $t - t_0$ *is proportional to this time.*

This is the *principle of the conservation of area* for plane motion.

Dynamically, equation (11) means that *if the resultant force passes constantly through a fixed point, the angular momentum about this point remains constant.* This proposition is called the *principle of the conservation of angular momentum.*

If v_0 be the initial velocity, p_0 the perpendicular to v_0 from the fixed point, equation (11) can also be written in the form

$$vp = v_0 p_0. \tag{13}$$

516. Exercise.

The equation (9) can be written $d(mvp)/dt = xY - yX$. Show that the two terms of $d(mvp)/dt = mpdv/dt + mvdp/dt$ represent the moments of the tangential and normal components of the resultant force R, respectively.

517. The Principle of d'Alembert. Let us consider a particle of mass m moving under the action of any forces $F_1, F_2, \cdots F_n$, whose resultant is R. The total acceleration j of the particle has the components d^2x/dt^2, d^2y/dt^2, d^2z/dt^2 parallel to the rectangular axes Ox, Oy, Oz. If the forces $F_1, F_2, \cdots F_n$ be imagined removed, a force equal to mj would be required to give the particle the same acceleration j that it had under the action of the forces $F_1, F_2, \cdots F_n$. This fictitious force, mj, whose components are md^2x/dt^2, md^2y/dt^2, md^2z/dt^2, is called the *effective* *force.* For the sake of distinction, the forces $F_1, F_2, \cdots F_n$, which actually produce the motion, are called the *impressed forces.*

The ordinary equations of motion of a particle,

$$m\frac{d^2x}{dt^2}=X,\ m\frac{d^2y}{dt^2}=Y,\ m\frac{d^2z}{dt^2}=Z, \tag{14}$$

where X, Y, Z are the components of the resultant R of the impressed forces, express merely the equality between the effective force mj and the resultant impressed force R. It follows that *if the reversed effective force* $-mj$, or its components, $-md^2x/dt^2$, $-md^2y/dt^2$, $-md^2z/dt^2$, *be combined with the impressed forces* $F_1, F_2, \cdots F_n$, *we have a system in equilibrium.* This is the fundamental idea of d'Alembert's principle.

518. The reversed effective force, $-mj$, is sometimes called the *force of inertia* of the particle. To understand the idea underlying this expression, imagine the impressed forces to be removed, and then push the particle, say with the hand, so as to give it the same motion that it had under the action of the impressed forces. The pressure of the hand on the particle must at every instant be equal to the resultant R, or to the effective force mj, while the equal and opposite pressure of the particle on the hand represents the force of inertia. It must, however, be clearly understood that this force of inertia, or inertia-resistance, is a force exerted on the hand and not on the particle.

519. Owing to the fact that, by combining with the impressed forces the reversed effective force, we obtain at any given instant a system in equilibrium, it becomes possible to apply to kinetical problems the statical conditions of equilibrium.

Since in the case of a single particle the forces are all concurrent, the conditions of equilibrium are obtained by equating to zero the sum of the components of the forces along each axis. This gives

$$X-m\frac{d^2x}{dt^2}=0,\ Y-m\frac{d^2y}{dt^2}=0,\ Z-m\frac{d^2z}{dt^2}=0,$$

and these are the ordinary dynamical equations of motion (see (14), Art. 517).

520. The conditions of equilibrium of a system of forces can also be expressed by means of the principle of virtual work (Art. 410). Thus, let δx, δy, δz be the components of any virtual displacement δs of the particle; then the principle of virtual work applied to our system of forces gives the single condition

$$\left(-m\frac{d^2x}{dt^2}+X\right)\delta x+\left(-m\frac{d^2y}{dt^2}+Y\right)\delta y+\left(-m\frac{d^2z}{dt^2}+Z\right)\delta z=0, \quad (15)$$

which is of course equivalent to the three equations (14) on account of the arbitrariness of the displacement δs.

The equation (15), which may also be written in the form

$$m\left(\frac{d^2x}{dt^2}\delta x+\frac{d^2y}{dt^2}\delta y+\frac{d^2z}{dt^2}\delta z\right)=X\delta x+Y\delta y+Z\delta z, \quad (16)$$

expresses d'Alembert's principle for a single particle: *for any virtual displacement the sum of the virtual works of the impressed forces is equal to that of the effective force.*

521. The advantage of using the equations of motion in the form given to them by d'Alembert arises mainly from the application of the principle of virtual work which thus becomes possible; this will be seen more clearly later on, in the treatment of constrained motion. For the present it may suffice to notice that, if the actual displacement ds of the particle in its path be selected as the virtual displacement δs, equation (16) becomes

$$m\left(\frac{d^2x}{dt^2}dx+\frac{d^2y}{dt^2}dy+\frac{d^2z}{dt^2}dz\right)=Xdx+Ydy+Zdz. \quad (17)$$

This is the equation of kinetic energy and work (Art. 504); for the left-hand member is the exact differential $d(\frac{1}{2}mv^2)$ of the kinetic energy, while the right-hand member represents the element of work of the impressed forces.

In the particular, but very common, case of *conservative* impressed forces, the right-hand member is likewise an exact differential, dU; hence, in this case a first integration can at once be performed, and we find, as in Art. 508,

$$\tfrac{1}{2}mv^2 - \tfrac{1}{2}mv_0^2 = \int(Xdx + Ydy + Zdz) = U - U_0. \qquad (18)$$

522. There is an essential distinction between the principle of d'Alembert on the one hand, and the principles of kinetic energy and of areas on the other. D'Alembert's principle merely gives a convenient form and interpretation to the dynamical equations of motion, through the application of the principle of virtual work; but it does not show how to integrate these equations.

The principle of kinetic energy and work and the principle of areas are really methods for integrating the equations of motion under certain conditions. The fact that these particular methods of combining the differential equations so frequently furnish the solution of physical problems, is the best proof of the adequacy of the fundamental definitions and assumptions of mechanics. It has led the physicist to ascribe real existence to the quantities whose exact differentials are introduced by the combination, viz., to force and work, to kinetic and potential energy, and to regard the conservation of energy as a law of nature. While this view may often be useful, it must not be forgotten that the question of the objective reality of these abstractions is beyond the ken of exact science.

2. CENTRAL FORCES.

523. We proceed to apply the general principles developed in the preceding articles to the motion of a particle under the action of a **central force**, *i. e.*, a force whose direction always passes through a fixed point called the *center of force* (see Art. 157). We shall here consider only *central forces whose magnitude is a function of the distance from the center alone.* Thus, let O be the center of force, P the position of the moving particle at any time t, m the mass of the particle, and $OP = r$ its distance from the center; then the general expression for such a central force F is

$$F = F(r) = mf(r),$$

where the function $F(r)$ represents the law of force, and the function $f(r)$ the law of the acceleration produced by this force in the particle m.

524. The most important special case is that of a force proportional to some power of the distance r, say

$$F(r) = \mu r^n,$$

where μ and n are constants. The constant μ, which represents the value of the force at unit distance from the center, is often called the *intensity* of the force, or of the center.

In the case of Newton's law of universal gravitation (Arts. 472, 473) we have $n = -2$, $\mu = \kappa m m'$, where κ is a constant, viz., the acceleration produced by a unit of mass acting on a unit of mass at unit distance, while m is the mass of the attracted particle, and m' that of the attracting center; that is, Newton's law is expressed by the formula

$$F = \kappa \frac{m m'}{r^2}.$$

525. From the physical point of view, attractions following Newton's law, and indeed, central forces generally, are usually regarded as due to the presence of mass (or an electric charge, etc.), not only in the moving particle, but also at the center of force; and the action between these two masses is then a mutual action, being of the nature of a *stress*, *i. e.*, consisting of two equal and opposite forces. It follows that what we have called the center of force is not a fixed point.

It will, however, be shown later (Arts. 567, 568) that a simple modification allows us to apply to this case the results deduced on the assumption that the center is fixed.

Again, the attracting or repelling masses will here be regarded as concentrated at points. It is shown in the theory of attraction that a homogeneous sphere, according to Newton's law, attracts a particle outside of its mass as if the whole mass of the sphere were concentrated at the center of the sphere. The attraction of *any* mass on a particle can, of course, always be reduced to a single force; but as the particle moves, the direction of this force will not in general pass through a fixed point; such a force is, therefore, not central.

526. If a particle P of mass m be acted upon by a single central force

$$F = m f(r),$$

its acceleration $j = F/m = f(r)$ will pass through the center of force and be a function of r alone. The problem reduces, therefore, at once to the kinematical problem of *central motion* (Art. 157). Although the leading ideas of the solution of this problem have been indicated in Kinematics (Arts. 157–174), the importance of the subject of central forces demands a restatement in this place of some of the results in the language of kinetics, and a more complete exposition of some special cases.

527. A particle of mass m acted upon by a single central force $F = mf(r)$ will describe a curvilinear path whenever the initial velocity is different from zero and does not pass through the center of force. For the case of *rectilinear* motion under a central force see Arts. 472–486.

All central motions, whatever may be the law of the force, have two properties in common: (a) the path of the particle, here often called the **orbit**, is a *plane* curve (Art. 158); (b) the sectorial velocity is constant (Arts. 159–161).

Taking the plane of motion as xy-plane and the center of force O as origin (Fig. 159), the direction cosines of the force F are $\mp x/r$, $\mp y/r$, the upper sign corresponding to an attractive force, the lower to a repulsion. Hence, the dynamical equations of motion are

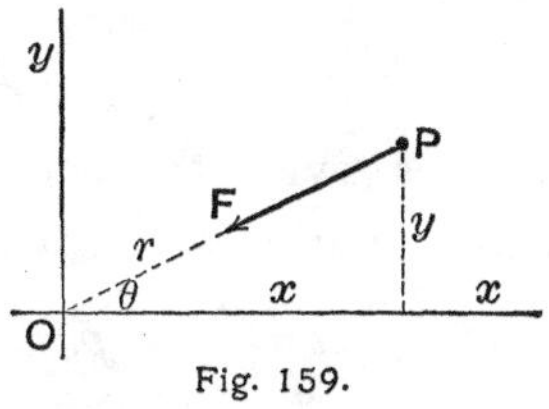

Fig. 159.

$$m\frac{d^2x}{dt^2} = \mp F\frac{x}{r}, \quad m\frac{d^2y}{dt^2} = \mp F\frac{y}{r}. \qquad (1)$$

If $mf(r)$ be substituted for F, the factor m disappears, and the equations become purely kinematical.

To avoid the use of the double sign, we shall give the equations in the form corresponding to the more important case of attraction; for a repulsive force it will only be necessary to change throughout the sign of F or $f(r)$. Thus the fundamental equations of motion are (comp. Art. 162):

$$\frac{d^2x}{dt^2} = -f(r)\frac{x}{r}, \quad \frac{d^2y}{dt^2} = -f(r)\frac{y}{r}. \qquad (2)$$

If polar co-ordinates r, θ (Fig. 159), with the center of force as pole, be used, the equations of motion are, since the total acceleration is along the radius vector (see Art. 114):

$$j_r \equiv \frac{d^2r}{dt^2} - r\left(\frac{d\theta}{dt}\right)^2 = -f(r), \quad j_\theta \equiv \frac{1}{r}\frac{d}{dt}\left(r^2\frac{d\theta}{dt}\right) = 0. \tag{3}$$

528. Two principal problems present themselves: (*a*) the problem of finding the orbit for a given law of force, and (*b*) the converse problem of determining the law of force, *i. e.*, the function $f(r)$, when the orbit is given. The solution of the former problem is effected by obtaining first integrals of the equations of motion from the principle of areas and from the principle of kinetic energy, and by combining these integrals so as to effect a second integration. Formulæ for the solution of the latter problem will be found incidentally.

529. As shown in Kinematics (Arts. 159–161) the second of the equations (3) gives the first integral

$$r^2\frac{d\theta}{dt} = c, \tag{4}$$

where c is twice the sectorial velocity; and with the notation indicated in Fig. 160 it readily follows that

$$c = pv = p_0v_0 = vr\sin\psi = v_0r_0\sin\psi_0; \tag{5}$$

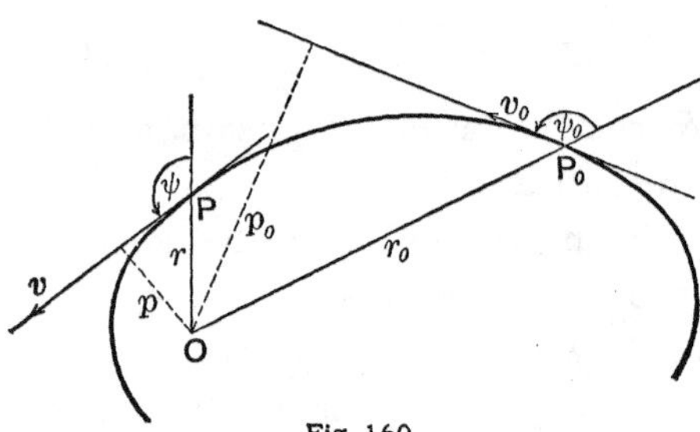

Fig. 160.

i. e., the velocity is inversely proportional to its perpendicular distance from the center, or, as it is sometimes expressed, the moment of the velocity about the center of force is constant.

The relation (4) can also be obtained from the equations (2) by applying the principle of areas (Art. 512, see Art. 163).

530. Another first integral of the equations of motion is obtained by combining the equations (1) according to the principle of kinetic energy and work (Art. 504, comp. Art. 164). This gives

$$d(\tfrac{1}{2}mv^2) = -Fdr, \text{ or } d(\tfrac{1}{2}v^2) = -f(r)dr, \tag{6}$$

whence

$$v^2 = v_0^2 - 2\int_{r_0}^{r} f(r)dr; \tag{7}$$

i. e., the velocity at any distance r depends only on this distance (besides the initial radius vector and velocity) and is independent of the path described, being the same as if the particle had been projected with the initial velocity along the straight line joining the initial position to the center.

531. To perform the second integration we have only to substitute in (7) for v its value in terms of r and t or r and θ. Now the general expression for the velocity in any curvilinear motion is (Art. 96)

$$v^2 = \left(\frac{dr}{dt}\right)^2 + r^2\left(\frac{d\theta}{dt}\right)^2 = \left(\frac{d\theta}{dt}\right)^2\left[\left(\frac{dr}{d\theta}\right)^2 + r^2\right].$$

From these expressions one of the variables θ and t can be eliminated by substituting for $d\theta/dt$ its value c/r^2 from (4); this gives

$$v^2 = \left(\frac{dr}{dt}\right)^2 + \frac{c^2}{r^2} = \frac{c^2}{r^4}\left[\left(\frac{dr}{d\theta}\right)^2 + r^2\right]. \tag{8}$$

It is often convenient to replace the radius vector r by its reciprocal $u = 1/r$; we then have

$$v^2 = \frac{1}{u^4}\left(\frac{du}{dt}\right)^2 + c^2u^2 = c^2\left[\left(\frac{du}{d\theta}\right)^2 + u^2\right]. \tag{9}$$

532. The formulæ (4) and (7), together with the expression (8) or (9), contain the complete solution of the two principal problems mentioned in Art. 528. Thus, if the law of force be given, the form of the function $f(r)$ is known, and v can be found from (7) in function of r or u; substituting this value of v in either (8) or (9), we have a differential equation of the first order between r and t, or between r and θ. The integration of the latter equation gives the integral equation of the orbit.

On the other hand, if the equation of the path be given, the expressions (8) or (9) furnish the value of v^2, which, substituted in (6), determines the law of force $f(r)$.

When the equation of the orbit is known, *i. e.*, when r is known as a function of θ, or *vice versa*, the time t of the motion can be found from (4), which gives

$$t = \frac{1}{c}\int r^2 d\theta.$$

533. If the second expression for v^2 in (9) be introduced in (6), we find, as shown in Art. 167,

$$f(r) = c^2u^2\left(\frac{d^2u}{d\theta^2} + u\right). \tag{10}$$

This will generally be found the most convenient form for finding the law of force when the polar equation of the orbit is given. Again, when $f(r)$ is given, the integration of this differential equation of the second order is often more convenient for finding the equation of the orbit than the method indicated in Art. 532.

It may be noted that the important relation (10) can be derived directly from the equations of motion (3), by eliminating t by means of (4) and introducing u for $1/r$. We have

$$\frac{dr}{dt} = \frac{dr}{d\theta}\frac{d\theta}{dt} = \frac{c}{r^2}\frac{dr}{d\theta} = -c\frac{du}{d\theta},$$

$$\frac{d^2r}{dt^2} = -c\frac{d^2u}{d\theta^2}\frac{d\theta}{dt} = -c^2u^2\frac{d^2u}{d\theta^2};$$

if these values be substituted in the first of the equations (3), the relation (10) will result.

534. When the equation of the orbit can be expressed conveniently in terms of the radius vector r and the perpendicular p from the center to the velocity, as is, for instance, the case for the conic sections, it is of advantage to combine the equation of kinetic energy (6) directly with the equation resulting from the principle of areas, $pv = c$ (Art. 529). This gives

$$-f(r) = \tfrac{1}{2}\frac{d(v^2)}{dr} = \frac{c^2}{2}\frac{d}{dr}\frac{1}{p^2} = -\frac{c^2}{p^3}\frac{dp}{dr}. \tag{11}$$

535. Exercises.

(1) Find the law of force when the equation of the orbit is $r^n = q^n/(1 + e\cos n\theta)$, e being positive, and investigate the particular cases $n = 1$, $n = 2$, $n = -1$, $n = -2$.

(2) Find the law of the central force directed to the origin under whose action a particle will describe the following curves: (*a*) the spiral of Archimedes $r = a\theta$; (*b*) the hyperbolic spiral $\theta r = a$; (*c*) the logarithmic or equiangular spiral $r = ae^{n\theta}$; (*d*) the curve $r = a\cos n\theta$.

(3) A particle moves in a circle under the action of a central force directed towards a point on the circumference. Find the law of force.

(4) A particle is acted upon by a force perpendicular to a given plane and inversely proportional to the cube of the distance from the plane. Determine its motion.

(5) A particle moves in a semi-ellipse under the action of a force perpendicular to the axis joining the ends of the semi-ellipse. Determine the law of force and the velocity at the ends.

536. Force Proportional to the Distance: $f(r) = \kappa^2 r$. The equations of motion (2) are in this case

$$\frac{d^2x}{dt^2} = \mp\kappa^2 x, \quad \frac{d^2y}{dt^2} = \mp\kappa^2 y,$$

the upper sign holding for attraction, the lower for repulsion. Their solution is very simple, because each equation can be integrated separately. We find, in the case of *attraction*,

$$x = a_1\cos\kappa t + a_2\sin\kappa t, \quad y = b_1\cos\kappa t + b_2\sin\kappa t,$$

and in the case of *repulsion*,

$$x = a_1 e^{\kappa t} + a_2 e^{-\kappa t}, \quad y = b_1 e^{\kappa t} + b_2 e^{-\kappa t};$$

a_1, a_2, b_1, b_2, being the constants of integration.

537. To find the equation of the orbit, it is only necessary to eliminate t in each case.

In the case of attraction, this elimination can be performed by solving for cos κt, sin κt, squaring and adding. The result is

$$(a_1 y - b_1 x)^2 + (a_2 y - b_2 x)^2 = (a_1 b_2 - a_2 b_1)^2,$$

and this represents an ellipse, since

$$(a_1^2 + a_2^2)(b_1^2 + b_2^2) - (a_1 b_1 + a_2 b_2)^2 \equiv (a_1 b_2 - a_2 b_1)^2$$

is always positive. The center of the ellipse is at the origin, and the lines $a_1 y = b_1 x$, $a_2 y = b_2 x$ are a pair of conjugate diameters.

538. In the case of repulsion, solve for $e^{\kappa t}$ and $e^{-\kappa t}$, and multiply. The resulting equation,

$$(a_1 y - b_1 x)(b_2 x - a_2 y) = (a_1 b_2 - a_2 b_1)^2,$$

represents a hyperbola whose asymptotes are the lines $a_1 y = b_1 x$, $a_2 y = b_2 x$.

539. It is worthy of notice that the more general problem of *the motion of a particle attracted by any number of fixed centers, with forces directly proportional to the distances from these centers,* can be reduced to the problem of Art. 536.

Let x, y, z be the co-ordinates of the particle, r_i its distance from the center O_i; x_i, y_i, z_i the co-ordinates of O_i; and $-\kappa_i^2 r_i$ the acceleration produced by O_i. Then the x-component of the resultant acceleration is

$$= -\Sigma \kappa_i^2 r_i \cdot \frac{x - x_i}{r_i} = -\Sigma \kappa_i^2 (x - x_i) = -x \Sigma \kappa_i^2 + \Sigma \kappa_i^2 x_i;$$

and similar expressions obtain for the y and z components. Hence, the equations of motion are

$$\frac{d^2 x}{dt^2} = -x \Sigma \kappa_i^2 + \Sigma \kappa_i^2 x_i, \quad \frac{d^2 y}{dt^2} = -y \Sigma \kappa_i^2 + \Sigma \kappa_i^2 y_i, \quad \frac{d^2 z}{dt^2} = -z \Sigma \kappa_i^2 + \Sigma \kappa_i^2 z_i.$$

As the right-hand members are linear in x, y, z, there is one, and only one, point at which the resultant acceleration is zero. Denoting its co-ordinates by $\bar{x}, \bar{y}, \bar{z}$, we have

$$\bar{x} = \frac{\Sigma \kappa_i^2 x_i}{\Sigma \kappa_i^2}, \quad \bar{y} = \frac{\Sigma \kappa_i^2 y}{\Sigma \kappa_i^2}, \quad \bar{z} = \frac{\Sigma \kappa_i^2 z_i}{\Sigma \kappa_i^2}.$$

The form of these equations shows that this point of zero acceleration, which is sometimes called the *mean center,* is the centroid of the centers of force, if these centers be regarded as containing masses equal to κ_i^2. It is evidently a fixed point.

540. By introducing the co-ordinates of the mean center, we can now reduce the equations of motion to the simple form

$$\frac{d^2x}{dt^2} = -\kappa^2(x-\bar{x}),\ \frac{d^2y}{dt^2} = -\kappa^2(y-\bar{y}),\ \frac{d^2z}{dt^2} = -\kappa^2(z-\bar{z}),$$

where $\kappa^2 = \Sigma\kappa_i^2$. Finally, taking the mean center as origin, we have

$$\frac{d^2x}{dt^2} = -\kappa^2 x,\ \frac{d^2y}{dt^2} = -\kappa^2 y,\ \frac{d^2z}{dt^2} = -\kappa^2 z.$$

It thus appears that *the motion of the particle is the same as if there were only a single center of force*, viz., the mean center $(\bar{x}, \bar{y}, \bar{z})$, *attracting with a force proportional to the distance from this center.*

The plane of the orbit is, of course, determined by the mean center and the initial velocity.

541. It is easy to see that most of the considerations of Art. 539 apply even when some or all of the centers *repel* the particle with forces proportional to the distance. It may, however, happen in this case that the mean center lies at infinity, in which case, of course, it can not be taken as origin.

Simple geometrical considerations can also be used to solve such problems. Thus, in the case of two attractive centers O_1, O_2 (Fig. 161) of equal intensity κ^2, the forces can evidently be represented by the distances $PO_1 = r_1$, $PO_2 = r_2$ of the particle P from the centers. Their resultant is therefore $= 2\,PO$, if O denotes the point midway between O_1 and O_2; and this resultant always passes through this fixed point O, and is proportional to the distance PO from this point.

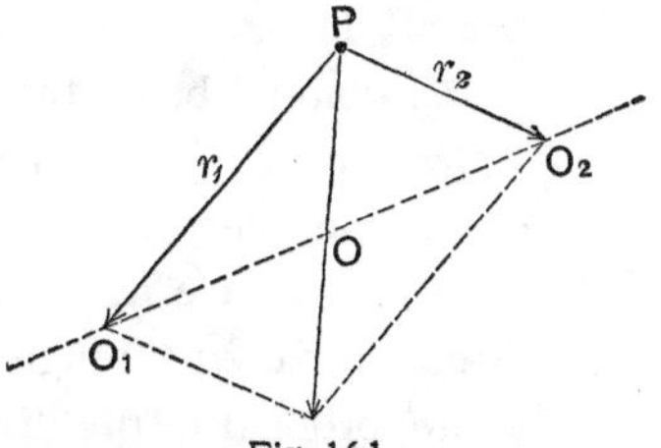

Fig. 161.

542. Exercises.

(1) Determine the constants of integration in Art. 535, if x_0, y_0 are the co-ordinates of the particle at the time $t = 0$ and v_1, v_2 the components of its velocity v_0 at the same time. The equation of the orbit will assume the form

$$\kappa^2(x_0 y - y_0 x)^2 + (v_1 y - v_2 x)^2 = (x_0 v_2 - y_0 v_1)^2$$

for attraction, and

$$\kappa^2(x_0 y - y_0 x)^2 - (v_1 y - v_2 x)^2 = -(x_0 v_2 - y_0 v_1)^2$$

for repulsion.

(2) Show that the semi-diameter conjugate to the initial radius vector has the length v_0/κ, where $v_0^2 = v_1^2 + v_2^2$. As any point of the orbit can be regarded as initial point, it follows that *the velocity at any point is proportional to the parallel diameter of the orbit.*

(3) Find what the initial velocity must be to make the orbit a circle in the case of attraction, and an equilateral hyperbola in the case of repulsion.

(4) The initial radius vector r_0 and the initial velocity v_0 being given geometrically, show how to construct the axes of the orbit described under the action of a central force (of given intensity κ^2) proportional to the distance from the origin.

(5) A particle describes an ellipse under the action of a central force proportional to the distance; show that the eccentric angle is proportional to the time, and find the corresponding relation for a hyperbolic orbit.

(6) A particle of mass m describes a conic under the action of a central force $F = \mp m\kappa^2 r$. Show that the sectorial velocity is $\frac{1}{2}c = \frac{1}{2}\kappa ab$, a and b being the semi-axes of the conic.

(7) In Ex. (6) show that the time of revolution is $T = 2\pi/\kappa$, if the conic is an ellipse.

(8) A particle describes a conic under the action of a force whose direction passes through the center of the conic. Show that the force is proportional to the distance from the center.

(9) A particle is acted upon by two central forces of the same intensity (κ^2), each proportional to the distance from a fixed center. Determine the orbit: (*a*) when both forces are attractive; (*b*) when both are repulsive; (*c*) when one is an attraction, the other a repulsion.

(10) A particle of mass m is attracted by two centers O_1, O_2 of equal mass m' and repelled by a third center O_3, whose mass is $m'' = 2m'$. If the forces are all directly proportional to the respective distances, determine and construct the orbit.

(11) When a particle moves in an ellipse under a force directed towards the center, find the time of moving from the end of the major axis to a point whose polar angle is θ.

(12) Prove that if, in the problem of Art. 541, the intensities of O_1 and O_2 are κ_1, κ_2, the resultant attraction F passes through the centroid G of two masses κ_1, κ_2, placed at O_1, O_2, and that $F = (\kappa_1 + \kappa_2)PG$.

(13) In Art. 536, in the case of attraction, the component motions are evidently simple harmonic oscillations. Show that the equation of the path can be put in the form (comp. Art. 142)

$$\frac{x^2}{a^2} - \frac{2\,xy}{ab}\sin\delta + \frac{y^2}{b^2} = \cos^2\delta.$$

(14) Show that the total energy of a particle of mass m describing an ellipse of semi-axes a, b under a force $m\kappa^2 r$ directed to the center is $= \frac{1}{2}\,m\kappa^2(a^2 + b^2)$.

543. Force Inversely Proportional to the Square of the Distance: $f(r) = \mu/r^2$ (Newton's law).

It has been shown in Kinematics (Arts. 157–169) how this law of acceleration can be deduced from Kepler's laws of planetary motion. From Kepler's first law Newton concluded that the acceleration of a planet (regarded as a point of mass m) is constantly directed towards the sun; from the second he found that this acceleration is inversely proportional to the square of the distance. The motion of a planet can therefore be explained on the hypothesis of an attractive force,

$$F = m\frac{\mu}{r^2},$$

issuing from the sun.

The value of μ, which represents the acceleration at unit distance or the so-called intensity of the force, was found to be (Art. 169; or below, Art. 556)

$$\mu = 4\,\pi^2\frac{a^3}{T^2};$$

and as, according to Kepler's third law, the quantity a^3/T^2 has the same value for all the planets, Newton inferred that the intensity of the attracting force is the same for all planets; in other words, that it is one and the same central force that keeps the different planets in their orbits.

544. It was further shown by Newton and Halley that the motions of the comets are due to the same attractive force. The orbits of the comets are generally ellipses of great eccentricity, with the sun at one of the foci. As a comet is within range of observation only while in that portion of its path which lies nearest to the sun, a portion of a parabola, with the same focus and vertex, can be substituted for this portion of the elliptic orbit, as a first approximation.

It is also found from observation that the motions of the moons or satellites around the planets follow very nearly Kepler's laws. A planet can therefore be regarded as attracting each of its satellites with a force proportional to the mass of the satellite and inversely proportional to the square of the distance.

545. All these facts led Newton to suspect that the force of terrestrial gravitation, as observed in the case of falling bodies on the earth's surface, might be the same as the force that keeps the moon in its orbit around the earth. This inference could easily be tested, since the acceleration g of falling bodies as well as the moon's distance and time of revolution were known.

Let m be the mass of the moon, a the major semi-axis of its orbit, T the time of revolution, r the distance between the centers of earth and moon; then the earth's attraction on the moon is (Art. 543)

$$F = 4\pi^2 m \frac{a^3}{T^2 r^2},$$

or, since the eccentricity of the moon's orbit is so small that the orbit can be regarded as nearly circular,

$$F = 4\pi^2 m \frac{a}{T^2}.$$

On the other hand, the attraction exerted by the earth on a mass m on its surface, *i. e.*, at the distance $R = 3963$ miles from the center, is

$$F' = mg.$$

Now, if these forces are actually in the inverse ratio of the squares of the distances, we must have

$$\frac{F'}{F} = \frac{a^2}{R^2},$$

or, since the distance of the moon is nearly $= 60 R$,

$$F' = 60^2 F.$$

Substituting the above values of F and F', we find

$$g = 4\pi^2 \cdot \frac{60^3 R}{T^2}.$$

With $R = 3963$ miles, $T = 27^d\, 7^h\, 43^m$, this gives

$$g = 32.0,$$

a value which agrees sufficiently with the observed value of g, considering the rough degree of approximation used.

546. In this way Newton was finally led to his **law of universal gravitation**, which asserts that *every particle of mass m attracts every other particle of mass m' with a force*

$$F = \kappa \frac{mm'}{r^2}, \qquad (12)$$

where r is the distance of the particles and κ a constant, viz., the acceleration produced by a unit of mass in a unit of mass at unit distance (see Arts. 472, 473).

The best test of this hypothesis as an actual law of physical nature is found in the close agreement of the results of theoretical astronomy based on this law with the observed celestial phenomena.

547. Taking Newton's law as a basis, let us now turn to the converse problem of *determining the motion of a particle acted upon by a single central force for which* $f(r) = \mu/r^2$ (problem of planetary motion).

It has been shown in Kinematics (Arts. 170–173) that *if the force be attractive*, the particle will describe a conic section with one of the foci at the center of force, the conic being an ellipse, parabola, or hyperbola, according as

$$v_0^2 \lesseqgtr \frac{2\mu}{r_0}. \tag{13}$$

If the force be repulsive, the same reasoning will apply, except that μ is then a negative quantity. The orbit is, therefore, in this case always hyperbolic; the branch of the hyperbola that forms the orbit must evidently turn its convex side towards the focus at which the center of force is situated, since the force always lies on the concave side of the path.

548. To exhibit fully the determination of the constants and the dependence of the nature of the orbit on the initial conditions, a solution somewhat different from that given in Kinematics will here be given for the problem of planetary motion in its simplest form.

With $f(r) = \mu/r^2$, the equation of kinetic energy and work, (7), Art. 530, gives

$$v^2 = v_0^2 - 2\mu \int_{r_0}^{r} \frac{dr}{r^2} = v_0^2 + 2\frac{\mu}{r} - \frac{2\mu}{r_0},$$

or, if the constant of integration be denoted briefly by h and $u = 1/r$ be introduced,

$$v^2 = 2\mu u + h, \text{ where } h = v_0^2 - \frac{2\mu}{r_0}. \tag{14}$$

Substituting this expression of v^2 in the equation (9), Art. 531, we find the differential equation of the orbit in the form

$$\left(\frac{du}{d\theta}\right)^2 + u^2 = \frac{1}{c^2}(2\mu u + h), \tag{15}$$

or

$$\left(\frac{du}{d\theta}\right)^2 = -\left(u - \frac{\mu}{c^2}\right)^2 + \frac{\mu^2}{c^4} + \frac{h}{c^2}.$$

To integrate, we introduce a new variable u' by putting

$$u - \frac{\mu}{c^2} = u'\sqrt{\frac{\mu^2}{c^4} + \frac{h}{c^2}};$$

the resulting equation,

$$\left(\frac{du'}{d\theta}\right)^2 = 1 - u'^2, \text{ or } d\theta = \pm\frac{du'}{\sqrt{1 - u'^2}},$$

has the general integral

$$\theta - \alpha = \mp \cos^{-1} u', \text{ or } u' = \cos(\theta - \alpha),$$

where α is the constant of integration. The orbit has, therefore, the equation

$$\frac{1}{r} = \frac{\mu}{c^2} + \sqrt{\frac{\mu^2}{c^4} + \frac{h}{c^2}} \cos(\theta - \alpha), \tag{16}$$

which agrees with the equation (47) given in Kinematics, Art. 173, excepting the different notation used for the constants.

549. The equation (16) represents a conic section referred to its focus as origin. The general focal equation of a conic is

$$\frac{1}{r} = \frac{1}{l} + \frac{e}{l} \cos(\theta - \alpha), \tag{17}$$

where l is the semi-latus rectum, or parameter, e the eccentricity, and α the angle made with the polar axis by the line joining the focus to the nearest vertex.

In a planetary orbit (Fig. 162), the sun S being at one of the foci, the nearest vertex A is called the *perihelion*, the other vertex A' the *aphelion*, and the angle $\theta - \alpha$ made by any radius vector $SP = r$ with the perihelion distance SA is called the *true anomaly*.

Comparing equations (17) and (16), we find, for the determination of the constants:

$$\frac{1}{l} = \frac{\mu}{c^2}, \quad \frac{e}{l} = \sqrt{\frac{\mu^2}{c^4} + \frac{h}{c^2}};$$

hence,

$$l = \frac{c^2}{\mu}, \quad e = \sqrt{1 + \frac{hc^2}{\mu^2}}, \tag{18}$$

or, solving for c and h,

$$c = \sqrt{\mu l}, \quad h = \mu\frac{e^2 - 1}{l}. \tag{19}$$

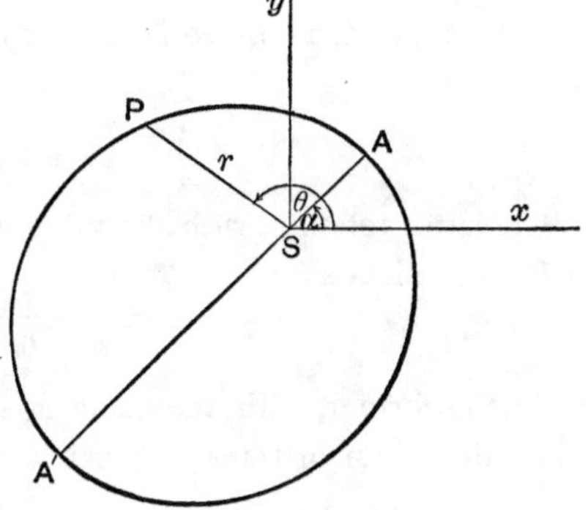

Fig. 162.

550. The expression for the eccentricity e in (18) determines the nature of the conic; the orbit is an ellipse, parabola, or hyperbola, according as $e \lesseqgtr 1$; hence, by (18), according as the constant h of the equation of kinetic energy is negative, zero, or positive. Owing to the value of h given in (14), this criterion agrees with the form (13), Art. 547.

It should be observed that it follows from (13) that *the nature* of the conic is independent of the *direction* of the initial velocity.

551. The criterion (13) can be given the following interpretation. Consider a particle attracted by a fixed center according to Newton's law. If it move in a straight line passing through the center, the principle of kinetic energy gives for its velocity, at the distance r,

$$v^2 = v_0^2 - 2\mu \int_{r_0}^{r} \frac{dr}{r^2} = \frac{2\mu}{r} + v_0^2 - \frac{2\mu}{r_0};$$

hence, if it start from rest at an infinite distance from the center, it would acquire the velocity $\sqrt{2\mu/r}$ at the distance r. The criterion (13) is therefore equivalent to saying that *the orbit is an ellipse, a parabola, or a hyperbola, according as the velocity at any point is less than, equal to, or greater than the velocity which the particle would have acquired at that point by falling towards the center from infinity* (comp. Art. 475).

552. For a *central conic*, whose axes are $2a$, $2b$, we have $l = b^2/a$, $e = \sqrt{a^2 \mp b^2}/a$ (the upper sign relating to the ellipse, the lower to the hyperbola), so that the equations (19) reduce to the following:

$$c = b\sqrt{\frac{\mu}{a}}, \quad h = \mp\frac{\mu}{a}. \tag{20}$$

The latter relation, with the value of h from (14), gives for the major or focal semi-axis a:

$$\pm\frac{1}{a} = \frac{2}{r_0} - \frac{v_0^2}{\mu}; \tag{21}$$

while the former, with the value of c as given in (5), Art. 529, determines the minor or transverse axis b:

$$b = c\sqrt{\frac{a}{\mu}} = r_0 v_0 \sin\psi_0 \sqrt{\frac{a}{\mu}}. \tag{22}$$

553. The magnitudes of the axes having thus been found, their directions can be determined by a simple construction which furnishes the second focus.

In the *ellipse*, the focal radii have a constant sum $= 2a$, and lie on the same side of the tangent, making equal angles with it. In the *hyperbola*, they have a constant difference $= 2a$, and lie on opposite sides of the tangent.

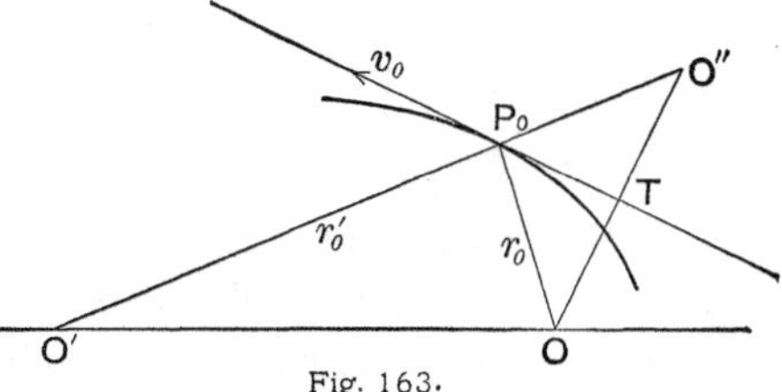

Fig. 163.

Hence, determining the point O'' (Fig. 163), which is symmetrical to the center of force O with respect to the initial velocity, and drawing the line P_0O'', we have only to lay off on this line from P_0 a length $P_0O' = \pm(2a - r_0)$; then O' is the second focus, which for an elliptic orbit must be taken with O on the same side of the tangent P_0T, and for a hyperbolic orbit on the opposite side.

554. For a *parabola*, since $e = 1$, we find, from (19),

$$h = 0, \quad l = \frac{c^2}{\mu} = \frac{v_0^2 r_0^2 \sin^2 \psi_0}{\mu}. \tag{23}$$

The axis of the parabola is readily found by remembering that the perpendicular let fall from the focus on the tangent bisects the tangent (*i. e.*, the segment of the tangent between the point of contact and the axis). Hence, if OT (Fig. 164) be the perpendicular let fall from the center O on the velocity v_0, it is only necessary to make $TT' = P_0T$, and T' will be a point of the axis. Moreover, the perpendicular let fall from T on OT' will meet the axis at the vertex A of the parabola, so that $OA = \frac{1}{2} l$.

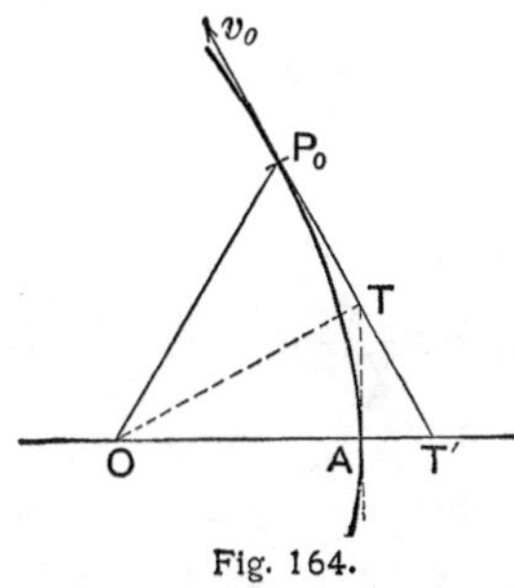

Fig. 164.

555. The relation (21), which must evidently hold at *any* point of the orbit, can be written in the form

$$v^2 = 2\mu\left(\frac{1}{r} \mp \frac{1}{2a}\right), \tag{24}$$

the upper sign relating to the ellipse, the lower to the hyperbola, while for the parabola, the second term in the parenthesis vanishes (since $a=\infty$).

This convenient expression for the velocity in terms of the radius vector might have been derived directly from the fundamental relation (5), $v=c/p$, the first of the equations (19), $c^2=\mu l$, and the geometrical properties of the conic sections ($r \pm r' = 2a$, $pp'=b^2$, $p'r=pr'$, where r, r' are the focal radii, and p, p' the perpendiculars let fall from the foci on the tangent). The proof is left to the student.

556. Time. In the case of an elliptic orbit, the time T of a complete revolution, usually called the **periodic time**, is found by remembering that the sectorial velocity is constant and $=\frac{1}{2}c$ (Art. 529), whence

$$T=\frac{2\pi ab}{c},$$

or, by (20),

$$T=2\pi\sqrt{\frac{a^3}{\mu}}=\frac{2\pi}{n}. \tag{25}$$

The constant

$$n=\sqrt{\frac{\mu}{a^3}},$$

which evidently represents the mean angular velocity about the center in one revolution, is called the **mean motion** of the planet. It should be noticed that it depends not only on the intensity of the force, but also on the major axis of the orbit, while in the case of a force directly proportional to the distance the periodic time is independent of the size of the orbit (see Art. 542, Ex. 7).

The periodic time T and the major axis a of a planetary orbit determine the intensity μ of the force:

$$\mu=4\pi^2\frac{a^3}{T^2}, \tag{26}$$

whence

$$F=mf(r)=m\frac{\mu}{r^2}=4\pi^2 m\frac{a^3}{T^2r^2}, \tag{27}$$

where m is the mass of the planet.

557. To find generally the time t in terms of θ or r, we can, of course, proceed as indicated in Art. 532; but the resulting expressions are somewhat complicated, and it is best to introduce the eccentric angle ϕ of the ellipse as a new variable, and to express t, r, and θ in terms of ϕ. In astronomy, the polar angle θ is known as the *true anomaly* and the eccentric angle ϕ as the *eccentric anomaly*.

558. The relation of the eccentric angle ϕ to the polar co-ordinates r, θ will appear from Fig. 165, in which P is the position of the planet

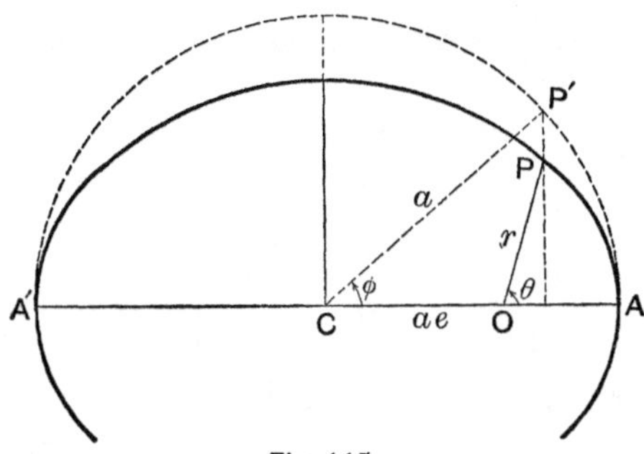

Fig. 165.

at the time t, P' the corresponding point on the circumscribed circle, $\measuredangle AOP = \theta$ the true anomaly, and $\measuredangle ACP' = \phi$ the eccentric anomaly. The focal equation of the ellipse

$$r = \frac{l}{1 + e\cos\theta} = \frac{a(1 - e^2)}{1 + e\cos\theta}$$

gives $r + er\cos\theta = a - ae^2$; and the figure shows that $r\cos\theta = a\cos\phi - ae$; hence,

$$r = a(1 - e\cos\phi), \text{ or } a - r = ae\cos\phi. \tag{28}$$

Equating this value of r to that given by the polar equation of the ellipse, we have

$$1 - e\cos\phi = \frac{1 - e^2}{1 + e\cos\theta}, \text{ or } \cos\theta = \frac{\cos\phi - e}{1 - e\cos\phi}.$$

A more symmetrical form can be given to this relation by computing

$$1 - \cos\theta \equiv 2\sin^2\tfrac{1}{2}\theta = (1 + e)\frac{1 - \cos\phi}{1 - e\cos\phi},$$

$$1 + \cos\theta \equiv 2\cos^2\tfrac{1}{2}\theta = (1 - e)\frac{1 + \cos\phi}{1 - e\cos\phi};$$

whence, by division,

$$\tan\tfrac{1}{2}\theta = \sqrt{\frac{1 + e}{1 - e}}\tan\tfrac{1}{2}\phi. \tag{29}$$

559. To find t in terms of r, we have only to substitute in (24) for v^2 its value from (8), Art. 531, and to integrate the resulting differential equation

$$\left(\frac{dr}{dt}\right)^2+\frac{c^2}{r^2}=\frac{2\mu}{r}-\frac{\mu}{a}.$$

As, by (20), Art. 552, $c^2=\mu b^2/a=\mu a(1-e^2)$, this equation becomes

$$r^2\left(\frac{dr}{dt}\right)^2=\frac{\mu}{a}[a^2e^2-(a-r)^2],$$

or

$$dt=\sqrt{\frac{a}{\mu}}\frac{rdr}{\sqrt{a^2e^2-(a-r)^2}}.$$

The integration is easily performed by introducing the eccentric angle ϕ as variable by means of (28); this gives

$$dt=\sqrt{\frac{a}{\mu}}\cdot a(1-e\cos\phi)d\phi.$$

If the time be counted from the perihelion passage of the planet, we have $t=0$ when $r=a-ae$, *i.e.*, when $\phi=0$; hence, putting $\sqrt{\mu/a^3}=n$, as in Art. 556, we find

$$nt=\phi-e\sin\phi. \tag{30}$$

This relation is known as *Kepler's equation;* the quantity nt is called the *mean anomaly*.

560. Kepler's equation (30) can be derived directly by considering that the ellipse APA' (Fig. 165) can be regarded as the projection of the circle $AP'A'$, after turning this circle about AA' through an angle $=\cos^{-1}(b/a)$. For it follows that the elliptic sector AOP is to the circular sector AOP' as b is to a. Now, for the circular sector we have

$$AOP'=ACP'-OCP'=\tfrac{1}{2}a^2\phi-\tfrac{1}{2}ae\cdot a\sin\phi=\tfrac{1}{2}a^2(\phi-e\sin\phi);$$

hence, the elliptic sector described in the time t is

$$AOP=\frac{b}{a}\cdot AOP'=\tfrac{1}{2}ab(\phi-e\sin\phi).$$

The sectorial velocity being constant by Kepler's first law, we have

$$\frac{AOP}{t}=\frac{\pi ab}{T};$$

hence,

$$t=\frac{T}{2\pi}(\phi-e\sin\phi),$$

and this agrees with (30) since, by (25), $2\pi/T=n$.

561. Kepler's equation (30) gives the time as a function of ϕ; by means of (28), it establishes the relation between t and r; by means of (29), it connects t with θ. It is, however, a transcendental equation and cannot be solved for ϕ in a finite form.

For orbits with a small eccentricity e, an approximate solution can be obtained by writing the equation in the form

$$\phi = nt + e \sin \phi,$$

and substituting under the sine for ϕ its approximate value nt:

$$\phi = nt + e \sin nt. \tag{31}$$

This amounts to neglecting terms containing powers of e above the first power.

Substituting this value of ϕ in (28), we have with the same approximation

$$r = a(1 - e \cos nt). \tag{32}$$

To find θ in terms of t, we have, from the equation of the ellipse, $r = a(1 - e^2)(1 + e \cos \theta)^{-1} = a(1 - e \cos \theta)$, neglecting again terms in e^2; hence, $r^2 = a^2(1 - 2e \cos \theta)$. Substituting this value in the equation of areas, $r^2 d\theta = c dt = \sqrt{\mu a(1 - e^2)}\, dt$, we find

$$(1 - 2e \cos \theta)\, d\theta = \sqrt{\frac{\mu}{a^3}}\, dt = n dt;$$

whence, by integration, since $\theta = 0$ for $t = 0$,

$$\theta - 2e \sin \theta = nt,$$

or finally,

$$\theta = nt + 2e \sin nt. \tag{33}$$

Thus we have in (31), (32), (33) approximate expressions for ϕ, r, and θ directly in terms of the time. The quantity $2e \sin nt$, by which the true anomaly θ exceeds the mean anomaly nt, is called the *equation of the center*.

562. Exercises.

(1) A particle describes an ellipse under the action of a central force. Determine the law of force by means of formula (11), Art. 534: (*a*) when the center of force is at the center of the ellipse; (*b*) when it is at a focus.

(2) A particle is attracted by a fixed center according to Newton's law. What must be the initial velocity if the orbit is to be circular?

(3) A number of particles are projected, from the same point in the field of a force following Newton's law, with the same velocity, but

in different directions. Show that the periodic times are the same for all the particles.

(4) The mean distance of Mars from the sun being 1.5237 times that of the earth, what is the time of revolution of Mars about the sun?

(5) A particle describes a conic under the action of a central force following Newton's law; if the intensity μ of the force be suddenly changed to μ', what is the effect on the orbit?

(6) In Ex. (5), if the original orbit was a parabola and the intensity be doubled, what is the new orbit?

(7) Regarding the moon's orbit about the earth as circular, what would it become: (*a*) if the earth's mass were suddenly doubled? (*b*) if it were reduced to one half?

(8) In Ex. (5), determine the effect on the major semi-axis (or "mean distance") a and on the periodic time T, of a *small* change in the intensity μ of the force.

(9) If the mass M of the sun be suddenly increased by M/n, n being very large, while the earth is at the end of the minor axis of its orbit, what would be the effect on the earth's mean distance and on the period of revolution T?

(10) Find the equation of the hodograph of planetary motion, derive from it the expression for the velocity in terms of the radius vector, and show that the velocity is a maximum in perihelion and a minimum in aphelion.

(11) Show that the greatest velocity of a planet in its orbit about the sun is to its least velocity as $1 + e$ is to $1 - e$; and find this ratio for the earth, whose orbit has the eccentricity $e = 0.016\ 771\ 2$.

(12) Find the time exactly as a function of θ, for a parabolic orbit.

(13) The latus rectum passing through the sun divides the earth's orbit into two different parts; in what time are these described if the whole time is $365\frac{1}{4}$ days?

(14) Show that the path of a projectile *in vacuo* is an ellipse, parabola, or hyperbola, according as $v_0 \lesseqqgtr 33{,}000$ ft. per second ($= 7$ miles per second, nearly). One of the foci lies at the center of the earth, and the ordinary assumption that the path is parabolic means that this center can be regarded as infinitely distant. Show also that the path becomes circular for $v_0 = 5$ miles per second, nearly.

563. The Problem of Two Bodies. In the preceding discussion of the motion of a particle under the action of a central force, it has been assumed that the center of force is fixed. In the applications of the theory of central forces this assumption is in general not satisfied. Thus, in considering the motion of a planet around the sun, the force of attraction is, according to Newton's law of universal gravitation (Art. 546), regarded as due to the presence of a mass M at the center (sun), and of a mass m at the attracted point (planet); and the action between these two masses is a mutual action, being of the nature of a *stress, i.e.*, consisting of two equal and opposite forces, each equal to

$$F = \kappa \frac{mM}{r^2}.$$

Hence, the mass m of the planet attracts the mass M of the sun with precisely the same force with which the mass M of the sun attracts the mass m of the planet. The attraction affects, therefore, the motions of both bodies.

564. The *accelerations* produced by the two forces are, of course, not equal. Indeed, the acceleration $F/m = \kappa M/r^2$, produced in the planet by the sun, is very much greater than the acceleration $F/M = \kappa m/r^2$, produced by the planet in the sun; for the mass of even the largest planet (Jupiter) is less than one thousandth of that of the sun. The assumption of a fixed center can therefore be regarded as a first approximation in the problem of the motion of a planet about the sun.

In the case of the earth and moon, the difference of the masses is not so great, the mass of the moon being nearly one eightieth of that of the earth.

It can be shown, however, that the results deduced on the assumption of a fixed center can, by a simple modification, be made available for the solution of *the general problem of the motions of two particles of masses m, M, subject to no forces besides their mutual attraction.* In astronomy, this is called the **problem of two bodies.** In the solution below we assume the attraction to follow Newton's law of the inverse square of the distance. It will be convenient to speak of the two particles, or bodies, as planet (m) and sun (M).

565. With regard to any fixed system of rectangular axes, let x, y, z be the co-ordinates of the planet (m), at the time t; x', y', z' those of the sun (M), at the same time; so that for their distance r we have

$$r^2 = (x - x')^2 + (y - y')^2 + (z - z')^2.$$

Then the equations of motion of the planet are

$$m\frac{d^2x}{dt^2} = \kappa\frac{Mm}{r^2} \cdot \frac{x' - x}{r},$$

$$m\frac{d^2y}{dt^2} = \kappa\frac{Mm}{r^2} \cdot \frac{y' - y}{r}, \qquad (1)$$

$$m\frac{d^2z}{dt^2} = \kappa\frac{Mm}{r^2} \cdot \frac{z' - z}{r};$$

while the equations of motion of the sun are

$$M\frac{d^2x'}{dt^2} = \kappa\frac{mM}{r^2} \cdot \frac{x - x'}{r},$$

$$M\frac{d^2y'}{dt^2} = \kappa\frac{mM}{r^2} \cdot \frac{y - y'}{r}, \qquad (2)$$

$$M\frac{d^2z'}{dt^2} = \kappa\frac{mM}{r^2} \cdot \frac{z - z'}{r}.$$

566. By adding the corresponding equations of the two sets, we find

$$\frac{d^2}{dt^2}(mx + Mx') = 0, \quad \frac{d^2}{dt^2}(my + My') = 0, \quad \frac{d^2}{dt^2}(mz + Mz') = 0.$$

If it be remembered that the centroid of the two masses m, M has the co-ordinates

$$\bar{x} = \frac{mx + Mx'}{m + M}, \quad \bar{y} = \frac{my + My'}{m + M}, \quad \bar{z} = \frac{mz + Mz'}{m + M},$$

it appears that these equations can be written in the form

$$\frac{d^2\bar{x}}{dt^2} = 0, \quad \frac{d^2\bar{y}}{dt^2} = 0, \quad \frac{d^2\bar{z}}{dt^2} = 0;$$

in words: *the acceleration of the common centroid of planet and sun is zero;* i. e., *this centroid moves with constant velocity in a straight line.*

567. The integration of the equations (1) would give the absolute path of the planet. But the constants could not be determined, because the absolute initial position and velocity of the planet are, of course, not known. The same holds for the absolute path of the sun. All we can do is to determine the *relative* motion, and we proceed to find the motion of the planet relative to the sun.

Taking the sun's center as new origin for parallel axes, we have for the co-ordinates ξ, η, ζ of the planet in this new system,

$$\xi = x - x', \ \eta = y - y', \ \zeta = z - z'.$$

Now, dividing the equations (1) by m, the equations (2) by M, and subtracting the equations of set (2) from the corresponding equations of set (1), we find for the relative acceleration of the planet

$$\begin{aligned} \frac{d^2\xi}{dt^2} &= -\kappa\frac{M+m}{r^2}\cdot\frac{\xi}{r}, \\ \frac{d^2\eta}{dt^2} &= -\kappa\frac{M+m}{r^2}\cdot\frac{\eta}{r}, \\ \frac{d^2\zeta}{dt^2} &= -\kappa\frac{M+m}{r^2}\cdot\frac{\zeta}{r}. \end{aligned} \tag{3}$$

The form of these equations shows that *the relative motion of the planet with respect to the sun is the same as if the sun were fixed and contained the mass* $M + m$. Thus the problem is reduced to that of a fixed center, the only modification being that the mass of the center M should be increased by that of the attracted particle m.

568. This result can also be obtained by the following simple consideration. The *relative* motion of the planet with respect to the sun would obviously not be altered if geometrically equal accelerations were applied to both. Let us, therefore, subject each body to an additional acceleration equal and opposite to the actual acceleration of the sun (whose components are obtained by dividing the equations (2) by M). Then the sun will be reduced to equilibrium, while the resulting acceleration of the planet, which is its relative acceleration with respect to the sun, will evidently be the sum of the acceleration exerted on it by the sun and the acceleration exerted on the sun by the planet. This is just the result expressed by the equations (3).

569. It can here only be mentioned in passing that, while the problem of two bodies thus leads to equations that can easily be integrated, *the problem of three bodies* is one of exceeding difficulty, and has been solved only in a few very special cases. Much less has it been possible to integrate the $3n$ equations of the problem of n bodies.

570. According to the equations (3), the first and second laws of Kepler can be said to hold for the *relative* motion of a planet about the sun (or of a satellite about its primary). The third law of Kepler requires some modification, since the intensity of the center μ should not be κM, but $\kappa(M+m)$. We have, by (26), Art. 556,

$$\mu = \kappa(M+m) = 4\pi^2 \frac{a^3}{T^2};$$

in other words, the quotient a^3/T^2 is not independent of the mass m of the planet.

Thus, if m_1, m_2 be the masses of two planets, a_1, a_2 the major semi-axes of their orbits, and T_1, T_2 their periodic times, we have

$$\frac{a_1^3/T_1^2}{a_2^3/T_2^2} = \frac{M+m_1}{M+m_2} = \frac{1+m_1/M}{1+m_2/M}.$$

This quotient is approximately equal to 1 if M is very large in comparison with both m_1 and m_2; hence, for the orbits of the planets about the sun, Kepler's third law is very nearly true.

571. Exercises.

(1) Two particles of masses m_1, m_2, attract each other with a force which is any function of the distance r between them, say $F = m_1 m_2 f(r)$. Show that their common centroid moves uniformly in a straight line, and find the equations of this line.

(2) In Ex. (1), write out the equations for the relative motion of either particle with respect to the common centroid.

IV. *Constrained Motion.*

1. INTRODUCTION.

572. The motion of a free particle is fully determined if all the forces acting upon the particle, as well as the so-called initial conditions, are given. The motion of a particle may, however, depend not only on given forces, but on other conditions not directly expressed in terms of forces. The motion is then said to be *constrained* (comp. Art. 396).

To mention some examples: a heavy particle sliding down a smooth inclined plane is subject not only to the force of gravity, but also to the condition that it cannot pass through the plane; a railway train running on the rails, a piece of machinery sliding in a groove or between guides, can, for many purposes, be regarded as a particle constrained to a curve; the bob of a pendulum, a stone attached to a cord and swung around by the hand, may be regarded as constrained to a surface.

573. Sometimes these constraining conditions can be easily replaced by forces. Thus, in the first illustration above, the condition that the particle cannot pass through the inclined plane can be expressed by introducing the reaction of the plane, *i. e.*, a force acting on the particle at right angles to the plane, so as to prevent it from passing through the plane. Similarly, in the case of the stone attached to the cord, we may imagine the cord cut and its tension introduced so as to replace the condition by a force.

Whenever the constraints to which a particle is subjected can thus be expressed by means of forces, these forces can be combined with the other impressed forces, and then, of course, the equations of motion for a free particle can be applied. Thus, let X', Y', Z' be the components of the resultant of all the constraints; X, Y, Z those of the resultant of all the other impressed forces. Then the equations of motion are:

$$m\frac{d^2x}{dt^2}=X+X', \quad m\frac{d^2y}{dt^2}=Y+Y', \quad m\frac{d^2z}{dt^2}=Z+Z'. \qquad (1)$$

It must, however, be noticed that the reactions representing the constraints, such as the tension of the string in the example referred to, are generally not given beforehand.

574. On the other hand, the constraints are often expressed more conveniently by conditional equations. Thus, if the motion of a particle be restricted to a surface, the equation of this surface, say

$$\phi(x, y, z) = 0, \tag{2}$$

may be given as a constraining condition to be fulfilled by the co-ordinates of the moving particle.

As a particle has but three degrees of freedom, it can be subjected to only one or two conditions of the form (2). One such condition confines it to a surface; two to the curve of intersection of the two surfaces represented by the two conditional equations; three conditions would evidently prevent it entirely from moving.

575. The curve or surface to which a particle is constrained may vary its position and even its shape in the course of time. In this case the conditional equations, referred to fixed axes, will contain not only the co-ordinates, but also the time. That is, they will be of the more general form

$$\phi(x, y, z, t) = 0. \tag{3}$$

576. To constrain a particle *completely* to a surface, we may imagine it confined between two very near impenetrable surfaces. The complete constraint to a curve may be imagined as due to a narrow tube having the shape of the curve, or by regarding the particle as a bead sliding along a wire.

In many cases, however, the constraint is not complete, but only partial, or one-sided. Thus, the rails compelling the train to move in a definite curve do not prevent its being lifted vertically out of this curve, nor does the cord that confines the motion of the stone to a sphere prevent it from moving towards the inside of the spherical surface.

While complete constraints are generally expressed by equations, one-sided constraints should be properly expressed by inequalities. Thus, in the case of the stone, the condition is really that its distance r from the hand is not greater than the length l of the cord, *i. e.*,

$$r \leqq l;$$

but as soon as r becomes less than l, the constraining action ceases, and the stone becomes free. It is, therefore, in general sufficient to consider conditional *equations;* but the nature of the constraint, whether complete or partial, must be taken into account to determine when and where the constraint ceases to exist.

We now proceed to consider briefly the motion of a particle constrained to a fixed curve and that of a particle constrained to a fixed surface.

2. MOTION ON A FIXED CURVE.

577. The condition that a particle should move on a given fixed curve can always be replaced by introducing a single additional force F' called the *constraining force*, or the *constraint.*

Consider, for instance, a particle of mass m, subject to the force of gravity $F = mg$ alone; in general it will describe a parabola whose equation can be found if the initial conditions are known. To compel the particle to describe some other curve, say a vertical circle, a constraining force F' (Fig. 166) must be introduced such that the resultant R of F and F' shall produce the acceleration required for motion in the circle. Thus, for instance, for *uniform* motion in the circle the resulting acceleration must be directed towards the center and must be $= \omega^2 a$, if a is the radius and ω the constant angular velocity. We have, therefore, in this case $R = m\omega^2 a$ along the radius, $F = mg$ vertically downward; and hence, denoting by θ the angle made by the radius CP with the vertical (Fig. 166),

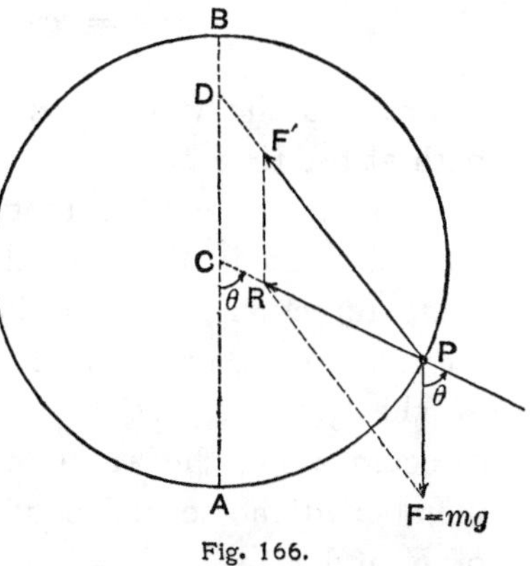

Fig. 166.

$$F'^2 = F^2 + R^2 + 2FR\cos\theta$$
$$= m^2(g^2 + \omega^4 a^2 + 2g\omega^2 a\cos\theta).$$

The constraint F', which is thus seen to vary with the angle θ, can be resolved into a tangential component F_t' and a normal component F_n'. As in our problem the velocity is to remain of constant magnitude, the tangential constraint must just counterbalance the tangential component of gravity $F_t = mg \sin\theta$. The normal constraint F_n' not only counterbalances the normal component of gravity $F_n = mg\cos\theta$, but also furnishes the centripetal force $R = m\omega^2 a$ required for motion in the circle; *i.e.*,

$$F_n' = R + F\cos\theta = m(\omega^2 a + g\cos\theta).$$

578. In the general case of a particle of mass m acted upon by any given forces and constrained to any fixed curve, it is convenient to resolve both the resultant F of the given forces and the constraint F' along the tangent and the normal plane. The equations of motion (see Art. 500) can then be written in the form

$$m\frac{dv}{dt} = F_t - F_t',$$

$$m\frac{v^2}{\rho} = \text{resultant of } F_n \text{ and } F_n',$$

where v is the velocity and ρ the radius of curvature of the path at the time t.

It should be noticed that the components F_n and F_n', though both situated in the normal plane, do not in general have the same direction. But in the important special case of plane motion, *i. e.*, when the path is a plane curve and the resultant F of the given forces lies in this plane, F_n and F_n' are both directed along the radius of curvature so that the right-hand member of the second equation becomes the sum or difference of F_n and F_n'.

579. The normal component F_n' of the constraining force is generally denoted by the letter N and is called the *resistance* or *reaction* of the curve; a force $-N$, equal and opposite to this reaction, represents the *pressure* exerted by the particle on the curve.

The tangential component F_t' of the constraint will exist only when the constraining curve is "rough," *i. e.*, offers frictional resistance; we have then, denoting the coefficient of friction by μ,

$$F_t' = \mu N.$$

We shall therefore write the equations of motion as follows:

$$m\frac{dv}{dt} = F_t - \mu N, \tag{1}$$

$$m\frac{v^2}{\rho} = \text{resultant of } F_n \text{ and } N. \tag{2}$$

580. The normal component, mv^2/ρ, of the effective force is sometimes called the *centripetal force*; it is directed along the principal normal of the path towards the center of curvature. A force equal and opposite to this centripetal force, *i. e.*, $= -mv^2/\rho$, is called **centrifugal force**. It should be noticed that this is a force exerted not *on* the moving particle, but *by* it.

It appears from equation (2) that the normal reaction N is the resultant of the centripetal force mv^2/ρ and the reversed normal component of the given forces, $-F_n$. Changing all the signs, we can express the same thing by saying that *the pressure on the curve*, $-N$, *is the resultant of the centrifugal force*, $-mv^2/\rho$, *and of the normal component* F_n *of the given forces.*

If, in particular, this normal component F_n is zero, the pressure on the curve is equal to the centrifugal force. This case is of frequent occurrence. Thus if a small stone attached to a cord be whirled around rapidly, the action of gravity on the stone can often be neglected in comparison with the centripetal force due to rotation; hence the centrifugal force measures approximately the tension of the cord, and may cause it to break. Again, when a railway train runs in a curve, the centrifugal force produces the horizontal pressure on the rails, which tends to displace and deform the rails.

581. It may happen that at a certain time t the pressure $-N$ vanishes. If the constraint be complete (Art. 576), this would merely indicate that the pressure in passing through zero inverts its sense. If, however, the constraint be one-sided, the consequence will be that the particle at this time leaves the constraining curve; for at the next moment the pressure will be exerted in a direction in which the particle is free to move.

Now N vanishes when its components $-F_n$ and mv^2/ρ become equal and opposite. The conditions under which the particle would leave the curve are, therefore, that the resultant F of the given forces should lie in the osculating plane of the path, and that $F_n = mv^2/\rho$.

582. To obtain the equations of motion expressed in rectangular cartesian co-ordinates, let X, Y, Z be the components of the resultant F of the given forces, and N_x, N_y, N_z those of the normal reaction N of the curve. If there be friction, the frictional resistance μN, being directed along the tangent to the path opposite to the sense of the motion, has the direction cosines $-dx/ds$, $-dy/ds$, $-dz/ds$, so that the components of the force of friction are $-\mu N dx/ds$, $-\mu N dy/ds$, $-\mu N dz/ds$. The general equations of motion are, therefore,

$$m\frac{d^2x}{dt^2} = X + N_x - \mu N\frac{dx}{ds},$$
$$m\frac{d^2y}{dt^2} = Y + N_y - \mu N\frac{dy}{ds}, \qquad (3)$$
$$m\frac{d^2z}{dt^2} = Z + N_z - \mu N\frac{dz}{ds}.$$

If the acceleration of the particle be zero, the left-hand members are all $= 0$, and the equations reduce to the conditions of equilibrium of a particle on a fixed curve.

In addition to the equations (3) we have of course the equations of the curve, say

$$\phi(x, y, z) = 0, \quad \psi(x, y, z) = 0, \qquad (4)$$

and the relations $N^2 = N_x^2 + N_y^2 + N_z^2, \qquad (5)$

$$N_x\frac{dx}{ds} + N_y\frac{dy}{ds} + N_z\frac{dz}{ds} = 0, \qquad (6)$$

the latter expressing that N is normal to the path.

583. Multiplying the equations (3) by dx, dy, dz, and adding, we find the equation of kinetic energy

$$d(\tfrac{1}{2}mv^2) = Xdx + Ydy + Zdz - \mu N ds. \qquad (7)$$

This relation might have been obtained directly from the consideration that for a displacement ds along the fixed curve the normal reaction N does no work, while the work of friction is $-\mu N ds$.

If there is no friction along the curve ($\mu = 0$), it follows from (7) that the velocity is independent of the reaction of the curve.

584. Exercises.

(1) Show that when the given forces are zero and there is no friction, the particle moves uniformly on the curve, and the pressure on the curve is proportional to the curvature of the path.

(2) A particle of mass m moves down a straight line inclined to the horizon at an angle θ, under the action of gravity alone.

(*a*) If there be no friction, we have by Art. 579, since $\rho = \infty$ (see Fig. 167):

$$m\frac{dv}{dt} = mg \sin \theta,$$

$$0 = mg \cos \theta - N.$$

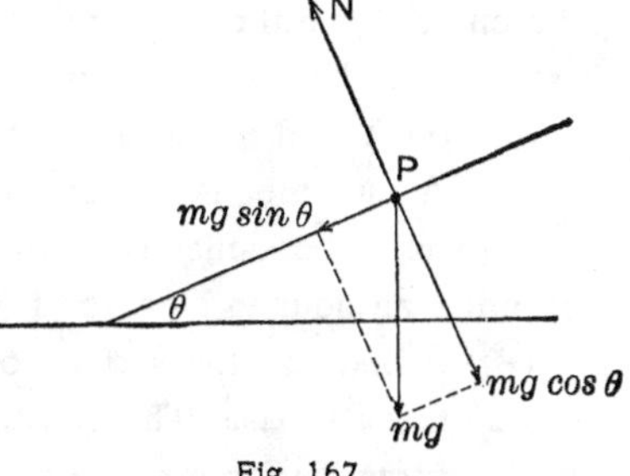

Fig. 167.

The first of these equations is the same as that derived in Kinematics for motion down an inclined plane (Arts. 116, 117). The second equation gives the normal reaction of the line $N = mg \cos \theta$; hence, the pressure on the line, $-N$, is constant.

(*b*) If the line be rough, the second equation remains the same, while the first must be replaced by the following:

$$m\frac{dv}{dt} = mg \sin \theta - \mu N = mg(\sin \theta - \mu \cos \theta).$$

As the acceleration is constant whether there be friction or not, the motion is uniformly accelerated, unless $\sin \theta - \mu \cos \theta = 0$, *i.e.*, $\mu = \tan \theta$.

Find v and s; show that in the exceptional case $\mu = \tan \theta$, the motion is uniform unless the initial velocity be zero; show that, for motion *up* the line, the first equation becomes $dv/dt = -g(\sin\theta + \mu\cos\theta)$, the motion being uniformly retarded until $t = v_0/g(\sin\theta + \mu\cos\theta)$, when the particle either begins to move down the line or remains at rest.

(3) A cord of length l (ft.) carries at one end a mass of m lbs., while the other end is fixed at a point O on a smooth horizontal table. A blow, at right angles to the cord, imparts to m an initial velocity v_0 ft. per second. Show that, as m describes the circle of radius l about O on the table, the velocity remains constant and the tension of the cord is $= mv_0^2/gl$ lbs.

(4) In Ex. (3), let $m = 2$ lbs.; $l = 3$ ft.; find the tension in pounds: (*a*) when the mass makes one revolution per second; (*b*) when it makes 8 revolutions per second. (*c*) If the cord cannot stand a tension of more than 300 lbs., what is the greatest allowable number of revolutions?

(5) A locomotive weighing 32 tons moves in a curve of 800 ft. radius with a velocity of 30 miles an hour; find the horizontal pressure on the rails.

(6) To prevent the lateral pressure on the rails in a curve, the track is inclined inwards. Determine the required elevation e (in inches) of the outer above the inner rail for a given velocity v and radius R if the gauge (*i. e.*, the distance between the rails) is 4 ft. 8 in. Show that with $R = 2000$ ft. and $v = 45$ miles per hour, $e = 3.8$ in.

(7) A plummet is suspended from the roof of a railroad car; how much will it be deflected from the vertical when the train is running 45 miles an hour in a curve of 300 yards radius?

(8) A body on the surface of the earth partakes of the earth's daily rotation on its axis. The constraint holding it in its circular path is due to the attractive force of the earth. Taking the earth's equatorial radius as 3963 miles, show that the centripetal acceleration of a particle at the equator is about $\frac{1}{9}$ ft. per second, or about $\frac{1}{290}$ of the actually observed acceleration $g = 32.09$ of a body falling *in vacuo*.

(9) If the earth were at rest, what would be the acceleration of a body falling *in vacuo* at the equator?

(10) Show that if the velocity of the earth's rotation were over 17 times as large as it actually is, the force of gravity would not be sufficient to detain a body near the surface at the equator (comp. Ex. (14), Art. 562).

(11) Show that in latitude ϕ the acceleration of a falling body, if the earth were at rest, would be $g_1 = g + j \cos^2 \phi$, where g is the observed acceleration of a falling body on the rotating earth and j the centripetal acceleration at the equator. Thus, in latitude $\phi = 45°$, $g = 980.6$ cm.; hence $g_1 = 982.3$.

(12) A chandelier weighing 80 lbs. is suspended from the ceiling of a hall by means of a chain 12 ft. long whose weight is neglected. By how much is the tension of the chain increased if it be set swinging so that the velocity at the lowest point is 6 ft. per second?

(13) A cord of 2 ft. length passes at its middle point through a hole in a smooth horizontal table. It carries at its lower end a mass of 2 lbs., at its other end a mass of 1 lb. The latter is set to revolve in a circle about the hole so as to stretch the cord and just prevent the mass of 2 lbs. from descending. (*a*) How many revolutions must it make? (*b*) If only one fourth of the cord lie on the table while three fourths hang down, how many revolutions must be made?

(14) Show that, when a particle moves with constant velocity in a vertical circle, the constraining force F' (Art. 577) is always directed towards a fixed point on the vertical diameter.

(15) A bicyclist is rounding a curve of 50 ft. radius at the rate of 10 miles an hour; find his inclination to the vertical.

(16) Owing to the earth's rotation on its axis the direction of a plumb-line does not pass through the center of the earth, even when the earth, as here assumed, is regarded as a homogeneous sphere. Express the angle δ of the deviation in latitude ϕ, and determine in what latitude δ is greatest.

(17) A cord, 4 ft. long, whose breaking strength is 24 lbs., is swung in a circle with a 3-lb. mass attached at its end. Find the greatest number of revolutions allowable: (*a*) when the circle is horizontal; (*b*) when it is vertical.

(18) At what speed would a locomotive whose centroid is 6 ft. above the track be upset in a curve of 400 ft. radius, the gauge being 4 ft. $8\frac{1}{4}$ in., if the rails are on a level?

585. *A particle of mass m subject to gravity alone is constrained to move in a vertical circle of radius l.* If there be no friction on the curve and the constraint be produced by a weightless rod or cord joining the particle to the center of the circle, we have the problem of the **simple mathematical pendulum.**

Equation (1), Art. 579, is readily seen to reduce in this case (see Fig. 168) to the form

$$l\frac{d^2\theta}{dt^2}+g\sin\theta=0. \qquad (8)$$

A first integration gives, as shown in Kinematics (Arts. 149, 150),

$$\tfrac{1}{2}v^2=g\left(l\cos\theta+\frac{v_0^{\,2}}{2g}-l\cos\theta_0\right), \qquad (9)$$

where v_0 is the velocity which the particle has at the time $t=0$ when its radius makes the angle $AOP_0=\theta_0$ with the vertical. Multiplying by m, we have, for the kinetic energy of the particle,

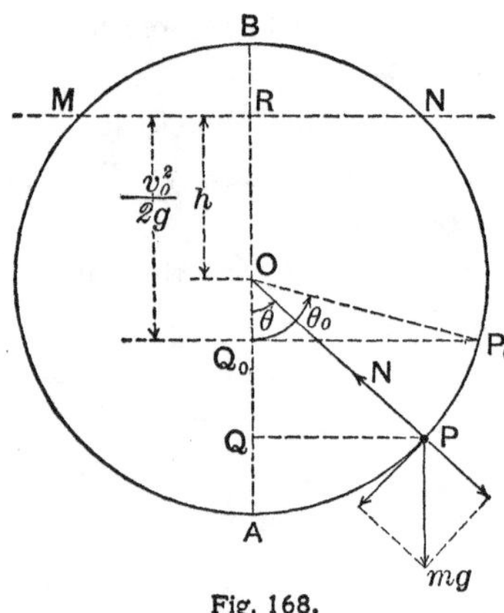

Fig. 168.

$$\tfrac{1}{2}mv^2=mg(l\cos\theta+h), \qquad (10)$$

where $h=v_0^2/2g-l\cos\theta_0$ is a constant. If the horizontal line MN, drawn at the height $v_0^2/2g$ above the initial point P_0, intersect the vertical diameter AB at R, it appears from the figure that $h=RO$.

586. Taking R as origin and the axis of z vertically downwards, we have $RQ=z=l\cos\theta+h$; hence the force-function U has the simple expression

$$U=mgz;$$

and the velocity $v=\sqrt{2gz}$ is seen to become zero when the particle reaches the horizontal line MN.

For the further treatment of the problem, three cases must be distinguished according as this line of zero-velocity MN

intersects the circle, touches it, or does not meet it at all; *i. e.*, according as

$$h \lesseqgtr l, \text{ or } \frac{v_0^2}{2g} \lesseqgtr 2l\cos^2 \tfrac{1}{2}\theta_0. \tag{11}$$

587. Equation (2), Art. 579, serves to determine the reaction N of the circle, or the pressure $-N$ on the circle. We have

$$m\frac{v^2}{l} = -mg\cos\theta + N,$$

whence

$$N = m\left(\frac{v^2}{l} + g\cos\theta\right).$$

Substituting for v^2 its value from (10), we find

$$N = mg\left(2\frac{h}{l} + 3\cos\theta\right). \tag{12}$$

The pressure on the curve has therefore its greatest value when $\theta = 0$, *i. e.*, at the lowest point A. It becomes zero for $l\cos\theta_1 = -\frac{2}{3}h$, which is easily constructed.

588. If the constraint be complete as for a bead sliding along a circular wire, or a small ball moving within a tube, the pressure merely changes sign at the point $\theta = \theta_1$. But if the constraint be one-sided, the particle may at this point leave the circle. The one-sided constraint may be such that $OP \leqq l$, as when the particle runs in a groove cut on the inside of a ring, or when it is joined to the center by a cord; in this case the particle may leave the circle at some point of its upper half. Again, the one-sided constraint may be such that $OP \geqq l$, as when the particle runs in a groove cut on the rim of a disk; in this case the particle can of course only move on the upper half of the circle.

589. Exercises.

(1) For $h = l$, equation (10) can be integrated in finite terms. Show that in this limiting case the particle approaches the highest point B of the circle asymptotically, reaching it only in an infinite time.

(2) Derive the equations of motion for the problem of the simple pendulum (Art. 585) from the general equations of Arts. 582, 583.

(3) For $\theta_0 = 60°$, $l = 1$ ft., $v_0 = 9$ ft. per second, show that the particle will leave the circle very nearly at the point $\theta_1 = 120°$, if the constraint be such that $OP \leqq l$ (Art. 588).

(4) For $v_0 = 10$ ft. per second, everything else being as in Ex. (3), show that the particle will leave the circle at the point $\theta_1 = 134\frac{1}{2}°$, nearly.

(5) A particle, subject to gravity and constrained to the inside of a vertical circle ($OP \leqq l$), makes complete revolutions. Show that it cannot leave the circle at any point, if $\frac{2}{3}h > l$; and that it will leave the circle at the point for which $\cos\theta = -\frac{2}{3}h/l$, if $\frac{2}{3}h < l$.

(6) In the experiment of swinging in a vertical circle a glass containing water, and suspended by means of a cord, if the cord be 2 ft. long, what must be the velocity at the lowest point if the experiment is to succeed?

(7) A particle subject to gravity moves on the outside of a vertical circle; determine where it will leave the circle: (*a*) if MN (Fig. 168) intersects the circle; (*b*) if MN touches the circle; (*c*) if MN does not meet the circle.

(8) A particle subject to gravity is compelled to move on any vertical curve $z = f(x)$ without friction. Show that the velocity at any point is $v = \sqrt{2gz}$ (comp. Art. 586) if the horizontal axis of x be taken at a height above the initial point equal to the "height due to the initial velocity," *i. e.*, $v_0^2/2g$.

(9) A particle slides on the outside of a smooth vertical circle, starting from rest at the highest point of the circle. Find where it will meet the horizontal plane through the lowest point of the circle.

(10) A heavy particle is constrained (without friction) to a common cycloid in a vertical plane, the axis of the cycloid being vertical and the concavity turned upward. Counting the arc s of the cycloid from the vertex, the equation of motion is $d^2s/dt^2 = -gs/4a$. If $v = 0$ when $s = s_0$, this gives $s = s_0 \cos\sqrt{g/4a} \cdot t$. Show that the motion is *isochronous*, *i. e.*, that the time t_1 of reaching the lowest point is independent of s_0.

(11) The involute of a cycloid being an equal cycloid, with its vertex at the cusp, its cusp on the axis, of the original cycloid, the particle in Ex. (10) can be constrained to the cycloid by means of a cord of

length $2a$, attached to the cusp of the involute, and wrapping itself on a cylinder erected on the involute as base. Show that, if the particle starts from rest at the cusp of the original cycloid, the tension of the cord is twice the normal component of the weight of the particle.

3. MOTION ON A FIXED SURFACE.

590. Just as for motion on a curve (Art. 582), we find the general equations of motion

$$m\frac{d^2x}{dt^2} = X + N_x - \mu N\frac{dx}{ds},$$

$$m\frac{d^2y}{dt^2} = Y + N_y - \mu N\frac{dy}{ds}, \qquad (1)$$

$$m\frac{d^2z}{dt^2} = Z + N_z - \mu N\frac{dz}{ds}.$$

The normal reaction

$$N = \sqrt{N_x^2 + N_y^2 + N_z^2} \qquad (2)$$

being at right angles to the constraining surface

$$\phi(x, y, z) = 0, \qquad (3)$$

the following conditions must be satisfied:

$$\frac{N_x}{\phi_x} = \frac{N_y}{\phi_y} = \frac{N_z}{\phi_z}, \qquad (4)$$

where ϕ_x, ϕ_y, ϕ_z, denote, as usual, the partial derivatives of $\phi(x, y, z)$ with regard to x, y, z, respectively.

If the acceleration of the particle be zero, the equations (1) reduce to the conditions of equilibrium of a particle on a surface.

591. *A particle of mass m, subject to gravity, is constrained to remain on the surface of a sphere of radius r.* If the constraint is produced by a weightless rod or cord joining the particle to the center of the sphere, the rod or cord describes a cone, and the apparatus is called a **conical** or **spherical pendulum.**

Taking the center O of the sphere as origin (Fig. 169), and the axis of z vertically downwards, we have for the equation of the sphere

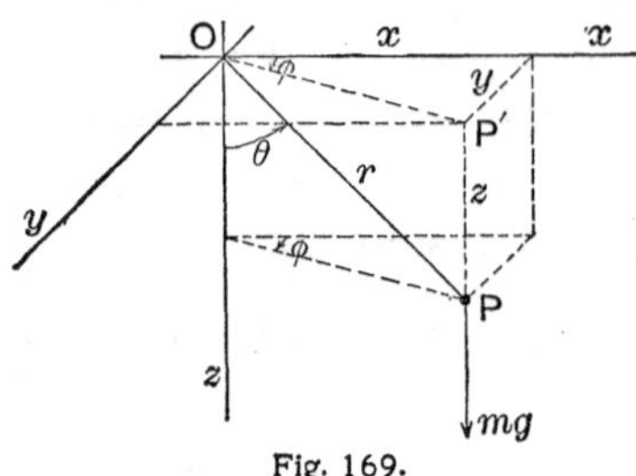

Fig. 169.

$$x^2+y^2+z^2-r^2=0, \qquad (5)$$

whence $\phi_x/x=\phi_y/y=\phi_z/z$. The direction cosines of N are $-x/r$, $-y/r$, $-z/r$. Hence, the equations of motion, as there is no friction:

$$m\frac{d^2x}{dt^2}=-N\frac{x}{r}, \quad m\frac{d^2y}{dt^2}=-N\frac{y}{r}, \quad m\frac{d^2z}{dt^2}=mg-N\frac{z}{r}. \qquad (6)$$

As the resistance N does no work, the principle of kinetic energy gives

$$\tfrac{1}{2}mv^2=mgz+C,$$

or, dividing by $\frac{1}{2}m$,

$$v^2=2(gz+h). \qquad (7)$$

To determine the constant of integration h, we have $v=v_0$ when $z=z_0$; hence

$$v^2=v_0^2+2g(z-z_0). \qquad (8)$$

592. That particular case of the problem of the conical pendulum in which the particle moves in a horizontal circle can be treated directly in an elementary manner. It finds its application in the theory of the *governor of a steam engine.*

Let OQ (Fig. 170) represent a vertical shaft turning about its axis with angular velocity ω; $OP=l$ a rigid arm hinged to the shaft at O so as to turn freely in the vertical plane POQ. This arm, whose weight is neglected, carries at P a heavy mass m. The forces acting on m are its weight mg and the tension N of the arm PO; the resultant R of these forces lies in the vertical plane.

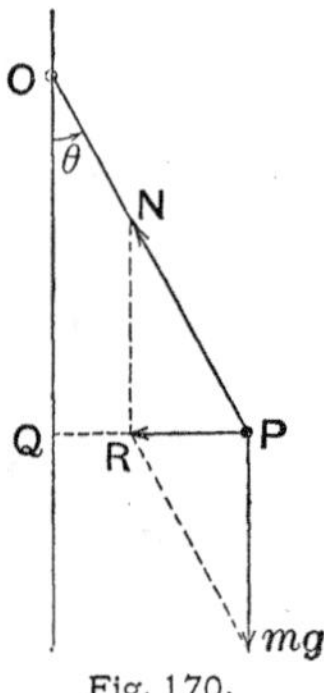

Fig. 170.

If the point P is to describe a horizontal circle about the axis OQ, the resultant R must be perpendicular to OQ, and the angular velocity ω and the angle $QOP=\theta$ must be such that

$$R=m\omega^2\cdot QP=m\omega^2 l\sin\theta.$$

The figure shows that this requires

$$\frac{R}{mg} = \tan\theta, \quad N\cos\theta = mg,$$

hence $$\omega^2 = \frac{g}{l\cos\theta} = \frac{g}{h}, \quad N = \frac{mg}{\cos\theta} = mg\cdot\frac{l}{h} = m\omega^2 l, \qquad (9)$$

where $h = QO$.

If ω is constant, the time T of one revolution is

$$T = \frac{2\pi}{\omega} = 2\pi\sqrt{\frac{h}{g}}, \qquad (10)$$

i. e., the same as the period of a simple pendulum of length $h = l\cos\theta$. If θ is small, *i. e.*, if QP is small in comparison with l and h, a slight change in θ will affect the period T but slightly; in other words, the revolutions of a long conical pendulum with small amplitude are nearly isochronous.

By (9), we have $\omega^2 = g/h$; as $h \leq l$, the angular velocity of the vertical shaft must be $> \sqrt{g/l}$ to make the bob fly out; for smaller velocities the pendulum would hang down vertically.

593. Exercises.

(1) A conical pendulum makes 60 revolutions per minute. (*a*) What is the height h of the cone? (*b*) If the bob weighs 1 lb., and the length of the arm is 1 ft., what is the tension of the arm? (*c*) What is the angle θ? Answer the same questions when the number of revolutions per minute is 180.

(2) A straight rod AB is fixed at an angle α to a vertical axis AC, the angle α opening upward. A bead of mass m slides without friction along AB. When AB turns about AC with angular velocity ω, what is the position of equilibrium of m?

(3) If the bead in Ex. (2) be prevented from sliding along AB by a projecting ring, find the pressure on this ring: (*a*) when α opens upward; (*b*) when α opens downward.

(4) If in Ex. (2) the straight rod AB be replaced by an arc of a circle, with its center on AC, we have a kind of centrifugal governor. In practice, however, the mass m is generally not made to slide along a material circular arc, but is constrained to a circle by being suspended from a point C on the axis by means of a rigid arm. For the sake of symmetry, two equal arms and masses are generally used.

(5) Determine the curve that should be substituted for the circular arc in Ex. (4) if the mass m is to be in equilibrium, for a given angular velocity ω, at every point of the arc.

(6) From the equations (5) and (6), Art. 591, derive the approximate path of the bob of a conical pendulum when the angle θ remains very small.

CHAPTER VI.

KINETICS OF THE RIGID BODY.

I. *General Principles.*

594. In kinetics the term *rigid body* means any system or aggregate of mass-particles whose mutual distances remain invariable. A rigid body may therefore consist of a finite number of rigidly connected particles or of a continuous mass of one, two, or three dimensions. Its motion depends not only on the forces acting on the body, but also on the way in which the mass is distributed throughout the body.

In the present section the rigid body is assumed to be free unless the contrary be stated explicitly. For the sake of simplicity the body is conceived as a rigidly connected system of a *finite* number of particles.

595. Let us consider any one particle m of the body; at any time t, let j be its acceleration and F the resultant of all the forces acting on the particle. Then the motion of this particle (see Art. 500) is determined by the equation

$$mj = F. \tag{1}$$

It should be noticed that among the forces acting on the particle are included not only those external forces acting on the rigid body that happen to be applied at m, but also the so-called *internal forces* which would replace the rigid connection of the particle m with the rest of the body.

596. If, at the time t, x, y, z are the co-ordinates of the particle m with respect to a fixed set of rectangular axes, then the components of its velocity v may be denoted by $\dot{x}, \dot{y}, \dot{z}$; those of its acceleration j by $\ddot{x}, \ddot{y}, \ddot{z}$.* And if the components of F along the same axes are X, Y, Z, the equation (1) can be replaced by the following three:

$$-m\ddot{x}+X=0, \quad -m\ddot{y}+Y=0, \quad -m\ddot{z}+Z=0. \qquad (2)$$

Such a set of three equations can be written down for each particle; hence, if the body consist of n particles, there would be in all $3n$ equations.

597. For the solution of particular problems these $3n$ equations are of little use, not only because their number would in general be very great, but mainly because the forces X, Y, Z include the unknown reactions between the particles. It is, however, possible to deduce certain general propositions from these equations.

The $3n$ equations express the equilibrium of the system formed by *all* the forces, both internal and external, acting on the particles, and the reversed effective forces. To apply the principle of virtual work to this system, let us multiply the three equations (2) by the components $\delta x, \delta y, \delta z$ of some virtual displacement of the particle m; let the same thing be done for every other particle of the body; and let all the resulting equations be added:

$$\Sigma(-m\ddot{x}+X)\delta x+\Sigma(-m\ddot{y}+Y)\delta y+\Sigma(-m\ddot{z}+Z)\delta z=0. \qquad (3)$$

598. It is important to notice that the internal reactions between the particles which make the body rigid occur in pairs of equal and opposite forces, and form, therefore, a system which is in equilibrium by itself. This may be regarded as an assumption which should be included in the definition of the rigid body. Hence, while these internal forces enter into

---

* This is the so-called *fluxional notation*, according to which derivatives with respect to the time are denoted by dots; thus $\dot{x}$ stands for dx/dt, $\ddot{x}$ for d^2x/dt^2, etc. Derivatives were called *fluxes* by Newton; thus the component of the acceleration of a point in any direction is the time-flux of its velocity in that direction; the component of its effective force in any direction is the time-flux of its momentum.

the equations (2), they do not appear in equation (3), since the equal and opposite forces cancel in the summation. Thus, equation (3) expresses that *the external forces acting on the rigid body and the reversed effective forces form a system in equilibrium* ; and this is **d'Alembert's Principle** for the rigid body.

It must, however, not be forgotten that the displacements δx, δy, δz should be so selected as to be compatible with the nature of the rigid body ; *i. e.*, with the conditions that the distances between the particles should not be disturbed.

599. The number of conditions expressing the invariability of the distances between n particles is $3n-6$. For if there were but 3 particles, the number of independent conditions would evidently be 3 ; for every additional particle, 3 additional conditions are required. Hence, the total number of conditions is $3+3(n-3)=3n-6$.

It follows that if a rigid body be subject to no other constraining conditions, the number of its equations of motion must be $3n-(3n-6)=6$. Hence, *a free rigid body has six independent equations of motion* (comp. Art. 31).

600. The six equations of motion of the rigid body can be obtained as follows.

Imagine the equations (2), viz.:

$$m\ddot{x}=X, \quad m\ddot{y}=Y, \quad m\ddot{z}=Z,$$

written down for every particle, and add the corresponding equations. This gives the first 3 of the 6 equations of motion:

$$\Sigma m\ddot{x}=\Sigma X, \quad \Sigma m\ddot{y}=\Sigma Y, \quad \Sigma m\ddot{z}=\Sigma Z. \qquad (4)$$

As the internal forces cancel in the summation, the right-hand members of these equations represent the components R_x, R_y, R_z of the resultant R of all the external forces acting on the body. The left-hand members can be written in the form $d(\Sigma m\dot{x})/dt$, $d(\Sigma m\dot{y})/dt$, $d(\Sigma m\dot{z})/dt$: these are the time derivatives or fluxes of the sums of the linear momenta of all the particles parallel to the axes. The equations (4) can therefore be written in the form

$$\frac{d}{dt}\Sigma m\dot{x} = R_x, \quad \frac{d}{dt}\Sigma m\dot{y} = R_y, \quad \frac{d}{dt}\Sigma m\dot{z} = R_z. \tag{5}$$

The axes of co-ordinates are arbitrary. Hence, if we agree to call *linear momentum of the body in any direction* the algebraic sum of the linear momenta of all the particles in that direction, the equations (5) express the proposition that *the rate at which the linear momentum of a rigid body in any direction changes with the time is equal to the sum of the components of all the external forces in that direction.*

601. Let us now combine the second and third of the equations (2) by multiplying the former by z, the latter by y, and subtracting the former from the latter. If this be done for each particle, and the resulting equations be added, we find $\Sigma m(y\ddot{z} - z\ddot{y}) = \Sigma(yZ - zY)$. Similarly, we can proceed with the third and first, and with the first and second of the equations (2). The result is:

$$\Sigma m(y\ddot{z} - z\ddot{y}) = \Sigma(yZ - zY), \quad \Sigma m(z\ddot{x} - x\ddot{z}) = \Sigma(zX - xZ),$$
$$\Sigma m(x\ddot{y} - y\ddot{x}) = \Sigma(xY - yX). \tag{6}$$

Here again the internal forces disappear in the summation, so that the right-hand members are the components H_x, H_y, H_z, of the vector H of the resultant couple, found by reducing all the external forces for the origin of co-ordinates. The left-hand members are the components of the resultant couple of the effective forces for the same origin.

We can also say that the right-hand members are the sums of the moments of the external forces about the co-ordinate axes (Art. 391), while the left-hand members represent the moments of the effective forces about the same axes. The latter quantities are exact derivatives, as shown in Art. 512. The equations (6) can therefore be written in the form

$$\frac{d}{dt}\Sigma m(y\dot{z} - z\dot{y}) = H_x, \quad \frac{d}{dt}\Sigma m(z\dot{x} - x\dot{z}) = H_y, \quad \frac{d}{dt}\Sigma m(x\dot{y} - y\dot{x}) = H_z. \tag{7}$$

As explained in Art. 514, the quantity $m(y\dot{z}-z\dot{y})$ is called the *angular momentum* (or the moment of momentum) of the particle m about the axis of x. We may now agree to call the quantity $\Sigma m(y\dot{z}-z\dot{y})$ the *angular momentum of the body* about the axis of x, just as $\Sigma m\dot{x}$ is the linear momentum of the body along this axis; and similarly for the other axes. The meaning of the equations (7) can then be stated as follows: *The rate at which the angular momentum of a rigid body about any axis changes with the time is equal to the sum of the moments of all the external forces about this line.*

The equations (4) and (6), or (5) and (7), are the **six equations of motion of the rigid body.** The three equations (4) or (5) may be called the *equations of linear momentum*, while (6) or (7) are the *equations of angular momentum.*

602. The equations (4) and (6) can also be derived from the equation (3), which expresses d'Alembert's principle, by selecting for δx, δy, δz convenient displacements.

Thus, the rigidity of the body will evidently not be disturbed if we give to all its points equal and parallel infinitesimal displacements, since this merely amounts to subjecting the whole body to an infinitesimal translation. Equation (3) can in this case be written

$$\delta x\Sigma(-m\ddot{x}+X)+\delta y\Sigma(-m\ddot{y}+Y)+\delta z\Sigma(-m\ddot{z}+Z)=0,$$

and is therefore equivalent to the three equations (4), since δx, δy, δz are independent and arbitrary.

Again, let the body be subjected to an infinitesimal rotation of angle $\delta\theta$ about any line l. As shown in Art. 185, the linear velocities of any point (x, y, z) of a rigid body, due to a rotation of angular velocity $\omega=\delta\theta/\delta t$ about any line l through the origin, are, if ω_x, ω_y, ω_z denote the components of ω:

$$\dot{x}=\omega_y z-\omega_z y,\quad \dot{y}=\omega_z x-\omega_x z,\quad \dot{z}=\omega_x y-\omega_y x.$$

Hence, putting $\omega_x\delta t=\delta\theta_x$, $\omega_y\delta t=\delta\theta_y$, $\omega_z\delta t=\delta\theta_z$, we have for the displacements of the point (x, y, z), due to a rotation of angle $\delta\theta$,

$$\delta x=z\,\delta\theta_y-y\,\delta\theta_z,\quad \delta y=x\,\delta\theta_z-z\,\delta\theta_x,\quad \delta z=y\,\delta\theta_x-x\,\delta\theta_y.$$

If these values be introduced in d'Alembert's equation (3) and the terms in $\delta\theta_x$, $\delta\theta_y$, $\delta\theta_z$ be collected, it assumes the form

$$\delta\theta_x\Sigma[-m(y\ddot{z}-z\ddot{y})+yZ-zY]+\delta\theta_y\Sigma[-m(z\ddot{x}-x\ddot{z})+zX-xZ]$$
$$+\delta\theta_z\Sigma[-m(x\ddot{y}-y\ddot{x})+xY-yX]=0;$$

as $\delta\theta_x$, $\delta\theta_y$, $\delta\theta_z$ are independent and arbitrary, their coefficients must vanish separately, and this gives the equations (6).

603. The equations of linear momentum, (4) or (5), admit of a further simplification, owing to the fundamental property of the centroid. By Art. 212, the co-ordinates $\bar{x}, \bar{y}, \bar{z}$ of the centroid satisfy the relations

$$M\bar{x}=\Sigma mx,\quad M\bar{y}=\Sigma my,\quad M\bar{z}=\Sigma mz,$$

where $M=\Sigma m$ is the whole mass of the body. Differentiating these equations, we find

$$M\dot{\bar{x}}=\Sigma m\dot{x},\quad M\dot{\bar{y}}=\Sigma m\dot{y},\quad M\dot{\bar{z}}=\Sigma m\dot{z},$$

and $$M\ddot{\bar{x}}=\Sigma m\ddot{x},\quad M\ddot{\bar{y}}=\Sigma m\ddot{y},\quad M\ddot{\bar{z}}=\Sigma m\ddot{z},$$

where $\dot{\bar{x}}, \dot{\bar{y}}, \dot{\bar{z}}$ are the components of the velocity $\bar{v}$, and $\ddot{\bar{x}}, \ddot{\bar{y}}, \ddot{\bar{z}}$ those of the acceleration $\bar{j}$, of the centroid.

The equations (4) or (5) can therefore be reduced to the form

$$M\ddot{\bar{x}}\equiv\frac{d}{dt}M\dot{\bar{x}}=R_x,\quad M\ddot{\bar{y}}\equiv\frac{d}{dt}M\dot{\bar{y}}=R_y,\quad M\ddot{\bar{z}}\equiv\frac{d}{dt}M\dot{\bar{z}}=R_z,\qquad(8)$$

whence $$M\bar{j}=\frac{d}{dt}M\bar{v}=R;\qquad(9)$$

i. e., if the whole mass of the body be regarded as concentrated at the centroid, the effective force of the centroid, or the time-rate of change of its momentum, is equal to the resultant of all the external forces. It follows that *the centroid of a rigid body moves as if it contained the whole mass, and all the external forces were applied at this point parallel to their original directions.*

604. If, in particular, the resultant R vanish (while there may be a couple H acting on the body), we have by (8) and

(9) $\bar{j} = 0$; hence $\bar{v} =$ const.; *i. e., if the resultant force be zero, the centroid moves uniformly in a straight line.*

This proposition, which can also be expressed by saying that, if $R = 0$, the momentum $M\bar{v}$ of the centroid remains constant, or, using the form (5) of the equations of motion, that the linear momentum of the body in any direction is constant, is known as the **principle of the conservation of linear momentum**, or the principle of the conservation of the motion of the centroid.

605. Let us next consider the equations of angular momentum, (6) or (7). To introduce the properties of the centroid, let us put $x - \bar{x} = \xi$, $y - \bar{y} = \eta$, $z - \bar{z} = \zeta$, so that ξ, η, ζ are the co-ordinates of the point (x, y, z) with respect to parallel axes through the centroid. The substitution of $x = \bar{x} + \xi$, $y = \bar{y} + \eta$, $z = \bar{z} + \zeta$ and their derivatives in the expression $y\dot{z} - z\dot{y}$ gives

$$y\dot{z} - z\dot{y} = \bar{y}\dot{\bar{z}} - \bar{z}\dot{\bar{y}} + \bar{y}\dot{\zeta} - \bar{z}\dot{\eta} + \eta\dot{\bar{z}} - \zeta\dot{\bar{y}} + \eta\dot{\zeta} - \zeta\dot{\eta}.$$

To form $\Sigma m(y\dot{z} - z\dot{y})$ we must multiply by m and sum throughout the body; in this summation, $\bar{y}$, $\bar{z}$, $\dot{\bar{y}}$, $\dot{\bar{z}}$ are constant and, by the property of the centroid, $\Sigma m\eta = 0$, $\Sigma m\zeta = 0$, $\Sigma m\dot{\eta} = 0$, $\Sigma m\dot{\zeta} = 0$. Hence we find

$$\Sigma m(y\dot{z} - z\dot{y}) = \Sigma m(\eta\dot{\zeta} - \zeta\dot{\eta}) + M(\bar{y}\dot{\bar{z}} - \bar{z}\dot{\bar{y}}).$$

The second term in the right-hand member is the angular momentum of the centroid about the axis of x (the whole mass M of the body being regarded as concentrated at this point), while the first term is the angular momentum of the body (in its motion relatively to the centroid) about a parallel to the axis of x, drawn through the centroid.

Similar relations hold for the angular momenta about the axes of y and z; and as these axes are arbitrary, we conclude that *the angular momentum of a rigid body about any line is equal to its angular momentum about a parallel through the centroid plus the angular momentum of the centroid about the former line.*

606. Differentiating the above expression, we find

$$\frac{d}{dt}\Sigma m(y\dot{z}-z\dot{y})=\frac{d}{dt}\Sigma m(\eta\dot{\zeta}-\zeta\dot{\eta})+M(\bar{y}\ddot{\bar{z}}-\bar{z}\ddot{\bar{y}}).$$

The first of the equations (7) can therefore be written

$$\frac{d}{dt}\Sigma m(\eta\dot{\zeta}-\zeta\dot{\eta})+M(\bar{y}\ddot{\bar{z}}-\bar{z}\ddot{\bar{y}})=H_x.$$

Now, if at any time t the centroid were taken as origin, so that $\bar{y}=0$, $\bar{z}=0$, this equation would reduce to the form

$$\frac{d}{dt}\Sigma m(\eta\dot{\zeta}-\zeta\dot{\eta})=H_x,$$

which is entirely independent of the co-ordinates of the centroid. On the other hand, wherever the origin is taken, if the centroid were a fixed point, the same equation would be obtained.

Similar considerations apply of course to the other two equations (7). It follows that *the motion of a rigid body relative to the centroid is the same as if the centroid were fixed.*

607. If, in particular, the resultant couple H be zero for any particular origin O (which will be the case not only when all external forces are zero, but whenever the directions of all the forces pass through the point O), the equations (7) can be integrated and give

$$\Sigma m(y\dot{z}-z\dot{y})=C_1,\ \Sigma m(z\dot{x}-x\dot{z})=C_2,\ \Sigma m(x\dot{y}-y\dot{x})=C_3, \quad (10)$$

where C_1, C_2, C_3 are constants of integration. Hence, *if the external forces pass through a fixed point, the angular momentum of the body about any line through this point is constant; if there are no external forces, the angular momentum is constant for any line whatever.* This is the **principle of the conservation of angular momentum.**

608. Another interpretation can be given to these equations. As shown in Arts. 97, 513, the quantities $y\dot{z}-z\dot{y}$, $z\dot{x}-x\dot{z}$,

$x\dot{y} - y\dot{x}$ can be regarded as sectorial velocities. Thus, if the radius vector, drawn from the origin to the particle m, be projected on the yz-plane, $y\dot{z} - z\dot{y}$ is twice the sectorial velocity of this radius vector in the yz-plane, $\frac{1}{2}(ydz - zdy)$ being the elementary sector described in the element of time dt. Let us denote by dS_x the sum of all these elementary sectors for the various particles, each multiplied by the mass of the particle; and similarly by dS_y, dS_z the corresponding sums of the projections on the other co-ordinate planes. Then the equations (10) can be written in the form

$$2\dot{S}_x = C_1,\ 2\dot{S}_y = C_2,\ 2\dot{S}_z = C_3. \tag{11}$$

Hence the proposition of Art. 607 might be called the principle of the conservation of sectorial velocities; it is more commonly called the **principle of the conservation of areas.**

The equations (11) can be integrated again and give, if the sectors be measured from the positions of the radii vectores at the time $t = 0$,

$$S_x = \tfrac{1}{2}C_1t,\ S_y = \tfrac{1}{2}C_2t,\ S_z = \tfrac{1}{2}C_3t.$$

609. If the radii vectores be projected on any plane through the origin whose normal has the direction cosines α, β, γ, the sum of the elementary sectors described in this plane, each multiplied by the mass, will be

$$dS = \tfrac{1}{2}(C_1\alpha + C_2\beta + C_3\gamma)dt;$$

hence
$$S = \tfrac{1}{2}(C_1\alpha + C_2\beta + C_3\gamma)t.$$

On the other hand, by (10), the angular momentum of the body about the normal of this plane has the expression $C_1\alpha + C_2\beta + C_3\gamma$, as it must be equal to the sum of the projections on this normal of the angular momenta about the axes of co-ordinates, which can be regarded as vectors laid off on these axes.

Now it is easy to see that this angular momentum $C_1\alpha + C_2\beta + C_3\gamma$, and hence the quantity S at a given time t, is greatest for the diagonal of the parallelepiped, whose edges are equal to C_1, C_2, C_3 along the axes, *i. e.*, for the normal to the plane

$$C_1x + C_2y + C_3z = 0. \tag{12}$$

For, the direction cosines of this normal are $\alpha' = C_1/D$, $\beta' = C_2/D$, $\gamma' = C_3/D$, where $D = \sqrt{C_1^2 + C_2^2 + C_3^2}$; and the quantity $C_1\alpha + C_2\beta + C_3\gamma$ can be written in the form

$$D\left(\frac{C_1}{D}\alpha + \frac{C_2}{D}\beta + \frac{C_3}{D}\gamma\right) = D(\alpha'\alpha + \beta'\beta + \gamma'\gamma),$$

where the quantity in parenthesis is the cosine of the angle between the directions (α', β', γ') and (α, β, γ), and is therefore greatest when these directions coincide.

The plane (12) about whose normal the angular momentum is greatest, and by projection on which the area S is made greatest, is called Laplace's **invariable plane.** As its equation is independent of t, it remains fixed. The normal of this plane is sometimes called the *invariable line or direction.*

610. Let us now return to the general case of the motion of a rigid body acted upon by any forces whatever.

The propositions of Arts. 603 and 606 together establish the so-called **principle of the independence of the motions of translation and rotation.** In studying the motion of a rigid body it is possible, according to this principle, to consider separately the motion of translation of the centroid, and the rotation of the body about the centroid.

By Art. 603, the motion of the centroid is the same as that of a particle of mass M acted upon by all the external forces transferred parallel to themselves to the centroid. As the motion of a particle has been discussed in Chapter V., nothing further need be said about this part of the problem.

By Art. 606, the motion of the body about the centroid is the same as if the centroid were fixed. The problem of the motion of a rigid body with a fixed point is therefore of great importance; it will, however, not be discussed in its generality in this elementary work. The more simple special case of a rigid body with a fixed axis is treated in Section III. The solution of both these problems depends on the equations (6) or (7).

611. In d'Alembert's equation (3) it is of course allowable to substitute for the virtual displacements δx, δy, δz the actual displacements Δx, Δy, Δz of the particles in any small motion of a free rigid body, since these actual displacements are certainly compatible with the condition of rigidity. The equation can then be written

$$\Sigma m(\ddot{x}\Delta x + \ddot{y}\Delta y + \ddot{z}\Delta z) = \Sigma(X\Delta x + Y\Delta y + Z\Delta z). \qquad (13)$$

After dividing by the time Δt of the small displacement $(\Delta x, \Delta y, \Delta z)$ and passing to the limit for $\Delta t = 0$, the left-hand member of this equation becomes the exact time-derivative of the kinetic energy

$$T = \Sigma \tfrac{1}{2} mv^2 = \Sigma \tfrac{1}{2} m(\dot{x}^2 + \dot{y}^2 + \dot{z}^2) \qquad (14)$$

of the body. The right-hand member of (13) represents approximately the *elementary work* done by the external forces during the small displacement. Hence integrating, say from $t = 0$ to $t = t$, we find the relation

$$T - T_0 \equiv \Sigma \tfrac{1}{2} mv^2 - \Sigma \tfrac{1}{2} mv_0^2 = \int_0^t \Sigma(Xdx + Xdy + Zdz), \quad (15)$$

where the right-hand member represents the work W done by the external forces on the body during the time t. This equation expresses the **principle of kinetic energy and work**, for a free rigid body: *in any motion of the body, the increase of the kinetic energy is equal to the work done by the external forces.*

The relation (15) is often written in the equivalent differential form:

$$dT \equiv d\Sigma \tfrac{1}{2} mv^2 = Xdx + Ydy + Zdz \equiv dW.$$

612. By introducing the co-ordinates of the centroid, *i.e.*, by putting $x = \bar{x} + \xi$, $y = \bar{y} + \eta$, $z = \bar{z} + \zeta$, as in Art. 605, the expression for the kinetic energy assumes the form (since $\Sigma m\dot{\xi} = 0$, $\Sigma m\dot{\eta} = 0$, $\Sigma m\dot{\zeta} = 0$):

$$T = \Sigma \tfrac{1}{2} m(\dot{\bar{x}}^2 + \dot{\bar{y}}^2 + \dot{\bar{z}}^2) + \Sigma \tfrac{1}{2} m(\dot{\xi}^2 + \dot{\eta}^2 + \dot{\zeta}^2)$$
$$= \tfrac{1}{2} M\bar{v}^2 + \Sigma \tfrac{1}{2} mu^2,$$

where $\bar{v}$ is the velocity of the centroid and u the *relative* velocity of any particle m with respect to the centroid.

Thus, it appears that *the kinetic energy of a free rigid body consists of two parts, one of which is the kinetic energy of the centroid* (the whole mass being regarded as concentrated at this point), *while the other may be called the relative kinetic energy with respect to the centroid.*

613. By the same substitution the right-hand member of equation (13), *i. e.*, the elementary work $\Sigma(X\Delta x + Y\Delta y + Z\Delta z)$, resolves itself into the two parts

$$(\Delta\bar{x}\Sigma X + \Delta\bar{y}\Sigma Y + \Delta\bar{z}\Sigma Z) + \Sigma(X\Delta\xi + Y\Delta\eta + Z\Delta\zeta).$$

The first parenthesis contains the work that would be done by all the external forces if they were applied at the centroid; it is therefore equal to the kinetic energy of the centroid, that is, to $\Delta(\frac{1}{2} M\bar{v}^2)$. The equation of kinetic energy (13) reduces, therefore, to the following:

$$\Delta(\Sigma \tfrac{1}{2} mu^2) = \Sigma(X\Delta\xi + Y\Delta\eta + Z\Delta\zeta); \qquad (16)$$

in other words, *the principle of kinetic energy holds for the relative motion with respect to the centroid.*

614. Impulses. The equations determining the effect of a system of impulses on a rigid body are readily obtained from the general equations of motion (4) and (6). We shall denote the impulse of a force F by $\underset{\cdot}{F}$. It will be remembered that the impulse $\underset{\cdot}{F}$ of a force F is its time integral; *i. e.*,

$$\underset{\cdot}{F} = \int_t^{t'} F dt.$$

We confine ourselves to the case when $t'-t$ is very small and F very large, in which case the action of the impulsive force F (Arts. 426, 427) is measured by its impulse $\underset{.}{F}$.

If all the forces acting on a rigid body are of this nature, and the impulses of X, Y, Z during the short interval $t'-t$ be denoted by $\underset{.}{X}$, $\underset{.}{Y}$, $\underset{.}{Z}$, the integration of the equations (4) from $t=t$ to $t=t'$ gives

$$\Sigma m(\dot{x}'-\dot{x})=\Sigma \underset{.}{X},\quad \Sigma m(\dot{y}'-\dot{y})=\Sigma \underset{.}{Y},\quad \Sigma m(\dot{z}'-\dot{z})=\Sigma \underset{.}{Z}, \qquad (17)$$

where $\dot{x}$, $\dot{y}$, $\dot{z}$ denote the velocities of the particle m at the time t just before the impulse, and $\dot{x}'$, $\dot{y}'$, $\dot{z}'$ those at the time t' just after the action of the impulse.

Similarly the equations (6) give

$$\begin{aligned} \Sigma m[y(\dot{z}'-\dot{z})-z(\dot{y}'-\dot{y})] &= \Sigma(y\underset{.}{Z}-z\underset{.}{Y}), \\ \Sigma m[z(\dot{x}'-\dot{x})-x(\dot{z}'-\dot{z})] &= \Sigma(z\underset{.}{X}-x\underset{.}{Z}), \\ \Sigma m[x(\dot{y}'-\dot{y})-y(\dot{x}'-\dot{x})] &= \Sigma(x\underset{.}{Y}-y\underset{.}{X}). \end{aligned} \qquad (18)$$

615. In determining the effect on a rigid body of a system of such impulses, any ordinary forces acting on the body at the same time are neglected because the changes of velocity produced by them during the very short time $t'-t$ are small in comparison with the changes of velocity $\dot{x}'-\dot{x}$, $\dot{y}'-\dot{y}$, $\dot{z}'-\dot{z}$ produced by the impulses. If the impulse $\underset{.}{F}$ of an impulsive force F be defined as the limit of the integral $\int_t^{t'} Fdt$ when $t'-t$ approaches zero and F approaches infinity, it is strictly true that the effect of ordinary forces can be neglected when impulsive forces act on the body.

If the rigid body be originally at rest, it will be convenient to denote by $\dot{x}$, $\dot{y}$, $\dot{z}$ the components of the velocity of the particle m just after the action of the impulses. We may also denote by $\underset{.}{R}$ the resultant of all the impulses, by $\underset{.}{H}$ the resultant impulsive couple for the reduction to the origin of coordinates, and mark the components of $\underset{.}{R}$ and $\underset{.}{H}$ by subscripts, as in the case of forces. With these notations the effect of a system of impulses on a body at rest is given by the equations

$$\Sigma m\dot{x} = \underset{\cdot}{R}_x, \quad \Sigma m\dot{y} = \underset{\cdot}{R}_y, \quad \Sigma m\dot{z} = \underset{\cdot}{R}_z, \tag{19}$$

$$\Sigma m(y\dot{z} - z\dot{y}) = \underset{\cdot}{H}_x, \; \Sigma m(z\dot{x} - x\dot{z}) = \underset{\cdot}{H}_y, \; \Sigma m(x\dot{y} - y\dot{x}) = \underset{\cdot}{H}_z. \tag{20}$$

In the equations (19) we have, of course, $\Sigma m\dot{x} = M\dot{\bar{x}}$, $\Sigma m\dot{y} = M\dot{\bar{y}}$, $\Sigma m\dot{z} = M\dot{\bar{z}}$, where $\dot{\bar{x}}$, $\dot{\bar{y}}$, $\dot{\bar{z}}$ are the components of the velocity of the centroid, and M is the mass of the body; *i. e.*, *the momentum of the centroid is equal to the resultant impulse.* The meaning of the equations (20) can be stated by saying that *the angular momentum of the body about any axis is equal to the moment of all the impulses about the same axis.*

II. *Moments of Inertia and Principal Axes.*

1. INTRODUCTION.

616. As will be shown in Section III., the rotation of a rigid body about any axis depends not only on the forces acting on the body, but also on the way in which the mass is distributed throughout the body. This distribution of mass is characterized by the position of the centroid and by that of certain lines in the body called *principal axes.*

It has been shown in Art. 212 that the centroid of a mass is found by determining the *moments*, or more precisely, the *moments of the first order*, Σmx, Σmy, Σmz, of the mass with respect to the co-ordinate planes, *i. e.*, the sums of all mass-particles m each multiplied by its distance from the co-ordinate plane.

The principal axes of a mass or body can be found by determining the *moments of the second order*, Σmx^2, Σmy^2, Σmz^2, Σmyz, Σmzx, Σmxy of the mass with respect to the same planes. We proceed, therefore, to study the theory of such moments.

617. If in a rigid body the mass m of each particle be multiplied by the square of its distance r from a given point, plane, or line, the sum

$$\Sigma mr^2 = m_1r_1^2 + m_2r_2^2 + \cdots,$$

extended over the whole body, is called the *quadratic moment*, or, more commonly, the **moment of inertia** of the body for that point, plane, or line.

If the body is not composed of discrete particles, but forms a continuous mass of one, two, or three dimensions, this mass can be resolved into elements of mass dm, and the sum Σmr^2 becomes a single, double, or triple integral $\int r^2 dm$.

Expressions of the form Σmr_1r_2, or $\int r_1r_2 dm$, where r_1, r_2 are the distances of m or of dm from two planes (usually at right angles), are called *moments of deviation* or **products of inertia**.

618. The determination of the moment of inertia of a continuous mass is a mere problem of integration; the methods are similar to those for finding the moments of mass of the first order required for determining centroids (Arts. 219–250), the only difference being that each element of mass must be multiplied by the square, instead of the first power, of the distance.

A moment of inertia is not a directed quantity; it is not a vector, but a scalar; indeed, it is a positive quantity, provided the masses are all positive, as we shall here assume.

If the mass is homogeneous, the density appears merely as a constant factor; regarding the density in this case as $= 1$, it is customary to speak of moments of inertia of volumes, areas, and lines.

The moment of inertia of any number of bodies or masses for any given point, plane, or line is obviously the sum of the moments of inertia of the separate bodies or masses for the same point, plane, or line.

619. The moment of inertia Σmr^2 of any body whose mass is $M = \Sigma m$ can always be expressed in the form

$$\Sigma mr^2 = M \cdot r_0^2,$$

where r_0 is a length called the **radius of inertia**, arm of inertia, or *radius of gyration.* This length r_0 is evidently a kind of average value of the distances r, its value being intermediate between the greatest r' and least r'' of these distances r. For we have $\Sigma mr'^2 > \Sigma mr^2 > \Sigma mr''^2$, or, since $\Sigma mr'^2 = Mr'^2$, $\Sigma mr^2 = Mr_0^2$, $\Sigma mr''^2 = Mr''^2$,

$$r' > r_0 > r''.$$

620. As an example, let us determine the moment of inertia of a *homogeneous rectilinear segment* (straight rod or wire of constant cross-section and density) for its middle point (or what amounts to the same thing, for a line or plane through this point at right angles to the segment).

Let l be the length of the rod (Fig. 171), O its middle point, ρ'' its density (*i. e.*, the mass of unit length), x the distance OP

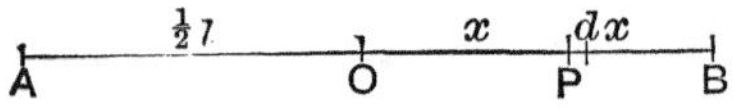

Fig. 171.

of any element $dm = \rho'' dx$ from the middle point. Observing that the moment of inertia for O of the whole rod AB is the sum of the moments of inertia of the halves AO and OB, and that the moments of inertia of these halves are equal, we have, for the moment of inertia I of AB,

$$I = 2\int_0^{\frac{1}{2}l} x^2 \cdot \rho'' dx = \tfrac{1}{12}\rho'' l^3,$$

and for the radius of inertia r_0, since the whose mass is $M = \rho'' l$,

$$r_0^2 = \frac{I}{M} = \tfrac{1}{12} l^2.$$

621. Exercises.

Determine the radius of inertia in the following cases. When nothing is said to the contrary, the masses are supposed to be homogeneous.

(1) Segment of straight line of length l, for a perpendicular through one end.

(2) Rectangular area of length l and width h: (*a*) for the side h (*b*) for the side l; (*c*) for a line through the centroid parallel to the side h; (*d*) for a line through the centroid parallel to the side l.

(3) Triangular area of base b and height h, for a line through the vertex parallel to the base.

(4) Square of side a, for a diagonal.

(5) Regular hexagon of side a, for a diagonal.

(6) Right cylinder or prism of height h, for the plane bisecting the height at right angles.

(7) Segment of straight line of length l, for one end, when the density is proportional to the nth power of the distance from this end. Deduce from this: (*a*) the result of Ex. (1); (*b*) that of Ex. (3); (*c*) the radius of inertia of a homogeneous pyramid or cone (right or oblique) of height h, for a plane through the vertex parallel to the base.

(8) Circular area (plate, disk, lamina) of radius a, for any diameter.

(9) Circular line (wire) of radius a, for a diameter.

(10) Solid sphere, for a diametral plane.

(11) Solid ellipsoid, for the three principal planes.

(12) Area of ring bounded by concentric circles of radii a_1, a_2, for a diameter.

(13) Area of the cross-section of a ⊥-iron: (*a*) for its line of symmetry; (*b*) for its base. (Dimensions as in Fig. 65, Art. 231.)

(14) A rectangular door of width b and height h has a thickness δ to a distance a from the edges, while the rectangular panel (whose dimensions are $b - 2a$, $h - 2a$) has half this thickness. Find the moment of inertia for a line through the centroid parallel to the side b.

622. The moment of inertia of any mass M *for a point* can easily be found if the moments of inertia of the same mass are known for any line passing through the point, and for the plane through the point perpendicular to the line. Let O (Fig. 172) be the point, l the line, π the plane; r, q, p the perpendicular distances of any particle of mass m from O, l, π, respectively. Then we have, evidently, $r^2 = q^2 + p^2$. Hence, multiplying by m, and summing over the whole mass M,

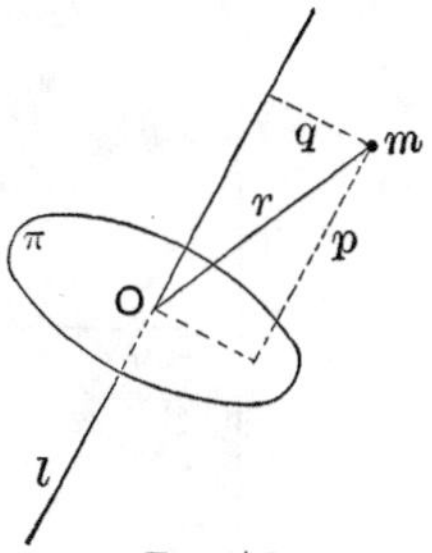

Fig. 172.

$$\Sigma mr^2 = \Sigma mq^2 + \Sigma mp^2; \qquad (1)$$

or, putting $\Sigma mr^2 = Mr_0^2$, $\Sigma mq^2 = Mq_0^2$, $\Sigma mp^2 = Mp_0^2$, where r_0, q_0, p_0 are the radii of inertia for O, l, π,

$$r_0^2 = q_0^2 + p_0^2. \tag{1'}$$

623. The moment of inertia of any mass *M for a line* is equal to the sum of the moments of inertia of the same mass for any two rectangular planes passing through the line. Thus, in particular, the moment of inertia for the axis of x in a rectangular system of co-ordinates is equal to the sum of the moments of inertia for the zx-plane and xy-plane. This follows at once by considering that the square of the distance of any point from the line is equal to the sum of the squares of the distances of the same point from the two planes. Thus, if q be the distance of any point (x, y, z) from the axis of x, we have $q^2 = y^2 + z^2$; whence

$$\Sigma mq^2 = \Sigma my^2 + \Sigma mz^2.$$

624. It follows, from the last article, that *the moment of inertia I_x of a plane area, for any line perpendicular to its plane*, is

$$I_x = I_y + I_z,$$

if I_y, I_z are the moments of inertia of the area for any two rectangular lines in the plane through the foot of the perpendicular line.

625. The problem of *finding the moment of inertia of a given mass for a line l', when it is known for a parallel line l*, is of great importance.

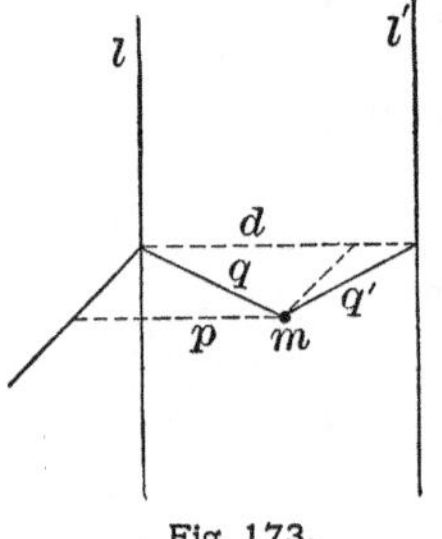

Fig. 173.

Let Σmq^2 be the moment of inertia of the given mass for the line l (Fig. 173), $\Sigma mq'^2$ that for a parallel line l' at the distance d from l. The distances q, q' of any particle m from l, l' form with d a triangle which gives the relation

$$q'^2 = q^2 + d^2 - 2qd \cos(q, d).$$

Multiplying by m, and summing over the whole mass M, we find

$$\Sigma mq'^2 = \Sigma mq^2 + Md^2 - 2\,d\,\Sigma mq \cos(q, d).$$

Now the figure shows that the product $q \cos(q, d)$ in the last term is the distance p of the particle m from a plane through l at right angles to the plane determined by l and l'. We have, therefore,

$$\Sigma mq'^2 = \Sigma mq^2 + Md^2 - 2\,d\Sigma mp, \tag{2}$$

where the last term contains the moment of the first order $\Sigma mp = M\bar{p}$ of the given mass M for the plane just mentioned.

If, in particular, this plane contains the centroid G of the mass M, we have $\Sigma mp = 0$, so that the formula reduces to

$$\Sigma mq'^2 = \Sigma mq^2 + Md^2. \tag{3}$$

Introducing the radii of inertia q_0', q_0, this can be written

$$q_0'^2 = q_0^2 + d^2. \tag{3'}$$

626. Similar considerations hold for the moments of inertia Σmp^2, $\Sigma mp'^2$ with respect to two parallel planes π, π' at the distance d from each other. We have, in this case, $p' = p - d$; hence,

$$\Sigma mp'^2 = \Sigma mp^2 + Md^2 - 2\,d\Sigma mp, \tag{4}$$

and if the plane π contain the centroid G,

$$\Sigma mp'^2 = \Sigma mp^2 + Md^2. \tag{5}$$

627. Of special importance is the case in which one of the lines (or planes), say $l\,(\pi)$, contains the centroid. The formulæ (3), (3'), and (5) hold in this case; and if we agree to designate any line (plane) passing through the centroid as a *centroidal* line (plane), our proposition can be expressed as follows: *The moment of inertia for any line (plane) is found from the moment of inertia for the parallel centroidal line (plane) by adding to the latter the product Md^2 of the whole mass into the square of the distance of the lines (planes).*

It will be noticed that of all parallel lines (planes) the centroidal line (plane) has the least moment of inertia.

628. Exercises.

Determine the radius of inertia of the following homogeneous masses:

(1) Rectangular plate of length l and width h, for a centroidal line perpendicular to its plane.

(2) Area of equilateral triangle of side a: (*a*) for a centroidal line parallel to the base; (*b*) for an altitude; (*c*) for a centroidal line perpendicular to its plane.

(3) Circular disk of radius a: (*a*) for a tangent; (*b*) for a line through the center perpendicular to the plane of the disk; (*c*) for a perpendicular to its plane through a point in the circumference.

(4) Solid sphere, for a diameter.

(5) Area of ring bounded by concentric circles of radii a_1, a_2, for a line through the center perpendicular to the plane of the ring.

(6) Right circular cylinder, of radius a and height h: (*a*) for its axis; (*b*) for a generating line; (*c*) for a centroidal line in the middle cross-section.

(7) By Ex. (3) (*b*), the moment of inertia of the area of a circle of radius a, for its *axis* (*i. e.*, the perpendicular to its plane, passing through the center), is $I = \frac{1}{2}\pi a^4$. Differentiating with respect to a, we find:

$$\frac{dI}{da} = 2\pi a^3 = 2\pi a \cdot a^2;$$

hence, approximately for small Δa:

$$\Delta I = 2\pi a^3 \Delta a = 2\pi a \Delta a \cdot a^2.$$

This is the moment of inertia of the thin ring, of thickness Δa, for its axis. (Comp. Ex. (5).)

If the constant *surface* density (Art. 214) of the circle be ρ', we have $I = \frac{1}{2}\rho' \pi a^4$; hence

$$\Delta I = 2\pi a \rho' \Delta a \cdot a^2,$$

where $\rho' \Delta a$ is the *linear* density ρ'' of the ring.

(8) Apply the method of Ex. (7) to derive from Ex. (4) the moment of inertia of a thin spherical shell, of radius a and thickness Δa, for a diameter.

(9) Area of ellipse: (*a*) for the major axis; (*b*) for the minor axis; (*c*) for the perpendicular to its plane through the center.

(10) Solid ellipsoid, for each of the three axes.

(11) Area of the cross-section of a T-iron, for a centroidal line parallel to the flange. (Comp. Art. 621, Ex. (13), and Art. 231.)

(12) Area of the cross-section of a symmetrical double T-iron, width of flanges $2b$, thickness of flanges δ, height of web a, thickness of web 2δ (Fig. 69, Art. 241, with $b' = b$); for the two axes of symmetry, and for a centroidal line perpendicular to its plane.

(13) Wire bent into an equilateral triangle of side a, for a centroidal line at right angles to the plane of the triangle.

(14) Paraboloid of revolution, bounded by the plane through the focus at right angles to the axis, for the axis.

(15) Anchor-ring, produced by the revolution of a circle of radius a about a line in its plane at the distance b ($> a$) from the center, for the axis of revolution.

(16) Prove that the moment of inertia of the solid generated by the revolution of any plane area about a line l, situated in its plane, but not intersecting it, is $I = M(3q^2 + b^2)$, where b is the distance of the centroid of the area from l, and q the radius of inertia of the area for the centroidal line parallel to l.

(17) Show that the moment of inertia of a homogeneous triangular plate for the centroidal line parallel to the base is equal to that of one eighth of the whole mass concentrated at the vertex, or to that of one half the whole mass concentrated at the base.

(18) Find the radius of inertia of a parallelogram whose sides a, b include an angle θ, for the centroidal line perpendicular to its plane.

(19) Find the radius of inertia of the area of a trapezoid (parallel sides a, b, height h) for the centroidal line parallel to the parallel sides.

(20) Prove that the radius of inertia q of any homogeneous right prism or cylinder, for the centroidal line perpendicular to the axis (*i. e.*, the line joining the centroids of the bases), can be found from the formula $q^2 = q_a^2 + q_c^2$, where q_a is the radius of inertia of the axis, q_c that of the middle cross-section, for the same centroidal line.

2. ELLIPSOIDS OF INERTIA.

629. The moments of inertia of a given mass for the different lines of space are not independent of each other. Several examples of this have already been given. It has been shown, in particular (Art. 625), that if the moment of inertia be known for any line, it can be found for any parallel line. It follows that if the moments be known for all lines through any given point, the moments for all lines of space can be found. We now proceed to study the relations between the moments of inertia for all the lines passing through any given point O.

630. It will here be convenient to refer the given mass M to a rectangular system of co-ordinates with the origin at the point O. Let x, y, z be the co-ordinates of any particle m of the mass; and let us denote by A, B, C the moments of inertia of M for the axes of x, y, z; by A', B', C' those for the planes yz, zx, xy; by D, E, F the products of inertia (Art. 617) for the co-ordinate planes; *i. e.*, let us put

$$\begin{aligned} A &= \Sigma m(y^2+z^2), & A' &= \Sigma mx^2, & D &= \Sigma myz, \\ B &= \Sigma m(z^2+x^2), & B' &= \Sigma my^2, & E &= \Sigma mzx, \\ C &= \Sigma m(x^2+y^2), & C' &= \Sigma mz^2, & F &= \Sigma mxy. \end{aligned} \tag{6}$$

631. These nine quantities are not independent of each other. We have evidently

$$A = B' + C', \qquad B = C' + A', \qquad C = A' + B';$$

hence, solving for A', B', C',

$$A' = \tfrac{1}{2}(B+C-A), \quad B' = \tfrac{1}{2}(C+A-B), \quad C' = \tfrac{1}{2}(A+B-C).$$

The moment of inertia for the origin O is

$$\Sigma mr^2 = \Sigma m(x^2+y^2+z^2) = A'+B'+C' = \tfrac{1}{2}(A+B+C). \tag{7}$$

632. The moment of inertia I for any line through O can be expressed by means of the six quantities A, B, C, D, E, F; and the moment of inertia I' for any plane through O can be expressed by means of A', B', C', D, E, F.

Let π (Fig. 174) be any plane passing through O; l its normal;

α, β, γ the direction cosines of l; and, as before (Art. 622), p, q, r the distances of any point (x, y, z) of the given mass from π, l, and O, respectively. Then, projecting the closed polygon formed by r, x, y, z on the line l, we have

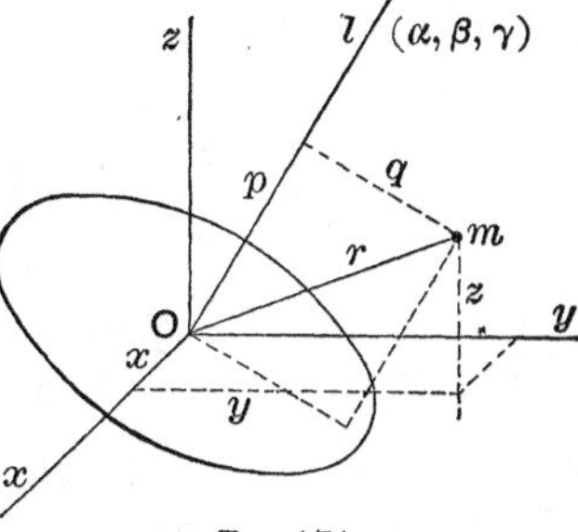

Fig. 174.

$$p = \alpha x + \beta y + \gamma z;$$

hence, squaring, multiplying by m, and summing over the whole mass, we find

$$\Sigma mp^2 =$$

$$\alpha^2 \Sigma mx^2 + \beta^2 \Sigma my^2 + \gamma^2 \Sigma mz^2 + 2\beta\gamma \Sigma myz + 2\gamma\alpha \Sigma mzx + 2\alpha\beta \Sigma mxy,$$

or, with the notations (6),

$$I' = A'\alpha^2 + B'\beta^2 + C'\gamma^2 + 2D\beta\gamma + 2E\gamma\alpha + 2F\alpha\beta. \quad (8)$$

Thus *the moment of inertia for any plane through the origin is expressed as a homogeneous quadratic function of the direction cosines of the normal of the plane.*

633. The moment of inertia $I = \Sigma mq^2$ for the line l can now be found from equation (1), Art. 622, by substituting for Σmr^2 and Σmp^2 their values from (7) and (8):

$$I = \Sigma mr^2 - I' = A' + B' + C' - I'$$

$$= A'(1-\alpha^2) + B'(1-\beta^2) + C'(1-\gamma^2) - 2D\beta\gamma - 2E\gamma\alpha - 2F\alpha\beta,$$

or, since $\alpha^2 + \beta^2 + \gamma^2 = 1$,

$$I = A'(\beta^2+\gamma^2) + B'(\gamma^2+\alpha^2) + C'(\alpha^2+\beta^2) - 2D\beta\gamma - 2E\gamma\alpha - 2F\alpha\beta$$

$$= \alpha^2(B'+C') + \beta^2(C'+A') + \gamma^2(A'+B') - 2D\beta\gamma - 2E\gamma\alpha - 2F\alpha\beta;$$

hence, finally, applying the relations of Art. 631,

$$I = A\alpha^2 + B\beta^2 + C\gamma^2 - 2D\beta\gamma - 2E\gamma\alpha - 2F\alpha\beta. \quad (9)$$

The moment of inertia for any line through the origin is, therefore, also *a homogeneous quadratic function of the direction cosines of the line.*

634. These results suggest a geometrical interpretation. Imagine an arbitrary length $OP = \rho$ laid off from the origin O on the line l whose direction cosines are α, β, γ; the co-ordinates of the extremity P of this length will be $x = \rho\alpha$, $y = \rho\beta$, $z = \rho\gamma$. Now, if equation (9) be multiplied by ρ^2, it assumes the form

$$Ax^2 + By^2 + Cz^2 - 2\,Dyz - 2\,Ezx - 2\,Fxy = \rho^2 I,$$

which represents a quadric surface provided that ρ be selected for the different lines through O so as to make $\rho^2 I$ constant, say $\rho^2 I = \kappa^2$. Hence, *if on every line l through the origin a length $OP = \rho = \kappa/\sqrt{I}$ be laid off*, i. e., *a length inversely proportional to the square root of the moment of inertia I for this line l, the points P will lie on the quadric surface*

$$Ax^2 + By^2 + Cz^2 - 2\,Dyz - 2\,Ezx - 2\,Fxy = \kappa^2.$$

The constant κ^2 may be selected arbitrarily; to preserve the homogeneity of the equation it will be convenient to take it in the form $\kappa^2 = M\epsilon^4$, where ϵ is still arbitrary.

635. As moments of inertia are essentially positive quantities, the radii vectores of the surface

$$Ax^2 + By^2 + Cz^2 - 2\,Dyz - 2\,Ezx - 2\,Fxy = M\epsilon^4 \qquad (10)$$

are all real, and the surface is an ellipsoid. It is called the *ellipsoid of inertia*, or the **momental ellipsoid,** of the point O. This point O is the center; the axes of the ellipsoid are called the **principal axes** at the point O; and the moments of inertia for these axes are called the *principal moments of inertia* at the point O. Among these there will evidently be the greatest and least of all the moments of the point O, the greatest moment corresponding to the shortest, the least to the longest axis of the ellipsoid.

It may be observed that, owing to the relations of Art. 631, which show that the sum of any two of the quantities A, B, C is always greater than the third, not every ellipsoid can be regarded as the momental ellipsoid of some mass. An ellipsoid can be a momental ellipsoid only when a triangle can be constructed of its semi-axes.

636. If the axes of the ellipsoid (10) be taken as axes of co-ordinates, the equation assumes the form

$$I_1x^2 + I_2y^2 + I_3z^2 = M\epsilon^4, \tag{11}$$

where I_1, I_2, I_3 are the principal moments at the point O.

By Art. 634 we have $\rho^2 = \kappa^2/I = M\epsilon^4/I$; hence $I = M\epsilon^4/\rho^2$. If, therefore, equation (11) be divided by ρ^2, the following simple expression is obtained for finding the moment of inertia, I, for a line whose direction cosines referred to the principal axes are α, β, γ,

$$I = I_1\alpha^2 + I_2\beta^2 + I_3\gamma^2. \tag{12}$$

637. To make use of this form for I, the principal axes at the point O, *i.e.*, the axes of the momental ellipsoid (10), must be known. The determination of the axes of an ellipsoid whose equation referred to the center is given is a well-known problem of analytic geometry. It can be solved by considering that the semi-axes are those radii vectores of the surface that are normal to it. The direction cosines of the normal of any surface $F(x, y, z) = 0$ are proportional to the partial derivatives $\partial F/\partial x$, $\partial F/\partial y$, $\partial F/\partial z$. If, therefore, the radius vector ρ is a semi-axis, its direction-cosines α, β, γ must be proportional to the partial derivatives of (10); *i.e.*, we must have

$$\frac{Ax - Fy - Ez}{\alpha} = \frac{-Fx + By - Dz}{\beta} = \frac{-Ex - Dy + Cz}{\gamma},$$

or dividing the numerators by ρ,

$$\frac{A\alpha - F\beta - E\gamma}{\alpha} = \frac{-F\alpha + B\beta - D\gamma}{\beta} = \frac{-E\alpha - D\beta + C\gamma}{\gamma}.$$

Denoting the common value of these fractions by I, we have

$$\alpha I = A\alpha - F\beta - E\gamma, \quad \beta I = -F\alpha + B\beta - D\gamma, \quad \gamma I = -E\alpha - D\beta + C\gamma;$$

multiplying these equations by α, β, γ, and adding, we find

$$I = A\alpha^2 + B\beta^2 + C\gamma^2 - 2D\beta\gamma - 2E\gamma\alpha - 2F\alpha\beta,$$

which, compared with (9), shows that I is the moment of inertia for the axis (α, β, γ). To obtain it in terms of A, B, C, D, E, F, we write the preceding three equations in the form

$$\begin{aligned}(I-A)\alpha + \quad F\beta + \quad E\gamma &= 0,\\ F\alpha + (I-B)\beta + \quad D\gamma &= 0,\\ E\alpha + \quad D\beta + (I-C)\gamma &= 0,\end{aligned} \tag{13}$$

whence, eliminating α, β, γ, we find I determined by the cubic equation

$$\begin{vmatrix} I-A, & F, & E \\ F, & I-B, & D \\ E, & D, & I-C \end{vmatrix} = 0. \tag{14}$$

The roots of this cubic are the three principal moments I_1, I_2, I_3 of the point O. The direction cosines of the principal axes are then found by substituting successively I_1, I_2, I_3 in (13) and solving for α, β, γ.

638. The geometrical representation of the moments of inertia for all lines passing through a point, by means of the radii vectores of the momental ellipsoid at the point, gives at once a number of propositions about these moments. It is only necessary to interpret properly the geometrical properties of the ellipsoid. Thus, it is known that the sum of the squares of the reciprocals of any three rectangular semi-diameters of an ellipsoid is constant. It follows that the sum of the three moments of inertia for any three rectangular lines passing through the same point has a constant value.

In general, the three principal moments of inertia I_1, I_2, I_3 at a point O are different. If, however, two of them are equal, say $I_2 = I_3$, the momental ellipsoid becomes an ellipsoid of revolution about the third, I_1, as axis; and it follows that the moments of inertia for all lines through O lying in the plane of the two equal axes are equal.

If $I_1 = I_2 = I_3$, the ellipsoid becomes a sphere, and the moments of inertia are the same for all lines passing through O.

639. If the equation of the momental ellipsoid at a point O be of the form $Ax^2 + By^2 + Cz^2 - 2\,Dyz = M\epsilon^4$, *i. e.*, if the two conditions

$$E \equiv \Sigma mzx = 0, \quad F \equiv \Sigma mxy = 0$$

be fulfilled, the axis of x coincides with one of the three axes of the ellipsoid, the surface being symmetrical with respect to the yz-plane. Hence, *if the conditions $E = 0$, $F = 0$ are satisfied, the axis of x is a principal axis at the origin.* The converse is evidently also true; *i. e.*, if a line is a principal axis at one of its points, then, taking this point as origin and the line as axis of x, the conditions $\Sigma mzx = 0$, $\Sigma mxy = 0$ must be satisfied.

It is easy to see that if a line be a principal axis at one of its points, say O, it will in general not be a principal axis at any other one of its points. For, taking the line as axis of x and O as origin, we have $\Sigma mzx = 0$, $\Sigma mxy = 0$. If now for a point O' on this line at the distance a from O the line is likewise a principal axis, the conditions

$$\Sigma mz(x-a) = 0, \quad \Sigma m(x-a)y = 0$$

must be fulfilled. These reduce to

$$\Sigma mz = 0, \quad \Sigma my = 0,$$

and show that the line must pass through the centroid. And as for a centroidal line these conditions are satisfied independently of the value of a, it appears that a centroidal principal axis is a principal axis at every one of its points. Hence, *a line cannot be principal axis at more than one of its points unless it pass through the centroid; in the latter case it is a principal axis at every one of its points.*

640. All those lines passing through a given point O for which the moments of inertia have the same value I can be shown to form a cone of the second order whose principal diameters coincide with the axes of the momental ellipsoid at O. This cone is called an **equimomental cone.** Its equation is obtained by regarding I as constant in equation (12) and introducing rectangular co-ordinates. Multiplying (12) by $\alpha^2 + \beta^2 + \gamma^2 = 1$, we find

$$(I_1 - I)\alpha^2 + (I_2 - I)\beta^2 + (I_3 - I)\gamma^2 = 0;$$

and multiplying by ρ^2, we obtain the equation of the equimomental cone in the form

$$(I_1 - I)x^2 + (I_2 - I)y^2 + (I_3 - I)z^2 = 0. \qquad (15)$$

641. A slightly different form of the equations (11), (12), (15) is often more convenient; it is obtained by introducing the three **principal radii of inertia** q_1, q_2, q_3 defined by the relations

$$I_1 = Mq_1^2, \quad I_2 = Mq_2^2, \quad I_3 = Mq_3^2.$$

The equation (11) of the momental ellipsoid at the point O then assumes the form

$$q_1^2x^2 + q_2^2y^2 + q_3^2z^2 = \epsilon^4. \qquad (11')$$

The expression of the radius of inertia q for any line (α, β, γ) through O becomes

$$q^2 = q_1^2\alpha^2 + q_2^2\beta^2 + q_3^2\gamma^2. \qquad (12')$$

Dividing (11′) by the square of the radius vector, ρ^2, and comparing with (12′), we find

$$q = \frac{\epsilon^2}{\rho}, \quad \rho = \frac{\epsilon^2}{q}, \qquad (16)$$

as is otherwise apparent from the fundamental property of the momental ellipsoid (Art. 634).

The equation of the equimomental cone for all whose generators the radius of inertia has the value q is obtained from (15) in the form

$$(q_1^2 - q^2)x^2 + (q_2^2 - q^2)y^2 + (q_3^2 - q^2)z^2 = 0. \qquad (15')$$

This cone meets any one of the momental ellipsoids (11′) in points whose radii vectores ρ are all equal; and if we select the arbitrary constant ϵ equal to the radius of inertia q of the generators of the equimomental cone, it follows from (16) that $\rho = q$. This particular ellipsoid has the equation

$$q_1^2x^2 + q_2^2y^2 + q_3^2z^2 = q^4,$$

and its intersection with the equimomental cone (15′) lies on the sphere

$$x^2 + y^2 + z^2 = q^2.$$

In other words, the equimomental cone (15′) passes through the spheroconic in which the ellipsoid meets the sphere. Multiplying the equation of the sphere by q^2 and subtracting it from the equation of the ellipsoid, we obtain the equation (15′) of the cone.

The semi-axes of the ellipsoid are q^2/q_1, q^2/q_2, q^2/q_3. If we assume $I_1 > I_2 > I_3$, and hence $q_1 > q_2 > q_3$, q must be $\geqq q^2/q_3$ and $\leqq q^2/q_1$. As long as q is less than the middle semi-axis q^2/q_2 of the ellipsoid, the axis of the cone coincides with the axis of x; but when $q > q^2/q_2$, the axis of z is the axis of the cone. For $q = q^2/q_2$, i. e., $q = q_2$, the cone (15′) degenerates into the pair of planes $(q_1^2 - q_2^2)x^2 - (q_2^2 - q_3^2)z^2 = 0$. These are the planes of the central circular (or *cyclic*) sections of the ellipsoid; they divide the ellipsoid into four wedges, of which one pair contains all the equimomental cones whose axes coincide with the greatest axis of the ellipsoid, while the other pair contains all those whose axes lie along the least axis of the ellipsoid.

642. There is another ellipsoid closely connected with the theory of principal axes; it is obtained from the momental ellipsoid by the process of reciprocation.

About any point O (Fig. 175) taken as center let us describe a sphere of radius ϵ, and construct for every point P its polar plane π with regard to the sphere. If P describe any surface, the plane π will envelop another surface which is called the *polar reciprocal* of the former surface with regard to the sphere.

Let Q be the intersection of OP with π, and put $OP=\rho$, $OQ=q$; then it appears from the figure that

$$\rho q=\epsilon^2. \tag{16}$$

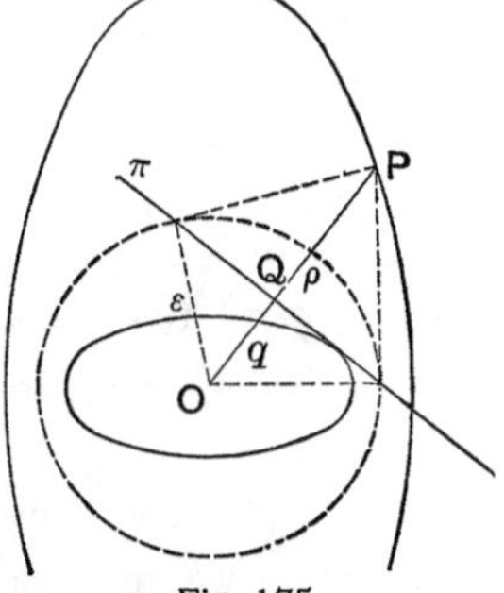

Fig. 175.

643. It is easy to see that the polar reciprocal of the momental ellipsoid (11′) with respect to the sphere of radius ϵ is the ellipsoid

$$\frac{x^2}{q_1^{\,2}}+\frac{y^2}{q_2^{\,2}}+\frac{z^2}{q_3^{\,2}}=1. \tag{17}$$

To prove this it is only necessary to show that the relation (16) is fulfilled for ρ as radius vector of (11′), and q as perpendicular to the tangent plane of (17). Now this tangent plane has the equation

$$\frac{x}{q_1^{\,2}}X+\frac{y}{q_2^{\,2}}Y+\frac{z}{q_3^{\,2}}Z=1;$$

hence we have for the direction cosines α, β, γ, and for the length q, of the perpendicular to the tangent plane

$$\frac{\alpha}{x/q_1^{\,2}}=\frac{\beta}{y/q_2^{\,2}}=\frac{\gamma}{z/q_3^{\,2}}=\frac{1}{[x^2/q_1^{\,4}+y^2/q_2^{\,4}+z^2/q_3^{\,4}]^{\frac{1}{2}}}=q.$$

These relations give $q_1\alpha=(x/q_1)q$, $q_2\beta=(y/q_2)q$, $q_3\gamma=(z/q_3)q$, whence

$$q_1^{\,2}\alpha^2+q_2^{\,2}\beta^2+q_3^{\,2}\gamma^2=\left(\frac{x^2}{q_1^{\,2}}+\frac{y^2}{q_2^{\,2}}+\frac{z^2}{q_3^{\,2}}\right)q^2=q^2. \tag{18}$$

For the radius vector ρ of (11′) whose direction cosines α, β, γ are the same as those of q, we have by (11′):

$$\rho^2=\frac{\epsilon^4}{q_1^{\,2}\alpha^2+q_2^{\,2}\beta^2+q_3^{\,2}\gamma^2}.$$

Hence $\rho^2q^2=\epsilon^4$; and this is what we wished to prove.

644. The surface (17) has variously been called the *ellipsoid of gyration*, the *ellipsoid of inertia*, the **reciprocal ellipsoid.** We shall adopt the last name. The semi-axes of this ellipsoid are equal to the principal radii of inertia at the point O. The directions of its axes coincide with those of the momental ellipsoid; but the greatest axis of the former coincides with the least of the latter, and *vice versa.*

By comparing the equations (12′) and (18) it will be seen that q is the radius of inertia of the line (α, β, γ) on which it lies. Thus, *while the radius vector* $OP = \rho$ *of the momental ellipsoid is inversely proportional to the radius of inertia*, i. e., $\rho = \epsilon^2/q$, *the reciprocal ellipsoid gives the radius of inertia* q *for a line* l *as the segment cut off on this line by the perpendicular tangent plane.*

645. We are now prepared to determine the moment of inertia for any line in space. Let us construct at the centroid G of the given mass or body both the momental ellipsoid and its polar reciprocal. The former is usually called the **central ellipsoid** of the body; the latter we may call the **fundamental ellipsoid** of the body. As soon as this fundamental ellipsoid

$$\frac{x^2}{q_1^{\,2}} + \frac{y^2}{q_2^{\,2}} + \frac{z^2}{q_3^{\,2}} = 1$$

is known, the moment of inertia of the body for any line whatever can readily be found. For, by Art. 644, the radius of inertia q for any line l_0 passing through the centroid is equal to the segment OQ cut off on the line l_0 by the perpendicular tangent plane of the fundamental ellipsoid; and for any line l not passing through the centroid, the square of the radius of inertia can be determined by first finding the square of the radius of inertia for the parallel centroidal line l_0, and then, by Art. 627, adding to it the square of the distance d of the centroid from the line l.

646. In the problem of determining the ellipsoids of inertia for a given body at any point, considerations of symmetry are of great assistance, just as in the problem of finding the centroid (comp. Art. 244).

Suppose a given mass to have a plane of symmetry; then taking this plane as the yz-plane, and a perpendicular to it as the axis of

x, there must be, for every particle of mass m, whose co-ordinates are x, y, z, another particle of equal mass m, whose co-ordinates are $-x, y, z$. It follows that the two products of inertia Σmzx and Σmxy both vanish, whatever the position of the other two co-ordinate planes. Hence, any perpendicular to the plane of symmetry is a principal axis at its point of intersection with this plane.

Let the mass have two planes of symmetry at right angles to each other; then taking one as yz-plane, the other as zx-plane, and hence their intersection as axis of x, it is evident that all three products of inertia vanish,

$$\Sigma myz = 0, \qquad \Sigma mzx = 0, \qquad \Sigma mxy = 0,$$

wherever the origin be taken on the intersection of the two planes. Hence, for any point on this intersection, the principal axes are the line of intersection of the two planes of symmetry, and the two perpendiculars to it, drawn in each plane.

If there be three planes of symmetry, their point of intersection is the centroid, and their lines of intersection are the principal axes at the centroid.

647. Exercises.

Determine the principal axes and radii at the centroid, the central and fundamental ellipsoids, and show how to find the moment of inertia for any line, in the following Exercises (1), (2), (3).

(1) Rectangular parallelepiped, the edges being $2a$, $2b$, $2c$. Find also the moments of inertia for the edges and diagonals, and specialize for the cube.

(2) Ellipsoid of semi-axes a, b, c. Determine also the radius of inertia for a parallel l to the shortest axis passing through the extremity of the longest axis.

(3) Right circular cone of height h and radius of base a. Find first the principal moments at the vertex; then transfer to the centroid.

(4) Determine the momental ellipsoid and the principal axes at a vertex of a cube whose edge is a.

(5) Determine the radius of inertia of a thin wire bent into a circle, for a line through the center inclined at an angle α to the plane of the circle.

(6) A peg-top is composed of a cone of height H and radius a, and a hemispherical cap of the same radius. The pointed end, to a distance h from the vertex of the cone, is made of a material three times as heavy as the rest. Find the moment of inertia for the axis of rotation; specialize for $h = a = \frac{1}{3} H$.

(7) Show that the principal axes at any point P, situated on one of the principal axes of a body, are parallel to the centroidal principal axes, and find their moments of inertia.

(8) For a given body of mass M find the points (*spherical points of inertia*) at which the momental ellipsoid reduces to a sphere.

(9) Determine a homogeneous ellipsoid having the same mass as a given body, and such that its moment of inertia for every line shall be the same as that of the given body.

(10) For a given body M, whose centroidal principal radii are q_1, q_2, q_3, determine three homogeneous straight rods intersecting at right angles, of such lengths $2a$, $2b$, $2c$, and such linear density ρ'', that they have the same mass and the same moment of inertia (for any line) as the given body.

3. DISTRIBUTION OF PRINCIPAL AXES IN SPACE.

648. It has been shown in the preceding articles how the principal axes can be determined at any particular point. The distribution of the principal axes throughout space and their position at the different points is brought out very graphically by means of the theory of confocal quadrics. It can be shown that the directions of the principal axes at any point are those of the principal diameters of the tangent cone drawn from this point as vertex to the fundamental ellipsoid; or, what amounts to the same thing, they are the directions of the normals of the three quadric surfaces passing through the point and confocal to the fundamental ellipsoid.

In order to explain and prove these propositions it will be necessary to give a short sketch of the theory of confocal conics and quadrics.

649. *Two conic sections are said to be* **confocal** *when they have the same foci.* The directions of the axes of all conics having the same two points S, S' as foci must evidently coincide, and the equation of such conics can be written in the form

$$\frac{x^2}{a^2+\lambda}+\frac{y^2}{b^2+\lambda}=1, \tag{19}$$

where λ is an arbitrary parameter. For, whatever value may be assigned in this equation to λ, the distance of the center O from either focus will always be $\sqrt{a^2+\lambda-(b^2+\lambda)}=\sqrt{a^2-b^2}$; it is therefore constant.

650. The individual curves of the whole system of confocal conics represented by (19) are obtained by giving to λ any particular value between $-\infty$ and $+\infty$; thus we may speak of the conic λ of the system.

For $\lambda=0$ we have the so-called fundamental conic $x^2/a^2+y^2/b^2=1$; this is an ellipse. To fix the ideas let us assume $a>b$. For all values of $\lambda>-b^2$, *i.e.*, as long as $-b^2<\lambda<\infty$, the conics (19) are ellipses, beginning with the rectilinear segment SS' (which may be regarded as a degenerated ellipse $\lambda=-b^2$ whose minor axis is 0), expanding gradually, passing through the fundamental ellipse $\lambda=0$, and finally verging into a circle of infinite radius for $\lambda=\infty$.

It is thus geometrically evident that through every point in the plane will pass one, and only one, of these ellipses.

651. Let us next consider what the equation (19) represents when λ is algebraically less than $-b^2$. The values of λ that are $<-a^2$ give imaginary curves, and are of no importance for our purpose. But as long as $-a^2<\lambda<-b^2$, the curves are hyperbolas. The curve $\lambda=-b^2$ may now be regarded as a degenerated hyperbola collapsed into the two rays issuing in opposite directions from S and S' along the line SS'. The degenerated ellipse together with this degenerated hyperbola thus represents the whole axis of x.

As λ decreases, the hyperbola expands, and finally, for $\lambda=-a^2$, verges into the axis of y, which may be regarded as another degenerated hyperbola.

The system of confocal hyperbolas is thus seen to cover likewise the whole plane so that one, and only one, hyperbola of the system passes through every point of the plane.

652. The fact that every point of the plane has one ellipse and one hyperbola of the confocal system (19) passing through it, enables us to regard the two values of the parameter λ that determine these two curves as co-ordinates of the point; they are called *elliptic co-ordinates.* If x, y be the rectangular cartesian co-ordinates of the point, its elliptic co-ordinates λ_1, λ_2 are found as the roots of the equation (19) which is quadratic in λ. Conversely, to transform from elliptic to cartesian co-ordinates, that is, to express x and y in terms of λ_1 and λ_2, we have only to solve for x and y the two equations

$$\frac{x^2}{a^2+\lambda_1}+\frac{y^2}{b^2+\lambda_1}=1, \quad \frac{x^2}{a^2+\lambda_2}+\frac{y^2}{b^2+\lambda_2}=1.$$

653. The two confocal conics that pass through the same point P intersect at right angles. For the tangent to the ellipse at P bisects the exterior angle at P in the triangle SPS', while the tangent to the hyperbola bisects the interior angle at the same point; in other words, the tangent to one curve is normal to the other, and *vice versa.* The elliptic system of co-ordinates is, therefore, an *orthogonal* system; the infinitesimal elements $d\lambda_1 \cdot d\lambda_2$ into which the two series of confocal conics (19) divide the plane are rectangular, though curvilinear.

654. These considerations are easily extended to space of three dimensions. An ellipsoid

$$\frac{x^2}{a^2}+\frac{y^2}{b^2}+\frac{z^2}{c^2}=1, \text{ where } a>b>c,$$

has six real foci in its principal planes; two, S_1, S_1', in the xy-plane, on the axis of x, at a distance $OS_1=\sqrt{a^2-b^2}$ from the center O; two, S_2, S_2', in the yz-plane, on the axis of y, at the distance $OS_2=\sqrt{b^2-c^2}$ from the center; and two, S_3, S_3', in the zx-plane, on the axis of x, at the distance $OS_3=\sqrt{a^2-c^2}$ from the center. It should be noticed that, since $b>c$, we have $OS_3>OS_1$; i. e., S_1, S_1' lie between S_3, S_3' on the axis of x.

The same holds for hyperboloids.

655. *Two quadric surfaces are said to be confocal when their principal sections are confocal conics.* Now this will be the case for two quadric surfaces whose semi-axes are a_1, b_1, c_1, and a_2, b_2, c_2, if the directions of their axes coincide and if

$$a_1^2-b_1^2=a_2^2-b_2^2, \quad b_1^2-c_1^2=b_2^2-c_2^2, \quad a_1^2-c_1^2=a_2^2-c_2^2.$$

Writing these conditions in the form

$$a_2^2 - a_1^2 = b_2^2 - b_1^2 = c_2^2 - c_1^2, \text{ say } = \lambda,$$

we find $a_2^2 = a_1^2 + \lambda$, $b_2^2 = b_1^2 + \lambda$, $c_2^2 = c_1^2 + \lambda$. Hence the equation

$$\frac{x^2}{a^2+\lambda} + \frac{y^2}{b^2+\lambda} + \frac{z^2}{c^2+\lambda} = 1, \tag{20}$$

where λ is a variable parameter, represents a system of confocal quadric surfaces.

656. As long as λ is algebraically greater than $-c^2$, the equation (20) represents ellipsoids. For $\lambda = -c^2$ the surface collapses into the interior area of the ellipse in the xy-plane whose vertices are the foci S_2, S_2' and S_3, S_3'. For as λ approaches the limit $-c^2$, the three semi-axes of (20) approach the limits $\sqrt{a^2-c^2}$, $\sqrt{b^2-c^2}$, 0, respectively. This limiting ellipse is called the *focal ellipse*. Its foci are the points S_1, S_1', since $a^2 - c^2 - (b^2 - c^2) = a^2 - b^2$.

When λ is algebraically $< -c^2$, but $> -a^2$, the equation (20) represents hyperboloids; for values of $\lambda < -a^2$ it is not satisfied by any real points. As long as $-b^2 < \lambda < -c^2$, the surfaces are hyperboloids of one sheet. The limiting surface $\lambda = -c^2$ now represents the exterior area of the focal ellipse in the xy-plane. The limiting hyperboloid of one sheet for $\lambda = -b^2$ is the area in the zx-plane bounded by the hyperbola whose vertices are S_1, S_1', and whose foci are S_3, S_3'. This is called the *focal hyperbola*.

Finally, when $-a^2 < \lambda < -b^2$, the surfaces are hyperboloids of two sheets, the limiting hyperboloid $\lambda = -a^2$ collapsing into the yz-plane.

657. It appears from these geometrical considerations, that there are passing through every point of space three surfaces confocal to the fundamental ellipsoid $x^2/a^2 + y^2/b^2 + z^2/c^2 = 1$ and to each other, viz.: an ellipsoid, a hyperboloid of one sheet, and a hyperboloid of two sheets. This can also be shown analytically, as there is no difficulty in proving that the equation (20) has three real roots, say λ_1, λ_2, λ_3, for every set of real values of x, y, z, and that these roots are confined between such limits as to give the three surfaces just mentioned.

The quantities λ_1, λ_2, λ_3 can therefore be taken as co-ordinates of the point (x, y, z); and these *elliptic co-ordinates* of the point are, geometrically, the parameters of the three quadric surfaces passing through the point and confocal to the fundamental ellipsoid; while, analytically, they are the three roots of the cubic (20). To express x, y, z in terms of the elliptic co-ordinates, it is only necessary to solve for x, y, z the three equations obtained by substituting in (20) successively $\lambda_1, \lambda_2, \lambda_3$ for λ.

658. The geometrical meaning of the parameter λ will appear by considering two parallel tangent planes π_0 and π_λ (on the same side of the origin), the former (π_0) tangent to the fundamental ellipsoid $x^2/a^2+y^2/b^2+z^2/c^2=1$, the latter ($\pi_\lambda$) tangent to any confocal surface λ or $x^2/(a^2+\lambda)+y^2/(b^2+\lambda)+z^2/(c^2+\lambda)=1$. The perpendiculars q_0, q_λ, let fall from the origin O on these tangent planes π_0, π_λ, are given by the relations (the proof being the same as in Art. 643)

$$q_0^2=a^2\alpha^2+b^2\beta^2+c^2\gamma^2, \tag{21}$$

$$q_\lambda^2=(a^2+\lambda)\alpha^2+(b^2+\lambda)\beta^2+(c^2+\lambda)\gamma^2, \tag{22}$$

where α, β, γ are the direction cosines of the common normal of the planes π_0, π_λ. Subtracting (21) from (22), we find, since $\alpha^2+\beta^2+\gamma^2=1$,

$$q_\lambda^2-q_0^2=\lambda; \tag{23}$$

i. e., *the parameter λ of any one of the confocal surfaces* (20) *is equal to the difference of the squares of the perpendiculars let fall from the common center on any tangent plane to the surface λ, and on the parallel tangent plane to the fundamental ellipsoid $\lambda=0$.*

659. Let us now apply these results to the question of the distribution of the principal axes throughout space.

We take the centroid G of the given body as origin, and select as fundamental ellipsoid of our confocal system the polar reciprocal of the central ellipsoid, *i.e.*, the ellipsoid (17) formed for the centroid, for which the name "fundamental ellipsoid of the body" was introduced in Art. 645. Its equation is

$$\frac{x^2}{q_1^2}+\frac{y^2}{q_2^2}+\frac{z^2}{q_3^2}=1,$$

if q_1, q_2, q_3 are the principal radii of inertia of the body.

The radius of inertia q_0 for any centroidal line l_0 can be constructed (Art. 644) by laying a tangent plane to this ellipsoid perpendicular to the line l_0; if this line meets the tangent plane at Q_0 (Fig. 176), then $q_0=GQ_0$. Analytically, if α, β, γ be the direction cosines of l_0, q_0 is given by formula (21) or (12').

660. To find the radius of inertia q for a line l, parallel to l_0, and passing through any point P, we lay through P a plane π_λ, perpendicular to l, and a parallel plane π_0, tangent to the fundamental ellipsoid; let Q_λ, Q_0 be the intersections of these planes with the centroidal line l_0. Then, putting $GQ_0 = q_0$, $GQ_\lambda = q_\lambda$, $GP = r$, $PQ_\lambda = d$, we have, by Art. 627,

$$q^2 = q_0^2 + d^2.$$

The figure gives the relation $d^2 = r^2 - q_\lambda^2$, which, in combination with (23), reduces the expression for the radius of inertia for the line l to the simple form

$$q^2 = r^2 - \lambda. \tag{24}$$

661. The value of $r^2 - \lambda$, and hence the value of q, remains the same for the perpendiculars to all planes through P, tangent to the same quadric surface λ: these perpendiculars form, therefore, an equimomental cone at P. By varying λ we thus obtain all the equimomental cones at P. The principal diameters of all these cones coincide in direction, since they coincide with the directions of the principal axes of the momental ellipsoid at P (see Art. 640); but they also coincide with the principal diameters of the cones enveloped by the tangent planes π_λ. It thus appears that *the principal axes at the point P coincide in direction with the principal diameters of the tangent cone from P as vertex to the fundamental ellipsoid* $x^2/q_1^2 + y^2/q_2^2 + z^2/q_3^2 = 1$.

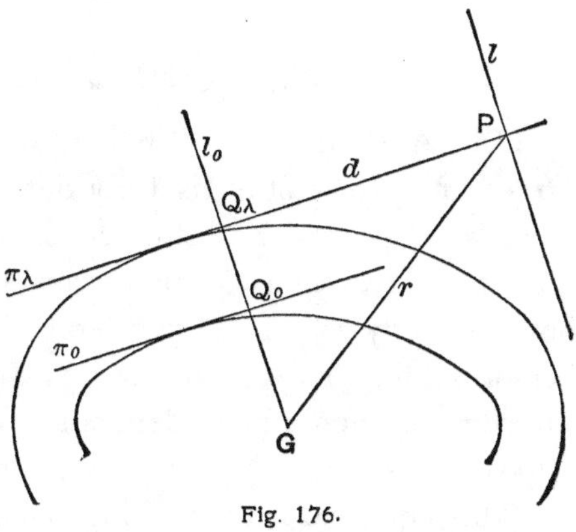

Fig. 176.

662. Instead of the fundamental ellipsoid, we might have used any quadric surface λ confocal to it. In particular, we may select the confocal surfaces λ_1, λ_2, λ_3 that pass through P. For each of these the cone of the tangent planes collapses into a plane, viz., the tangent plane to the surface at P, while the cone of the perpendiculars reduces to a single line, viz., the normal to the surface at P. Thus we find that *the prin-*

cipal axes at any point P coincide in direction with the normals to the three quadric surfaces, confocal to the fundamental ellipsoid and passing through P.

For the magnitudes of the principal radii q_x, q_y, q_z at P, we evidently have

$$q_x^2 = r^2 - \lambda_1, \qquad q_y^2 = r^2 - \lambda_2, \qquad q_z^2 = r^2 - \lambda_3.$$

663. Exercise.

(1) The principal radii q_1, q_2, q_3 of a body being given, find the equation of the momental ellipsoid at any point P, referred to axes through this point P parallel to the principal axes of the body; determine the directions of the principal axes at P, and show that these directions coincide with the normals of the three surfaces passing through P and confocal to the fundamental ellipsoid of the body.

III. *Rigid Body with a Fixed Axis.*

664. A rigid body with a fixed axis has but one degree of freedom. Its motion is fully determined by the motion of any one of its points (not situated on the axis), and any such point must move in a circle about the axis. Any particular position of the body is, therefore, determined by a single variable, or co-ordinate, such as the angle of rotation. Just as the equilibrium of such a body depends on a single condition (see Art. 399), so its motion is given by a single equation.

This equation is obtained at once by "taking moments about the fixed axis." For, according to the proposition of angular momentum (Art. 601), the time-rate of change of angular momentum about any axis is equal to the moment of the external forces about this axis. Hence, denoting this moment by H and taking the fixed axis as axis of z, we have as equation of motion the last of the equations (7), Art. 601, viz.,

$$\frac{d}{dt}\Sigma m(x\dot{y} - y\dot{x}) = H. \qquad (1)$$

665. The *angular momentum*, $\Sigma m(x\dot{y} - y\dot{x})$, about the fixed axis can be reduced to a more simple form. For rotation of angular velocity ω about the z-axis we have (Art. 175) $\dot{x} = -\omega y$, $\dot{y} = \omega x$, so that

$$\Sigma m(x\dot{y} - y\dot{x}) = \omega \Sigma m(x^2 + y^2) = \omega \cdot \Sigma mr^2 = I\omega,$$

where r is the distance of the particle m from the axis and $I = \Sigma mr^2$ the moment of inertia of the body for this axis.

This expression for the angular momentum can be derived without reference to any co-ordinate system. For evidently $m\omega r$ is the linear momentum of the particle m, $m\omega r^2$ is its moment, *i. e.*, the angular momentum of the particle, about the axis; and $\Sigma m\omega r^2 = \omega \Sigma mr^2 = I\omega$ is the angular momentum of the body about the axis.

It thus appears that, just as in translation the linear momentum of a body is the product of its mass into its linear velocity, so in the case of rotation *the angular momentum of the body about the axis of rotation is the product of its moment of inertia* (for this axis) *into the angular velocity.*

As regards the right-hand member of equation (1), the reactions of the axis need not be taken into account in forming the moment H; for as these reactions meet the axis, their moments about this axis are zero.

666. Substituting $I\omega$ for $\Sigma m(x\dot{y} - y\dot{x})$ in equation (1), and observing that the moment of inertia I about a *fixed* axis remains constant, we find the **equation of motion** in the form

$$I\frac{d\omega}{dt} = H; \qquad (2)$$

i. e., for rotation about a fixed axis, *the product of the moment of inertia for this axis into the angular acceleration equals the moment of the external forces about this axis;* just as, in the case of rectilinear translation, the product of the mass of the body into the linear acceleration equals the resultant force R along the line of motion:

$$m\frac{dv}{dt} = R.$$

And just as the latter equation may serve to determine experimentally the mass of a body by observing the acceleration produced in it by a given force R, *e. g.*, the force of gravity (as in the gravitation system, Art. 262), so the former equation, (2), may serve to determine experimentally the moment of inertia of a body about a line l, by observing the angular acceleration produced in the body when rotating about l under given forces.

667. For the **kinetic energy** of a body rotating with angular velocity ω about any axis, we have

$$T = \Sigma \tfrac{1}{2} mv^2 = \Sigma \tfrac{1}{2} m\omega^2 r^2 = \tfrac{1}{2} I\omega^2, \tag{3}$$

an expression which is again similar in form to that for the kinetic energy of a body in translation, viz., $T = \frac{1}{2} mv^2$.

668. When the axis is fixed so that I is constant, the equation of motion (2), multiplied by ω and integrated, say from $t = 0$ to $t = t$, gives the relation

$$\tfrac{1}{2} I\omega^2 - \tfrac{1}{2} I\omega_0^2 = \int_0^t H\omega dt, \tag{4}$$

which expresses the *principle of kinetic energy and work*, and might have been derived from the general formula (15), Art. 611. For, taking the axis of rotation as axis of z, we have in the present case $dx = -\omega y dt$, $dy = \omega x dt$, $dz = 0$, so that $\Sigma (Xdx + Ydy + Zdz) = \Sigma (xY - yX)\omega dt = H\omega dt$; this, with the value (3) for T, reduces equation (15) of Art. 611 to the above equation (4).

Conversely, the equation of motion (2) can be derived from the principle of kinetic energy and work, $dT = dW$ (Art. 611) or $d \frac{1}{2} I\omega^2 = \Sigma(Xdx + Ydy + Zdz) = H\omega dt$.

If, in particular, H is constant, and we put $\omega = d\theta/dt$, (4) reduces to

$$\tfrac{1}{2} I(\omega^2 - \omega_0^2) = H(\theta - \theta_0),$$

where $\theta - \theta_0$ is the angle through which the body turns in the time t in which the angular velocity increases from ω_0 to ω.

669. A rigid body with a fixed *horizontal* axis is called a **compound pendulum** if the only external force acting is the weight of the body.

The plane through axis and centroid will make, with the vertical plane (downwards) through the axis, an angle θ, which we may take as angle of rotation, so that $\omega = d\theta / dt$ (Fig. 177). The weights of the particles, being all parallel and proportional to their masses, have a single resultant Mg passing through the centroid G. Hence, if h be the perpendicular distance OG of the centroid from the axis, the moment of the external forces is $H = -Mgh \sin\theta$; and if the radius of inertia of the body for the centroidal axis parallel to the axis of rotation be q, the moment of inertia for the latter axis is $I = M(q^2 + h^2)$.

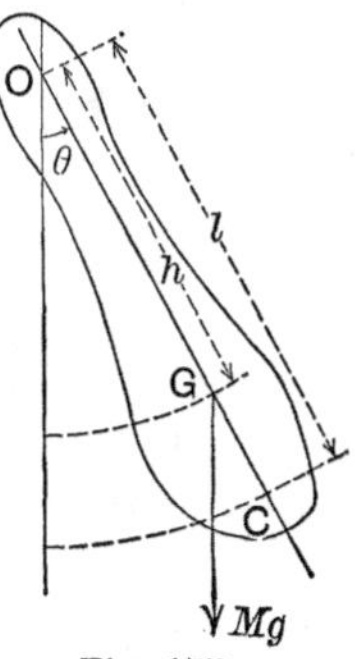

Fig. 177.

With these values the equation of motion (2) assumes the simple form

$$\frac{d^2\theta}{dt^2} = -\frac{gh}{q^2 + h^2} \sin\theta. \tag{5}$$

As shown in Art. 585, the equation of the *simple* pendulum of length l is

$$\frac{d^2\theta}{dt^2} = -\frac{g}{l} \sin\theta.$$

The two equations differ only in the constant factor of $\sin\theta$, and it appears that *the motion of a compound pendulum is the same as that of a simple pendulum whose length is*

$$l = h + \frac{q^2}{h}. \tag{6}$$

670. The problem of the compound pendulum has thus been reduced to that of the simple pendulum. The length l is called *the length of the equivalent simple pendulum.* The foot O (Fig. 177) of the perpendicular let fall from the centroid on the axis is called *the center of suspension.* If on the line OG a length $OC = l$ be laid off, the point C is called *the center of* *oscillation.* It appears, from (6), that G lies between O and C.

The relation (6) can be written in the form

$$h(l-h)=q^2, \text{ or } OG \cdot GC = \text{const.}$$

As this relation is not altered by interchanging O and C, it follows that *the centers of oscillation and suspension are interchangeable; i. e.*, the period of a compound pendulum remains the same if it be made to swing about a parallel axis through the center of oscillation.

671. Exercises.

(1) A pendulum, formed of a cylindrical rod of radius a and length L, swings about a diameter of one of the bases. Find the time of a small oscillation.

(2) A cube, whose edge is a, swings as a pendulum about an edge. Find the length of the equivalent simple pendulum.

(3) A circular disk of radius r revolves uniformly about its axis, making 100 rev./min. What is its kinetic energy?

(4) A fly-wheel of radius r, in which a mass, equal to that of the disk in Ex. (3), is distributed uniformly along the rim, has the same angular velocity as the disk. Neglecting the mass of the nave and spokes, determine its kinetic energy, and compare it with that of the disk.

(5) A fly-wheel of 12 ft. diameter, whose rim weighs 10 tons, makes 50 rev./min. Find its kinetic energy in foot-tons.

(6) A fly-wheel of 10 ft. diameter, weighing 5 tons, is making 40 revolutions when thrown out of gear. In what time does it come to rest if the diameter of the axle is 6 in. and the coefficient of friction $\mu = 0.05$?

(7) A fly-wheel of radius r and mass m is making N rev./min. when the steam is shut off. If the radius of the shaft be r', and the coefficient of friction μ, find after how many revolutions the wheel will come to rest owing to the axle friction.

(8) A homogeneous straight rod of length l is hinged at one end so as to turn freely in a vertical plane. If it be dropped from a horizontal position, with what angular velocity does it pass through the vertical position? (Equate the kinetic energy to the work of gravity.)

(9) A homogeneous plate whose shape is that of the segment of a parabola bounded by the curve and its latus rectum swings about the latus rectum which is horizontal. Find the length of the equivalent simple pendulum.

(10) A fly-wheel of 10 ft. radius makes 45 rev./min. Its rim (regarded as a circular line) weighs 8000 lbs., while each of the 10 spokes (regarded as a straight line) weighs 200 lbs. Find the kinetic energy stored in the wheel.

(11) Find the work that would have to be done by the engine to increase the number of revolutions to 60 per minute for the fly-wheel in Ex. (10).

(12) When q is given while l and h vary, the equation (6) represents a hyperbola whose asymptotes are the axis of l and the bisector of the angle between the (positive) axes of h and l. Show that $l_{\min} = 2q$ for $h=q$; also that l, and hence the period of oscillation, can be made very large by taking h either very large or very small. The latter case occurs for a ship whose *metacenter* (which plays the part of the point of suspension) lies very near its centroid.

(13) Explain how to determine experimentally the moment of inertia of a body for any line l by observing its small oscillations about l as axis.

(14) Find the horse-power required to keep a wheel weighing k tons rotating with N rev./min., the radius of the axle being r' ft. and the coefficient of friction μ.

(15) If q be the radius of inertia of the wheel in Ex. (14) (including the shaft and other attachments), for its axis, determine: (*a*) after how many revolutions, (*b*) in what time, the wheel will come to rest if left to itself.

(16) A homogeneous circular disk, 1 ft. in diameter and weighing 25 lbs., is making 240 rev./min. when left to itself. Determine the constant tangential force applied to its rim that would bring it to rest in 1 min.

(17) A cord is wrapped around the horizontal axle of a heavy wheel; the free end of the cord passes over a fixed vertical pulley and carries a mass of 75 lbs. It is found that when the mass has descended 10 ft. the wheel is making 67 rev./min. What is the moment of inertia of the wheel?

(18) A homogeneous circular hoop of radius a is suspended by a point in its rim. Find the length of the equivalent simple pendulum when the hoop swings: (a) in its own plane, (b) at right angles to this plane; (c) determine the ratio of the periods in the two cases.

672. Small Oscillations Due to Torsion. Let a homogeneous straight bar be suspended horizontally by means of a stout wire attached to its middle point. If the bar be turned, in a horizontal plane, out of its position of equilibrium, the torsion produced in the wire tends to bring the bar back to its original position; thus, vibrations about this position are set up. The torsional stress of the wire acts on the bar as a couple in the horizontal plane, and within certain limits, even for oscillations that are not very small, the moment of this couple can be taken proportional to the angle θ through which the bar is turned, say $=\mu\theta$. The motion of the bar about the vertical axis of the wire is therefore given by the equation

$$I\frac{d\omega}{dt}=-\mu\theta, \tag{7}$$

where I is the moment of inertia of the bar about the vertical centroidal axis, and μ the moment of the torsional couple that would turn the bar through 1 radian.

Comparing with the equation for the small oscillations of a pendulum, it appears that the length l of the equivalent simple pendulum is

$$l=\frac{gI}{\mu},$$

and hence the time of a complete oscillation

$$T=2\pi\sqrt{\frac{I}{\mu}}. \tag{8}$$

673. This formula can be used to determine experimentally the moment of inertia of the bar. To eliminate the unit torsion moment μ, the time of oscillation is generally observed first for the bar alone, then for the bar loaded, *i. e.*, in connection with pieces whose moment of inertia is known or easily determined. If the moment of inertia of the added pieces be I', the time of oscillation of the loaded bar T', we have

$$T'=2\pi\sqrt{\frac{I+I'}{\mu}}.$$

Dividing this equation by (8), we find:

$$\frac{T'^2}{T^2} = \frac{I+I'}{I},$$

whence, $$I = I' \cdot \frac{T^2}{T'^2 - T^2}. \tag{9}$$

674. Magnetic Needle. The *moment* M of a magnetic needle is defined as the moment of the couple acting on the needle when placed in a uniform field of unit strength, at right angles to the lines of force. The forces of this couple can be regarded as applied at the poles and as equal to the pole-strengths. If the strength of the field is not 1, but H, the moment of the couple acting on the needle is MH; and if the needle is placed at an angle θ with the lines of force, the moment of the so-called *restoring*, or *directing*, couple that tends to place the axis of the needle parallel to the lines of force is easily seen to be $= MH \sin\theta$.

The equation of motion for the oscillations of a magnetic needle, pivoted so as to turn freely in a horizontal plane, when the needle is placed in any position in the earth's magnetic field, is therefore

$$I \frac{d^2\theta}{dt^2} = -MH \sin\theta, \tag{10}$$

where H is the strength of the earth's horizontal field.

675. When the oscillations are small, $\sin\theta$ can be replaced by θ, and we find for the length of the equivalent simple pendulum

$$l = \frac{gI}{MH},$$

and for the time of one complete oscillation

$$T = 2\pi \sqrt{\frac{I}{MH}}. \tag{11}$$

This formula is used for determining MH by observing T; as the quotient M/H can be found from deflection observations, the formula serves ultimately for the determination of the earth's horizontal intensity H.

676. The *dynamical meaning of the radius of inertia* of a body for any line l appears by considering that, according to equation (2), if the body be set revolving about the line l as a fixed axis, under the action of any forces, the motion, *i. e.*, the variation of the angular velocity, depends only on the moment H of the forces and the moment of inertia I. The motion remains, therefore, the same when the body is replaced by any other body having the same moment of inertia for l, provided the moment H of the forces remains the same. Thus, in studying the motion of a rigid body about a fixed axis l, if the mass of the body be M and its moment of inertia $I = Mq^2$, the body might be replaced by a ring, or a cylindrical shell, of mass M and radius q about the axis l, or by a single particle of mass M placed at the distance q from the axis. Thus, the *radius of inertia* q receives its dynamical interpretation as *that distance from the axis at which the mass M of the body must be concentrated to produce the same motion of rotation as the actual motion.*

It should be carefully observed, however, that, in replacing the body by another of equal moment of inertia, the reactions of the axis, and hence the pressure on the axis, are in general changed, as will be explained later (Art. 699).

677. Reduced Mass. In substituting for a rotating body another of equal moment of inertia for the axis of rotation, it is not even necessary to keep the mass the same (as was done in Art. 676). Thus, a body of mass M and moment of inertia $I = Mq^2$ for the fixed axis of rotation l may be replaced by a mass M_r distributed uniformly along a circle of radius r, about l as axis, provided that M_r and r be selected so that

$$M_r \cdot r^2 = Mq^2 = I.$$

This equivalent mass M_r is called the *mass reduced to the distance* r from the axis.

As regards the external forces, since only their moment H about the fixed axis is essential, they can be replaced by a single force F, perpendicular to the axis l, at such a distance p from it that its moment Fp is equal to H. By thus putting $H = Fp$ *the external forces* are said to be *reduced to the distance* p from the axis.

678. If both the mass M of the rotating body and the external forces acting on it be reduced to the *same* distance r from the axis l, the equation of motion (2), $Id\omega/dt = H$, assumes the simple form

$$M_r \cdot r\frac{d\omega}{dt} = F;$$

and as $rd\omega/dt$ is the linear (tangential) acceleration of a point at the distance r from the axis, the equation is exactly the same as that for rectilinear translation: $mdv/dt = F$.

The following exercises will show in what way the idea of reduced mass can be used to advantage.

679. Exercises.

(1) Reduce the mass of a homogeneous circular plate to its circumference, for rotation about the axis of the plate.

(2) Reduce the mass of a homogeneous straight rod of length $2a$ to its middle point, for rotation about the perpendicular through one end.

(3) Reduce the mass of a homogeneous solid sphere of radius a: (*a*) to the surface, for rotation about a diameter; (*b*) to the center, for rotation about a tangent.

(4) Two masses m_1, m_2 hang from a weightless cord slung over a fixed pulley of mass m; show how to determine the effect of the inertia of the pulley on the motion.

Denoting by j the acceleration of the cord, by T_1, T_2 its tensions above m_1 and m_2, we have as equations of motion of m_1 and m_2 separately:

$$m_1 j = m_1 g - T_1, \quad m_2 j = -m_2 g + T_2,$$

and for the motion of the pulley:

$$mq^2 \cdot \frac{d\omega}{dt} = T_1 r - T_2 r,$$

where q is the radius of inertia, r the radius, of the pulley. If m_r be the mass of the pulley reduced to its circumference, the latter equation becomes

$$m_r r\frac{d\omega}{dt} = m_r j = T_1 - T_2.$$

Substituting for T_1, T_2 their values from the first two equations, and solving for j, we find :

$$j = \frac{m_1 - m_2}{m_1 + m_2 + m_r} g,$$

where $m_r = \frac{1}{2} m$ if the pulley can be regarded as a homogeneous disk, and $m_r = m$ if its mass be regarded as concentrated in the rim. For the tensions we find :

$$T_1 = \frac{2\, m_2 + m_r}{m_1 + m_2 + m_r} \cdot m_1 g, \quad T_2 = \frac{2\, m_1 + m_r}{m_1 + m_2 + m_r} \cdot m_2 g.$$

Compare the results of Art. 464, where the mass of the pulley was neglected.

(5) In Ex. (4), show how to take account of axle-friction.

The total pressure on the axle being $T_1 + T_2 + mg$, the frictional force is $F = \mu (T_1 + T_2 + mg)$; it acts at an arm equal to the radius ρ of the axle. To reduce this force to the circumference of the pulley we have to find F_r from

$$F_r \cdot r = F \cdot \rho = \mu\rho (T_1 + T_2 + mg).$$

We then have (see Ex. (4)) :

$$m_r j = T_1 - T_2 - F_r = T_1 - T_2 - \mu \frac{\rho}{r} (T_1 + T_2 + mg) ;$$

whence, substituting for T_1 and T_2 and solving for j:

$$j = \frac{m_1 - m_2 - \mu (m_1 + m_2 + m) \rho / r}{m_1 + m_2 + m_r - \mu (m_1 - m_2) \rho / r} g.$$

(6) The apparatus called "wheel and axle" can be regarded as consisting of two equal masses m suspended over two rigidly connected coaxial pulleys, with different radii r and $\rho < r$; find the acceleration j of the mass hanging from the larger pulley.

Neglecting axle friction, we have $T_1 = mg - mj$, $T_2 = mg + m \cdot (\rho / r) j$, since the accelerations are evidently as $r : \rho$. Hence,

$$m_r r^2 \cdot \frac{d\omega}{dt} = T_1 r - T_2 \rho,$$

or since $r d\omega / dt = j$,

$$j = \frac{m (1 - \rho / r)}{m (1 + \rho^2 / r^2) + m_r} g.$$

In particular, for $r = 2\, \rho$, we find $j = 2\, mg / (5\, m + 4\, m_r)$.

(7) A wheel, weighing 40 lbs., radius of inertia for its axis 1 ft., has a cord wrapped around its axle which is 6 in. in diameter; this cord passes over a pulley, and has a weight of 12 lbs. suspended from its end. Find in what time the weight will descend 15 ft.

680. Many pieces of machinery are bodies turning about fixed axes. Thus in most machines we find shafts, or axles, kept in nearly uniform rotation by a **driving force** supplied by a *prime mover* (steam engine, water wheel, turbine, electric motor, etc.); to the shaft we find rigidly attached, and turning with it, cranks, eccentrics, wheels, pulleys, etc., which, at a certain distance from the axis of the shaft, have to overcome a **resistance** and thus do useful work, such as lifting a hammer, or a rack, when the work is done against gravity, or shaping material, as in a planer, when the work is done against the cohesion of the material to be planed, and so forth.

The resistance is generally directed at right angles to the axis of the shaft, this being evidently the most effective way of doing work in this case.

681. The **work** done in one revolution by a uniformly revolving shaft against a constant perpendicular resistance R at the distance r from the axis of the shaft (Fig. 178) is $= 2\pi r \cdot R$, where r is called the *leverage* of the force R. The work done as the shaft turns through any angle θ is $\theta r \cdot R$; and as Rr is the moment of R about the axis, we may say that *the work done against the resistance is equal to the moment of the resistance multiplied by the angle of rotation.*

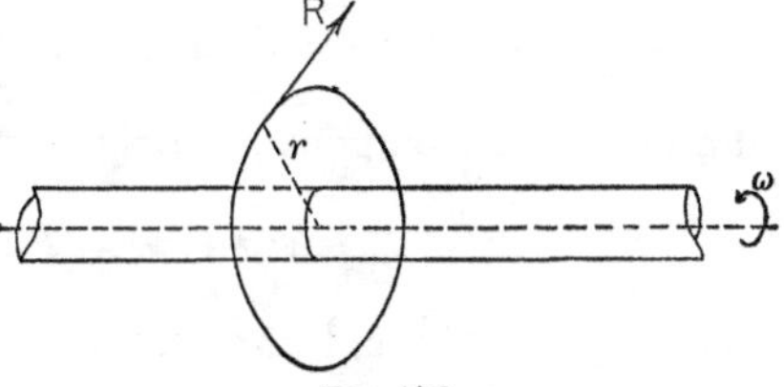

Fig. 178.

The rate of work, or **power**, is then equal to *the moment of the resistance multiplied by the angular velocity.*

This presupposes that the resistance is of constant magnitude, always at right angles to and at the same distance from the axis, and that the rotation is uniform. Otherwise we have to fall back on the general expressions of Arts. 664–668.

682. Steam Engine. In the steam engine the pressure P of the steam on the piston (Fig. 179) is first resolved at the cross-head P into a component $P' = P \sec \phi$ along the connecting rod PQ and a component $P'' = P \tan \phi$ at right angles to the guides of the cross-head.

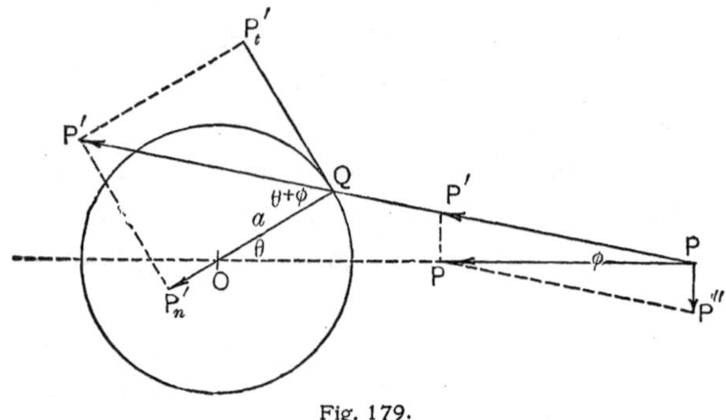

Fig. 179.

Only the former of these components P' acts on the crank; it can be resolved at the crank pin Q into a normal component along the crank QO:

$$P_n' = P' \cos(\theta + \phi) = P \sec \phi \cos(\theta + \phi) = P(\cos\theta - \sin\theta \tan\phi),$$

and a tangential component, at right angles to the crank,

$$P_t' = P' \sin(\theta + \phi) = P(\sin\theta + \cos\theta \tan\phi).$$

As P_n' passes through the center of rotation O it has no turning effect. The rotation of the crank is therefore due entirely to the force P_t'; its moment about O is

$$H = P_t' a = a(\sin\theta + \cos\theta\tan\phi) P,$$

where $a = OQ$ is the length of the crank.

Even if P were constant (which it is not, see Arts. 469–471), this moment H would vary with the angles θ and ϕ.

The angle ϕ can be eliminated, its variation depending on that of the angle θ. For, with $OQ = a$, $PQ = l$, the triangle OPQ gives

$$\frac{\sin\phi}{\sin\theta} = \frac{a}{l}, \text{ or } \sin\phi = \frac{1}{m}\sin\theta,$$

if we put $l/a = m$.

As the angle ϕ is generally small, the length of the connecting rod being usually at least 3 or 4 times that of the crank, we can substitute $\sin\phi$ for $\tan\phi$ so that we obtain for the *turning moment* the simple approximate expression

$$H = a\left(\sin\theta + \frac{1}{2m}\sin 2\theta\right)P.$$

683. Fly-Wheel. As the turning moment H varies in the course of the stroke, it follows that even if the resistance were constant, the angular velocity ω of the crank, or the linear velocity u of its end Q, will not remain constant. In order to diminish this irregularity in the rotation of the crank as much as possible, a heavy *fly-wheel* is fixed on the crank shaft.

The function of the fly-wheel is not to create energy, but to store and distribute it. During that part of the stroke during which the turning moment H is greater than its mean value for one stroke, energy is being accumulated in the mass of the fly-wheel; and when H is less than its mean value, part of this energy is consumed in doing useful work and thus making up for the lack of turning moment.

684. It must here suffice to discuss a simple ideal case. Assuming the connecting rod of infinite length so that $\phi = 0$, and the driving force $P' = P$ constant, its tangential component (Fig. 180) is $P_t' = P\sin\theta$, and the turning moment is

$$H = Pa\sin\theta.$$

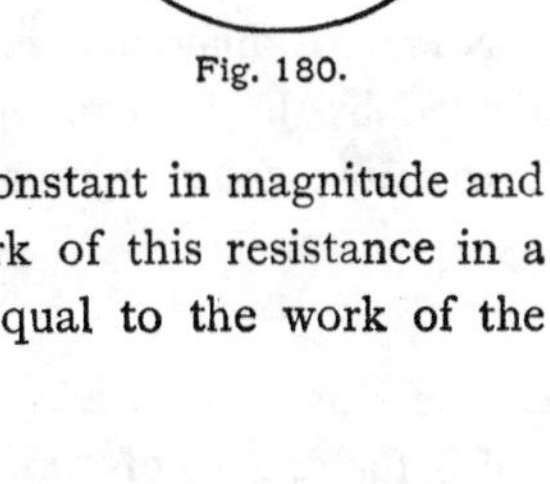

The work done by the driving force in a *double* (forth and back) stroke is evidently

$$W = 2\cdot P\cdot 2a = 4Pa.$$

Fig. 180.

If, for the sake of simplicity, we assume the resistance R to be overcome by the crank as constant in magnitude and always tangential to the crank circle, the work of this resistance in a double stroke is $= Q\cdot 2\pi a$. This must be equal to the work of the driving force, so that

$$4Pa = 2\pi Qa,$$

whence

$$Q = \frac{2}{\pi}P = 0.637\,P.$$

685. It is easy to determine the angle θ_1 for which the effective driving force $P_t' = P \sin \theta_1$ is just equal to the resistance $Q = 2P/\pi$; we find approximately

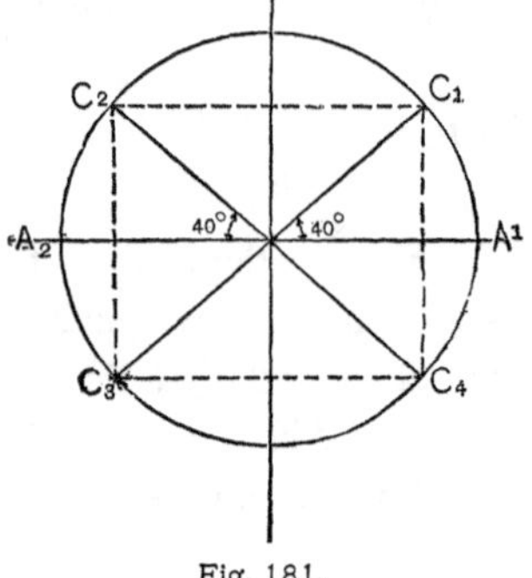

Fig. 181.

$$\theta_1 = 40°, 140°, 220°, 320°.$$

As long as $P_t' > Q$, *i.e.*, from C_1 to C_2 and from C_3 to C_4 (Fig. 181), the rotation is accelerated; from C_2 to C_3 and from C_4 to C_1 it is retarded. The velocity is therefore greatest at C_2 and C_4, least at C_3 and C_1.

Now, in the interval C_1C_2 the work W_{12} of the driving force is

$$W_{12} = P \cdot \text{chord } C_1C_2 = P \cdot 2a\cos\theta_1 = \pi Qa \cos\theta_1,$$

since $Q = 2P/\pi$, while the work of the resistance Q is

$$W_{12}' = Q \cdot \text{arc } C_1C_2 = Q \cdot a(\pi - 2\theta_1);$$

hence, the energy stored in the fly-wheel in this interval C_1C_2 is

$$W_{12} - W_{12}' = \pi Qa\left(\cos\theta_1 - 1 + 2\frac{\theta_1}{\pi}\right) = .21\,\pi Qa = \tfrac{2}{3}Qa,$$

approximately.

686. On the other hand, if ω_1, ω_2 be the angular velocities of the crank, v_1, v_2 the linear velocities of the crank-pin, at C_1, C_2, respectively, and if the mass of the fly-wheel reduced to the crank-pin be denoted by m_a, we find for the same energy stored in the fly-wheel in the period C_1C_2 the other expression:

$$T_2 - T_1 = \tfrac{1}{2} m_a a^2(\omega_2^2 - \omega_1^2) = \tfrac{1}{2} m_a(v_2^2 - v_1^2).$$

Now, the difference $v_2 - v_1$ between the greatest and least velocity, divided by their mean $\frac{1}{2}(v_1 + v_2) = v$, measures the relative *fluctuation* in velocity; the reciprocal of this quotient,

$$\kappa = \frac{v}{v_2 - v_1} = \frac{v_1 + v_2}{2(v_2 - v_1)},$$

can be regarded as a measure of the uniformity of the rotation; it is called the **degree of uniformity.** Introducing this quantity κ, we have

$$T_2 - T_1 = \frac{1}{\kappa} m_a v^2.$$

Equating the two expressions found for the kinetic energy stored in the fly-wheel, we find for the reduced mass of the wheel,

$$m_a = \tfrac{2}{3}\kappa \cdot \frac{Qa}{v^2}.$$

If the resistance Q be expressed in pounds, the mass m_a in pounds will be

$$m_a = \tfrac{2}{3}\kappa g \frac{Qa}{v^2}.$$

The coefficient κ is selected differently according to the nature of the engine and the object for which it is used. For slow-running engines it is between 10 and 20; for very fast engines it may reach 100 or more.

687. Exercises.

(1) Show that $m_a = 112{,}000\ \kappa \text{HP}/v^2N$, if HP is the horse-power, N the number of revolutions per minute.

(2) Find m_a when HP $= 100$, $N = 25$, $v = 56$ ft. per second, $\kappa = 50$.

688. Reactions of the Fixed Axis. A rigid body that turns about a fixed axis exerts an action on the fixed axis that tends partly to shift it bodily and partly to turn it out of its position. The axis must, therefore, be kept fixed by certain forces, called the **reactions** of the axis. As a straight line is determined by two of its points, to fix the axis it suffices to fix two of its points, say A and B. The reactions of the fixed axis can, therefore, be regarded as two (in general unknown and variable) forces, A at A and B at B.

Like any system of forces acting on a rigid body, these two forces can be replaced by a single force R together with a couple H. Thus, if we apply at A two equal and opposite forces, each equal and parallel to B, the resultant of A and B at A will form a single force R, while B at B with $-B$ at A forms a couple H.

689. In certain particular cases the reactions of the fixed axis may be zero. It is obvious that for a revolving piece of machinery it is desirable to avoid as far as possible any action on the bearings that keep the axis of the piece in a fixed position. It is, therefore, of importance to know under what conditions the reactions of the axis vanish.

Before discussing the general case of the determination of the reactions for a body turning about a fixed axis (for which see Arts. 697–705), it will be well to treat some simple cases commonly occurring in machines by the most direct methods.

690. Consider a revolving shaft, with wheels, pulleys, cranks, etc., mounted on it. Every particle m rigidly attached to the revolving body describes a circle about the axis of the shaft. *If the rotation be uniform*, the acceleration of the particle m, at the distance r from the axis, is the centripetal acceleration $\omega^2 r$ required for uniform circular motion (see Art. 120); it is directed along r towards the axis. The effective force of the particle is therefore $m\omega^2 r$; and by d'Alembert's principle (Art. 598) these effective forces, reversed in sense, that is, the **centrifugal forces** $m\omega^2 r$, directed along r away from the axis, *must be in equilibrium with the external forces.*

691. If the mass of the revolving body be distributed symmetrically about the axis of rotation, the centrifugal forces $m\omega^2 r$ are in equilibrium by themselves. For, this symmetry means that for every particle of mass m at the distance r from the axis there exists one of equal mass, at the same distance on the opposite side of the axis; and the centrifugal forces of these particles, being equal and opposite and acting along the same line, are in equilibrium.

Now, by Art. 690, if the centrifugal forces are in equilibrium by themselves, so must be the external forces. And if there be no external forces except the reactions of the fixed axis, these reactions will be zero, so that there is no tendency to either shift or turn the axis. This axis is then called a **permanent** axis of rotation.

Among the external forces the most important is generally the weight of the revolving body. But if the speed of rotation is very high, the centrifugal forces may be so large that the weight is insignificant in comparison with them. Unless this is the case, the weight will produce a pressure on the bearings which is easily determined.

692. The centrifugal forces may, however, be in equilibrium, and hence the reactions may be zero, even when the symmetry of mass distribution is not as perfect as assumed in Art. 691. Thus, let two particles of unequal masses m_1, m_2 be rigidly attached to the shaft, on opposite sides, on a line meeting the axis at right angles (Fig. 182). If their distances r_1, r_2 from the axis be so chosen that

$$m_1 r_1 = m_2 r_2,$$

their centrifugal forces, $m_1 \omega^2 r_1$ and $m_2 \omega^2 r_2$, will be equal and opposite and in the same line.

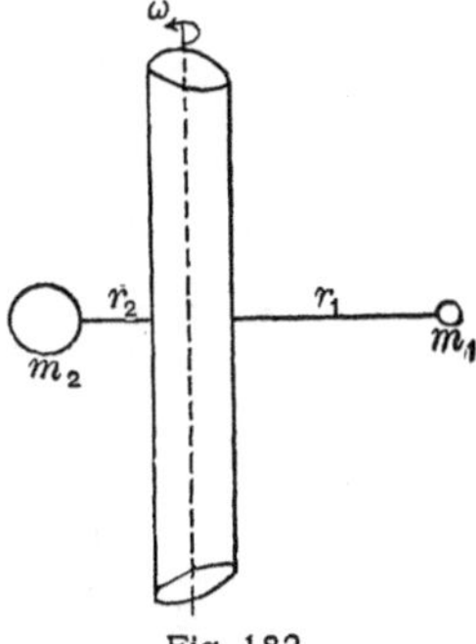

Fig. 182.

Any number of pairs of particles may be arranged in this way about the axis without disturbing the equilibrium of the centrifugal forces. The reactions are therefore zero if the other external forces can be neglected.

693. When placed, with its axis horizontal, on two supports, the shaft with the two masses attached will remain in equilibrium in any position, without any tendency to turn about its axis, even when the weight is taken into account. Such a body is said to have **standing balance.** It is clear that if the condition $m_1 r_1 = m_2 r_2$ were not satisfied, the shaft when placed on two supports would roll over until the heavier mass (m_2) lies vertically below the axis. And this tendency to shift the axis parallel to itself exists also when the shaft revolves and manifests itself in a pressure on the bearings. The amount of this pressure will be equal to the difference of the centrifugal forces, viz., $= (m_2 r_2 - m_1 r_1)\omega^2$, and can be regarded as applied at the centroid of m_1 and m_2, which does not lie on the axis, since $m_1 r_1 \neq m_2 r_2$.

These considerations are easily generalized; they show that any rigid body turning about a fixed axis exerts on this axis a pressure $= M\omega^2 \bar{r}$, directed along the perpendicular $\bar{r}$ let fall from the centroid on the axis. This pressure vanishes only if $\bar{r} = 0$, *i. e.*, if the centroid lies on the axis.

694. If two masses m_1, m_2 are not situated on the same perpendicular to the axis (Fig. 183), but still in the same plane with the axis and so that $m_2r_2 = m_1r_1$, their centroid lies on the axis and their centrifugal forces are equal and opposite, but not in the same line. These forces form, therefore, a couple whose plane contains the axis, and the tendency of this couple will be to turn the axis in this plane, *i. e.*, to change its direction.

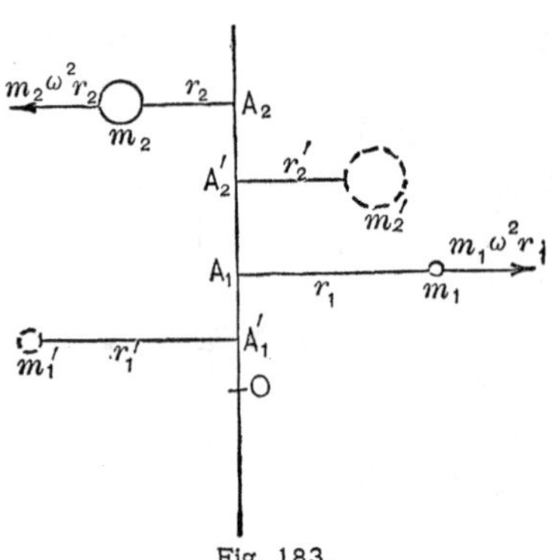

Fig. 183.

Thus the condition for "standing balance" is satisfied, and yet the axis is not "permanent."

Let O be an arbitrary point on the axis taken as origin; A_1, A_2 the points where r_1, r_2 meet the axis; and put $OA_1 = z_1$, $OA_2 = z_2$; then the moment of the couple that tends to turn the axis out of its direction is

$$m_2\omega^2r_2 \cdot z_2 - m_1\omega^2r_1 \cdot z_1 = m_1\omega^2r_1(z_2 - z_1).$$

This couple is called the **centrifugal couple.**

To obtain **running balance,** *i. e.*, to counterbalance this couple without disturbing the standing balance, two masses m_1', m_2' may be introduced, in the same plane through the axis in which m_1, m_2 are situated, and placed so that not only $m_2'r_2' = m_1'r_1'$ but also

$$m_1'r_1'(z_2' - z_1') = m_1r_1(z_2 - z_1).$$

695. For any number of masses m_i, attached to the shaft at distances r_i from the axis, but all in the same plane through the axis (Fig. 184), the condition for standing balance is evidently

$$\Sigma m_ir_i = 0,$$

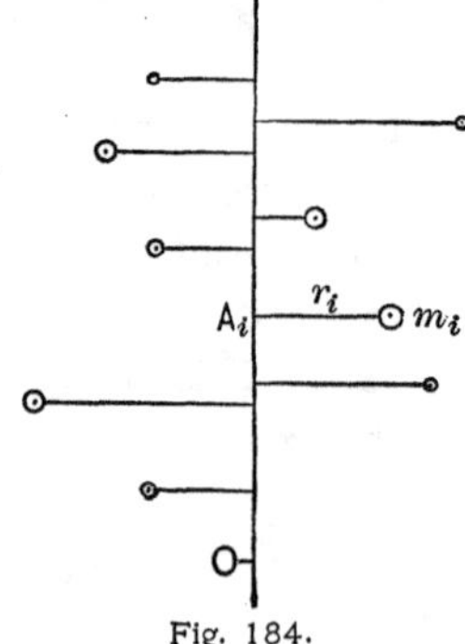

Fig. 184.

the distances r_i being taken positive on one side, negative on the opposite side of the axis. This condition means that the centroid of all the masses m_i must lie on the axis. The sum of the masses on one side of the axis need not be the same as that on the other.

For running balance the additional condition

$$\Sigma m_ir_iz_i = 0$$

must be satisfied, z_i being the distance OA_i of the plane in which m_i revolves from an arbitrary origin O, selected on the axis, the same for all the masses. It should be observed that there always exists on the axis one point for which $\Sigma m_i r_i z_i = 0$, viz., the centroid of masses equal to $m_i r_i$ placed at the points A_i on the axis; the origin O must be selected different from this particular point. The condition for running balance should be satisfied for *any* origin O on the axis, and this will be the case if it is satisfied for any one point besides the particular point just mentioned for which it is satisfied identically.

696. The theory of balancing rotating masses has assumed considerable importance in modern engineering practice, owing to the high speeds of revolution now often used. The student is referred for further details to W. E. Dalby, *The Balancing of Engines*, London, Arnold, 1902; P. H. Lorenz, *Dynamik der Kurbelgetriebe, mit besonderer Berücksichtigung der Schiffsmaschinen*, Leipzig, Teubner, 1901, where further references will be found.

697. Let us now proceed to *the general case of the motion of any rigid body with a fixed axis.* As shown in Arts. 664–666, the actual motion is given by a single equation, viz., the equation (2), Art. 666, obtained by taking moments about the fixed axis. To determine the **reactions** (comp. Art. 688) it is necessary to write out and discuss the other five equations of motion of a rigid body.

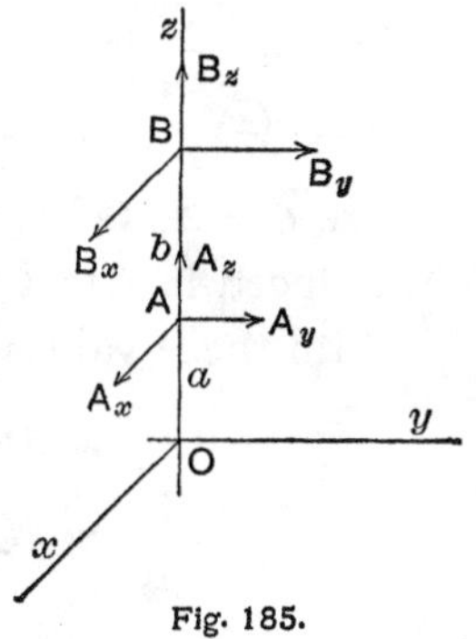

Fig. 185.

The axis will be fixed if any two of its points A, B are fixed. The reaction of the fixed point A can be resolved into three components A_x, A_y, A_z, that of B into B_x, B_y, B_z. By introducing these reactions the body becomes free; and the system composed of these reactions, of the external forces, and of the reversed effective forces must be in equilibrium. We take the axis of rotation as axis of z (Fig. 185) so that the z-co-ordinates of the particles are constant, and hence $\dot{z} = 0$, $\ddot{z} = 0$; and we put $OA = a$, $OB = b$. Then the six equations of motion are (see Art. 600 (4) and Art. 601 (6)):

$$\Sigma m\ddot{x} = \Sigma X + A_x + B_x,$$
$$\Sigma m\ddot{y} = \Sigma Y + A_y + B_y,$$
$$0 = \Sigma Z + A_z + B_z,$$
$$-\Sigma mz\ddot{y} = \Sigma(yZ - zY) - aA_y - bB_y,$$
$$\Sigma mz\ddot{x} = \Sigma(zX - xZ) + aA_x + bB_x,$$
$$\Sigma m(x\ddot{y} - y\ddot{x}) = \Sigma(xY - yX).$$

698. It remains to introduce into these equations the values for $\ddot{x}$, $\ddot{y}$. As the motion is a pure rotation, we have (see Art. 175) $\dot{x} = -\omega y$, $\dot{y} = \omega x$; hence, $\ddot{x} = -\dot{\omega} y - \omega^2 x$, $\ddot{y} = \dot{\omega} x - \omega^2 y$. Summing over the whole body, we find

$$\Sigma m\ddot{x} = -\dot{\omega}\Sigma my - \omega^2\Sigma mx = -M\dot{\omega}\bar{y} - M\omega^2\bar{x},$$
$$\Sigma m\ddot{y} = \dot{\omega}\Sigma mx - \omega^2\Sigma my = M\dot{\omega}\bar{x} - M\omega^2\bar{y},$$

where $\bar{x}$, $\bar{y}$ are the co-ordinates of the centroid; and

$$-\Sigma mz\ddot{y} = -\dot{\omega}\Sigma mzx + \omega^2\Sigma myz = -E\dot{\omega} + D\omega^2,$$
$$\Sigma mz\ddot{x} = -\dot{\omega}\Sigma myz - \omega^2\Sigma mzx = -D\dot{\omega} - E\omega^2,$$
$$\Sigma m(x\ddot{y} - y\ddot{x}) = \dot{\omega}\Sigma mx^2 - \omega^2\Sigma mxy + \dot{\omega}\Sigma my^2 + \omega^2\Sigma mxy = C\dot{\omega},$$

where $C = \Sigma m(x^2 + y^2)$, $D = \Sigma myz$, $E = \Sigma mzx$ are the notations introduced in Art. 630.

With these values the equations of motion assume the form:

$$\begin{aligned} -M\bar{x}\omega^2 - M\bar{y}\dot{\omega} &= \Sigma X + A_x + B_x,\\ -M\bar{y}\omega^2 + M\bar{x}\dot{\omega} &= \Sigma Y + A_y + B_y,\\ 0 &= \Sigma Z + A_z + B_z,\\ D\omega^2 - E\dot{\omega} &= \Sigma(yZ - zY) - aA_y - bB_y,\\ -E\omega^2 - D\dot{\omega} &= \Sigma(zX - xZ) + aA_x + bB_x,\\ C\dot{\omega} &= \Sigma(xY - yX). \end{aligned} \qquad (12)$$

699. The last equation is identical with equation (2), Art. 666.

The components of the reactions along the axis of rotation occur only in the third equation, and can therefore not be found separately. The longitudinal pressure on the axis is

$$=-A_z-B_z=\Sigma Z.$$

The remaining four equations are sufficient to determine A_x, A_y, B_x, B_y.

The total stress to which the axis is subject, instead of being represented by the two forces, at A and B, can be reduced for the origin O to a force and a couple (comp. Art. 688). The equations (12) give for the components of the force

$$\begin{aligned} -A_x-B_x &= \Sigma X+M\bar{x}\omega^2+M\bar{y}\dot{\omega}, \\ -A_y-B_y &= \Sigma Y+M\bar{y}\omega^2-M\bar{x}\dot{\omega}, \\ -A_z-B_z &= \Sigma Z. \end{aligned} \qquad (13)$$

This force consists of the resultant of the external forces,

$$R=\sqrt{(\Sigma X)^2+(\Sigma Y)^2+(\Sigma Z)^2},$$

and two forces in the xy-plane which form the reversed effective force of the centroid; for $M\bar{x}\omega^2$ and $M\bar{y}\omega^2$ give as resultant the centrifugal force $M\omega^2\sqrt{\bar{x}^2+\bar{y}^2}=M\omega^2\bar{r}$, directed from the origin towards the projection of the centroid on the xy-plane, while $M\bar{y}\dot{\omega}$, $-M\bar{x}\dot{\omega}$ form the tangential resultant $M\dot{\omega}\bar{r}$, perpendicular to the plane through axis and centroid.

The couple has a component in the yz-plane, and one in the zx-plane, viz.:

$$\begin{aligned} aA_y+bB_y &= \Sigma(yZ-zY)-D\omega^2+E\dot{\omega}, \\ -aA_x-bB_x &= \Sigma(zX-xZ)+E\omega^2+D\dot{\omega}, \end{aligned} \qquad (14)$$

while the component in the xy-plane is zero. The resultant couple lies, therefore, in a plane passing through the axis of rotation.

700. In the particular case *when no forces X, Y, Z are acting on the body*, the last of the equations (12), or equation (2), shows that *the angular velocity* ω *remains constant.* The stress on the axis of rotation will, however, exist; and the axis will in general tend to change both its direction, owing to the couple (14), and its position, owing to the force (13).

If the axis be not fixed as a whole, but only one of its points, the origin, be fixed, the force (13) is taken up by the fixed point, while the couple (14) will change the direction of the axis. Now this couple vanishes if, in addition to the absence of external forces, the conditions

$$D \equiv \Sigma myz = 0, \quad E \equiv \Sigma mzx = 0 \tag{15}$$

are fulfilled. In this case the body would continue to rotate about the axis of z even if this axis were not fixed, provided that the origin is a fixed point. A line having this property is called a **permanent axis of rotation.**

As the meaning of the conditions (15) is that the axis of z is a principal axis of inertia at the origin (see Art. 639), we have the proposition that *if a rigid body with a fixed point, not acted upon by any forces, begin to rotate about one of the principal axes at this point, it will continue to rotate uniformly about the same axis.* In other words, the principal axes at any point are always, and are the only, permanent axes of rotation. This can be regarded as the dynamical definition of principal axes.

701. It appears from the equations (13) that the *position* of the axis of rotation will remain the same if, in addition to the absence of external forces, the conditions

$$\bar{x} = 0, \bar{y} = 0 \tag{16}$$

be fulfilled; for in this case the components of the force (13) all vanish. If, moreover, the axis of rotation be a principal axis, the rotation will continue to take place about the same line even when the body has no fixed point.

The conditions (16) mean that the centroid lies on the axis of z; and it is known (Art. 639) that a centroidal principal axis is a principal axis at every one of its points. The axis of z must therefore be a principal axis of the body, *i. e.*, a principal axis at the centroid. We have thus the proposition: *If a free rigid body, not acted upon by any forces, begin to rotate about one of its centroidal principal axes, it will continue to rotate uniformly about the same line.*

702. The determination of the reactions is much simplified in the case *when both the mass of the body and the external forces are distributed symmetrically with respect to the centroidal plane perpendicular to the fixed axis.* For it is then sufficient to support the axis at a single point, viz., the point O (Fig. 186) where the axis meets the plane of symmetry; this point O is called the **center of suspension.** The reaction of the axis is therefore a single force P, passing through O, in the plane of symmetry; $-P$ is the pressure on the axis. The whole problem becomes plane; we take the fixed axis as axis of z, the plane of symmetry as xy-plane.

Owing to the symmetry, the impressed forces will reduce for any point in the xy-plane, say for the centroid G of the body, to a single resultant R and a couple H in this plane. It will generally be convenient to resolve the forces along OG and at right angles to it; thus the reaction P is replaced by its components P_r, P_θ, and the resultant R of the impressed forces by R_r, R_θ.

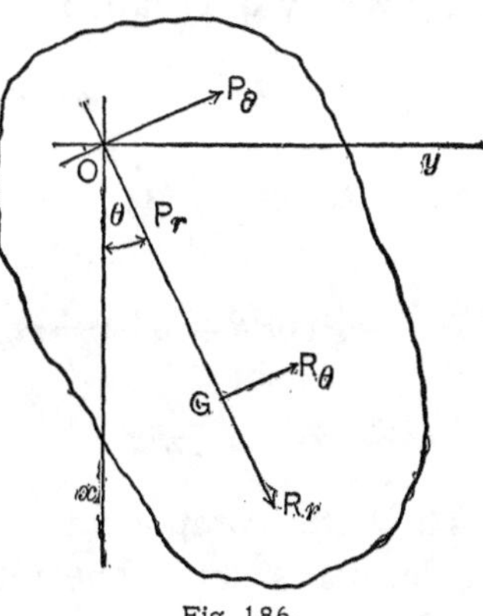

Fig. 186.

703. The centroid G describes a circle about O with radius $OG = h$; its accelerations are therefore $h\dot{\theta}^2$ along GO and $h\ddot{\theta}$ at right angles to GO, θ being the angle of rotation counted from the arbitrary axis Ox in the plane of symmetry.

The centroid moves as if all the forces were applied at this point (Art. 603). Hence resolving along OG and at right angles to it, and taking moments about O, we find the three equations of motion:

$$-mh\dot{\theta}^2 = R_r + P_r, \tag{17}$$

$$mh\ddot{\theta} = R_\theta + P_\theta, \tag{18}$$

$$m(q^2 + h^2)\ddot{\theta} = H, \tag{19}$$

where q is the radius of inertia for the centroidal axis parallel to the fixed axis.

The equation (19), which is the same as equation (2), Art. 666, determines θ and its derivatives; substituting their values in (17) and (18), the reactions P_r, P_θ are found.

704. If the fixed axis be assumed horizontal and gravity as the only impressed force, we have the case of the compound pendulum (Art. 669). Taking the axis of x vertically downward, we have

$$R_r = mg\cos\theta, \quad R_\theta = -mg\sin\theta, \quad H = -mgh\sin\theta.$$

Equation (19) becomes

$$\ddot{\theta} = -\frac{gh}{q^2 + h^2}\sin\theta,$$

and gives by integration if $\omega = \omega_0$ for $\theta = \theta_0$:

$$\dot{\theta}^2 \equiv \omega^2 = \omega_0^2 + \frac{2gh}{q^2 + h^2}(\cos\theta - \cos\theta_0).$$

Substituting in (17) and (18), we find for the components of the pressure on the axis:

$$-P_r = mg\cos\theta + mh\omega^2 = mg\left(\omega_0^2\frac{h}{g} - \frac{2h^2}{q^2+h^2}\cos\theta_0 + \frac{q^2+3h^2}{q^2+h^2}\cos\theta\right),$$

$$-P_\theta = -mg.\frac{q^2}{q^2+h^2}\sin\theta.$$

The latter component is independent of the initial conditions, the former is not. The total pressure on the axis, $\sqrt{P_r^2 + P_\theta^2}$, varies in general in the course of the motion both in magnitude and in its direction in the body. In the particular case when

$\omega_0^2 = 2gh\cos\theta_0/(q^2+h^2)$, *i. e.*, when the initial kinetic energy $\frac{1}{2}m(q^2+h^2)\omega_0^2 = mgh\cos\theta_0$ (which means that ω is zero when OG is horizontal), we have

$$-P_r = mg\frac{q^2+3h^2}{q^2+h^2}\cos\theta, \quad -P_\theta = -mg\frac{q^2}{q^2+h^2}\sin\theta,$$

and the inclination ϕ of the total pressure to OG is given by

$$\tan\phi = \frac{-P_\theta}{-P_r} = -\frac{q^2}{q^2+3h^2}\tan\theta.$$

705. Exercises.

(1) A homogeneous straight rod of mass m and length l, hinged at one end so as to swing in a vertical plane, is let go from a horizontal position. Determine the pressure on the axis of the hinge (Fig. 187).

With $h=\frac{1}{2}l$, $q^2=\frac{1}{12}l^2$, $\theta_0=-\frac{1}{2}\pi$, $\omega_0=0$, we have, by Art. 704,

$$\omega^2 = \frac{3g}{l}\cos\theta, \quad -P_r = \frac{5}{2}mg\cos\theta, \quad -P_\theta = -\frac{1}{4}mg\sin\theta.$$

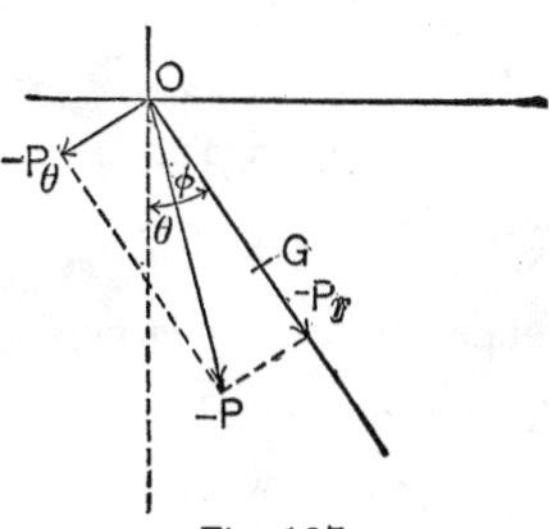

Fig. 187.

The component pressure along the rod, $-P_r$, is always positive, *i. e.*, directed from O toward G; as θ varies from $-\frac{1}{2}\pi$ through zero to $\frac{1}{2}\pi$, $-P_r$ varies from zero to $\frac{5}{2}$ of the weight of the rod to zero. The component perpendicular to the rod, $-P_\theta$, is positive while the rod descends and negative while it ascends; it is therefore always inclined downward; its maximum value, equal to $\frac{1}{4}$ of the weight of the rod, occurs when the rod is horizontal. The total pressure on the axis is

$$P = \sqrt{P_r^2+P_\theta^2} = \frac{1}{4}mg\sqrt{100\cos^2\theta+\sin^2\theta} = \frac{1}{4}mg\sqrt{99\cos^2\theta+1}.$$

This pressure is greatest, viz., $=\frac{5}{2}mg$ when the rod is vertical; it is least, viz., $=\frac{1}{4}mg$, when the rod is horizontal. The inclination ϕ of P to the rod is given by

$$\tan\phi = \frac{P_\theta}{P_r} = -\frac{1}{10}\tan\theta.$$

Discuss in a similar way the horizontal and vertical components of P.

(2) A rapidly revolving axle whose axis is Az (Fig. 188) has the straight homogeneous rod AC attached to it at a constant angle, by means of the link BC, at right angles to Az. The mass of the axle and of the link BC are neglected; also the action of gravity on the whole revolving mass. If $AB = b$, $BC = c$, determine the pressure on the axis.

As of the six equations of motion (12) (Art. 698) only the last, which does not involve the reactions, requires integration, so that the determination of the reactions, after ω has been found from the sixth equation, is a mere algebraical process, we may select the zx-plane so as to pass through the centroid. Taking A as origin, and A and B as fixed points, we have in the equations (12) of Art. 698:

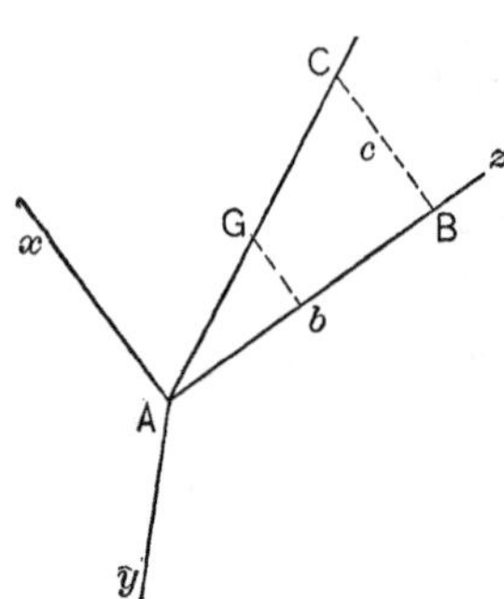

Fig. 188.

$$a = 0, \ b = b, \ \bar{x} = \tfrac{1}{2}c, \ \bar{y} = 0, \ \bar{z} = \tfrac{1}{2}b;$$

and the sixth equation gives, as there are no impressed forces, $\omega =$ const. It is easy to see that $D = 0$, $E = \frac{1}{3}mbc$. Hence the equations reduce to

$$-\tfrac{1}{2}mc\omega^2 = A_x + B_x, \quad 0 = A_y + B_y, \quad 0 = A_z + B_z,$$

$$0 = -bB_y, \quad -\tfrac{1}{3}mbc\omega^2 = bB_x, \quad C \cdot \dot{\omega} = 0,$$

whence $A_x = -\frac{1}{6}m\omega^2 c$, $B_x = -\frac{1}{3}m\omega^2 c$, $A_y = 0$, $B_y = 0$.

The two parallel forces $-A_x$, $-B_x$ are equivalent to a single resultant $= \frac{1}{2}m\omega^2 c$, which is the centrifugal force of the centroid; but it should be noticed that this resultant is not applied at the centroid G, but at the distance $z_1 = \frac{2}{3}b$ from the xy-plane; in other words, it passes through the centroid of the triangle ABC.

(3) Solve (2) when the axis Az is vertical (downward) and the weight of the rod AC is taken into account.

(4) A homogeneous plate of the shape of a right-angled triangle of sides b, c, revolves about the side b, under no impressed forces; determine the pressure on the axis.

(5) Solve (4) when the axis b is vertical and the weight of the plate is taken into account.

706. Impulses. Suppose a rigid body with a fixed axis is acted upon, when at rest, by a single impulse F, in a plane perpendicular to the axis and at the distance p from the axis. It is required to determine the initial motion of the body just after impact.

As the impulsive reactions of the fixed axis have no moment about this axis, the initial angular momentum of the body about the fixed axis must be equal to the moment of the impulse $\underset{.}{F}$ about the same axis; *i. e.*, to $\underset{.}{F}p$. If ω is the initial angular velocity, the angular momentum of the body about the fixed axis is, by Art. 665, $= \omega I$, where I is the moment of inertia of the body for the fixed axis. Hence we have

$$\omega = \frac{\underset{.}{F}p}{I}. \tag{20}$$

707. Let the impulse $\underset{.}{F}$ be produced by a particle of mass m impinging on the body with the fixed axis l; the mass of this body we shall here denote by m'. The idea of reduced mass (Art. 677) can often be used to advantage to determine $\underset{.}{F}$.

Let u be the projection of the velocity of the particle m on the plane perpendicular to the axis (the component parallel to l is evidently ineffective as regards rotation about l); and let the particle m strike the body at the point P where the direction of u meets the plane through the axis l perpendicular to u. Then we have only to reduce the mass m' of the body to the point P by putting

$$I = m'_p \cdot p^2,$$

where I is the moment of inertia of the body for the fixed axis l, p the distance of P from l, *i. e.*, the shortest distance of u and l, and m'_p the mass of the body reduced to P. The impact can now be treated like that of two homogeneous spheres (Arts. 432 sq.), except that the mass m' of the body impinged upon is replaced by m'_p. The velocity v of m and the velocity v' of the point P are therefore given by the equations

$$mv + m'_p v' = mu + m'_p u',$$

$$v' - v = e(u - u'),$$

where e is the coefficient of restitution and u' is the linear velocity of the point P just before impact.

708. In the case of inelastic impact we have $e = 0$, and hence $v' = v$; the former of the two equations gives, therefore, if the body m' is initially at rest,

$$v' = v = \frac{mu}{m + m'_p},$$

whence

$$\omega = \frac{v'}{p} = \frac{mup}{p^2(m + m'_p)} = \frac{mup}{mp^2 + I}. \qquad (21)$$

The momentum imparted to the body by the impact of the particle is

$$\underset{\cdot}{F} = m'_p v' = \frac{mm'_p u}{m + m'_p} = \frac{I}{I + mp^2} \cdot mu. \qquad (22)$$

This value of $\underset{\cdot}{F}$ reduces the formula (20) to (21).

The result (22) can be expressed by saying that, in the case of inelastic impact where the particle after impact moves along with the body, we may take for $\underset{\cdot}{F}$ the whole momentum of the impinging mass if at the same time we increase the moment of inertia I of the body by that of the particle, viz., mp^2.

The same result can be derived without using the idea of reduced mass. As the particle after impact moves along with the body, with the velocity v' of the point P, the momentum given up by it to the body is $\underset{\cdot}{F} = m(u - v') = mu - m\omega p = mu - m\underset{\cdot}{F}p^2/I$, by (20); hence $\underset{\cdot}{F} = muI/(I + mp^2)$, which agrees with (22).

709. To determine the impulsive stress produced on the axis by a single impulse $\underset{\cdot}{F}$, we have to apply the general equations (17), (18) of Art. 614, or (19), (20) of Art. 615. But we can also proceed directly as follows:

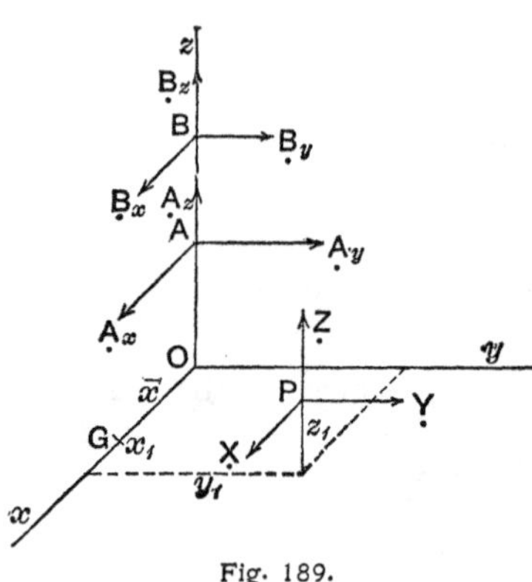

Fig. 189.

Take the fixed axis as the axis of z and the zx-plane through the centroid G (Fig. 189) and let $\bar{x}$, 0, 0 be the co-ordinates of G, and x_1, y_1, z_1 those of the point of application P of the impulse $\underset{\cdot}{F}$. The components of $\underset{\cdot}{F}$ may be denoted by $\underset{\cdot}{X}, \underset{\cdot}{Y}, \underset{\cdot}{Z}$; those of the reactions of the axis by $\underset{\cdot}{A}_x, \underset{\cdot}{A}_y, \underset{\cdot}{A}_z, \underset{\cdot}{B}_x, \underset{\cdot}{B}_y, \underset{\cdot}{B}_z$ (comp. Art. 697).

As the initial motion after impact is a rotation about the axis of z, we have $\dot{x} = -\omega y$, $\dot{y} = \omega x$, $\dot{z} = 0$, so that the momentum of a particle of mass m has the components $-m\omega y$, $m\omega x$, 0. Reducing these momenta to the origin O, we find a resultant momentum whose components are $-\omega\Sigma my = 0$, $\omega\Sigma mx = M\omega\bar{x}$, 0; and a resulting couple whose vector has the components $-\omega\Sigma mzx = -E\omega$, $-\omega\Sigma myz = -D\omega$, $\omega\Sigma m(x^2+y^2) = C\omega$, where C, D, E have the same meaning as in Art. 698.

The six equations of motion just after the impact are therefore, if the body was originally at rest,

$$
\begin{aligned}
0 &= \underset{\cdot}{X} + \underset{\cdot}{A}_x + \underset{\cdot}{B}_x, \\
M\bar{x}\omega &= \underset{\cdot}{Y} + \underset{\cdot}{A}_y + \underset{\cdot}{B}_y, \\
0 &= \underset{\cdot}{Z} + \underset{\cdot}{A}_z + \underset{\cdot}{B}_z, \\
-E\omega &= y_1\underset{\cdot}{Z} - z_1\underset{\cdot}{Y} - a\underset{\cdot}{A}_y - b\underset{\cdot}{B}_y, \\
-D\omega &= z_1\underset{\cdot}{X} - x_1\underset{\cdot}{Z} + a\underset{\cdot}{A}_x + \underset{\cdot}{B}b_x, \\
C\omega &= x_1\underset{\cdot}{Y} - y_1\underset{\cdot}{X}.
\end{aligned}
\qquad (23)
$$

It is easily verified that this is the form assumed by the equations (19), (20) of Art. 615 in the present case.

710. The last of these equations is nothing but the equation (20) of Art. 706. The components $\underset{\cdot}{A}_z$, $\underset{\cdot}{B}_z$ along the axis cannot be determined separately; the other components of the reactions can be found from the first, second, fourth, and fifth equations.

The impulsive stress to which the axis is subjected by the impulse, or the so-called *percussion of the axis*, instead of being represented by two impulses, $\underset{\cdot}{A}$, $\underset{\cdot}{B}$ as above, can also be regarded as composed of an impulsive pressure at O whose components are

$$
-\underset{\cdot}{A}_x - \underset{\cdot}{B}_x = \underset{\cdot}{X}, \quad -\underset{\cdot}{A}_y - \underset{\cdot}{B}_y = Y_b - M\bar{x}\omega, \quad -\underset{\cdot}{A}_z - \underset{\cdot}{B}_z = \underset{\cdot}{Z},
$$

and an impulsive couple whose vector has the components

$$
a\underset{\cdot}{A}_y + b\underset{\cdot}{B}_y = y_1\underset{\cdot}{Z} - z_1\underset{\cdot}{Y} + E\omega, \quad -a\underset{\cdot}{A}_x - b_x\underset{\cdot}{B}_x = z_1\underset{\cdot}{X} - x_1\underset{\cdot}{Z} + D\omega, \quad 0.
$$

The last component being zero, the resulting couple lies in a plane passing through the axis of z.

If there were any number of impulses acting on the body simultaneously, the effect on the axis could be determined in the same way, except that the quantities $\underset{.}{X}$, $\underset{.}{Y}$, $\underset{.}{Z}$, $y_1\underset{.}{Z} - z_1\underset{.}{Y}$, $z_1\underset{.}{X} - x_1\underset{.}{Z}$, must be replaced by the corresponding sums.

711. It follows from the preceding article that the conditions under which a single impulse acting on a rigid body with a fixed axis will produce no stress on the axis are

$$\underset{.}{X}=0, \quad \underset{.}{Y}=M\bar{x}\omega, \quad \underset{.}{Z}=0, \quad -z_1M\bar{x}+E=0, \quad D=0. \quad (24)$$

If these conditions are fulfilled, the resulting motion will be the same even when the axis is free.

The first and third equations show that *the impulse must be perpendicular to the plane passing through axis and centroid.* The meaning of the fourth and fifth conditions becomes apparent if the xy-plane be taken so as to pass through the point of application P of the impulse. The new origin O' is the foot of the perpendicular let fall from P on the fixed axis. To transform the conditions (24) to the new system it is only necessary to substitute $z+z_1$ for z; the first three conditions are not affected, and the last two become

$$-z_1M\bar{x} + \Sigma mzx + z_1\Sigma mx = 0, \quad \Sigma myz + z_1\Sigma my = 0,$$

or, since $\Sigma mx = M\bar{x}$, $\Sigma my = 0$,

$$E' = 0, \quad D' = 0,$$

where E', D' are the products of inertia at O'.

It thus appears that *the axis of z must be a principal axis at the foot of the perpendicular let fall on this axis from the point of application of the impulse.*

712. It should be noticed that a line taken at random in a body is not necessarily a principal axis at any one of its points. But if a line is a principal axis at a point O', then it is always possible to determine an impulse that will produce no stress on this line so that the body will begin to rotate about it as axis even though it be not fixed. As shown in the last article, the impulse must be $=M\bar{x}\omega$, and must be directed at right angles to the plane through axis and centroid. The point where it meets this plane is called the **center of percussion** for the line. Its distance x_1 from the axis is found from the equation of motion, viz., the last of the equations (23), which, owing to the conditions (24), reduces to

$$C = M\bar{x}x_1.$$

If q' be the radius of inertia of the body for a parallel centroidal axis, we have $C = M(q'^2 + \bar{x}^2)$; hence

$$x_1 = \bar{x} + \frac{q'^2}{\bar{x}}. \tag{25}$$

Hence, if a given line l be a principal axis for one of its points O', there exists a center of percussion; it lies on the intersection of the plane (l, G) with the plane through O' perpendicular to l, at the distance x_1, given by (25), from the line l. An impulse $M\bar{x}\omega$ through the center of percussion at right angles to the plane through axis and centroid, while producing no percussion on the axis, sets the body rotating with angular velocity ω if it was originally at rest; on the other hand, if the body was originally in rotation about the axis, such an impulse can bring the body to rest without affecting the axis.

713. Exercises.

(1) A homogeneous straight rod of length l and mass m is suspended vertically from a horizontal axis through its end. At what point and in what direction must it be struck to produce no shock on the axis?

(2) Show that the center of percussion of a compound pendulum coincides with the center of oscillation (Art. 670).

(3) A homogeneous straight rod of length l lies on a smooth horizontal table; how should it be struck to begin turning about a vertical axis through one end?

(4) A homogeneous circular disk, of radius a and mass m, while turning about its axis with angular velocity ω, is suddenly stopped by a peg (which projects radially from its circumference) striking a fixed obstacle. Determine the momentum of the impact and the stress on the bearings which are at distances a, b from the disk on opposite sides.

(5) "A pendulum is constructed of a sphere (radius a, mass M) attached to the end of a thin rod (length b, mass m). Where should it be struck at each oscillation that there may be no impulsive pressure to wear out the point of support?" (Routh.)

(6) Show that if a body with a fixed *centroidal* axis is struck a blow, the axis is always subject to an impulsive stress.

(7) A *trip-hammer* consists essentially of a heavy mass carried by an arm which can turn about a fixed axis c'. A cam projecting from a revolving shaft strikes the arm of the hammer at each revolution, raising it a certain distance so as to let the head of the hammer drop on the anvil. Let m be the mass of this shaft (with its fly-wheel), I its moment of inertia for its axis c, p the distance of the point of impact P from the axis c of the shaft; then $m_p = I/p^2$ is the mass reduced to P. Let m' be the mass of the hammer and arm, I' their moment of inertia for the axis c', and p' the distance of the point of impact P from c', so that $m'_p = I'/p'^2$ is the mass of the hammer reduced to P. Regarding the impact as inelastic, we then have for the velocity of the point P just after impact (see Art. 708),

$$v = \frac{m_p}{m_p + m'_p} u,$$

where $u = \omega p$, ω being the angular velocity of the shaft c. The lost kinetic energy is (by (10), Art. 444),

$$E = \frac{m'_p}{m_p + m'_p} \cdot \tfrac{1}{2} m_p u^2.$$

To make this as small as possible, m_p/m'_p should be large; *i. e.*, the *reduced* mass of the fly-wheel should be large in comparison with that of the hammer.

To reduce as much as possible the percussion of the axis c' of the hammer, the point of impact P should be near the center of percussion (Art. 712) of the hammer; that is, p' should satisfy the relation

$$I' = m'\bar{x}p',$$

where $\bar{x}$ is the distance of the centroid of the hammer (with its arm) from the axis c'; this gives

$$p' = \frac{I'}{m'\bar{x}}.$$

(8) In the trip-hammer, Ex. (7), let m_1 be the mass of the hammer-head regarded as concentrated at the distance a from the axis c'; m_2 the mass of the arm regarded as a straight rod of length $l (> a)$. Find

the center of percussion P, and show that with $m_1/m_2 = 8$, $a/l = 0.75$, we find $p'/l = 0.74$.

(9) In the problem of Arts. 707, 708, determine F for perfectly elastic impact ($e = 1$).

(10) The *ballistic pendulum* consists essentially of a heavy wooden block (or a case filled with sand) suspended from a horizontal axis. A bullet is shot into the block at rest, at right angles to the vertical plane through the axis of the pendulum. From the observed rise of the block due to the impact of the bullet, the velocity of the bullet is computed.

The bullet enters a certain distance into the wood; the time of this motion is neglected, *i. e.*, the impact is supposed to take place at the point A where the bullet stops. Let a be the distance of this point from the axis, α the angle it makes with the vertical; then the moment of the momentum mu of the bullet is $mu \cdot a \cos \alpha$; equating this to the angular momentum of pendulum and bullet (since after the impact the bullet moves with the pendulum), we have the equation of impact, which gives the initial angular velocity of the pendulum. If m' be the mass of the pendulum, q its radius of inertia for the axis, h the distance of its centroid from the axis, the equation of motion of the pendulum (with the bullet) can be written down and integrated. This gives

$$(m'q^2 + ma^2)\,\omega^2 = 2\,m'gh\,(1 - \cos \theta_1) + 2\,mga\,[\cos \alpha - \cos(\theta_1 - \alpha)]$$
$$= 2\,(m' + m)gh\,(1 - \cos \theta_1), \text{ nearly.}$$

The angle θ_1 through which the pendulum swings from the vertical, can be measured by a tape attached to the block, say at the distance b from the axis, and drawn out of a reel as the pendulum moves; if the length of tape drawn out is c, we evidently have $c = 2\,b \sin \frac{1}{2}\theta_1$.

Combining the results, we find finally

$$u = \left(1 + \frac{m'}{m}\right)\frac{hc}{ab \cos \alpha}\sqrt{gl},$$

where l is the length of the equivalent simple pendulum.

(11) A homogeneous cube whose edges are a rests on a horizontal plane; one of its base edges being a fixed axis. Determine the impulse F of the least blow that will upset the cube about the fixed axis. Observe that the kinetic energy due to the blow must be sufficient to raise the centroid to the position vertically above the fixed axis. This work,

which must be done before the upsetting takes place, is sometimes called the *dynamic stability*.

(12) In Ex. (11), if the edge of the cube is 1 ft., determine with what velocity a mass equal to one third of that of the cube would have to strike it (at the middle of the edge opposite the axis, at right angles to the diagonal plane) to upset the cube. (See Art. 708.)

IV. *Plane Motion.*

714. As explained in Art. 610, the general problem of the motion of a rigid body resolves itself into two problems which may be treated separately: the motion of the centroid regarded as a particle containing the whole mass of the body and having all the forces acting on the body applied to it parallel to their actual directions, and the motion of the body about the centroid regarded as a fixed point. The former problem is solved by integrating the three equations (4) of Art. 600, which can also be written in the form (5), Art. 600, or (8) Art. 603; the latter by integrating the three equations (6) of Art. 601, which can also be written in the form (7), Art. 601.

715. If the centroid moves in a *plane* curve and the rotation about the centroid always takes place about axes perpendicular to the plane of this curve, the motion of the body is called *plane motion*. It suffices to study the motion of the cross-section of the body in this plane. The motion of the centroid $G(\bar{x}, \bar{y})$ is given by the two **equations of linear momentum**:

$$M\ddot{\bar{x}} = \Sigma X, \ M\ddot{\bar{y}} = \Sigma Y, \tag{1}$$

or

$$\frac{d}{dt}M\dot{\bar{x}} = R_x, \ \frac{d}{dt}M\dot{\bar{y}} = R_y. \tag{1'}$$

The motion about the centroid is given by a single equation of the form

$$\Sigma m(x\ddot{y} - y\ddot{x}) = \Sigma(xY - yX), \tag{2}$$

or

$$\frac{d}{dt}\Sigma m(x\dot{y} - y\dot{x}) = H, \tag{2'}$$

the **equation of angular momentum**, obtained by "taking moments" about any point of the plane. As the motion about the centroid takes place as if the centroid were fixed (Art. 606), it will often be convenient to take moments about the centroid. The equation (2′) then assumes the form

$$I\frac{d\omega}{dt} = H, \tag{2''}$$

where I is the moment of inertia of the body about the centroidal axis perpendicular to the plane, H the sum of the moments of the external forces about the same axis, ω the angular velocity about the axis.

Analogous considerations hold for the equations of the change of motion due to impulses (Arts. 614, 615).

716. As an instructive example of plane motion consider the motion of a *homogeneous circular cylinder down an inclined plane, the axis of the cylinder remaining horizontal.* Let m be the mass, a the radius of

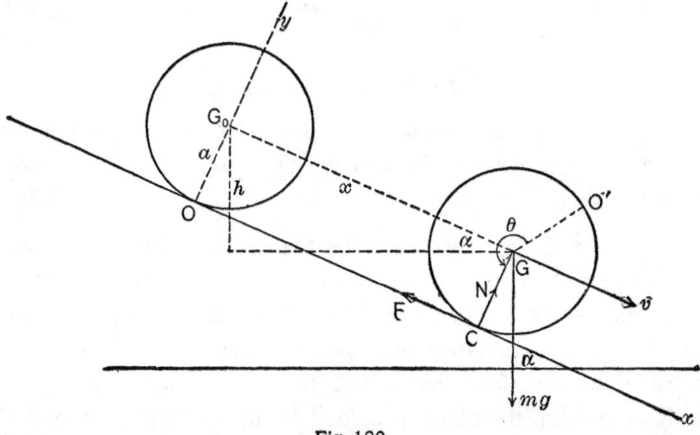

Fig. 190.

the cylinder. In Fig. 190, G_0 is the initial position of the centroid (at the time $t = 0$) when the linear velocity v of the centroid and the angular velocity ω about it are both zero; G is the position of the centroid at any time t. The inclination of the plane to the horizon is α.

If that point of the circumference which is initially in contact with the plane (at O) has at the time t the position O', then $O'GC = \theta$ is the angle through which the cylinder has turned about its axis in the time t. This angle θ and the distance $OC = G_0G = x$ through which the centroid has moved are the variables that determine the position of the cylinder at any time.

717. The motion of a cylinder on a fixed plane is called pure *rolling* if at every instant the line of contact is the instantaneous axis. Rolling is to be clearly distinguished not only from *sliding*, when the cylinder has a motion of translation parallel to the plane so that the instantaneous axis lies at infinity, but also from *spinning*, when the line of contact turns about one of its points in the plane. In the most general case the motion of the cylinder on the plane would be a combination of these three kinds of motion.

It follows from the definition that in pure rolling the length of the circular arc $O'C = a\theta$ must equal the distance $OC = G_0G = x$, through which the centroid has moved. The *kinematical condition for pure rolling* is therefore:

$$x = a\theta; \tag{3}$$

hence

$$\frac{dx}{dt} = a\frac{d\theta}{dt}, \quad \text{or } v = a\omega, \tag{4}$$

and

$$\frac{d^2x}{dt^2} = a\frac{d^2\theta}{dt^2} = a\frac{d\omega}{dt}, \quad \text{or } j \equiv \frac{dv}{dt} = a\frac{d\omega}{dt}, \tag{5}$$

where v and j are the (linear) velocity and acceleration of the centroid, ω and $d\omega/dt$ the angular velocity and acceleration about the centroid, *i. e.*, about the axis of the cylinder.

718. Let the only impressed force be the weight mg of the cylinder. The reaction of the plane at the point of contact C can be resolved into its normal component N, at right angles to the plane, and the frictional resistance F, along the greatest slope of the plane.

If the plane were perfectly smooth so that $F = 0$, the cylinder would necessarily *slide* down the plane, provided its initial angular velocity be zero. For, as the only forces acting, mg and N, both pass through the centroid, no angular momentum about the axis can be generated. In this case the principle of kinetic energy and work gives, if the velocity at the time t for this limiting case be denoted by v_1,

$$\tfrac{1}{2} m v_1^2 = mgh, \quad \text{or } v_1 = \sqrt{2\,gh}, \tag{6}$$

where $h = x \sin\alpha$ is the vertical height through which the centroid G has descended.

719. *Pure rolling* occurs only if the frictional force F is of sufficient amount to prevent sliding (or *slipping*, as it is called). The moment Fa of the friction produces angular momentum, giving the cylinder an angular velocity ω (clockwise in Fig. 190) about its axis. Mechanical means might be substituted for friction to produce the same effect, *e. g.*, gearing, or a flexible band wrapped around the cylinder and stretched up the plane.

In the case of pure rolling the principle of kinetic energy and work gives (rolling friction being neglected)

$$\tfrac{1}{2} mv^2 + \tfrac{1}{2} I\omega^2 = mgh, \tag{7}$$

the work of N as well as that of F being zero because the point of application C is instantaneously at rest.

The left-hand member which represents the kinetic energy of the cylinder can be simplified by introducing the mass reduced to the point of contact C, *i. e.*, to the distance a from the centroid (Art. 677): $m' = I/a^2$; as in the case of pure rolling (Art. 717) $a\dot{\omega} = v$, (7) reduces to

$$\tfrac{1}{2}(m + m')\, v^2 = mgh. \tag{8}$$

Denoting, as in Art. 718, the velocity of frictionless sliding by v_1, we find:

$$v = \frac{1}{\sqrt{1 + m'/m}}\, v_1. \tag{9}$$

For the homogeneous cylinder, $m' = \frac{1}{2} m$ (see Ex. (1), Art. 679), so that

$$v = \sqrt{\tfrac{2}{3}}\, v_1.$$

The same reasoning applies when the cylinder is replaced by any other solid of revolution whose mass is distributed symmetrically both with respect to the axis of revolution and with respect to the centroidal plane perpendicular to this axis; the formulæ (6) to (9) hold without change in this more general case. But the quotient m'/m will not always be a pure number; it will depend on the dimensions of the body and on the distance a of the centroid from the inclined plane.

720. Exercises.

(1) A homogeneous cylinder of 1 ft. diameter rolls down a plane inclined at 30° to the horizon, over a distance of 20 ft.; find the final linear and angular velocity.

(2) A sphere, a circular disk or cylinder, and a hoop or thin cylindrical shell, all three homogeneous, roll down the same inclined plane through the same distance; show that the velocities acquired are

$$v_s = \sqrt{\tfrac{5}{7}}\, v_1 = 0.845\, v_1,\quad v_c = \sqrt{\tfrac{2}{3}}\, v_1 = 0.816\, v_1,\quad v_h = \sqrt{\tfrac{1}{2}}\, v_1 = 0.707\, v_1.$$

(3) Two equal circular disks, each of radius R and mass m, are rigidly connected by a short cylindrical shaft, of radius $a < R$, the axis of the shaft coinciding with the axes of the disks. The shaft rolls down an inclined plane formed by a board set on edge. Neglecting the mass of the shaft, find the linear velocity acquired after descending through a vertical height h. Show that when a is small in comparison with R, we have approximately $v/v_1 = a/R$.

(4) A solid sphere of radius a and a hollow sphere of outer radius a and inner radius a', both homogeneous, are rolled down a plane; compare the final linear velocities: (a) when $a' = \frac{1}{2}a$, (b) when a' is nearly equal to a. The hollow sphere could thus be distinguished from the solid sphere even if their weights were equal.

(5) For the three typical cases of Ex. (2) determine what portion of the total energy is rotational.

721. The three equations of motion (Art. 715) assume the following form for the problem of Art. 716, if O (Fig. 190) be taken as origin, the axis of y at right angles to the inclined plane, the axis of x down its greatest slope:

$$m\frac{dv}{dt} = mg\sin\alpha - F,\quad 0 = N - mg\cos\alpha,$$
$$I\frac{d\omega}{dt} = Fa,\quad \text{or}\quad m'a\frac{d\omega}{dt} = F, \qquad (10)$$

where $m' = I/a^2$ is the mass reduced to the point of contact C.

These equations hold not only for a cylinder, but also in the more general case mentioned at the end of Art. 719, provided the centroid has the distance a from the plane.

722. The second of the equations (10) merely determines the pressure on the plane. The third, in the case of pure rolling when the condition (5) holds, becomes

$$I\frac{dv}{dt} = Fa^2,\quad \text{or}\quad m'\frac{dv}{dt} = F. \qquad (11)$$

Combining this with the first of the equations (10) and putting $I=mq^2$, we find for the acceleration of the centroid G and for the reaction F along the plane:

$$\frac{dv}{dt}=\frac{a^2}{a^2+q^2}g\sin\alpha=\frac{m}{m+m'}g\sin\alpha, \tag{12}$$

$$F=\frac{q^2}{a^2+q^2}mg\sin\alpha=\frac{m'}{m+m'}mg\sin\alpha. \tag{13}$$

It appears from (12) that in pure rolling the acceleration of the centroid is constant but always less than in frictionless sliding; indeed, of the component of the weight along the plane, $mg\sin\alpha$, only the fraction $m/(m+m')$ produces linear acceleration; the remaining portion, viz., $m'/(m+m')$ of $mg\sin\alpha$, is used to produce angular acceleration.

The acceleration dv/dt of the centroid being constant, the motion of the centroid is uniformly accelerated. And as, by (5), $ad\omega/dt=dv/dt$ the rotation about the centroid is likewise uniformly accelerated.

723. The value of the frictional resistance F is given by (13). It is important to notice the difference in the effect of friction in the case of sliding from that in the case of rolling. In sliding motion, the whole amount of friction comes into play, the resistance along the plane being $F=\mu N$, where μ is the coefficient of friction. Not so in the case of pure rolling, where only as much of this force becomes active as is necessary to prevent sliding; hence, in the case of rolling, the resistance F along the plane is in general $<\mu N$.

It follows that if rolling takes place and is kept up by friction alone, we must have for the resistance F along the plane

$$F\leqq\mu mg\cos\alpha.$$

Owing to the value (13) of F, this gives the following *dynamical condition for the possibility of pure rolling:*

$$\frac{m'}{m+m'}\tan\alpha\leqq\mu,\text{ or }\tan\alpha\leqq\left(1+\frac{m}{m'}\right)\mu. \tag{14}$$

724. Suppose that the condition (14) is not satisfied so that

$$\tan\alpha>\left(1+\frac{m}{m'}\right)\mu. \tag{15}$$

The motion will be a combination of rolling and sliding. The distance $x=G_0G=OC$ (Fig. 190) through which the centroid moves in any time t is not equal to, but greater than, the arc $O'C = a\theta$ through which the body turns about the centroid in the same time; hence the relations (3), (4), (5) between x and θ, v and ω, dv/dt and $d\omega/dt$ do not hold.

The equations of motion still have the form (10); but as actual sliding takes place at the point of contact C, we now have

$$F = \mu N. \tag{16}$$

Combining this relation with the first two of the equations (10), we find

$$\frac{dv}{dt} = g(\sin\alpha - \mu\cos\alpha). \tag{17}$$

Hence the motion of the centroid is again uniformly accelerated and, owing to (15), the acceleration is positive.

The third of the equations (10) gives

$$\frac{d\omega}{dt} = \frac{\mu mga\cos\alpha}{I}; \tag{18}$$

hence the rotation about the centroid is also uniformly accelerated.

725. Exercises.

(1) For the three types of bodies mentioned in Ex. (2), Art. 720, the centroid having in each case the distance a from the inclined plane, determine the linear and angular accelerations: (*a*) in the case of pure rolling; (*b*) in the case of rolling combined with sliding; (*c*) determine the condition for pure rolling if the angle of friction is ϕ.

(2) A homogeneous circular disk, 2 ft. in diameter, rolls down a plane sloping 1 : 5. Starting from rest, how far will it go in 10 sec., (*a*) if $\mu = 0.1$? (*b*) if $\mu = 0.05$?

(3) Discuss the motion of a homogeneous cylinder with horizontal axis *up* an inclined plane, if initially $v = v_0$, $\omega = 0$: (*a*) when there is no friction; (*b*) when there is friction, but not sufficient to produce rolling.

(4) In Ex. (3), if the friction is sufficient to produce rolling, the cylinder can be made to roll up the plane under the action of gravity alone by giving it an initial angular velocity ω_0 about its axis. This can, *e. g.*, be done by a blow directed up the plane and striking the cylinder *above* the centroid (*i. e.*, at a distance from the plane $> a$). For such a blow is equivalent to an equal and parallel blow through the centroid, giving the centroid an initial velocity v_0, together with an impulsive couple tending to make the body roll up the plane with

initial angular velocity ω_0. It should be noticed that the frictional resistance F, in the case of rolling up the plane, is directed *up* the plane, just as in the case of rolling down the plane. For, what produces, or in the present case destroys, angular momentum is, properly speaking, not the force F, but the component $mg \sin \alpha$ of gravity which is directed down the plane. Write down the equations of motion and solve them.

(5) The effect of *rolling friction*, hitherto neglected, can be expressed by a couple of moment $\mu' mga \cos \alpha$, where μ' is the coefficient of rolling friction (see Arts. 378, 379). Its sense agrees with that of the moment of F in downward motion, but is opposite to it in upward motion. Thus, for rolling down the plane we have

$$m\frac{dv}{dt} = mg \sin \alpha - F,\ I\frac{d\omega}{dt} = Fa + \mu' mg a \cos \alpha;$$

the condition for pure rolling (5) gives

$$F = \frac{m'}{m+m'}(\sin \alpha - \mu' \frac{m}{m'} \cos \alpha) mg.$$

726. A homogeneous circular cylinder, of radius a and mass m, placed on a rough horizontal plane, is acted upon by gravity and a horizontal force X, passing through the centroid at right angles to the axis of the cylinder.

The equations of motion,

$$m\frac{dv}{dt} = X - F,\quad 0 = N - mg,\quad I\frac{d\omega}{dt} = Fa, \tag{19}$$

give in the case of pure rolling, owing to (5),

$$X = m\frac{dv}{dt} + \frac{I}{a}\frac{d\omega}{dt} = (m+m')\frac{dv}{dt}, \tag{20}$$

where for the cylinder $m' = \frac{1}{2}m$. The same solution holds, with the proper value of m', for the more general solid of revolution mentioned at the end of Art. 719.

The expression for X shows that the centroid of the body moves like a particle, not of mass m, but of mass $m + m'$, under the force X alone. This result is used in studying the motion of a carriage or train to take into account the "rotary inertia" of the wheels by adding to the total mass the sum $\Sigma m'$ of the masses of the wheels reduced to their circumference. Thus for a train running down an incline of angle α, under the action of gravity alone, if M be the mass of the train (including the wheels), m' the sum of the reduced masses of the wheels, the acceleration will be

$$\frac{M}{M+m'} g \sin \alpha.$$

727. Exercise.

(1) Determine the height at which the buffers and coupling chain of a railroad car should be placed to prevent "pitching," M being the mass of the car, m that of the wheels, $m' = mq^2/a^2$ their mass reduced to the point of contact with the rails, h the height of the centroid of the car above the axles.

728. A homogeneous sphere, of radius a and mass m, is placed on a rough horizontal plane, the sphere having initially a velocity of translation v_0 parallel to the plane and an angular velocity ω_0 about the horizontal diameter perpendicular to v_0. Gravity and the reaction of the plane are the only forces acting on the sphere. As these cannot change the *directions* of either v_0 or the axis of ω_0, the motion must be plane.

The following discussion applies without change to the case of a homogeneous cylinder and of a hoop; it is only necessary to give the proper value to the reduced mass m'.

Two cases may be distinguished according as the initial angular velocity ω_0 gives to the point of contact A a linear velocity $a\omega_0$ of the same sense as v_0 or of the opposite sense.

729. If $a\omega_0$ agrees in sense with v_0 (Fig. 191), the frictional resistance $F = \mu mg$ is opposite to v_0 as well as to $a\omega_0$, and tends to diminish both v_0 and ω_0. Hence the equations of motion are, if the sense of v_0 be taken as positive,

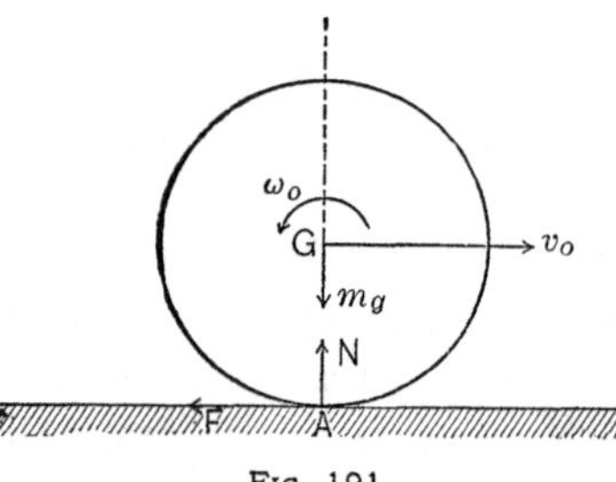

Fig. 191.

$$m\frac{dv}{dt} = -\mu mg, \quad 0 = N - mg,$$

$$m'\frac{d\omega}{dt} = -\mu m\frac{g}{a},$$

where $m' = I/a^2$ is the mass reduced to the point of contact. The first and last of these equations show that the motion of the centroid as well as the rotation about the centroid is uniformly retarded. Integrating these equations, we find

$$v = v_0 - \mu g t, \quad \omega = \omega_0 - \mu \frac{m}{m'} \frac{g}{a} t.$$

Hence v would vanish at the time $t_1 = v_0/\mu g$, ω at the time $t_2 = m' a\omega_0/\mu m g$. But the motion changes its character as soon as the velocity of the point of contact A becomes zero. Initially, this velocity is $v_0 + a\omega_0 > 0$, so that the initial motion is rolling combined with sliding, the instantaneous axis lying (parallel to the axis of ω) at the distance v_0/ω_0 above the centroid. At the time t, the velocity of A is

$$v + a\omega = v_0 + a\omega_0 - \mu g\left(1 + \frac{m}{m'}\right) t;$$

it vanishes at the time

$$t_3 = \frac{v_0 + a\omega_0}{\mu g (1 + m/m')},$$

after which the motion becomes pure rolling.

As v_0 and $a\omega_0$ are both positive, this time t_3 must lie between the times t_1 and t_2. If $t_1 > t_2$, so that ω vanishes first and has become negative at the time t_3 when pure rolling sets in, the sphere will, after the time t_3, *roll forward* (*i. e.*, in the sense of v_0). If, however, $t_1 < t_2$, v first reduces to zero, and is, therefore, negative when pure rolling begins; the sphere will, therefore, after the time t_3, *roll backward*. If $t_1 = t_2$, the sphere comes to a stop at the time $t_3 = t_1 = t_2$.

The values of t_1 and t_2 show that we have $t_1 \gtreqless t_2$ according as $v_0 \gtreqless \frac{m'}{m} a\omega_0$.

730. Exercises.

(1) Discuss that case of the problem of Art. 728, when the initial angular velocity ω_0 gives to the lowest point of the sphere a velocity $a\omega_0$ opposite in sense to v_0; distinguish the three cases $v_0 \gtreqless |a\omega_0|$, taking the sense of v_0 as positive so that ω_0 is a negative quantity.

(2) What must be the initial motion to produce rolling backward: (*a*) for a sphere? (*b*) for a cylinder or disk? (*c*) for a hoop?

(3) A homogeneous circular disk 2 ft. in diameter is set spinning at the rate of 5 rev./sec.; it is thus placed, with its axis horizontal, on

a horizontal plane whose coefficient of friction is $\frac{1}{80}$. How long will it slip (or "skid"), and what will be its linear and angular velocity when the motion becomes pure rolling?

(4) At what height above the table (cushion height) must a billiard ball be struck (horizontally) if its motion is to be pure rolling from the beginning?

(5) A billiard ball is started as a "dead ball," *i.e.*, with no initial rotation, and with $v_0 = 7$ ft. per second. If it goes 10 ft. before the motion becomes pure rolling, what is the coefficient of friction?

ANSWERS.

Page 16.

(1) Join the point P to the instantaneous center C; the direction of motion is perpendicular to CP.

(3) See Art. 24.

(4) With O as origin and the diameter perpendicular to l as x-axis, the space centrode is $y^2 = cx \pm a\sqrt{x^2+y^2}$, where a is the radius of the circle, c the distance of O from l. With A as origin and l' as x-axis the body centrode is $x^2 = ay \pm c\sqrt{x^2+y^2}$. The upper sign corresponds to l' sliding over the first and second quadrants of the circle, the lower sign to l' sliding over the third and fourth quadrants. If $c>a$, the complete fixed centrode has a node at O with the tangents $ay = \pm\sqrt{c^2-a^2}x$. The polar equations of the centrodes are $r\sin^2\theta = c\cos\theta + a$ and $r'\cos^2\theta' = a\sin\theta' + c$. The body centrode for $c>a$ is (apart from position) the fixed centrode for $c<a$, and *vice versa*.

(5) $y^2 = 2a(x+\frac{1}{2}a)$.

(6) The fixed centrode is a circle passing through O', O''; the body centrode is a circle of twice the radius of the fixed centrode. The path of any point in the fixed plane is a Pascal limaçon; the points of the body centrode describe cardioids.

(8) Two equal parabolas; the motion is the same as that of Ex. (5).

(10) With O as pole and OB as polar axis the equation of the *fixed centrode* is $r^2\cos^2\theta - 2ar\cos^2\theta + a^2 = l^2$. With O as origin and OB as x-axis the equation in cartesian co-ordinates is

$$(x^2+a^2-l^2)\sqrt{x^2+y^2} = 2ax^2,$$

or

$$(x^2+a^2-l^2)^2(x^2+y^2) = 4a^2x^4.$$

This last equation represents, however, not only the centrode of AB as B moves on the positive x-axis, but also the centrode of AB when B

moves on the negative x-axis. With $l+a=s$, $l-a=d$, the cartesian equation can also be written:

$$y=\pm\frac{x}{x^2-sd}\sqrt{(x^2-s^2)(x^2-d^2)}.$$

The equation of the *body centrode*, with A as pole and AB as polar axis, $AC=r'$, $\measuredangle BAC=\theta'$, is found by observing that $r=r'+a$, $l\sin\theta'=OB\sin\theta=r\cos\theta\sin\theta$; substituting the resulting values of r and θ in the equation of the fixed centrode, we find

$$(a^2-l^2\cos^2\theta')r'^2-2\,al^2r'\sin^2\theta'+l^2(l^2-a^2\cos\theta')=0,$$

which breaks up into the two equations

$$r_1'=l\cdot\frac{l+a\cos\theta'}{a+l\cos\theta'},\quad r_2'=l\cdot\frac{l-a\cos\theta'}{a-l\cos\theta'}.$$

These relations can be read off directly from the figure if perpendiculars be dropped from O on AB and from B on AC. In cartesian co-ordinates each of these equations is of the fourth degree.

For the *path of any point P* whose body co-ordinates, with A as origin and AB as x'-axis, are x', y', we have

$$x=a\cos\theta+x'\cos\phi+y'\sin\phi,\quad y=a\sin\theta-x'\sin\phi+y'\cos\phi.$$

After eliminating ϕ by means of the relation $l/a=\sin\theta/\sin\phi$, it remains to eliminate θ; the result is somewhat complicated.

For the *path of the midpoint of AB* the equations reduce to

$$x=a\cos\theta+\tfrac{1}{2}l\cos\phi,\quad y=a\sin\theta-\tfrac{1}{2}l\sin\phi,$$

whence
$$x=\sqrt{a^2-4y^2}+\tfrac{1}{2}\sqrt{l^2-4y^2},$$

which is of the fourth degree.

Page 27.

(3) $\sqrt{3}$.

(4) 9.3 miles; N. 10° E.

(6) $2\,a\cos\frac{1}{2}\alpha$.

(7) (*b*) 160° 48'.7.

(8) $(\sqrt{3}-1)a$, $\frac{1}{2}\sqrt{2}(\sqrt{3}-1)a$.

(10) Inclination to vertical: (*a*) $9\frac{1}{2}°$; (*b*) $22\frac{2}{3}°$; (*c*) $36\frac{5}{6}°$; (*d*) $68\frac{1}{5}°$.

Page 33.

(5) (*a*) 5.9; (*b*) 40.6; (*c*) 73.3; (*d*) 35.2; (*e*) 1093.

(6) $t=2\,(a+b)/(v_1+v_2)$.

(8) 184, 200 miles per second $=2.964\times10^{10}$ cm. per second. More exact measurements give 2.9989×10^{10} cm. per second.

(9) (*a*) 2 hr.; (*b*) $13\frac{1}{3}$ min.

(10) About 31°.

(11) 24 miles per hour.

Page 37.

(1) $\frac{11}{15}$ ft./sec^2.

(2) 32.186.

(3) 0.11 ft./sec^2.

(4) 0.0034.

Page 40.

(1) (*c*) 145 ft. (2) 0.275 ft./sec^2.

(4) $h = c\left[t + \frac{c}{g} - \sqrt{\frac{c}{g}\left(2t + \frac{c}{g}\right)}\right]$. An approximate value is $h = \frac{1}{2}gct^2/(c+gt)$. For a direct numerical computation the *method of successive approximations* can be used. Thus, neglecting the time t_2, required by the sound, find the depth s approximately from $s = \frac{1}{2}gt^2$, with $t = 4$; with this value of s find t_2; hence the time of fall t_1, with which correct s; etc. Result: $s = 70.4$ meters.

(5) (*a*) $4\frac{5}{8}$ min. (*b*) 0.18 ft./sec^2. (*c*) $33\frac{7}{11}$ miles per hour. (*d*) 4 min. $4\frac{1}{2}$ sec.

(8) (*a*) $4\frac{1}{4}$ miles. (*b*) 645 ft./sec. (*c*) $1\frac{1}{4}$ min. (*d*) 1200 ft./sec. (*e*) 58 and 17 sec. The resistance of the air would modify these results quite appreciably.

(9) 80 ft./sec.

(10) (*a*) $t = h/v_0$. (*b*) $h - s = \frac{1}{2}gh^2/v_0^2$. (*c*) $v_0 = \sqrt{gh}$.

(11) 20 min. (12) $\frac{1}{260}$ sec. (13) 426 ft. (14) 30 miles per hour.

(16) $h - h' = \tau\sqrt{2gh}$ approximately; 0.6 ft.

Page 45.

(1) $v = 26{,}000$ ft. per second, $t = 34$ min. 50 sec., approximately.

(2) It represents a cycloid.

(4) (*a*) $v = 7$ miles per second, $t = \infty$. (*b*) $v = 7$ miles per second, $t = 116$ hr.

(5) $v = -\sqrt{2gR}\sqrt{\frac{R}{s} \mp \frac{1}{\kappa^2}}$, where $\frac{v_0^2}{2gR} - \frac{R}{s_0} = \mp\frac{1}{\kappa^2}$.

(*a*) If $v_0 < \sqrt{2gR}\sqrt{\frac{R}{s_0}}$, $t = \frac{\kappa}{\sqrt{2gR}}\Big[\sqrt{s(\kappa^2R - s)} - \sqrt{s_0(\kappa^2R - s_0)}$

$$+ \kappa^2R\left(\cos^{-1}\frac{1}{\kappa}\sqrt{\frac{s}{R}} - \cos^{-1}\frac{1}{\kappa}\sqrt{\frac{s_0}{R}}\right)\Big].$$

(*b*) If $v_0 = \sqrt{2gR}\sqrt{\frac{R}{s_0}}$, $t = \frac{1}{3}\sqrt{\frac{2}{g}}\frac{1}{R}(s_0^{\frac{3}{2}} - s^{\frac{3}{2}})$. (*c*) If $v_0 > \sqrt{2gR}\sqrt{\frac{R}{s_0}}$,

$$t = \frac{\kappa}{\sqrt{2gR}}\left[\kappa^2R\log\frac{\sqrt{s + \kappa^2R} + \sqrt{s}}{\sqrt{s_0 + \kappa^2R} + \sqrt{s_0}} + \sqrt{s_0(s_0 + \kappa^2R)} - \sqrt{s(s + \kappa^2R)}\right].$$

(6) (*a*) $v=5$ miles per second, $t=26$ min. (*b*) $v=7$ miles per second, $t=27$ min. (*c*) $v=12$ miles per second, $t=19$ min.

(7) If $v_0<\sqrt{2gR}$, the height above the earth's surface to which the particle rises is $h=v_0^2R/(2gR-v_0^2)$; the time of rising to this height is

$$\frac{R}{2gR-v_0^2}\left[v_0+\frac{2gR}{\sqrt{2gR-v_0^2}}\sin^{-1}\frac{v_0}{\sqrt{2gR}}\right].$$

If $v_0=\sqrt{2gR}$, the time of rising to the distance s from the center is

$$t=\frac{2}{3R\sqrt{2g}}(s^{\frac{3}{2}}-R^{\frac{3}{2}}),$$

and the particle does not return.

If $v_0>\sqrt{2gR}$,

$$t=\frac{\kappa}{R\sqrt{2g}}\left[\sqrt{s(s+\kappa^2)}-\sqrt{R(R+\kappa^2)}+\kappa^2\log\frac{\sqrt{R}+\sqrt{R+\kappa^2}}{\sqrt{s}+\sqrt{s+\kappa^2}}\right],$$

where

$$\kappa^2=\frac{2gR^2}{v_0^2-2gR}.$$

(8) $h=R$, $t=(1+\frac{1}{2}\pi)\sqrt{R/g}=34$ min. 50 sec., approximately. The time of falling back is the same as the time of rising; comp. Ex. (1).

(9) About 7 miles per second.

Page 48.

(1) Differentiating $s=C_1\sin\mu t+C_2\cos\mu t$, we find the velocity $ds/dt=v$ as a function of the time

$$v=\mu C_1\cos\mu t-\mu C_2\sin\mu t.$$

As $s=R$ when $t=0$ (in the problem of Art. 82), the former equation gives

$$R=C_2\cos 0,\ \therefore\ C_2=R,$$

while the latter gives, with $v=0$ for $t=0$,

$$0=\mu C_1,\ \therefore\ C_1=0.$$

With these values of C_1 and C_2 the two equations reduce to

$$s=R\cos\mu t,\ v=-\mu R\sin\mu t,$$

of which the latter can also be derived from the former by differentiation.

(2) $v=5$ miles per second; $T=2\pi\sqrt{R/g}=1$ hr. 25 min.

(3) $s=\frac{1}{2}R(e^{\mu t}+e^{-\mu t})=R\cosh\mu t$.

(4) $\sqrt{s_0^2+\left(\frac{v_0}{\mu}\right)^2}$.

Page 51.

(1) $\omega = \pi$ radians; $v = 15.7$ ft./sec. (2) (*a*) 3; (*b*) $28\frac{7}{11}$.

(3) -0.157 rad./sec^2. (5) (*a*) 402; (*b*) 25 sec.

(4) 5. (7) 31 ft./sec.

(8) (*a*) 0.022 rad./sec^2; (*b*) 15.7 ft./sec.; (*c*) 7.8 ft./sec.

Page 56.

(3) $v = 21$ ft./sec., at 22° to the train.

(5) $v = 24.2$ ft./sec., at $24\frac{1}{2}°$ to the track.

(6) About 20′′. (7) 36 miles per hour. (8) $v_1 = v_2 \sin \theta$.

(9) Resolve v into v_0 parallel to the track, and v_1 along the tangent to the wheel; show that v bisects the angle between these components; it follows that $v_0 = v_1$, and hence $v = 2\,v_0 \cos OCP$, where O is the center of the wheel and C its point of contact.

(11) $r = v_0 t$, $\theta = \omega t$; hence, eliminating t, $r = (v_0/\omega)\,\theta$, a spiral of Archimedes.

(12) At the pole O erect a perpendicular $OP' = a$ to the radius vector $OP = r$; then $P'P$ is the normal. Proof by Ex. (10).

(13) For the ellipse, $r_1 + r_2 = 2\,a$, whence $dr_2/dt = -\,dr_1/dt$. Notice that dr_1/dt, dr_2/dt are the *projections* (not the *components*) of the velocity v (with which the curve is described) on the radii vectores r_1, r_2. This is seen by observing that v can be resolved into components in two ways: (*a*) into dr_1/dt along r_1 and a component $\perp r_1$; (*b*) into dr_2/dt along r_2 and a component $\perp r_2$. Hence the perpendiculars erected at the ends of dr_1/dt and dr_2/dt (laid off from P in the proper sense) must meet at the end of v.

(14) The projections of the velocity on the radius vector and on the focal axis are in the constant ratio e of the focal radius vector to the distance to the directrix. It follows that the tangent meets the directrix at the same point as does the perpendicular to the radius vector through the focus.

Page 62.

(4) $N = 210$, $\omega = 22$, $V = 14$ ft./sec.

(5) 16.5 knots; $9\frac{1}{3}$ ft./sec.

(6) 55°, 66°, $2\frac{2}{3}$ in. (8) $5\frac{3}{4}$ ft./sec.

(7) 0.174, 0.119, 0.146 of the stroke. (9) At 0.447 of the stroke.

(10) $\dfrac{dx}{dt} = -a\dfrac{d\theta}{dt}\left(\sin\theta + \dfrac{n}{2}\dfrac{\sin 2\theta}{\sqrt{m^2-\sin^2\theta}}\right)$, $\dfrac{dy}{dt} = (1-n)a\dfrac{d\theta}{dt}\cos\theta$,

$$\frac{d^2x}{dt^2} = -a\left(\frac{d\theta}{dt}\right)^2\left[\cos\theta + n\cdot\frac{m^2\cos 2\theta + \sin^4\theta}{(m^2-\sin^2\theta)^{\frac{3}{2}}}\right],$$

$$\frac{d^2y}{dt^2} = -(1-n)a\left(\frac{d\theta}{dt}\right)^2\sin\theta.$$

If $m = l/a$ is large, we have approximately

$$\frac{d^2x}{dt^2} = -a\left(\frac{d\theta}{dt}\right)^2\left(\cos\theta + \frac{n}{m}\cos 2\theta\right),\ \frac{d^2y}{dt^2} = -(1-n)a\left(\frac{d\theta}{dt}\right)^2\sin\theta.$$

Page 67.

(2) Follows from the last of the equations (6), Art. 114.

(3) By (2), Art. 112, $j_n = \dfrac{v^2}{\rho} = \rho\cdot\left(\dfrac{v}{\rho}\right)^2$.

(4) By Art. 112, $j_n = j\sin\psi = v^2/\rho$; hence, $v^2 = j\cdot\rho\sin\psi$.

(6) Since j is directed towards A, we have, with A as origin, $j_\theta = 0$, *i.e.*, $r^2\dfrac{d\theta}{dt} = \text{const.}$

(7) $\dfrac{d\alpha}{dt} = \omega = \text{const.}$; $r = \text{const.}$; hence, by (6), Art. 114, $j = j_r = -r\omega^2$.

(9) Differentiate twice with respect to t the equations of the cycloid $x = a(\theta - \sin\theta)$, $y = a(1-\cos\theta)$. At the lowest point: $\dfrac{d^2x}{d^2t} = 0$, $\dfrac{d^2y}{dt^2} = a\left(\dfrac{d\theta}{dt}\right)^2$; at the highest point: $\dfrac{d^2x}{dt^2} = 2a\dfrac{d^2\theta}{dt^2}$, $\dfrac{d^2y}{dt^2} = -a\left(\dfrac{d\theta}{dt}\right)^2$.

Page 69.

(1) (*a*) 2360 ft. above the point; (*b*) after 3 min. 2 sec.; (*c*) 2670 ft. behind the train; (*d*) 25 miles per hour.

(4) $45°$.

(6) Construct a vertical circle having the given point as its highest point and touching, (*a*) the straight line, (*b*) the circle.

Page 72.

(9) (*a*) $137\frac{1}{2}$ ft. from the vertical of the starting point; (*b*) $6\frac{1}{4}$ sec.; (*c*) 201 ft./sec., at $6\frac{1}{3}°$ to the vertical.

(10) 227 ft./sec. (11) $4°\,21'$ or $86°\,48'$.

(13) Let $OV = v_0$ be the given initial velocity. On the vertical through O lay off $OD = H = v_0^2/2g$; then the horizontal through D is the directrix. Double the angle DOV, making $\measuredangle VOF = \measuredangle DOV$, and lay off $OF = OD = H$; then F is the focus.

(14) With $v_0^2/2g = H$, the locus is $x^2 = -4H(y-H)$, a parabola.

(17) A horizontal line.

(18) (*a*) 1.5 sec.; (*b*) 25.1 ft. from the building; (*c*) 597 ft./sec., at $16\frac{1}{2}°$ to the vertical.

(19) 300 ft. from tee, in 1 sec. (20) At a distance of 6260 ft.

Page 78.

(3) (*a*) 0, -4.94; (*b*) -2.72, -2.47; (*c*) -3.14, 0.

Page 82.

(1) $x = 10.81 \cos(\frac{1}{5}\pi t + 27\frac{1}{3}°)$. (2) $x = 2a \cos\frac{1}{2}\delta \cdot \cos(\omega t + \frac{1}{2}\delta)$.

(3) (*a*) $x = 2a\cos\omega t$; (*b*) $x = 0$, the case known in physics as *interference*.

(4) $x_1 = -5.18 \cos \pi t$, $x_2 = 14.14 \cos(\pi t + 30°)$.

Page 97.

(1) 0.99672; 86116. (3) 32.16. (5) 980.4.

(2) 3.2595 ft. (4) 28.8 ft.

(8) The pendulum should be lengthened by $\frac{1}{144}$ of its length.

(9) It will lose about 67 sec. (10) About a mile.

Page 99.

(3) 1.0038. (5) Use equation (24), Art. 150.

(6) Determining the constant from the condition $\theta = \pi$ for $v = 0$, we find instead of (24), Art. 150, $\frac{1}{2}v^2 = 2gl\cos^2\frac{1}{2}\theta$. Substituting $v = -l d\theta/dt$ and integrating, we find $t = \sqrt{l/g}\log\tan\frac{1}{4}(\pi + \theta)$, if $\theta = 0$ when $t = 0$. This shows that the point approaches the highest point of the circle asymptotically, *i. e.*, without reaching it in any finite time.

Page 109.

(2) Let $x^2 + y^2 = a^2$ be the circle, the acceleration being parallel to the axis of y; then $j = -a^2 v_1^2/y^3$, where v_1 is the x-component of the initial velocity.

(3) $f(r) = v_0^2/a$.

(4) Let $j = \mu^2 r$ be the acceleration; x_1, y_1 the initial position; v_1, v_2 the components of the initial velocity; then the path is the hyperbola

$$(v_2^2 - \mu^2 y_1^2)x^2 + 2(\mu^2 x_1 y_1 - v_1 v_2)xy + (v_1^2 - \mu^2 x_1^2)y^2 = (v_2 x_1 - v_1 y_1)^2.$$

(6) $a = \frac{\mu}{\epsilon^2}$, $b = \frac{v_0 r_0 \sin\psi_0}{\epsilon}$, $\tan\alpha = -\frac{v_0^2 \sin 2\psi_0}{\epsilon^2 + v_0^2 \cos 2\psi_0}$, where $\epsilon^2 = \frac{2\mu}{r_0} - v_0^2$.

Page 112.

(3) Find first the relative velocity of A' with respect to A, whence ω is obtained; determine the distances CA, CA'.

(5) $v_B = c \cot\phi$, $v_0 = \frac{1}{2}c/\sin\phi$, where $-c$ is the velocity of A.

(6) $v_B = 28.3$, $v_M = 22.4$ ft./sec.

Page 140.

(3) Based on the proposition that the bisector of an angle of a triangle divides the opposite side into segments proportional to the adjacent sides.

(5) The center of the incircle of the triangle formed by the mid-points of the sides.

(6) About 1000 miles below the earth's surface.

(8) $\bar{x} = \bar{y} = (2/\pi)r$.

(9) Taking OA as axis of x,

$$\bar{x} = \frac{2r}{\alpha^2}(\alpha \sin\alpha + \cos\alpha - 1), \quad \bar{y} = \frac{2r}{\alpha^2}(\sin\alpha - \alpha\cos\alpha).$$

(10) $\bar{x} = \frac{3\sqrt{2} - \log(1 + \sqrt{2})}{\sqrt{2} + \log(1 + \sqrt{2})} \cdot \frac{a}{4}$, $\bar{y} = \frac{2\sqrt{2} - 1}{\sqrt{2} + \log(1 + \sqrt{2})} \cdot \frac{4}{3}a$.

(11) $\bar{x} = \pi a$, $\bar{y} = \frac{4}{3}a$. (12) $\bar{x} = \bar{y} = \frac{4}{5}a$.

(13) $\bar{x} = 0$, $\bar{y} = \frac{cx}{s} + \frac{1}{2}y$, where $s = c(e^{\frac{x}{c}} - e^{-\frac{x}{c}})$.

(14) $\bar{x} = r \cdot \frac{\sin\theta}{\theta}$, $\bar{y} = r \cdot \frac{1 - \cos\theta}{\theta}$, $\bar{z} = \frac{1}{2}kr\theta$.

Page 150.

(1) $\frac{1}{3}\sqrt{a^2 + b^2}$, $\frac{1}{3}\sqrt{a^2 + 4b^2}$, $\frac{1}{3}\sqrt{4a^2 + b^2}$.

(2) With AE, AF as axes, $\bar{x} = \frac{61}{116}a$, $\bar{y} = \frac{63}{116}a$.

(3) With the sides of the triangle as axes, $\bar{x}=\bar{y}=\frac{3\pi-2}{6(\pi-1)}a$; distance from center $=\frac{\sqrt{2}}{6(\pi-1)}a$.

(5) Resolve the area into elements parallel to *BD*.

(6) With the lower base and the perpendicular side as axes, $\bar{x}=\frac{1}{3}(a^2+ab+b^2)/(a+b)$, $\bar{y}=\frac{1}{3}(a+2b)h/(a+b)$.

(7) Compare Art. 231.

(8) $\bar{x}=\frac{1}{2}\frac{(a+a')^2\beta+4a'(b-\beta)\alpha}{(a+a')\beta+2(b-\beta)\alpha}$.

(9)
$$\bar{x}=\frac{1}{2}\frac{a^2+b\delta-\delta^2}{a+b-\delta}=4.90\text{ in.};$$

first approximation, $\bar{x}=\frac{1}{2}\frac{a^2+b\delta}{a+b-\delta}=4.93$ in.;

second approximation, $\bar{x}=\frac{1}{2}\frac{a^2}{a+b}=4.50$ in.

(10) $\bar{x}=\frac{1}{2}\frac{a(a+2b')-(2a-b+b')\delta}{a+b+b'-2\delta}=4.5$ in.

(11) $\bar{x}=\frac{1}{2}\frac{a^2\beta+b\alpha^2-\alpha^2\beta}{a\beta+b\alpha-\alpha\beta}$, $\frac{1}{2}\frac{a^2\beta+b\alpha^2}{a\beta+b\alpha-\alpha\beta}$, $\frac{1}{2}\frac{a^2\beta}{a\beta+b\alpha}$; $0.32\,a$, $0.33\,a$, $0.25\,a$.

(14) $\bar{x}=-b^2c/(a^2-b^2)$.

(15) For a segment of a ring of angle 2α and radii r_1, r_2, the distance of the centroid from the center is $\bar{x}=\frac{2}{3}\cdot\frac{\sin\alpha}{\alpha}\cdot\frac{r_1^2+r_1r_2+r_2^2}{r_1+r_2}$. Hence, $\bar{x}=\frac{2}{147\pi}(740+73\sqrt{2})=3.65$ ft.; *i.e.*, the centroid lies about 1 in. above the lower arc.

(16) $\bar{x}=\frac{3}{5}x$, $\bar{y}=\frac{3}{8}y$.

(17) $\bar{x}=\frac{\pi}{2}$, $\bar{y}=\frac{\pi}{8}$.

(18) $\bar{x}=\frac{4}{3\pi}a=0.4053\,a$, $\bar{y}=\frac{4}{3\pi}b$.

(19) $\bar{x}=\frac{2}{3(\pi-2)}a=0.584\,a$, $\bar{y}=\frac{2}{3(\pi-2)}b$.

(21) Take the vertex as origin, the axis of the cone as axis of x, and one of the bounding planes as plane of xy. Then, if a be the radius of the base, h the height, and 2α the angle at the vertex of the cone, the formulæ of Art. 236 give

$$\eta=(a/h)x=\tan\alpha\cdot x,\quad S=\tfrac{1}{2}ah\sec\alpha\cdot\phi,$$
$$\bar{x}=\tfrac{3}{4}h,\quad \bar{y}=\tfrac{2}{3}a\cdot\frac{\sin\phi}{\phi},\quad \bar{z}=\tfrac{2}{3}a\cdot\frac{1-\cos\phi}{\phi}.$$

(22) About 2600 miles from the center.

(23) At $\frac{1}{5}r$ from the lid.

Page 158.

(1) $\bar{x} = \frac{3}{8}a$.

(2) $\bar{x} = \frac{1}{4}h \cdot \frac{3r_1^2 + 2r_1r_2 + r_2^2}{r_1^2 + r_1r_2 + r_2^2}$.

(3) Let V_1 be the volume of the supplementary pyramid, V_2 that of the whole pyramid, V that of the frustum; $\bar{x}_1$, $\bar{x}_2$, $\bar{x}$ the distances of their centroids from the lower base; h_1, h_2, h their heights. Then the equation of moments is $(V_2 - V_1)\bar{x} = V_2\bar{x}_2 - V_1\bar{x}_1$. By geometry, $V_2/V_1 = r_2^3/r_1^3$; hence $(r_2^3 - r_1^3)\bar{x} = r_2^3\bar{x}_2 - r_1^3\bar{x}_1$. Also $h_2/h_1 = r_2/r_1$, $h_2 - h_1 = h$, $\bar{x}_2 = \frac{1}{4}h_2$, $\bar{x}_1 = h + \frac{1}{4}h_1$; whence finally $\bar{x}$ is found as in (2).

(4) $\bar{x} = \frac{3}{4}\frac{(2a - h)^2}{3a - h}$.

(5) $\bar{x} = \frac{2}{3}h$.

(6) $\bar{y} = \frac{5}{12}y_1$.

(7) $\bar{y} = \frac{5}{16}y_1$.

(8) $\bar{x} = \frac{3}{8}a$, $\bar{y} = \frac{3}{8}b$, $\bar{z} = \frac{3}{8}c$.

(9) (*a*) $\bar{x} = \bar{y} = \frac{4}{3}a$;

(*b*) $\bar{x} = \frac{9\pi^2 + 16}{18\pi}a$, $\bar{y} = \frac{5}{6}a$;

(*c*) $\bar{x} = \frac{26}{15}a$;

(*d*) $\bar{x} = \frac{45\pi^2 + 128}{90\pi}a$;

(*e*) $\bar{x} = \frac{2(15\pi - 8)}{15(3\pi - 4)}a$;

(*f*) $\bar{x} = \frac{63\pi^2 - 64}{6(9\pi^2 - 16)}a$.

(10) Take as element a hemispherical shell of radius r and thickness dr; $\bar{x} = \frac{n+3}{2(n+4)}a$.

(11) Let P_1, P_2, P_3 be the vertices of the *triangle*, P_1P_4 a median, G the centroid; then

$$\frac{P_1G}{P_1P_4} = \frac{2}{3}, \quad \frac{P_1G}{P_1P_4} = \frac{\bar{x} - x_1}{x_4 - x_1}; \text{ whence, } \bar{x} = x_1 + \frac{2}{3}(x_4 - x_1) = \frac{1}{3}(x_1 + x_2 + x_3).$$

For the tetrahedron $P_1P_2P_3P_4$ let P_5 be the centroid of the base opposite P_1 so that $x_5 = \frac{1}{3}(x_2 + x_3 + x_4)$. Then, proceeding as above, we find $\bar{x} = \frac{1}{4}(x_1 + x_2 + x_3 + x_4)$.

(12) About 0.2 mile.

(13) $\bar{x} = \frac{1}{4}(H + h)$.

(14) Compare Art. 225, Ex. (5), Art. 224, and Ex. (11), Art. 238.

$V = \frac{2}{3}\pi(p + q + r)A$, where A is the area of the triangle;

$S = \pi[a(q + r) + b(r + p) + c(p + q)]$.

(15) Taking the axis of the cup as axis of y, let (x_1, y_1) be the centroid of the mass m of cup and handle, $(0, y_2)$ that of the water whose mass $m' = ky_2$. Then the co-ordinates x, y of the centroid of the cup, handle, and water satisfy the equation

$$(m + ky_1)x^2 - kx_1xy - 2\,mx_1x + m_1x_1^2 = 0.$$

(16) Taking the axis of z parallel to the axis of the cylinder, and the origin in the line of intersection of the bases, we have $V = \int\int z dx dy$, or if ϕ be the angle of inclination of the bases,

$$V = \tan\phi \int\int y dx dy = \tan\phi \cdot \bar{y} \int\int dx dy.$$

(17) Apply (16) twice.

Page 161.

(2) $34\frac{1}{11}$ miles per hour. (3) 32,000 lb.-ft.

Page 165.

(1) 6.4×10^5 poundals, 8.9×10^9 dynes.

(2) 4.5 lb. (3) 0.14. (5) 60.

Page 174.

(3) 120°. (4) 218 lb. at 36° 35′ to the force of 100 lbs.

(5) 10.35, 14.64. (7) $Q = \frac{1}{2}(-P + \sqrt{4R^2 - 3P^2})$.

(8) 569, inclination to horizon = 99° 26′.

(9) Twice the focal distance.

(10) 124° 14′. (11) 90°. (12) 18.48.

(13) $R = 6$ and acts along 5.

(14) The resultant acts along the diameter through A, and is in magnitude equal to the perimeter.

(15) $(1 + \sqrt{2})P$.

(16) (*a*) $W\sin\theta$, $W\cos\theta$; (*b*) $W\tan\theta$, $W/\cos\theta$; (*c*) $W\sin\theta/\cos\alpha$, $W\cos(\theta + \alpha)/\cos\alpha$.

(19) Produce BO to the intersection D with the circumscribed circle; then DA is equal and parallel to the resultant of OA, OB; $DAO'C$ is a parallelogram; hence, $DA = CO'$.

(20) Resolve the components P_1, P_2 along the bisectors of θ.

Page 179.

(2) $T = W \cdot a/c$, $P = W \cdot b/c$.

(3) AC must bisect the angle BCW.

(4) $R^2 = A^2 + B^2 + C^2 + 2\,BC\cos\alpha + 2\,CA\cos\beta + 2\,AB\cos\gamma$.

(5) See Arts. 286–288.

(7) The sum of their moments must vanish for two points in the plane not in line with their point of intersection.

(10) $P = \frac{4}{3}W$; $T = \frac{5}{3}W$.

(11) $T = W$; $P = 0.89\,W$ along the bisector of $\measuredangle BCW$.

(12) $P = W\sin(\alpha+\beta)\sin\beta$ becomes a maximum for $\beta = \frac{1}{2}(\pi - \alpha)$, *i. e.*, when the sail bisects the angle between boat and wind.

(13) $W\dfrac{\sin\beta}{\sin(\alpha+\beta)}$; $W\dfrac{\sin\alpha}{\sin(\alpha+\beta)}$.

(14) Tension in AB and $CD = W\cdot l/d$, tension in $BC = W\cdot(c-l)/2d$, where $d = \sqrt{l^2 - \frac{1}{4}(c-l)^2}$.

(15) $P_{max.} = 0.26\,R$.

(18) $\frac{1}{2}F\tan\alpha$; 13.4, 28.9, 50, 86.6, 137.4, 186.6, 283.6, 5715, ∞.

(19) 848, 282; 1000, 600. (20) $0.640\,W$.

(21) (*a*) $\sqrt{2}\,W$; (*b*) $2\,W\cos\frac{1}{2}(\frac{1}{2}\pi \pm \theta)$.

(23) $2\,W\cos\frac{1}{2}(\frac{1}{2}\pi + \alpha)$, etc. (*a*) $\alpha = 30°$, $\beta = 120°$, $\gamma = 30°$; (*b*) impossible.

(25) $T = 0.56\,W$; $A = C = 0.72\,W$, $B = 0.67\,W$.

(26) $T = A = C = \frac{1}{3}\sqrt{3}\,W$, $B = W$.

(27) $\delta = \pi - 3\beta$, $30° < \beta < 60°$.

Page 192.

(1) Take moments about the fulcrum. (*a*) 4.32; (*b*) 3.94.

(2) (*a*) $A = 8\frac{2}{3}$, $B = 7\frac{1}{3}$; (*c*) $A = 14\frac{1}{6}$, $B = 14\frac{5}{6}$.

(3) (*a*) $P = W$; (*b*) $P = (1 + \sqrt{2})W$.

(6) (*a*) 19.4 tons and 21.1 tons; (*b*) 30.5, 9.9.

(8) Let α be the angle subtended at the center by the side 12, and θ the angle at which the diagonal 13 is inclined to the horizon; then

$$\tan\theta = \frac{W_3 - W_1}{W_4 - W_2}\csc\alpha + \cot\alpha.$$

(9) $x = F_2 l\sin\alpha_2/(F_1\sin\alpha_1 + F_2\sin\alpha_2)$.

(11) 49 lbs. per square inch. (12) 63.1 in.

Page 201.

(1) $C=1$, $D=1\frac{1}{3}$, $E=6\frac{2}{3}$, $AB=4.5$, $BC=4.1$, $CD=4.0$, $DE=4.2$, $EF=8.9$; reaction at $A=4.5$, at $F=8.9$.

(2) $H=75$, $T=81$.

(3) $c=\sqrt{\dfrac{x^3}{6(s-x)}}$; $H=wc$.

(4) 1185 lbs. per square inch; 3.5 ft.

(5) $H=47.4$ lbs., $T=57.2$ lbs., $y-c=9.8$ ft.

Page 215.

(1) $T=7.68$, $A=9.76$, $E=12.80$ lbs.

(2) $T=2\,mW$, $A=\sqrt{4\,m^2-2\,m+1}\,W$, where $m=c/l$.

(3) The three forces W, T, A, must pass through a point; $\cos\phi = 2\sqrt{\frac{1}{3}(1-m^2)}$, where $m=l/b$; $T=W\sec\phi$, $A=W\tan\phi$.

(4) $T=\frac{1}{2}W\cos\theta/\sin(\theta-\phi)$.

(5) $A_x=-B=-\dfrac{ac}{l\sqrt{l^2-a^2}}W$, $A_y=W$.

(6) $B=\dfrac{ac}{l^2}W$, $A_x=-\dfrac{ac}{l^3}\sqrt{l^2-a^2}\,W$, $A_y=\left(1-\dfrac{a^2c}{l^3}\right)W$.

(7) (*a*) Equilibrium impossible; (*b*) $F=E=\dfrac{l\cos\theta}{e-f}W$, $A=W$.

(8) $x=am$, $A=\sqrt{m^2-1}\,W$, $C=mW$, where $m=(l/a)^{\frac{1}{3}}$.

(9) $B=\frac{1}{2}(3\,w+W)\tan\theta$, $A=(3\,w+W)\sqrt{\frac{1}{4}\tan^2\theta+1}$.

(10) $\cos\theta=\frac{1}{8}(m+\sqrt{m^2+32})$, where $m=l/a$.

(11) $P=\frac{1}{2}W\dfrac{\cos\theta}{\sin(\theta+\phi)}$, $A=\frac{1}{2}W\dfrac{\cos\theta\sqrt{1+\sin^2\phi}}{\sin(\theta+\phi)}$, $\cot\phi=\dfrac{c-l\cos\theta}{l\sin\theta}$, ϕ being the angle at which BC is inclined to the horizon.

(12) $m=(W+2\,F'\sin\alpha')/(W+2\,F\sin\alpha)$, $C_x=F\cos\alpha-F'\cos\alpha'$, $C_y=W+F\sin\alpha+F'\sin\alpha'$.

(13) $A=\frac{1}{2}W$, $B=\frac{1}{2}W\cos\alpha$, $P=\frac{1}{2}W\sin\alpha$.

(14) $P=\dfrac{\sqrt{2\,m-1}}{m-1}W$, $P'=\dfrac{m}{m-1}W$.

(15) $P=\dfrac{2\sqrt{n}}{(n-1)\cos\alpha+2\sqrt{n}\sin\alpha}W$, $P_{\text{min.}}=W\sin\alpha$.

Page 219.

(3) 60°.

(5) $\cot\theta = \frac{3}{8}$ for hemisphere, $h/4\,a$ for cone, $h/2\,a$ for pyramid.

(6) $h = \sqrt{3\,\rho_1/\rho_2}\,a$. (7) $h = \sqrt{\rho_1/2\,\rho_2}\,a$.

Page 222.

(3) Let h be the altitude of ABC, through C, and a, b the segments into which this altitude divides AB; then $C_x = -\dfrac{h}{a+b}W$, $C_y = \dfrac{1}{2}\dfrac{b-a}{b+a}W$.

(4) $A_x = \dfrac{\cos\alpha\cos\beta}{\sin(\alpha+\beta)}W = -B_x$, $A_y = \left[\dfrac{1}{2} + \dfrac{\sin\alpha\cos\beta}{\sin(\alpha+\beta)}\right]W = 2W - B_y$.

Page 235.

(1) 4 tons. (2) $P = \dfrac{\sin\phi}{\cos(\phi-\alpha)}W$.

(3) (*a*) $\dfrac{\sin(\theta-\phi)}{\cos\phi}W < P < \dfrac{\sin(\theta+\phi)}{\cos\phi}W$; (*b*) $P = 0$, or $P \gtreqless 2W\sin\theta$; (*c*) if P act up the plane, $P \gtreqless \dfrac{\sin(\phi+\theta)}{\cos\phi}W$; if P act down the plane, $P \gtreqless \dfrac{\sin(\phi-\theta)}{\cos\phi}W$.

(5) $226\frac{1}{4}$ lbs.; $56\frac{1}{2}$ lbs.

(6) (*a*) $P = \dfrac{\sin(\theta-\phi)}{\cos(\alpha+\phi)}W$; (*b*) $P = \dfrac{\sin(\theta+\phi)}{\cos(\alpha-\phi)}W$.

(7) $\theta = \dfrac{\pi}{2} - 2\phi$. (8) $\theta = \tan^{-1}\dfrac{\mu W + \mu' W'}{W + W'}$.

(9) $A = (1 - m\cos^2\theta)W$, $C = m\cos\theta W$, $\mu = \dfrac{m\cos\theta\sin\theta}{1 - m\cos^2\theta}$, where $m = l/c$.

(10) $A = m\dfrac{\sin(\theta-\phi)\cos\theta}{\sin\phi}W$, $C = m\cos\theta W$, $\sin 2\theta = \dfrac{1}{m}\tan 2\phi$, where $m = l/c$.

(11) $\sin\theta \leqq \frac{3}{8}$.

Page 282.

(1) 3.5 ft. (2) 1380 ft./sec.

(3) $8\frac{6}{7}$ ft./sec. (4) $36\frac{3}{7}$ ft./sec.

(5) (*a*) $v = 4\frac{3}{13}$, $v' = 5\frac{3}{13}$; (*b*) $v = -1\frac{12}{13}$, $v' = 7\frac{1}{13}$.

(6) If the original velocities are of the same sense, $v = 42\frac{4}{9}$, $v' = 46\frac{1}{9}$; if not, $v = -19\frac{7}{9}$, $v' = 10\frac{5}{9}$.

(7) $e=0$ gives (a) $44\frac{7}{12}$, (b) $-2\frac{1}{12}$; $e=1$ gives (a) $v=38\frac{1}{6}$, $v'=$ $9\frac{1}{6}$, (b) $v=-55\frac{1}{6}$, $v'=35\frac{5}{6}$.

(8) $v=-eu$. (10) $8\frac{1}{3}$ ft.

(11) (a) 0.31 ft.; (b) $9\frac{1}{2}$ sec.; (c) $66\frac{6}{11}$ ft. (13) $\left(\frac{1+e}{2}\right)^n u$.

(14) (a) $4\frac{4}{5}$ ft./sec.; (b) $38\frac{2}{5}$ ft./sec.

(15) For $e=0$: (a) $v=\frac{m}{m+m'}u$; (b) $\lim v=0$; (c) $\lim v=u'$.
For $e=1$: (a) $v=\frac{m-m'}{m+m'}u$, $v'=\frac{2m}{m+m'}u$; (b) $\lim v=-u$, $\lim v'=0$; (c) $\lim v=2u'-u$, $\lim v'=u'$. Interpret these results.

Page 288.

(1) The momenta are as 20 : 1; the kinetic energy is the same.

(2) 3125 lbs. (3) 9 lbs. (4) 6250 lbs.

(5) About 450 lbs.; about $\frac{1}{49}$ of the available energy is wasted.

(6) 4.9 ft.-lbs. (7) 363 ft.-tons; 9 miles per hour. (10) 3 tons.

(11) $13\frac{1}{2}$ and 2 ft.-tons. (12) 144,000 lbs. (13) About 3300 lbs.

(14) About 60 lbs. per sq. in. (15) About 400 lbs. per sq. in.

(16) 449 lb.-ft. (17) 26 lbs.

Page 292.

(1) 57 lb.-ft. (2) 16 ft./sec.

(3) $v=8.8$, $\beta=59°$; $v'=14.4$, $\beta'=17\frac{1}{2}°$.

(4) $v=6\frac{2}{5}$ ft. per second; $F=160{,}000$ lbs.

(5) 5 : 1. (6) 17 : 1.

(7) The impinging sphere is brought to rest;
$v'=\sqrt{u^2+u'^2}$, $\tan\beta'=u'/u$.

Page 300.

(1) (b) 250 lbs. (2) (a) 8 ft./sec.; (b) 20 ft.

(3) (a) 764 lbs.; (b) $1\frac{1}{2}$ mile; (c) 1146 lbs. (4) 4.9 sec.

(5) 35.8 and 4.2 ft. above the ground; $50\frac{1}{2}$ lbs.; 11.2 sec.

(6) $j=(m_1\sin\theta_1-m_2\sin\theta_2-\mu_1 m_1\cos\theta_1-\mu_2 m_2\cos\theta_2)g/(m_1+m_2)$,
$T=(\sin\theta_1+\sin\theta_2-\mu_1\cos\theta_1+\mu_2\cos\theta_2)\,m_1 m_2 g/(m_1+m_2)$.

(7) $j=5.8$ ft./sec.^{2}; $T=1\frac{7}{11}$ lbs.

(9) 0.036. (10) 288 ft.

(11) (*a*) About 1600 lbs; (*b*) about 3750 lbs.

(12) 589 ft. (Find first $\sin\theta$ by successive approximation.)

(13) (*a*) 120; (*b*) 200 lbs. (14) 4125 lbs. (15) $562\frac{1}{2}$ lbs.

(16) (*a*) 2625, 3058; (*b*) 1375, 1602; (*a*) 217; (*b*) 113.

(17) (*a*) 1267.2 ft.-tons; (*b*) 4435.2 ft.-tons; (*c*) 5 : 2; (*d*) about $1\frac{1}{4}$ mile.

(18) (*a*) 1146; (*b*) 1946; (*c*) 800; (*d*) 2050 lbs.

(19) 2016 ft.-tons. (21) About 7 in. (23) 513,274 ft.-tons.

(20) $397\frac{1}{4}$ ft.-tons. (22) 30 ft.

Page 313.

(1) (*a*) 11,133 ft.-lbs.; (*b*) $22\frac{1}{4}$ ft.

(2) 6 lbs./in.^{2}; 83,400 ft.-lbs. (4) 864 ft./sec.^{2}

(5) Time $= \dfrac{2}{\kappa}\left(\pi + 2\dfrac{l}{x_0}\right)$, work $= \frac{1}{2} m\kappa^2 x_0^2$, where $\kappa = \sqrt{\dfrac{m_1 g}{m(l_1 - l)}}$.

(6) (*a*) $\frac{1}{2}$ lb.; (*b*) 11.3 ft./sec.; (*c*) 0.6 sec.

(7) If $x_0 \leqq 2\mu m_1(l_1 - l)m_1$, the particle comes to rest between P_0 and Q (Fig. 155). If $x_0 > 2\mu m_1(l_1 - l)/m_1$, but $x_0(\kappa^2 x_0 - 2\mu g) \leqq 2l/\mu m g$, the particle comes to rest between Q and Q'; etc.

(9) If $x_0 < e$, nothing is changed; if $x_0 > e$, the particle performs simple harmonic oscillations about Q_1, just as in the case $x_0 < e$.

(10) The length l is increased to $l + e + \sqrt{e(e + 2h)}$.

(11) Take $x_0 = l_2 - l$ in Ex. (8). (12) $42\frac{1}{2}$ min.

(14) $x_1 = h(1 + \sqrt{1 + 2h_1/h})$.

Page 317.

(2) The equation of motion $m\dfrac{d^2s}{dt^2} = -mg - mkv^2$ gives, with $k = \mu^2/g$,

$$v = \frac{g}{\mu}\frac{\mu v_0 \cos\mu t - g\sin\mu t}{\mu v_0 \sin\mu t + g\cos\mu t}, \quad s = \frac{g}{\mu^2}\log\left(\frac{\mu v_0}{g}\sin\mu t + \cos\mu t\right),$$

$$s = \frac{g}{2\mu^2}\log\frac{g^2 + \mu^2 v_0^2}{g^2 + \mu^2 v^2} = \frac{1}{2k}\log\frac{g + kv_0^2}{g + kv^2}.$$

(3) Time of ascent $= \dfrac{1}{\sqrt{kg}}\tan^{-1}\left(\sqrt{\dfrac{k}{g}}\cdot v_0\right)$, height $= \dfrac{1}{2k}\log\left(1 + \dfrac{k}{g}v_0^2\right)$.

to $-$.

$f(x)$ is a minimum if $f'(x) = 0$ and

to $+$. Granville's Calculus page 12

(4) $\frac{v_1}{v_0} = \frac{\sqrt{g}}{\sqrt{g + kv_0^2}}$.

(5) *In vacuo* $v_1 = 17$ ft. per second; in the air $v_1 = 122$ ft./sec.

(6) $s = \frac{v_0}{k}(1 - e^{-kt}),\ v = v_0 e^{-kt} = v_0 - ks$.

(7) $v = \frac{g}{k}(1 - e^{-kt}),\ s = \frac{g}{k}\left[t + \frac{1}{k}(e^{-kt} - 1)\right] = -\frac{v}{k} + \frac{g}{k^2}\log\frac{g}{g - kv}$.

Page 321.

(2) The logarithmic decrement is $\log e^{-\lambda t} = -\lambda t$.

(4) If $\mu \neq \kappa$, $s = c_1 \cos \kappa t + c_2 \sin \kappa t + \frac{a}{\kappa^2 - \mu^2}\sin \mu t$; if $\mu = \kappa$, $s = c_1 \cos \kappa t + c_2 \sin \kappa t + \frac{at}{2\kappa}\sin \kappa t$.

(5) The term due to the forced oscillation is

$$\frac{a}{\sqrt{(\kappa^2 - \mu^2)^2 + 4\lambda^2\mu^2}}\cos \mu(t - t_0);$$

hence, this oscillation lags behind the force by the phase difference μt_0; the amplitude is less than for undamped oscillations. The free oscillations (if any) will rapidly die out so that the motion soon approaches a state of motion given by the above term.

Page 324.

(1) 1 watt = 0.00134 H.P., 1 H.P. = 746 watts (with $g = 981$).

(2) 1 metric H.P. = 736 watts = 0.986 British H.P.

(3) $20\frac{1}{4}$.

(4) $41\frac{1}{4}$.

(5) (*a*) 64; (*b*) 224; (*c*) 384.

(6) 188 gals.

(7) 35,200 gals.

(8) About 1 hr.

(9) (*a*) 177,000 ft.-tons; (*b*) 51 hr.; (*c*) about 74 days (of 8 hr.).

(10) $25\frac{5}{16}$ lbs.

(11) 15.6 lbs.

(12) 12.8.

(13) $1\frac{1}{3}$.

(14) $21\frac{1}{4}$.

Page 331.

(1) $U = cz + C$.

(2) $V = mg(z - z_0)$.

(4) $U = -\int f(r)dr = -F(r)$.

(5) As $r^2 = (x - x_0)^2 + (y - y_0)^2 + (z - z_0)^2$, $rdr = (x - x_0)\,dx + (y - y_0)\,dy + (z - z_0)\,dz$; hence the direction cosines of $R = f(r)$ are $\pm\frac{x - x_0}{r} = \pm\frac{\partial x}{\partial r}$, etc. Hence $X = \pm f(r)\frac{\partial x}{\partial r}$, etc., and putting $\int f(r)\,dr = F(r)$, we find $U = \pm F(r)$.

Page 343.

(1) Put $r = 1/u$, and find $d^2u/d\theta^2$ in terms of u alone:

$$\frac{d^2u}{d\theta^2} = -u + (n-1)(1-e^2)q^{-2n}u^{-2n+1} - (n-2)q^{-n}u^{-n+1}.$$

Substituting in (10), Art. 533, we find

$$f(r) = \frac{c^2}{q^n}[(n-1)(1-e^2)q^{-n}u^{-2n+3} - (n-2)u^{-n+3}].$$

$n = 1$ gives an ellipse if $e < 1$, a parabola if $e = 1$, a hyperbola if $e > 1$, all referred to focus and focal axis; $n = 2$ gives conics referred to their axes; $n = -1$ gives Pascalian limaçons (cardioids for $e = \pm 1$); $n = -2$ gives a lemniscate if $e = \pm 1$.

(2) (a) $c^2\left(\frac{2a^2}{r^5} + \frac{1}{r^3}\right)$; (b) $\frac{c^2}{r^3}$; (c) $\frac{c^2(1+n^2)}{r^3}$; (d) $c^2\left(\frac{2n^2a^2}{r^5} + \frac{1-n^2}{r^3}\right)$.

(3) $8a^2c^2/r^5$.

(4) Ellipse, parabola, or hyperbola according as $\mu \gtreqless v_2^2 y_0^2$, where y_0 is the initial distance of the particle from the plane, v_2 the component of its initial velocity at right angles to the plane.

(5) $f(r) = -\frac{b^4}{a^2}\frac{v_1^2}{y^3}$.

Page 346.

(2) The equation of the orbit given in Ex. (1) is satisfied not only by (x_0, y_0), but also by $(v_1/\kappa, v_2/\kappa)$; *i. e.*, the orbit passes not only through the initial point P_0, but also through the point Q, which is the extremity of the radius vector $OQ = v_0/\kappa$ parallel to v_0; OP_0 and OQ are the conjugate semi-diameters whose equations are $x_0 y = y_0 x$, $v_1 y = v_2 x$.

(4) The problem reduces to that of constructing the axes of a conic from a pair of conjugate diameters.

(5) The curve being referred to its axes, the equations of motion are $x = a\cos\kappa t$, $y = b\sin\kappa t$ for the ellipse, and $x = \frac{1}{2}a(e^{\kappa t} + e^{-\kappa t}) = a\cosh\kappa t$, $y = \frac{1}{2}b(e^{\kappa t} - e^{-\kappa t}) = b\sinh\kappa t$ for the hyperbola.

(6) From the equations of Ex. (5) it follows that for the ellipse $\tan\theta = (b/a)\tan\kappa t$; hence $d\theta/dt = \kappa ab/r^2$; then apply (4), Art. 529.

(8) Use the equations of the conic in terms of the eccentric angle ϕ; or the equation

$$\frac{1}{p^2} = \frac{1}{a^2} \pm \frac{1}{b^2} \mp \frac{1}{a^2b^2}r^2,$$

where p is the perpendicular from the center to the tangent; the upper sign gives the ellipse, the lower the hyperbola; apply (11), Art. 534.

(9) (*a*) Ellipse; (*b*) hyperbola; (*c*) parabola.

(10) The parabola $x - x_0 = \frac{v_1}{v_2}(y-y_0) - \frac{2\,\kappa a}{v_2^2}(y-y_0)^2$, where $2\,a$ is the distance of O_3 from the point O that bisects O_1O_2; the point midway between O and O_3 is taken as origin, and OO_3 as axis of x.

(11) $t = \frac{1}{\kappa}\tan^{-1}\left(\frac{a}{b}\tan\theta\right)$.

Page 357.

(1) (*a*) $f(r) = \frac{c^2}{a^2b^2}\cdot r$; (*b*) $f(r) = \frac{c^2a}{b^2}\cdot\frac{1}{r^2}$.

(2) $v_0 = \sqrt{\mu/r_0}$.

(4) 687 days.

(5) By (24), Art. 555, $v^2 = \frac{2\,\mu}{r} \mp \frac{\mu}{a}$; as the velocity is not changed instantaneously, we must have $\frac{2\,\mu}{r} \mp \frac{\mu}{a} = \frac{2\,\mu'}{r} \mp \frac{\mu'}{a'}$, whence the new major semi-axis a' can be found.

(6) An ellipse with the end of its minor axis at the point where the change takes place.

(7) (*a*) Ellipse with $a = \frac{2}{3}r$; (*b*) parabola.

(8) Differentiate (24), Art. 555, with respect to μ and a.

(9) The periodic time T would be diminished by $\frac{2}{n}T$.

(10) $r = \frac{l}{1 + e\cos\theta}$; hence, $x = \frac{l\sin\theta}{1+e\cos\theta}$, $y = \frac{l\sin\theta}{1+e\cos\theta}$; differentiating and remembering that $r^2d\theta/dt = c$, we find

$$\frac{dx}{dt} = -\frac{c}{l}\sin\theta,\ \frac{dy}{dt} = \frac{c}{l}(\cos\theta + e);$$

eliminating θ, we find the equation of the hodograph

$$x^2 + \left(y - \frac{ec}{l}\right)^2 = \left(\frac{c}{l}\right)^2, \text{ or since } c = \sqrt{\mu l},\ x^2 + \left(y - \frac{\mu e}{c}\right)^2 = \left(\frac{\mu}{c}\right)^2.$$

(11) 1.034114. (12) $t = \sqrt{\frac{2\,a^3}{\mu}}(\tan\frac{1}{2}\theta + \frac{1}{3}\tan^3\frac{1}{2}\theta)$.

Page 362.

(2) Let ρ_1, ρ_2 be the distances of m_1, m_2 from the common centroid at any time t; $\xi_1 = x_1 - \bar{x}$, etc.; then the equations of the relative motion are

$$\frac{d^2\xi_1}{dt^2} = m_2 f\left(\frac{m_1+m_2}{m_2}\rho_1\right)\cdot\frac{\xi_1}{\rho_1}, \text{ etc.};\quad \frac{d^2\xi_2}{dt^2} = m_1 f\left(\frac{m_1+m_2}{m_1}\rho_2\right)\cdot\frac{\xi_2}{\rho_2}, \text{ etc.}$$

Page 370.

(4) (*a*) $7\frac{1}{2}$ lbs.; (*b*) 480 lbs.; (*c*) 6.4 rev./sec.

(5) 4840 lbs. (6) $e = \frac{7}{4}\frac{v^2}{R}$ in.

(7) $8\frac{1}{2}°$. (9) 32.20.

(12) $7\frac{1}{2}$ lbs. (13) (*a*) 76 per min.; (*b*) 108.

(14) In Fig. 166, $CD : RF' = PC : PR$, hence $CD = \frac{mg \cdot a}{\omega^2 a} =$ const.

(15) $7\frac{2}{3}°$. (16) $\tan \delta = \frac{R\omega^2 \sin \phi \cos \phi}{g - R\omega^2 \cos^2 \phi}$; $\delta_{max.}$ in latitude 44° 57′.

(17) (*a*) 76.4 rev./min., (*b*) 71.4. (18) About 48 miles per hour.

Page 373.

(1) The integration gives $\tan \frac{1}{4}(\pi + \theta) = \tan \frac{1}{4}(\pi + \theta_0) \cdot e^{\sqrt{\frac{g}{l}}t}$; as θ approaches π, t approaches infinity.

(5) The particle remains on the circle as long as $\cos \theta + \frac{2}{3} h/l$ is positive.

(6) $v_0 > 22$ ft./sec.

(7) To count the angles from the highest point of the circle, put $\pi - \theta = \phi$; then, putting $h - l = h'$, where h' is the height to which the velocity at the highest point is due, we have

$$N = -3\,mg\left(\cos \phi - \frac{2}{3}\frac{l + h'}{l}\right).$$

The particle remains on the curve as long as $\cos \phi > \frac{2}{3}(l + h')/l$. Distinguish the cases $h' > 0$, $h' = 0$, $h' < 0$.

(9) 1.4617 a. (10) $t_1 = \pi\sqrt{a/g}$.

Page 377.

(1) (*a*) 9.8 in.; (*b*) 1.23 lbs.; (*c*) $35\frac{1}{3}°$; 1.1 in., 11.1 lbs., $84\frac{4}{5}°$.

(2) Distance from axis $r = (g/\omega^2) \cot \theta$.

(3) $P = mg \cos \theta\,(r\omega^2/g \cot \theta - 1)$.

(5) Taking AC as axis of x, its intersection with the required curve as axis of y, the equation of the curve is $y^2 = (2\,g/\omega^2)\,x$.

(6) An ellipse.

Page 394.

(1) $r_0^2 = \frac{1}{3} l^2$. (2) (a) $\frac{1}{3} l^2$; (b) $\frac{1}{3} h^2$; (c) $\frac{1}{12} l^2$; (d) $\frac{1}{12} h^2$.

(3) $\frac{1}{2} h^2$. (4) $\frac{1}{12} a^2$. (5) $\frac{5}{24} a^2$. (6) $\frac{1}{12} h^2$.

(7) $\frac{n+1}{n+3} l^2$. (8) $\frac{1}{4} a^2$. (9) $\frac{1}{2} a^2$.

(10) $\frac{1}{5} a^2$. (11) $\frac{1}{5} a^2$, $\frac{1}{5} b^2$, $\frac{1}{5} c^2$. (12) $\frac{1}{4}(a_1^2 + a_2^2)$.

(13) (a) $I = \frac{2}{3}[b^3\alpha + (a - \alpha)\beta^3]$; (b) $I = \frac{2}{3}[a^3\beta + (b - \beta)\alpha^3]$. When $\beta(\alpha)$ is small, the second term within the bracket can be neglected.

(14) $I = \frac{1}{24} \rho\delta[2\, bh^3 - (b - 2\, a)(h - 2\, a)^3]$.

Page 398.

(1) $r_0^2 = \frac{1}{12}(h^2 + l^2)$. (2) (a) $\frac{1}{24} a^2$; (b) $\frac{1}{24} a^2$; (c) $\frac{1}{12} a^2$.

(3) (a) $\frac{5}{4} a^2$; (b) $\frac{1}{2} a^2$; (c) $\frac{3}{2} a^2$. (4) $\frac{2}{5} a^2$. (5) $\frac{1}{2}(a_1^2 + a_2^2)$.

(6) (a) $\frac{1}{2} a^2$; (b) $\frac{3}{2} a^2$; (c) $\frac{1}{12}(h^2 + 3\, a^2)$. (8) $\frac{2}{3} a^2$.

(9) (a) $\frac{1}{4} b^2$; (b) $\frac{1}{4} a^2$; (c) $\frac{1}{4}(a^2 + b^2)$.

(10) $\frac{1}{5}(b^2 + c^2)$, $\frac{1}{5}(c^2 + a^2)$, $\frac{1}{5}(a^2 + b^2)$.

(11) $I = \frac{2}{3}[(a^3\beta + (b - \beta)\, \alpha^3] - \frac{1}{2} \frac{(a^2\beta + b\alpha^2 - \alpha^2\beta)^2}{a\beta + (b - \beta)\, \alpha}$.

(12)
$$I_x = \frac{1}{6}[a^3 b - (a - 2\, \delta)^3(b - \delta)]$$
$$= \frac{1}{6}\, \delta[a^2(a + 6\, b) - 6\, a(a + 2\, b)\delta + 4(3\, a + 2\, b)\delta^2 - 8\, \delta^3];$$
$$I_y = \frac{2}{3}\, \delta[2\, b^3 - a\delta^2 - 2\, \delta^3];$$
$$I_z = I_x + I_y = \frac{1}{6}\, \delta[(a^3 + 6\, a^2 b + 8\, b^3) - 6\, a(a + 2\, b)\delta + 8(2\, a + b)\delta^2 - 16\, \delta^3].$$

(13) $\frac{1}{6} a^2$. (14) $\frac{4}{3} a^2$. (15) $\frac{3}{4} a^2 + b^2$. (18) $\frac{1}{12}(a^2 + b^2 \sin^2 \theta)$.

(19) For the line joining the midpoints of the non-parallel sides the moment of inertia of the trapezoid is the same (by symmetry) as that of the rectangle having this side as base and an altitude equal to that of the trapezoid. Transferring to the centroidal line, we find

$$r_0^2 = \frac{1}{18} h^2\left[1 + \frac{2\, ab}{(a + b)^2}\right].$$

Page 409.

(1) The centroidal principal axes are perpendicular to the faces. The moments for these axes are $\frac{1}{3}M(b^2+c^2)$, $\frac{1}{3}M(c^2+a^2)$, $\frac{1}{3}M(a^2+b^2)$. The central ellipsoid is $(b^2+c^2)x^2+(c^2+a^2)y^2+(a^2+b^2)z^2=3\,\epsilon^4$. For an edge $2\,a$, $I=\frac{4}{3}M(b^2+c^2)$; for a diagonal $I=\frac{2}{3}M(b^2c^2+c^2a^2+a^2b^2)/(a^2+b^2+c^2)$.

For the cube the fundamental ellipsoid becomes a sphere of radius $\frac{1}{3}\sqrt{6}\,a$; for an edge of the cube, $q^2=\frac{8}{3}a^2$; for a diagonal, $q^2=\frac{2}{3}a^2$.

(2) Central ellipsoid: $(b^2+c^2)x^2+(c^2+a^2)y^2+(a^2+b^2)z^2=5\,\epsilon^4$; for l, $q^2=\frac{1}{5}(6\,a^2+b^2)$.

(3) Take the vertex as origin, the axis of the cone as axis of x; then $I_1=\frac{3}{10}Ma^2$; I_1', *i.e.*, the moment of inertia for the yz-plane, $=\frac{3}{5}Mh^2$. As for a solid of revolution about the axis of x $B'=C'$ and $B=C$, we have $I_2'=I_3'=\frac{1}{2}I_1$, and $I_2=I_3=I_1'+\frac{1}{2}I_1$. Hence, $I_2=I_3=\frac{3}{5}M(h^2+\frac{1}{4}a^2)$. At the centroid the squares of the principal radii are $\frac{3}{10}a^2$, $\frac{3}{80}(4\,a^2+h^2)$.

(4) $A=B=C=\frac{2}{3}Ma^2$, $D=E=F=\frac{1}{4}Ma^2$; hence momental ellipsoid: $4(x^2+y^2+z^2)-3(yz+zx+xy)=6\,\epsilon^4/a^2$; squares of principal radii: $\frac{1}{6}a^2$, $\frac{11}{12}a^2$, $\frac{11}{12}a^2$.

(5) $q^2=\frac{1}{2}a^2(1+\sin^2\alpha)$.

(6) $I=\frac{1}{10}\rho\pi a^4(\frac{8}{3}a+H+2\,h^5/H^4)$; for $h=a=\frac{1}{3}H$, $q^2=\frac{461}{1410}a^2$.

(7) $A=I_1$, $B=I_2+Mx_1^2$, $C=I_3+Mx_1^2$.

(8) The centroid may be such a point; if the central ellipsoid be an oblate spheroid, the two points on the axis of revolution at the distance $\pm\sqrt{(I_1-I_2)/M}$ from the centroid are such points.

(9) The ellipsoid must have the same central ellipsoid as the given body; its equation is $x^2/A'+y^2/B'+z^2/C'=5/M$, where M is the mass and A', B', C' are the moments of inertia for the principal *planes* of the body at the centroid.

(10) $\rho''=M/N$, where

$$N=\sqrt{6}\,[(q_2^2+q_3^2-q_1^2)^{\frac{1}{3}}+(q_3^2+q_1^2-q_2^2)^{\frac{1}{3}}+(q_1^2+q_2^2-q_3^2)^{\frac{1}{3}}]^{\frac{3}{2}},$$

$$a^3=\frac{3}{4}\frac{M}{\rho''}(q_2^2+q_3^2-q_1^2), \text{ etc.}$$

Page 420.

(1) $2\pi\sqrt{\frac{4L^2+3a^2}{6gL}}$.

(2) $\frac{2}{3}\sqrt{2}\,a$.

(3) $m\left(\frac{5\pi r}{3}\right)^2$.

(5) 156 ft.-tons.

(6) 4 min. 22 sec.

(7) $\frac{\pi N^2 r^2}{3600\,g\mu r'}$.

(8) $\omega=\sqrt{\frac{3g}{l}}$.

(9) $\frac{4}{7}\,a$.

(10) 300,000 ft.-lbs.

(11) 234,000 ft.-lbs.

(14) $\frac{8}{21}\,\mu r'kN$.

(15) (*a*) $\frac{\pi}{3600\,g}\cdot\frac{q^2N^2}{\mu r'}=\frac{1}{36,900}\,\frac{q^2N^2}{\mu r'}$ rev.; (*b*) $\frac{1}{306}\,\frac{q^2N}{\mu r'}$ sec.

(16) $1\frac{1}{3}$ oz. (17) 1000 lb.-ft. (18) (*c*) $2/\sqrt{3}=1.155$.

Page 425.

(1) $m_r=\frac{1}{2}m$. (4) $\frac{4}{3}m$. (3) (*a*) $\frac{2}{5}m$; (*b*) $\frac{7}{5}m$.

(7) As the motion is uniformly accelerated, the final velocity is twice the average velocity, *i. e.*, $=2\times 15/t$, where t is the required time. The principle of kinetic energy and work then gives $t=7.1$ sec.

Page 431.

(2) 7143 lbs.

Page 447.

(1) $\frac{2}{3}\,l$. (4) $F=\frac{1}{2}\,ma\omega$; reactions: $-\frac{b}{a+b}F$, $-\frac{a}{a+b}F$.

(5) $x_1=\frac{M[\frac{2}{5}a^2+(a+b)^2]+\frac{1}{3}mb^2}{M(a+b)+\frac{1}{2}mb}$. (8) $p'=\frac{m_1a^2+\frac{1}{3}m_2l^2}{m_1a+\frac{1}{2}m_2l}$.

(9) F has twice the value found for inelastic impact.

(11) $F=\sqrt{\frac{1}{3}(\sqrt{2}-1)}\,m\sqrt{ga}$. (12) 12.6 ft./sec.

Page 453.

(1) $v=20.6$ ft. per second.

(4) (*a*) $0.833\,v_1$; (*b*) $0.775\,v_1$.

(5) $\frac{2}{7}$, $\frac{1}{3}$, $\frac{1}{2}$.

Page 456.

(1) (*a*) $j_s = \frac{5}{7} g \sin\alpha$, $j_c = \frac{2}{3} g \sin\alpha$, $j_h = \frac{1}{2} g \sin\alpha$.

(*b*) $\dot{\omega}_s = \frac{5}{2}\mu \frac{g}{a}\cos\alpha$, $\dot{\omega}_c = 2\mu \frac{g}{a}\cos\alpha$, $\dot{\omega}_h = \mu \frac{g}{a}\cos\alpha$.

(*c*) $\frac{\tan\alpha}{\tan\phi} \leqq \frac{7}{2}$, 3, 2, respectively.

(2) (*a*) 210.5 ft.; (*b*) 236.8 ft.

Page 458.

(1) $\bar{x} = \frac{M}{M+m+m'} h$.

Page 459.

(1) If $v_0 > |a\omega_0|$, F has the same sense as in the case of Art. 729; the equations are therefore the same; but ω_0 is a negative quantity. If $v_0 < |a\omega_0|$, the sense of F is reversed. In both cases the sphere rolls forward, *i. e.*, in the sense of v_0.

(3) $t_3 = 9\frac{3}{4}$ sec.; $v = -10.5$ ft./sec.; number of revolutions per second $= \frac{5}{8}$. (4) $\frac{7}{5} a$. (5) $\frac{3}{80}$.

INDEX.

[The Numbers refer to the Pages.]

Aberration of light, 57.
Absolute motion, 25.
—— units, 163–166.
Acceleration (in rectilinear motion), 36–49.
—— (in curvilinear motion), 62–68.
——, angular, 50.
—— in cartesian co-ordinates, 65–66.
—— directly proportional to the distance, 46–49.
—— of gravity, 37, 39.
—— inversely proportional to square of distance, 42–46.
——, normal, 64–65, 67.
—— in polar co-ordinates, 66–67.
—— in simple harmonic motion, 76, 78.
——, tangential, 64–65.
—— in uniform circular motion, 74.
—— as vector, 63.
Activity, 322.
Advantage, mechanical, 268.
d'Alembert, principle of, 334–337, 381.
Amplitude, 74.
——, correction for, 99.
Anchor ring, moment of inertia of, 399.
Angle of friction, 232.
—— of incidence and reflection, 291.
—— of repose, 233.
Angle-iron, centroid of cross-section, 151.
Angular acceleration, 50.
—— momentum, 333.
—— ——, conservation of, 386.
—— ——, equations of, 383.
—— —— about fixed axis, 416–417.
—— —— in plane motion, 450.
—— —— of rigid body, 383.
Angular velocity, 49, 50,
—— —— as rotor, 113.
—— —— resolved along axes, 117.
—— velocities, composition of, 114–117.
—— ——, parallelogram of, 116–117.
Anomaly, eccentric, 354.
Anomaly, mean, 356.
——, true, 351, 354.
Anti-parallelogram, 123–124.
Aperiodic motion, 319.
Aphelion, 351.
Apparent solar day, 29.
Appell, P., 179.
Archimedean spiral, 57, 343.
Arcs of curves, centroids, 135, 138–141.
Areal acceleration, 51.
—— density, 135.
—— velocity, 51.
Areas, centroids of, 141–153.
——, conservation of, 387.
——, principle of, 104.
Arm of a couple, 201.
—— of inertia, 394.
Astatic equilibrium, 218.
Attraction and repulsion, 102, 305–309, 343–344, 350.
Atwood's machine, 297–300, 425–426.
Available work, 268.
Average angular velocity, 50.
—— force, 303.
—— piston pressure, 305.
—— velocity, 34.
Axis of rotation, 5.
Axle-friction, 239.
—— in Atwood's machine, 426.

Balance, common, 218.
——, running, 434.
——, standing, 433.
Ballistic pendulum, 449.
Beat, 95.
Belt-friction, 240–242.
Belt on pulleys, 51, 52.
Bending moment, 228–230.
Binding screw, 271–272.
Body centrode, 11, 112.
—— falling *in vacuo*, 39–41.

Body falling toward the earth, 44–46.
—— —— through the earth, 47–49.
—— projected upward, 40–41, 46.
—— of reference, 25.
See also Rigid body.
Bole, 160.
Boyle's law, 292, 304.
Breaking strength of cord, 270–271.
Buffers, height of, 458.
Bullet, 34, 40–41, 70–73.
Buoyancy, 314.
Byerly, W. E., 86.

CARDIOID, centroid of arc, 141.
Catenary, centroid of arc, 141.
——, common, 198–201.
Catenaries, 193–201.
Center of displacement, 8, 9.
—— —— force, 100, 337.
—— —— gravity, 136, 191–192.
—— —— inertia, 136.
—— —— mass, 131, 134, 135–136.
—— —— oscillation, 419–420.
—— —— parallel forces, 189.
—— —— percussion, 447.
—— —— rotation, 6, 11.
—— —— suspension, 419–420, 439.
Centimeter, 7.
Central axis, 20–21, 209, 210–211, 250–251.
—— ellipsoid, 408.
—— forces, 351–362.
—— motion, 99–100.
Centrifugal couple, 434.
—— force, 367, 370–371, 432.
Centripetal force, 367.
Centrode, 10–11, 111–112.
Centroid, 131, 134, 135–136, 192.
—— of arcs of curves, 135, 138–141.
—— —— areas, 141–153, 193.
—— —— any solid, 155–156.
—— —— rigid body, motion of, 384.
—— —— solid of revolution, 155, 158.
—— —— volumes, 153–159.
Centroidal line, 397.
—— principal axes, 438–439.
C. G. S. system, 6, 32, 36, 132, 160, 163, 260, 322.
Chain, kinematic, 119.
—— raised by capstan, 302–303.
Circle, centroid of area, 141.
Circle, motion in vertical, 92–99, 365–366, 371–375.
——, motion under central forces, 343, 357.
Circular arc, centroid of, 139–140, 141.
—— disk, centroid of, 146–147.
—— —— on horizontal plane, 459–460.
—— ——, impact, 447.
—— —— as pendulum, 420.
—— ——, reduced mass, 425.
—— —— rotating, 421, 459–460.
—— line or wire, moment of inertia, 395, 398, 409.
—— orbit, 357.
—— ring, moment of inertia, 395, 398.
—— sections of ellipsoid, 406.
—— sector, centroid of, 145, 152.
—— segment, centroid of, 152.
Cissoid, centroid of area, 153.
Clamp, 271–272.
Coaster, 300–301.
Coefficient of friction, 231.
—— —— restitution, 281, 283.
—— —— rolling friction, 243.
Complete constraint, 117–119, 364.
Component, 4, 23, 56–58, 170.
Composition, 23.
—— of angular velocities, 114–117.
—— —— concurrent forces, 176.
—— —— couples, 205–206, 207–208.
—— —— force and couple, 206–207.
—— —— forces acting on the same particle, 170–175.
—— —— forces in the same plane, 208, 209–211.
—— —— intersecting rotors, 116–117.
—— —— parallel forces, 183–184, 185–190.
—— —— parallel rotors, 114–116.
—— —— simple harmonic motions, 79–92.
—— —— velocities, 53.
Compound harmonic motion, 78–92.
—— —— wave motion, 87.
—— pendulum, 419–422, 439–442, 447, 448–450.
Compression, 222.
Conchoidal motion, 14–15, 16.
Concurrent forces, 175–183.
Conditions of equilibrium for forces acting on the same particle, 172–174.
—— —— —— for concurrent forces, 176

Conditions of equilibrium for parallel forces, 190–191, 192–193.
—— —— —— for forces in a plane, 208–209, 210–211, 211–217.
—— —— —— for forces acting on any rigid body, 244, 250.
Condition for pure rolling, 452.
Cone, centroid of, 139, 153.
——, equimomental, 405.
—— of friction, 233.
——, moment of inertia of, 395.
——, principal axes of, 409.
Confocal conics, 410–412.
—— quadrics, 412–414.
Conical pendulum, 375–378.
Conic sections as orbits, 343–347, 348–358.
Conic, tangent to, 57.
Connecting rod, motion of, 14, 16, 58–62, 112.
—— —— and crank, forces acting on, 181, 272–274, 428–429.
Conservation of angular momentum, 334, 386.
—— —— areas, 334, 387.
—— —— energy, 286, 308, 308–309, 320, 323, 330.
—— —— linear momentum, 385.
—— —— motion of centroid, 280, 385.
Conservative forces, 329.
Constant, elastic, 311.
—— of gravitation, 305–306, 313.
Constrained motion, 362–378.
Constraining force, 365–366.
Constraints, 18–19, 117–119, 255–258, 263, 297, 363–366.
Constraint, complete, 117–119.
Cord, equilibrium of, 193–201.
—— running over pulley, 297–300, 425–426.
Correction for amplitude in pendulum, 99.
Cotterill, J. H., 61.
Couple of forces, 189–1[illegible] 201–208.
——, centrifugal, 434.
—— represented by vector, 204–205.
—— of rolling friction, 343.
Crane, 179–180.
Crank and connecting rod, 272, 274, 428–429.
Cube, impact on, 449–450.
—— as pendulum, 420.
——, principal axes of, 409.
Curvilinear motion, 325–378.
Cyclic sections of ellipsoid, 406.
Cycloid, 11, 57.
Cycloid, centroid of arc, 141.
——, centroid of solids generated by revolution of, 158.
——, motion on, 374–375.
Cylinder, centroid of, 153, 159.
——, moment of inertia of, 395, 398, 399.
——, moving down inclined plane, 451–456.
——, moving up inclined plane, 456–457.
—— on horizontal plane, 457–458.
Cylindrical rod as pendulum, 420.

Dalby, W. E., 435.
Damped oscillations, 318–320.
Damping ratio, 321.
Decrement, logarithmic, 321.
Degree of uniformity, 430.
Degrees of freedom, 18–19, 119, 255–258, 381.
Density, 131–132, 135.
Derived units, 31, 130.
Deviation due to obliquity of connecting rod, 60.
Dimensions, 31–32, 37, 160, 163, 260, 322.
Direct impact of spheres, 278–293.
Directing couple, 423.
Displacement, 3, 4, 8, 17, 19.
—— in simple harmonic motion, 74.
Distribution of principal axes in space, 410–416.
Door, moment of inertia, 395.
—— on hinges, 253–255.
Driving force, 268, 427.
Dynamic stability, 449.
Dynamical meaning of principal axes, 438.
—— —— —— radius of inertia, 424.
Dynamics, 1, 129–169.
Dyne, 163, 165–166.

Earth and moon, 141, 348–349, 358.
Eccentric anomaly, 354.
Effective force, 334–335.
Efficiency of machines, 268, 322–325.
Elastic constant, 311.
—— stress or tension, 310.
—— strings and springs, 310–315.
Elasticity, perfect and imperfect, in impact, 280–281.

Elasticity, perfect, 311.
Elevation of outer rail, 370.
—— of projectile, 70.
Elevator, 301.
Ellipse, centroid of area, 141, 152, 153.
——, focal, 413.
——, moment of inertia, 398.
—— as orbit, 343–347, 348–358.
——, tangent to, 57.
Ellipsograph, 123.
Ellipsoid, central, 408.
——, centroid of octant, 158.
——, equivalent, 410.
——, fundamental, 408.
—— of gyration or inertia, 408.
——, moment of inertia, 395, 398.
——, momental, 402.
——, principal axes, 409.
——, reciprocal, 408.
Ellipsoids of inertia, 399–410.
Elliptic co-ordinates, 412, 413–414.
—— harmonic motion, 88.
—— integral, 98–99.
—— motion, 12–14, 16, 112.
Energy, kinetic, 161–167.
——, potential, 308, 308–309, 330.
——, total, 312, 323.
Epicycloid, Epitrochoid, 11.
Epoch, epoch-angle, 75.
Equation of the center, 357.
Equation of motion of a particle, 293.
—— —— —— about a fixed axis, 417.
Equations of linear and angular momentum, 383, 450.
Equations of motion of a rigid body, 381–383.
Equiangular spiral, 343.
Equilibrium, 170, 172.
—— of cord or chain, 193–201.
—— of forces in a plane, 208–209, 210–211, 211–217.
—— on inclined plane, 175, 233, 235–236.
—— of rigid body, 244, 250.
——, stable, unstable, neutral (astatic), 217–219.
Equimomental cone, 405.
Equipotential surfaces, 330.
Equivalence of couples, 202–204, 207–208.
Equivalent ellipsoid, 410.
—— simple pendulum, 419, 420–423.
Erg, 260.
Everett, J. D., 6, 32, 83, 163.
Expanding gas, 313.
FALLING BODY, 39–41, 295–296.
—— —— in resisting medium, 316–318.
Field of force, 308.
First integrals of equations of motion, 104.
Fixed axis, body with, 416–450.
—— ——, reactions of, 431–442.
Fixed centrode, 10, 111.
—— curve, motion on, 365–375.
—— surface, motion on, 375–378.
Flux, fluxional notation, 380.
Fly-wheel, 51, 52, 420, 421, 429–431.
Focal ellipse and hyperbola, 413.
Foot, 7.
Foot-pound, foot-poundal, 260.
Force, 161–169, 275, 284.
——, attractive and repulsive, 305–309, 343–344, 350.
——, average or mean, 303.
——, central, 337–362.
——, centrifugal and centripetal, 367.
——, conservative, 329.
——, constant, 294–302.
——, effective, 334–335.
——, impressed, 335.
—— of inertia, 160, 335.
——, intensity of, 338.
——, internal, 379.
—— inversely proportional to square of distance, 305–309, 347–362.
——, law of, 168–169.
——, normal, 326.
—— proportional to distance, 309, 343–347.
—— reduced to distance from axis, 425.
—— of restitution, 280.
——, tangential, 326.
——, variable, 302–325.
Forced oscillations, 321–322.
Force-function, 329–331.
Force polygon, 172, 185, 225–228.
Four bar linkage, 119–121.
Fourier's theorem, 86.
F. P. S. system, 6, 32, 36, 132, 161, 260, 322.
Free curvilinear motion, 325–363.
Free oscillations, 309–315.
Freedom, degrees of, 18–19, 119, 255–258, 381.
Frequency, 75.
Friction, 236–243, 300, 314.

Friction angle, 232.
—— circle, 238.
—— cone, 233.
—— in pure rolling, 453, 455–457.
——, rolling, 242–243.
Frustum of cone or pyramid, centroid of, 158.
Fulcrum, 192.
Fundamental units, 31, 130.
—— ellipsoid, 408.
Funicular polygon, 186–188, 193–201, 225–228, 229–230.

GALILEO's laws of falling bodies, 40.
Gas engine, 325.
——, expanding, 313.
Gases, kinetic theory of, 291–293.
Geometric addition, 23.
—— derivative, 63.
—— difference, 25.
—— subtraction, 24.
—— sum, 23.
Geometry of motion, 1, 3–28.
Governor, 376–378.
Grain, gram, 130.
Graphical methods in statics, 171, 172–173, 183–184, 185–188, 191, 193–195, 201, 224–228, 234, 236.
—— time-table, 33.
Gravitation, constant of, 305–306, 313.
——, terrestrial, 348–349.
—— units, 163–166.
——, universal, 43–44, 349.
Gravity, acceleration of, 37, 39.
—— and centrifugal force, 370–371.
Guldinus, first proposition of, 140.
——, second proposition of, 148–149.
Gyration, ellipsoid of, 408.
——, radius of, 394.

HALLEY, 348.
Hammer and nail, 285–286, 288.
Harmonic motion, 74.
Hart's inversor, 125.
Head or height due to a velocity, 40.
Helix, centroid of arc, 141.
Hemisphere, centroid of volume, 158.
Heterogeneous mass, 131.
Hexagon, moment of inertia, 395.
Higher pairs, 118.
Hodograph, 63–64, 67, 73, 358.
Homogeneous mass, 131.
Hooke's law of elastic stress, 310.
Hoop, equivalent simple pendulum, 422.
—— on inclined plane, 454–457.
Horse-power, 322, 324–325, 421.
Hyperbola, as orbit, 343–347, 348–358.
——, focal, 413.
Hyperbolic spiral, 343.
Hypocycloid, Hypotrochoid, 11.

IMPACT, direct, 275–290, 291–293.
——, oblique, 290–291.
—— of water in pipe, 289.
Impressed force, 334.
Impulse, 161–162, 275, 390–392.
—— acting on body with fixed axis, 442–450.
Impulsive force, 276.
—— reactions, 444–450.
Inclined plane, 68–70, 175, 296, 299–300, 369–370, 451–457.
Independence of translation and rotation, 388.
Indicator, 304–305.
—— diagram, 305.
Inertia, 160.
——, ellipsoids of, 399–410.
——, force of, 335.
——, law of, 168.
——, moment of, 393.
——, product of, 393.
—— of pulley, 425–426.
——, radius of, 394.
——, spherical points of, 410.
Instantaneous axis, 15, 21.
—— center, 10.
—— force, 276.
Intensity of force, 338.
Internal forces, 379.
Invariable plane, line, direction, 388.
Invariant of forces acting on rigid body, 246, 248.
Inverse of a curve with respect to a circle, 124–125.
Inversors, 124–127.
Isochronous motions, 75, 374.

JACK, 235.
Jointed frames, 219–228.
Joule, 261.
Journal friction, 236–239.

KELVIN, 19, 82, 83.
Kennedy, 118, 119.
Kepler's equation, 356–357.
—— laws, 101–106, 347–348, 362.
Kilogram (force), 165.
Kilogram-meter, 260.
Kinematical conditions for pure rolling, 452.
Kinematic chain, 119.
Kinematics, 1, 29–128.
—— of machinery, 117–128.
—— and statics, 205.
Kinetic energy, 166–167.
—— ——, change from impact, 283–289.
—— —— of centroid, 390.
—— —— relative to centroidal axes, 390.
—— —— of rotation, 418.
—— —— and work, principle of, 104, 295, 303, 327–331, 389.
—— friction, 231.
—— theory of gases, 291–293.
Kinetics, 1, 275–460.
—— of the particle, 275–378.
—— of the rigid body, 379–460.
Knot, 62.

LAPLACE'S invariable plane, 388.
Law of universal gravitation, 43–44, 349.
Laws of force, inertia, stress, 168–169.
—— of friction, 231–232.
—— of motion, Newton's, 167–169.
Length of equivalent simple pendulum, 419.
—— of wave, 84.
Level surfaces, 330.
Lever, 192, 193, 269–270.
—— crank, 119, 121.
Limiting static friction, 231.
Line of quickest descent, 69–70.
—— of force, 331.
Linear density, 135, 398.
—— kinematics, 30–52.
—— momentum, conservation of, 385.
—— ——, equations of, 383.
—— —— of rigid body, 382.
—— —— in plane motion, 450.
—— motion, 4, 6–7.
—— velocity, 51.
Link, linkage, linkwork, 118–119.
Lissajous's curves, 90–92.
Load, 268.
Logarithmic decrement, 321.
Logarithmic spiral, 343.
Lorenz, P. H., 435.
Lost kinetic energy in impact, 288.
—— work, 286, 323.
Lower pairs, 118.

MACGREGOR, J. G., 89.
Mach, E., 169.
Machine, 118.
Magnetic needle, 423.
Mass, 129–133.
—— moment, 133.
——, reduced, 424, 425–426.
—— and weight, 130.
Material lines and surfaces, 131.
—— particle, 131, 159, 293–294.
Mathematical pendulum, 92–99.
Mean angular velocity, 50.
—— anomaly, 356.
—— force, 303.
—— motion, 354.
—— piston pressure, 305.
—— solar day, 29–30.
—— sun, 29.
—— time, 30.
—— velocity, 34.
Mechanical advantage, 268.
Mechanics, 1.
Mechanism, 119, 268.
—— for compound harmonic motion, 81.
—— for simple harmonic motion, 77, 78.
Menelaus, proposition of, 120.
Metacenter, 421.
Minchin, G. M., 86.
Modulus of a machine, 324.
Moment, bending, 228–230.
—— of a couple, 202.
—— of first order, 392.
—— of a force about an axis, 250.
—— of a force about a point, 177–178.
—— of inertia, 393, 392–416.
—— of inertia, determined experimentally, 421, 422–423.
—— of mass, 133.
—— of momentum, 333.
—— polygon, 186–188.
—— of second order, 392.
Momental ellipsoid, 402.
Momentum, 159–161, 167, 275, 277.
——, angular, 333.
—— of centroid, 277–278.

Momentum, conservation of angular, 386.
——, conservation of linear, 385.
——, equations of linear and angular, 383.
—— of rigid body, 277, 382, 383.
Moon and earth, 141, 348–349, 358.
Motion, 3.
—— on a fixed curve, 365–375.
—— on a fixed surface, 375–378.
——, mean, 354.

NEUTRAL equilibrium, 218–219.
Newton, 106, 347–348.
Newton's laws of motion, 167–169.
—— law of universal gravitation, 43–44, 349.
Newtonian forces, 305–309.
Normal acceleration, 64–65.
—— force, 326.

OBLIQUE impact, 290–291.
Octant of ellipsoid, centroid of volume, 158.
One-sided constraint, 364.
Orbit, 339, 348.
Oscillations, damped, 318–320.
——, due to torsion, 422–423.
——, forced, 321–322.
——, free, 309–315.

PAIRS, lower and higher, 118.
Pantograph, 122–123.
Pappus, first proposition of, 140, 141.
——, second proposition of, 148–149.
Parabola as catenary, 197.
——, centroid of arc, 141.
——, centroid of area, 152.
—— as orbit, 348–358.
Parabolic segment as pendulum, 421.
Paraboloid, centroid of volume, 158.
——, moment of inertia, 399.
Parallel forces, 183–201.
—— motion, 127–128.
Parallelopiped, centroid of volume, 153.
——, principal axes, 409.
Parallelogram, centroid of area, 141.
—— (mechanism), 122.
——, moment of inertia, 399.
—— of angular velocities, 116–117.
—— of forces, 171.
—— law, 23, 116.
Parallelogram law for couples, 205–206.
Particle, 131, 159, 293–294.
Particles, centroid of, 137–138, 140.
Peaucellier's cell, 125–127.
Peg-top, moment of inertia, 410.
Pendulum, 92–99.
——, compound, 419–422.
——, simple mathematical, 372–375.
——, spherical or conical, 375–378.
Percussion of axis, 445.
——, center of, 447.
Perfect elasticity, 311.
Perfectly smooth, 236.
Pericycloid, peritrochoid, 11.
Perihelion, 351.
Period, periodic time, 74, 106, 354–358.
Permanent axis of rotation, 432, 438.
—— set, 310.
Perry, J., 289, 325.
Phase, phase-angle, 75.
Pile-driver, 288–289.
Pin-friction, 239.
Piston-pressure, 303–305.
Piston-rod motion, 58–62.
Pivot-friction, 239–240.
Plane area, centroid of, 145–146.
—— kinematics, 52–128.
—— motion, 4, 7–16, 450–460.
—— statics, 208–243.
Planetary motion, 106–109, 347–362.
Plumb-line, 371.
Point of application, 172.
Polar co-ordinates, 156.
—— reciprocal of momental ellipsoid, 407–408.
Potential, 307, 329.
—— energy, 308, 308–309, 330.
Pound (force), 165.
—— (standard), 130.
Poundal, 163, 165–166.
Power, 268, 322, 427.
Pressure, 179.
—— on curve, 366–367.
—— of piston, 303–305.
—— on rails, 370.
Principal axes, 402, 404–405, 409–410.
—— ——, distribution in space, 410–416.
—— ——, dynamical meaning, 438.
—— moments of inertia, 402.
—— radii of inertia, 405.

Principle of angular momentum or of areas, 104, 332–334.
—— of conservation of angular momentum or of areas, 334, 386, 387.
—— —— —— of energy, 286, 308, 308–309, 320, 323, 330.
—— —— —— of linear momentum or of motion of centroid, 385.
—— of d'Alembert, 334–337.
—— of independence of translation and rotation, 388.
—— of kinetic energy and work, 104, 295, 303, 327–331, 389, 418.
—— of the lever, 185, 189.
—— of virtual velocities, 263.
—— —— —— work, 258–274, 262, 264–265, 336, 380.
—— of work, 261.
Prism, centroid of volume, 153.
——, moment of inertia, 395.
Problem of two bodies, 359–362.
—— of three bodies, 361–362.
Product of inertia, 393, 400.
Projectile motion, 70–73, 358.
Propagation, velocity of, 84.
Pyramid, centroid of volume, 139, 153.
——, moment of inertia, 395.

QUADRIC surfaces, confocal, 412–414.
Quadrant of circle, centroid of area, 141.
Quadrilateral, centroid of area, 142–143.
Quantity of motion, 159–161.
Quickest descent, line of, 69–70.

RADIAN, 7.
Radius of gyration, 394.
—— of inertia, 394.
—— —— ——, dynamical meaning of, 424.
Railroad train, 33, 34, 40–41, 69, 235, 288, 293, 300–302, 370–371, 457–458.
Raindrop falling, 27.
Range of projectile, 72.
Reaction, 179, 192, 211, 255, 263, 366.
——, total, 232.
Reactions in compound pendulum, 439–442.
—— of fixed axis, 431–442.
——, impulsive, 444–450.
Reciprocal ellipsoid, 408.
Recoil, 289–290, 293.
Rectangle, moment of inertia, 394, 398.
Rectangular door, moment of inertia, 395.
Rectilinear motion of particle, 293–325.
—— segment, centroid, 138–139.
—— ——, moment of inertia, 394, 395.
Reduced mass, 424, 425–426.
Reduction of forces in plane, 208, 209–211.
—— —— —— in space, 243–244, 244–249.
Relative motion, 25–26.
—— —— of planet with respect to sun, 361–362.
—— velocity, 56, 56–58.
Repulsion and attraction, 343–344, 350.
Resistance, 179, 268, 366, 427.
—— of a medium, 315–318.
Resolution of angular velocity along the axes, 117.
—— of a force into parallel components, 193.
—— of velocity, 53–55, 56–58.
Restitution, 279–281.
Restoring couple, 423.
Resultant, 4, 23, 170, 172, 183, 186–190, 245.
Resulting couple, 245.
Reuleaux, 117, 119.
Revolving shaft, work of, 427.
Rifle-ball, 301.
Rigid body, 4, 175–176, 379.
—— ——, conditions of equilibrium, 250.
—— ——, kinetics of, 379–460.
—— —— supported at three points, 252–253, 257–258.
—— —— with fixed axis, 256–257, 416–450.
—— —— with fixed plane, 257–258.
—— —— with fixed point, 255–256.
Ring, circular, moment of inertia, 395, 398.
Rod, equilibrium of, 211–217, 265–267.
——, oscillating vertically in water, 314–315.
——, reduced mass, 425.
——, swinging about one end, 441–442, 447.
Rods, equilibrium of jointed, 219–228, 267.
Rolling, 452.
—— cones, 18.
—— friction, 242–243, 457.

Rotation, 4, 5, 49–52, 113.
Rotor, 113, 167, 205.
Rough, 236.
Running balance, 434.

SAFETY-VALVE, 193.
Sagging of telegraph wire, 201.
Sanborn, F. B., 325.
Screw, binding, 271–272.
—— motion, 18–22.
Second, 29–30.
Seconds pendulum, 95–97.
Sector of circle, centroid, 145, 152.
—— of sphere, centroid, 154–155.
Sectorial acceleration, 51.
—— velocity, 51, 67, 100–103, 324.
Segment of circle, centroid, 152.
—— of sphere, centroid, 158.
——, rectilinear, moment of inertia, 394, 395.
Semi-circle, centroid of arc, 141.
Sense, 3, 22, 33, 202.
Set, permanent, 310.
Shearing force, 228.
Sidereal day, 29–30.
Similar curves, 122–123.
Simple harmonic motion, 74–78, 347.
—— —— wave motion, 85–87.
—— mathematical pendulum, 372–375.
Sine-curve, centroid of area, 152.
Skew forces, 207, 247–248.
Sleigh, 300.
Slide valve, 61, 62.
Slider crank, 119, 272–274.
Sliding, 452.
—— friction, 236–242.
—— pair, 118.
Small oscillations due to torsion, 422–423.
Solid statics, 243–274.
—— —— of revolution, centroid, 155, 158.
Specific density, specific gravity, 132–133.
Speed, 33.
Sphere, centroid of volume, 154–155.
—— on horizontal plane, 458–460.
—— on inclined plane, 454–457.
——, moment of inertia, 395, 398.
——, reduced mass, 425.
Spheres, impact of, 278–293.
Spherical motion, 16–18.
—— pendulum, 375–378.
—— points of inertia, 410.
Spherical sector, centroid, 154–155.
—— segment, centroid, 158.
—— shell, moment of inertia, 398.
—— surface, centroid of area, 148.
Spinning, 452.
Spiral of Archimedes, 57, 343.
——, equiangular, hyperbolic, logarithmic, 343.
—— spring, 314.
Square, moment of inertia, 395.
Stability, 217–219.
——, dynamic, 449.
Stable equilibrium, 217–219.
Standard mass, 130.
Standards, 7.
Standing balance, 433.
Static friction, 231.
Statics, 1, 170–274.
—— and kinematics compared, 205.
Steam engine, 313, 324–325.
—— —— indicator, 304–305.
—— hammer, 289.
Strain, strain energy, 312.
Stress, 195, 222, 338, 359.
—— diagram, 172, 185.
—— during impact, 279.
——, elastic, 310.
——, law of, 169.
Stresses in a frame, 222–228.
Stroke, 58.
Strut, 222.
Surface area, centroid, 149–150.
—— density, 135, 398.
—— of revolution, centroid, 147–148.
Suspension bridge, 197, 201.
Swing, 95.
Sylvester, 119.
Symmetry, 136, 153–154, 408–409.

T-IRON, centroid of cross-section, 144–145, 151–152.
——, moment of inertia of cross-section, 395, 399.
Tacking against the wind, 180–181.
Tangential acceleration, 64–65.
—— force, 326.
Tension, 179, 195, 222, 297.
——, elastic, 310.
Terrestrial gravitation, 348–349.
Thomson and Tait, 19, 83, 86.
Three bodies, problem of, 361–362.

Tie, 222.
Time, 29–30.
—— of flight, 72.
—— in planetary motion, 354–358.
Toggle-joint press, 181.
Top, moment of inertia, 410.
Torque, 202.
Torsion, oscillations due to, 422–423.
Total energy, 312, 323.
—— reaction, 232.
—— work, 268, 323.
Translation, 4, 9, 22–28, 112.
Trapezoid, centroid of area, 143–144, 151.
——, moment of inertia, 399.
Triangle, centroid of area, 138–139, 141–142, 150–151.
—— of forces, 171.
——, moment of inertia, 395, 398, 399.
Triangular frame, centroid, 141.
—— plate swinging, 442.
Trip-hammer, 448.
Trochoid, 11.
True anomaly, 351, 354.
—— solar day, 29.
Truncated cylinder, centroid, 159.
Turning pair, 118.
Twist, 21.
Twisting pair, 118.
Two bodies, problem of, 359–362.

U-IRON, centroid of cross-section, 151.
Uniform circular motion, 67–68, 73–74.
—— mass distribution, 131.
—— rectilinear motion, 30, 30–34.
—— rotation, 49.
Uniformly accelerated motion, 36, 37–41.
—— —— rotation, 50.
Unit of acceleration, 36–37.
—— of angle, 7.
—— of density, 132.
—— of force, 163–166.
—— of length, 7.
—— of mass, 130.
—— of momentum, 160–161.
—— of power, 322, 324.
—— of velocity, 31.
—— of work, 260–261.
Units, 6.
—— of force and work, 286–287.
——, fundamental and derived, 130.
Universal gravitation, 43–44, 349.
Unstable equilibrium, 218–219.
Useful work, 268, 286, 323.

VALVE-GEAR motion, 62.
Variable force, 302–325.
—— rectilinear motion, 34, 34–49.
Varignon's theorem, 177–179, 184–185.
Vector, 23, 23–28, 53, 204–205.
Velocity, 34–35.
—— of light, 34, 57.
—— in plane motion, 51–52, 52–62.
—— of propagation of wave, 84.
——, sectorial or areal, 51, 55, 334.
—— of uniform motion, 30.
Velocities in the rigid body, 109–117.
Virtual moment, 263.
—— velocity, 263.
—— work, 262.
Volume density, 135.

WASTED work, 286, 323.
Water in a pipe, impact of, 289.
Water-wheel, 325.
Watt, 324.
Watt's parallel motion, 127–128.
Wave length, 84.
—— motion, 83–87.
Weber, H., 86.
Wedge, 270–271.
Weight, 164, 179, 191–192, 268.
Wheel, pulled over obstacle, 216–217.
——, rolling, 16, 57, 68.
——, rotating, 51, 52.
—— and axle, 251–252, 421, 426.
Wire, stretched between two points, 201.
Work, 166–167, 258–260, 284.
—— against gravity, 297, 301.
——, available or total, 268, 323.
——, lost or wasted, 323.
—— of piston pressure, 303–305.
—— of revolving shaft, 427.
——, useful, 268, 323.

www.ingramcontent.com/pod-product-compliance
Lightning Source LLC
LaVergne TN
LVHW011256110826
845149LV00001B/158
9781418184896